Applications of Artificial Intelligence in Food Industry

Applications of Artificial Intelligence in Food Industry

Guest Editors

Weiqing Min
Zhiming Guo

MDPI

Basel • Beijing • Wuhan • Barcelona • Belgrade • Novi Sad • Cluj • Manchester

Guest Editors

Weiqing Min
Institute of Computing Technology
Chinese Academy of Science
Beijing
China

Zhiming Guo
School of Food and Biological Engineering
Jiangsu University
Zhenjiang
China

Editorial Office
MDPI AG
Grosspeteranlage 5
4052 Basel, Switzerland

This is a reprint of the Special Issue, published open access by the journal *Foods* (ISSN 2304-8158), freely accessible at: https://www.mdpi.com/journal/foods/special_issues/DMP6O77HO4.

For citation purposes, cite each article independently as indicated on the article page online and as indicated below:

Lastname, A.A.; Lastname, B.B. Article Title. *Journal Name* **Year**, *Volume Number*, Page Range.

ISBN 978-3-7258-4835-5 (Hbk)
ISBN 978-3-7258-4836-2 (PDF)
https://doi.org/10.3390/books978-3-7258-4836-2

Contents

About the Editors

Weiqing Min

Weiqing Min is currently an Associate Professor with the Key Laboratory of Intelligent Information Processing, Institute of Computing Technology, Chinese Academy of Sciences. His research interests include multimedia content analysis and food computing. He has authored and co-authored more than 70 peer-referenced papers in relevant journals and conferences, including *IEEE Trans. on Pattern Analysis and Machine Intelligence, Patterns (Cell Press), ACM Computing Surveys, Trends in Food Science & Technology, IEEE Trans. on Image Processing, Food Chemistry, ACM MM, AAAI, IJCAI,* etc. He has served as the Associate Editor of *IEEE TMM* and as the *ACM MM2021/ICME* 2022–2025 Area Chair. He is the reviewer of many prestigious international journals including *Nature Food, Patterns (Cell Press), Advanced Science, IEEE TIP/TMM,* etc. He is a distinguished member of *CCF*, and a senior member of *IEEE*, CSIG and Chinese Institute of Food Science and Technology. He is the recipient of the 2024 Beijing Outstanding Youth Fund, 2023 CSIG Young Scientists Award, Second Prize of 2020 Beijing Science and Technology Progress Award, and 2020 ACM CHINA SIGMM RISING STAR AWARD. He is also the recipient of the 2016 ACM TOMM Nicolas D. Georganas Best Paper Award and 2017 IEEE Multimedia Magazine Best Paper Award.

Zhiming Guo

Zhiming Guo is a Professor in the School of Food and Biological Engineering at Jiangsu University, China, where he also serves as a principal investigator in the field of food analysis and detection sciences. His research focuses on the rapid and non-destructive detection of food quality and safety, optical sensing and automation for food quality evaluations, applications of the Internet of Things (IoT) for assessing food quality and safety, and advanced artificial intelligence algorithms.

He has been awarded funding from more than 11 national projects, including 5 National Key R&D projects and 4 National Natural Science Foundation of China (NSFC) projects. His significant research achievements have earned him prestigious awards such as the Second Prize of Natural Science from the Ministry of Education of China, the First Prize of Science and Technology from Jiangsu Province, and the Special Prize of the China Business Federation.

Dr. Guo has published more than 160 peer-reviewed journal articles and holds 45 patents. He is actively involved in his leadership roles as the Deputy Secretary-General of both the Jiangsu Institute of Food Science and Technology (JIFST) and the Agricultural Products Processing Machinery Branch of the Chinese Society of Agricultural Engineering (CSAE). He is also a member of the American Society of Agricultural and Biological Engineers (ASABE).

Additionally, Dr. Guo's excellence in academia has been recognized by several awards, including his inclusion within the first "Shennong Young Talents" list announced by the Ministry of Agriculture and Rural Affairs of China and his recognition as an outstanding young teacher in the Jiangsu universities' "Qinglan Project". He was also named as part of the national excellent university teacher team by the Ministry of Education of China.

foods

MDPI

Editorial

Applications of Artificial Intelligence in Food Industry

Heera Jayan [1,2], Weiqing Min [3,*] and Zhiming Guo [1,2,*]

1 School of Food and Biological Engineering, Jiangsu University, Zhenjiang 212013, China; heerajayan93@outlook.com
2 China Light Industry Key Laboratory of Food Intelligent Detection & Processing, Jiangsu University, Zhenjiang 212013, China
3 Key Laboratory of Intelligent Information Processing, Institute of Computing Technology, Chinese Academy of Sciences, Beijing 100086, China
* Correspondence: minweiqing@ict.ac.cn (W.M.); guozhiming@ujs.edu.cn (Z.G.)

With breakthroughs in artificial intelligence (AI) brought by the fourth industrial revolution, intelligent applications are providing innovative solutions across food industry [1,2]. The powerful capabilities of AI in perceiving, reasoning, and decision-making are supporting the transformation and upgrading of food technologies, enabling automation, digitization and intelligence throughout the sector [3–5]. Novel AI based applications are improving every aspect of the food industry from production, processing and packaging to transportation, storage, consumption and waste management, thus benefiting all stakeholders in the food industry [6,7]. When integrated with complimentary techniques such as the Internet of Things (IoT) and Big Data, AI facilitates advancement such as development of new product, food process simulation and optimization, and enhanced monitoring of food quality and safety (Figure 1) [8–10].

Development of cutting-edge methodologies for assessing food quality and nutritional content remains a significant focus for researchers, particularly in fruits and vegetables [11,12]. Ropelewska & Szwejda-Grzybowska [13,14] demonstrated strong correlations between image texture parameters and chemical properties of red and yellow pepper, highlighting the potential of multispectral imaging for broader applications in food analysis. Another notable contribution is to integrate deep learning to evaluate the chemical composition and nutritional content of food product [15–17]. Additionally, the combination of deep learning and near-infrared spectroscopy has proven highly effective for determining soluble solid content (SSC) in apples [18,19]. The use of ant colony optimization algorithm achieved superior results for SSC prediction while also excelling in maturity classification based on apple color characteristics. Furthermore, the integrative technique was applied to assess the quality of dried kiwifruit, emphasizing the importance of visual attributes in predicting moisture content and optimising drying processes [20]. These studies display the significant potential of integrating deep learning with advanced imaging and spectroscopic technique for evaluating food quality and nutritional content, paving the way for more efficient and accurate food analysis [21,22].

The classification of fruits and vegetables based on the quality, authenticity maturity is also vital for informed consumer choices and efficient food processing. Noutfia & Ropelewska [23] highlighted the application of image analysis to classify date fruits into categories based on the drying processes, achieving over 94% accuracy for fresh, convective-dried, and infrared-dried samples across tested cultivars. Despite the availability of various methods for classification based on quality parameters, automating the entire process remains crucial for achieving real-time, on-line classification at an industrial scale [24,25]. Granados-Vega et al. [26] addressed this need by developing a prototype for Classification of Persian lemon artificial using neural network (ANN) trained on digital images. The ANN successfully classified 96.60% of samples, and simulations confirmed the feasibility of implementing this prototype in practical settings. Zhu et al. [27] introduced an online classification system for maize seeds based on the vitality, utilizing near-infrared spectroscopy data with machine learning models. Their approach achieved a classification accuracy exceeding 91%, highlighting the potential of NIR-based devices for real-time classification of agricultural products. These studies represent a significant leap forward in the automation of fruit and vegetable classification by harnessing the power of advanced machine learning and computer vision techniques [28,29].

Beyond component analysis and classification, machine learning and AI excel in optimizing food industry processes, enabling increased efficiency, better yields, and cost reductions [30,31]. Pao-la-or et al. [32] employed Artificial Neural Fuzzy Intelligent System (ANFIS) to determine the optimal extraction condition for supercritical fluid extraction of oil from pomegranate seeds. In a related study, Pao-la-or et al. [33] utilized deep learning to optimize the drying process of fingerroot extract. Their model successfully preserved bioactive compounds, enhancing the potential applications. This work demonstrated the effectiveness of AI in refining extraction process. On the other hand, AI based systems can track the movement of food products through the supply chain, enabling rapid identification and recall of contaminated products, ensuring food safety [34]. Deep learning techniques have been proven to provide information regarding the risk associated with food products and their potential recalls by relying on historical data [35].

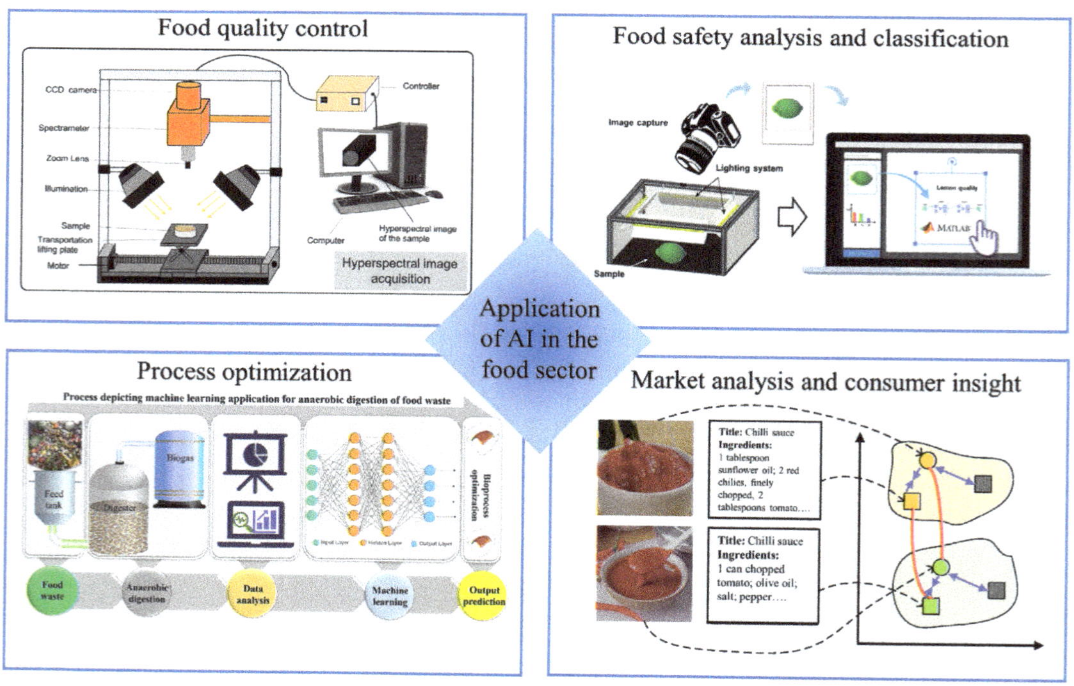

Figure 1. Application of AI in various field of food industry [26,36–38].

AI has the unique ability to process vast amounts of unstructured data from social media platforms, online reviews, and surveys, which will help in understanding emerging preferences and market gaps [39–41]. For example, Tao et al. [42] developed a tool to mine social media data for detecting foodborne illness trends, which could also provide insights into food safety and quality concerns of consumers. Similarly, Zuo et al. [38] proposed a Multi-Modal Alignment Method to retrieve recipes based on food images, highlighting the integration of cross-modal data to understand consumer preferences better and support tailored product development. Additionally, Chen et al. [43] introduced a health-aware food recommendation system leveraging a collaborative recipe knowledge graph and multi-task learning to provide personalized recommendations that balance dietary preferences with health requirements. This approach ensures consumers receive options tailored to both taste and health needs [44–46]. AI revolutionizes the food industry by empowering businesses to identify emerging trends, optimize formulations, and create personalized food recommendations [47]. This accelerates innovation and provides a strategic advantage in the rapidly evolving market.

Although AI based tools have been applied across various areas of food industry throughout the food supply chain, a significant gap remains between the development of AI technologies and their practical application [48–50]. This is especially evident in the fields of crop and agricultural sciences, where the integration of AI remains limited. Zhang et al. [51] recognized the potential of a multimodal technique that integrates AI with common non-destructive methods, demonstrating enhanced comprehensiveness and accuracy in food safety analyses. For example, Guo, et al. [52] combined SERS sensors with intelligent algorithms to achieve highly sensitive and in-situ detection of pesticides on apple. Further, extraction of useful information from temporal data is an extremely relevant research area, presenting significant opportunities for further exploitation in the food domain. Such advancement would allow AI to address complex challenges in the food industry, paving the way for effective and innovative solutions [53–55].

The future of AI applications in the food industry expected involve enhanced integration of technologies such as IoT, blockchain, and machine learning [56–58]. This integrative approach potentially enhance supply chain transparency, improve food safety, and optimize production processes [59]. In the future, AI is estimated to play a pivotal role in advancing sustainability by optimizing resource utilization, minimizing waste, and promoting environmentally sustainable practices across the entire food supply chain [60–62]. Further AI will remain as an integral part of developing personalized nutrition strategies and the development of products catered to meet the individual dietary requirement and choices [63–65]. For instance, deep generative networks can be employed to model user-specific information to generate accurate recommendations according to the guidelines. Additionally, AI-powered chatbots can engage in conversation regarding customer dietary preference and provide immediate suggestions and feedbacks [66,67].

Author Contributions: Conceptualization, Z.G. and W.M.; methodology, Z.G.; investigation, H.J.; resources, Z.G. and W.M.; writing—original draft preparation, H.J.; writing—review and editing, Z.G.; visualization, H.J.; supervision, Z.G.; project administration, Z.G. and W.M.; funding acquisition, Z.G. and W.M. All authors have read and agreed to the published version of the manuscript.

Funding: This research was funded by National Natural Science Foundation of China (32350410403, W2412103), Jiangsu Funding Program for Excellent Postdoctoral Talent (2023ZB868), Zhenjiang Science & Technology Program (GJ2024009).

Institutional Review Board Statement: Not applicable.

Informed Consent Statement: Not applicable.

Data Availability Statement: No new data were created or analyzed in this study. Data sharing is not applicable to this article.

Acknowledgments: We acknowledge academic editor for handling, editing, and promoting this Special Issue.

Conflicts of Interest: The authors declare no conflict of interest.

References

1. Hassoun, A.; Aït-Kaddour, A.; Abu-Mahfouz, A.M.; Rathod, N.B.; Bader, F.; Barba, F.J.; Biancolillo, A.; Cropotova, J.; Galanakis, C.M.; Jambrak, A.R.; et al. The Fourth Industrial Revolution in the Food Industry—Part I: Industry 4.0 Technologies. *Crit. Rev. Food Sci. Nutr.* **2023**, *63*, 6547–6563. [CrossRef] [PubMed]
2. Guo, Z.M.; Barimah, A.O.; Shujat, A.; Zhang, Z.Z.; Qin, O.Y.; Shi, J.Y.; El-Seedi, H.R.; Zou, X.B.; Chen, Q.S. Simultaneous Quantification of Active Constituents and Antioxidant Capability of Green Tea Using NIR Spectroscopy Coupled with Swarm Intelligence Algorithm. *Lwt-Food Sci. Technol.* **2020**, *129*, 109510. [CrossRef]
3. Li, H.; Geng, W.; Hassan, M.M.; Zuo, M.; Wei, W.; Wu, X.; Ouyang, Q.; Chen, Q. Rapid Detection of Chloramphenicol in Food Using SERS Flexible Sensor Coupled Artificial Intelligent Tools. *Food Control* **2021**, *128*, 108186. [CrossRef]
4. Chen, J.; Zhang, M.; Xu, B.; Sun, J.; Mujumdar, A.S. Artificial Intelligence Assisted Technologies for Controlling the Drying of Fruits and Vegetables Using Physical Fields: A Review. *Trends Food Sci. Technol.* **2020**, *105*, 251–260. [CrossRef]
5. Lu, B.; Han, F.; Aheto, J.H.; Rashed, M.M.A.; Pan, Z. Artificial Bionic Taste Sensors Coupled with Chemometrics for Rapid Detection of Beef Adulteration. *Food Sci. Nutr.* **2021**, *9*, 5220–5228. [CrossRef]
6. Esmaeily, R.; Razavi, M.A.; Razavi, S.H. A Step Forward in Food Science, Technology and Industry Using Artificial Intelligence. *Trends Food Sci. Technol.* **2024**, *143*, 104286. [CrossRef]
7. Chen, Q.; Sun, C.; Ouyang, Q.; Wang, Y.; Liu, A.; Li, H.; Zhao, J. Classification of Different Varieties of Oolong Tea Using Novel Artificial Sensing Tools and Data Fusion. *LWT-Food Sci. Technol.* **2015**, *60*, 781–787. [CrossRef]
8. Wang, C.; Gu, C.; Zhao, X.; Yu, S.; Zhang, X.; Xu, F.; Ding, L.; Huang, X.; Qian, J. Self-Designed Portable Dual-Mode Fluorescence Device with Custom Python-Based Analysis Software for Rapid Detection via Dual-Color FRET Aptasensor with IoT Capabilities. *Food Chem.* **2024**, *457*, 140190. [CrossRef]
9. Wang, W.; Jayan, H.; Majeed, U.; Zou, X.; Hu, Q.; Guo, Z. Visual detection of Auramine O using dual-signal ratiometric fluorescent nanopaper sensor combined portable smartphone. *Food Biosci.* **2025**, *65*, 106135. [CrossRef]
10. Gu, W.; Wen, W.; Wu, S.; Zheng, C.; Lu, X.; Chang, W.; Xiao, P.; Guo, X. 3D Reconstruction of Wheat Plants by Integrating Point Cloud Data and Virtual Design Optimization. *Agriculture* **2024**, *14*, 391. [CrossRef]
11. Nirere, A.; Sun, J.; Atindana, V.A.; Hussain, A.; Zhou, X.; Yao, K. A Comparative Analysis of Hybrid SVM and LS-SVM Classification Algorithms to Identify Dried Wolfberry Fruits Quality Based on Hyperspectral Imaging Technology. *J. Food Process. Preserv.* **2022**, *46*, e16320. [CrossRef]
12. Zhao, Q.; Shi, Y.; Xu, C.; Jiang, Z.; Liu, J.; Sui, Y.; Zhang, H. Control of Postharvest Blue and Gray Mold in Kiwifruit by Wickerhamomyces Anomalus and Its Mechanism of Antifungal Activity. *Postharvest Biol. Technol.* **2023**, *201*, 112345. [CrossRef]
13. Ropelewska, E.; Szwejda-Grzybowska, J. The Estimation of Chemical Properties of Pepper Treated with Natural Fertilizers Based on Image Texture Parameters. *Foods* **2023**, *12*, 2123. [CrossRef]
14. Xu, B.; Sylvain Tiliwa, E.; Yan, W.; Roknul Azam, S.M.; Wei, B.; Zhou, C.; Ma, H.; Bhandari, B. Recent Development in High Quality Drying of Fruits and Vegetables Assisted by Ultrasound: A Review. *Food Res. Int.* **2022**, *152*, 110744. [CrossRef]
15. Xu, M.; Sun, J.; Cheng, J.; Yao, K.; Wu, X.; Zhou, X. Non-destructive Prediction of Total Soluble Solids and Titratable Acidity in Kyoho Grape Using Hyperspectral Imaging and Deep Learning Algorithm. *Int. J. Food Sci. Technol.* **2023**, *58*, 9–21. [CrossRef]
16. Zhang, W.; Luo, Z.; Wang, A.; Gu, X.; Lv, Z. Kinetic Models Applied to Quality Change and Shelf Life Prediction of Kiwifruits. *LWT* **2021**, *138*, 110610. [CrossRef]
17. Du, C.; Han, D.; Song, Z.; Chen, Y.; Chen, X.; Wang, X. Calibration of Contact Parameters for Complex Shaped Fruits Based on Discrete Element Method: The Case of Pod Pepper (*Capsicum annuum*). *Biosyst. Eng.* **2023**, *226*, 43–54. [CrossRef]
18. Guo, Z.; Zhang, Y.; Wang, J.; Liu, Y.; Jayan, H.; El-Seedi, H.R.; Alzamora, S.M.; Gómez, P.L.; Zou, X. Detection Model Transfer of Apple Soluble Solids Content Based on NIR Spectroscopy and Deep Learning. *Comput. Electron. Agric.* **2023**, *212*, 108127. [CrossRef]
19. Tian, Y.; Sun, J.; Zhou, X.; Yao, K.; Tang, N. Detection of Soluble Solid Content in Apples Based on Hyperspectral Technology Combined with Deep Learning Algorithm. *J. Food Process. Preserv.* **2022**, *46*, e16414. [CrossRef]
20. Yu, S.; Zheng, H.; Wilson, D.I.; Yu, W.; Young, B.R. Integrating Image Analysis and Machine Learning for Moisture Prediction and Appearance Quality Evaluation: A Case Study of Kiwifruit Drying Pretreatment. *Foods* **2024**, *13*, 1789. [CrossRef]
21. Zhou, X.; Sun, J.; Tian, Y.; Lu, B.; Hang, Y.; Chen, Q. Hyperspectral Technique Combined with Deep Learning Algorithm for Detection of Compound Heavy Metals in Lettuce. *Food Chem.* **2020**, *321*, 126503. [CrossRef]
22. Zhou, X.; Zhao, C.; Sun, J.; Cao, Y.; Yao, K.; Xu, M. A Deep Learning Method for Predicting Lead Content in Oilseed Rape Leaves Using Fluorescence Hyperspectral Imaging. *Food Chem.* **2023**, *409*, 135251. [CrossRef] [PubMed]
23. Noutfia, Y.; Ropelewska, E. Exploration of Convective and Infrared Drying Effect on Image Texture Parameters of 'Mejhoul' and 'Boufeggous' Date Palm Fruit Using Machine Learning Models. *Foods* **2024**, *13*, 1602. [CrossRef]

24. Wang, J.; Gao, Z.; Zhang, Y.; Zhou, J.; Wu, J.; Li, P. Real-Time Detection and Location of Potted Flowers Based on a ZED Camera and a YOLO V4-Tiny Deep Learning Algorithm. *Horticulturae* **2021**, *8*, 21. [CrossRef]
25. El-Mesery, H.S.; Qenawy, M.; Li, J.; El-Sharkawy, M.; Du, D. Predictive Modeling of Garlic Quality in Hybrid Infrared-Convective Drying Using Artificial Neural Networks. *Food Bioprod. Process.* **2024**, *145*, 226–238. [CrossRef]
26. Granados-Vega, B.V.; Maldonado-Flores, C.; Gómez-Navarro, C.S.; Warren-Vega, W.M.; Campos-Rodríguez, A.; Romero-Cano, L.A. Development of a Low-Cost Artificial Vision System as an Alternative for the Automatic Classification of Persian Lemon: Prototype Test Simulation. *Foods* **2023**, *12*, 3829. [CrossRef]
27. Zhu, Y.; Fan, S.; Zuo, M.; Zhang, B.; Zhu, Q.; Kong, J. Discrimination of New and Aged Seeds Based on On-Line Near-Infrared Spectroscopy Technology Combined with Machine Learning. *Foods* **2024**, *13*, 1570. [CrossRef]
28. Cai, Z.; Sun, C.; Zhang, H.; Zhang, Y.; Li, J. Developing Universal Classification Models for the Detection of Early Decayed Citrus by Structured-Illumination Reflectance Imaging Coupling with Deep Learning Methods. *Postharvest Biol. Technol.* **2024**, *210*, 112788. [CrossRef]
29. Ukwuoma, C.C.; Zhiguang, Q.; Bin Heyat, M.B.; Ali, L.; Almaspoor, Z.; Monday, H.N. Recent Advancements in Fruit Detection and Classification Using Deep Learning Techniques. *Math. Probl. Eng.* **2022**, *2022*, 9210947. [CrossRef]
30. You, J.; Li, D.; Wang, Z.; Chen, Q.; Ouyang, Q. Prediction and Visualization of Moisture Content in Tencha Drying Processes by Computer Vision and Deep Learning. *J. Sci. Food Agric.* **2024**, *104*, 5486–5494. [CrossRef]
31. Deng, J.; Ni, L.; Bai, X.; Jiang, H.; Xu, L. Simultaneous Analysis of Mildew Degree and Aflatoxin B1 of Wheat by a Multi-Task Deep Learning Strategy Based on Microwave Detection Technology. *LWT* **2023**, *184*, 115047. [CrossRef]
32. Pao-la-or, P.; Marungsri, B.; Chirinang, P.; Posridee, K.; Oonsivilai, R.; Oonsivilai, A. Boosting *Purnica granatum* L. Seed Oil Yield: An Adaptive Neuro-Fuzzy Interference System Fuels SC-CO_2 Extraction Breakthrough. *Foods* **2024**, *13*, 161. [CrossRef] [PubMed]
33. Pao-la-or, P.; Posridee, K.; Buranakon, P.; Singthong, J.; Oonmetta-Aree, J.; Oonsivilai, R.; Oonsivilai, A. Beyond Traditional Methods: Deep-Learning Machines Empower Fingerroot (Boesenbergia Rotunda)-Extract Production with Superior Antioxidant Activity. *Foods* **2024**, *13*, 2676. [CrossRef]
34. Sharifmousavi, M.; Kayvanfar, V.; Baldacci, R. Distributed Artificial Intelligence Application in Agri-Food Supply Chains 4.0. *Procedia Comput. Sci.* **2024**, *232*, 211–220. [CrossRef]
35. Makridis, G.; Mavrepis, P.; Kyriazis, D. A Deep Learning Approach Using Natural Language Processing and Time-Series Forecasting towards Enhanced Food Safety. *Mach. Learn.* **2023**, *112*, 1287–1313. [CrossRef]
36. Li, T.; Wei, W.; Xing, S.; Min, W.; Zhang, C.; Jiang, S. Deep Learning-Based Near-Infrared Hyperspectral Imaging for Food Nutrition Estimation. *Foods* **2023**, *12*, 3145. [CrossRef]
37. Said, Z.; Sharma, P.; Thi Bich Nhuong, Q.; Bora, B.J.; Lichtfouse, E.; Khalid, H.M.; Luque, R.; Nguyen, X.P.; Hoang, A.T. Intelligent Approaches for Sustainable Management and Valorisation of Food Waste. *Bioresour. Technol.* **2023**, *377*, 128952. [CrossRef]
38. Zou, Z.; Zhu, X.; Zhu, Q.; Zhang, H.; Zhu, L. Disambiguity and Alignment: An Effective Multi-Modal Alignment Method for Cross-Modal Recipe Retrieval. *Foods* **2024**, *13*, 1628. [CrossRef]
39. Mu, W.; Kleter, G.A.; Bouzembrak, Y.; Dupouy, E.; Frewer, L.J.; Radwan Al Natour, F.N.; Marvin, H.J.P. Making Food Systems More Resilient to Food Safety Risks by Including Artificial Intelligence, Big Data, and Internet of Things into Food Safety Early Warning and Emerging Risk Identification Tools. *Compr. Rev. Food Sci. Food Saf.* **2024**, *23*, e13296. [CrossRef]
40. Lau, E.A.; Rukmana, A.Y.; Uhai, S.; Mokodenseho, S.; Tapaningsih, W.I.D.A. Mapping Research on the Influence of Social Media on Consumer Food Behavior a Bibliometric Approach. *Eastasouth J. Soc. Sci. Humanit.* **2024**, *1*, 84–94. [CrossRef]
41. Molenaar, A.; Lukose, D.; Brennan, L.; Jenkins, E.L.; McCaffrey, T.A. Using Natural Language Processing to Explore Social Media Opinions on Food Security: Sentiment Analysis and Topic Modeling Study. *J. Med. Internet Res.* **2024**, *26*, e47826. [CrossRef] [PubMed]
42. Tao, D.; Hu, R.; Zhang, D.; Laber, J.; Lapsley, A.; Kwan, T.; Rathke, L.; Rundensteiner, E.; Feng, H. A Novel Foodborne Illness Detection and Web Application Tool Based on Social Media. *Foods* **2023**, *12*, 2769. [CrossRef] [PubMed]
43. Chen, Y.; Guo, Y.; Fan, Q.; Zhang, Q.; Dong, Y. Health-Aware Food Recommendation Based on Knowledge Graph and Multi-Task Learning. *Foods* **2023**, *12*, 2079. [CrossRef] [PubMed]
44. Samad, S.; Ahmed, F.; Naher, S.; Kabir, M.A.; Das, A.; Amin, S.; Islam, S.M.S. Smartphone Apps for Tracking Food Consumption and Recommendations: Evaluating Artificial Intelligence-Based Functionalities, Features and Quality of Current Apps. *Intell. Syst. Appl.* **2022**, *15*, 200103. [CrossRef]
45. Patra, E.; Kokkinopoulou, A.; Wilson-Barnes, S.; Hart, K.; Gymnopoulos, L.P.; Tsatsou, D.; Solachidis, V.; Dimitropoulos, K.; Rouskas, K.; Argiriou, A.; et al. Personal Goals, User Engagement, and Meal Adherence within a Personalised AI-Based Mobile Application for Nutrition and Physical Activity. *Life* **2024**, *14*, 1238. [CrossRef]
46. Hart, K.H.; Wilson-Barnes, S.; Stefanidis, K.; Tsatsou, D.; Gymnopoulos, L.; Dimitropoulos, K.; Rouskas, K.; Argiriou, N.; Leoni, R.; Russell, D.; et al. The Suitability of Dietary Recommendations Suggested By Artificial Intelligence Technology via a Novel Personalised Nutrition Mobile Application. *Proc. Nutr. Soc.* **2022**, *81*, E37. [CrossRef]

47. Joachim, S.; Forkan, A.R.M.; Jayaraman, P.P.; Morshed, A.; Wickramasinghe, N. A Nudge-Inspired AI-Driven Health Platform for Self-Management of Diabetes. *Sensors* **2022**, *22*, 4620. [CrossRef]
48. Sun, J.; Cheng, J.; Xu, M.; Yao, K. A Method for Freshness Detection of Pork Using Two-Dimensional Correlation Spectroscopy Images Combined with Dual-Branch Deep Learning. *J. Food Compos. Anal.* **2024**, *129*, 106144. [CrossRef]
49. Ariyo Okaiyeto, S.; Bai, J.; Xiao, H. Generative AI in Education: To Embrace It or Not? *Int. J. Agric. Biol. Eng.* **2023**, *16*, 285–286. [CrossRef]
50. Ma, Z.; Zhu, Y.; Wu, Z.; Nfamoussa Traore, S.; Chen, D.; Xing, L. BP Neural Network Model for Material Distribution Prediction Based on Variable Amplitude Anti-Blocking Screening DEM Simulations. *Int. J. Agric. Biol. Eng.* **2023**, *16*, 190–199. [CrossRef]
51. Zhang, L.; Yang, Q.; Zhu, Z. The Application of Multi-Parameter Multi-Modal Technology Integrating Biological Sensors and Artificial Intelligence in the Rapid Detection of Food Contaminants. *Foods* **2024**, *13*, 1936. [CrossRef]
52. Guo, Z.; Zheng, Y.; Yin, L.; Xue, S.; Ma, L.; Zhou, R.; El-Seedi, H.R.; Zhang, Y.; Yosri, N.; Jayan, H.; et al. Flexible Au@AgNRs/MAA/PDMS-Based SERS Sensor Coupled with Intelligent Algorithms for in-Situ Detection of Thiram on Apple. *Sens. Actuators B Chem.* **2024**, *404*, 135303. [CrossRef]
53. Bai, J.-W.; Xiao, H.-W.; Ma, H.-L.; Zhou, C.-S. Artificial Neural Network Modeling of Drying Kinetics and Color Changes of Ginkgo Biloba Seeds during Microwave Drying Process. *J. Food Qual.* **2018**, *2018*, 3278595. [CrossRef]
54. Zhao, Z.; Jin, M.; Tian, C.; Yang, S.X. Prediction of Seed Distribution in Rectangular Vibrating Tray Using Grey Model and Artificial Neural Network. *Biosyst. Eng.* **2018**, *175*, 194–205. [CrossRef]
55. Huan, J.; Cao, W.; Liu, X. A Dissolved Oxygen Prediction Method Based on K-Means Clustering and the ELM Neural Network: A Case Study of the Changdang Lake, China. *Appl. Eng. Agric.* **2017**, *33*, 461–469. [CrossRef]
56. Ding, C.; Wang, L.; Chen, X.; Yang, H.; Huang, L.; Song, X. A Blockchain-Based Wide-Area Agricultural Machinery Resource Scheduling System. *Appl. Eng. Agric.* **2023**, *39*, 1–12. [CrossRef]
57. Zhang, Y.; Chen, L.; Battino, M.; Farag, M.A.; Xiao, J.; Simal-Gandara, J.; Gao, H.; Jiang, W. Blockchain: An Emerging Novel Technology to Upgrade the Current Fresh Fruit Supply Chain. *Trends Food Sci. Technol.* **2022**, *124*, 1–12. [CrossRef]
58. Sun, J.; Nirere, A.; Dusabe, K.D.; Yuhao, Z.; Adrien, G. Rapid and Nondestructive Watermelon (*Citrullus lanatus*) Seed Viability Detection Based on Visible Near-infrared Hyperspectral Imaging Technology and Machine Learning Algorithms. *J. Food Sci.* **2024**, *89*, 4403–4418. [CrossRef]
59. Chauhdary, J.N.; Li, H.; Akbar, N.; Javaid, M.; Rizwan, M.; Akhlaq, M. Evaluating Corn Production under Different Plant Spacings through Integrated Modeling Approach and Simulating Its Future Response under Climate Change Scenarios. *Agric. Water Manag.* **2024**, *293*, 108691. [CrossRef]
60. Tuly, J.A.; Ma, H. Bioconversion of Food Industrial Waste Okara by Microbial Fermentation: Scope of Omics Study and Possibility. *Trends Food Sci. Technol.* **2024**, *146*, 104391. [CrossRef]
61. Cayambe, J.; Heredia-R, M.; Torres, E.; Puhl, L.; Torres, B.; Barreto, D.; Heredia, B.N.; Vaca-Lucero, A.; Diaz-Ambrona, C.G.H. Evaluation of Sustainability in Strawberry Crops Production under Greenhouse and Open-Field Systems in the Andes. *Int. J. Agric. Sustain.* **2023**, *21*, 2255449. [CrossRef]
62. Wu, H.; Li, X.; Lu, H.; Tong, L.; Kang, S. Crop Acreage Planning for Economy- Resource- Efficiency Coordination: Grey Information Entropy Based Uncertain Model. *Agric. Water Manag.* **2023**, *289*, 108557. [CrossRef]
63. McCarthy, D.I. Nutritional Intelligence in the Food System: Combining Food, Health, Data and AI Expertise. *Nutr. Bull.* **2025**, *50*, 142–150. [CrossRef] [PubMed]
64. Cassotta, M.; Forbes-Hernández, T.Y.; Calderón Iglesias, R.; Ruiz, R.; Elexpuru Zabaleta, M.; Giampieri, F.; Battino, M. Links between Nutrition, Infectious Diseases, and Microbiota: Emerging Technologies and Opportunities for Human-Focused Research. *Nutrients* **2020**, *12*, 1827. [CrossRef]
65. Papastratis, I.; Konstantinidis, D.; Daras, P.; Dimitropoulos, K. AI Nutrition Recommendation Using a Deep Generative Model and ChatGPT. *Sci. Rep.* **2024**, *14*, 14620. [CrossRef]
66. Ariyo Okaiyeto, S.; Mujumdar, A.S.; Prakash Sutar, P.; Liu, W.; Bai, J.; Xiao, H. Chatbots: A Critical Look into the Future of the Academia. *Int. J. Agric. Biol. Eng.* **2024**, *17*, 287–288. [CrossRef]
67. Suresh Babu, C.V.; Simon, P.A.; Santhosh, S.; Shameem Ahamed, S.M.; Haran, J.H. Enhancing Support and Monitoring in Eating Disorders Through AI-Powered Chatbots and Neural Networks. In *Neuroscientific Insights and Therapeutic Approaches to Eating Disorders*; IGI Global: New York, NY, USA, 2024; pp. 82–100.

foods

MDPI

Article

Beyond Traditional Methods: Deep-Learning Machines Empower Fingerroot (*Boesenbergia rotunda*)-Extract Production with Superior Antioxidant Activity

Padej Pao-la-or [1], Kakanang Posridee [2], Pussarat Buranakon [2], Jittra Singthong [3], Jirawan Oonmetta-Aree [4], Ratchadaporn Oonsivilai [2,5,*] and Anant Oonsivilai [1,*]

1 School of Electrical Engineering, Institute of Engineering, Suranaree University of Technology, Nakhon Ratchasima 30000, Thailand; padej@sut.ac.th

2 School of Food Technology, Institute of Agricultural Technology, Suranaree University of Technology, Nakhon Ratchasima 30000, Thailand; posridee.ka@gmail.com (K.P.); p.buranakon@gmail.com (P.B.)

3 Department of Agro-Industry, Faculty of Agriculture, Ubon Ratchathani University, Warinchamrap, Ubon Ratchathani 34190, Thailand; jittrawara@gmail.com

4 Food Science and Technology Program, Faculty of Science and Technology, Nakhon Ratchasima Rajabhat University, Nakhon Ratchasima 30000, Thailand; ojirawan@gmail.com

5 Health and Wellness Research Unit, Suranaree University of Technology, Nakhon Ratchasima 30000, Thailand

* Correspondence: roonsivi@g.sut.ac.th (R.O.); anant@sut.ac.th (A.O.); Tel.: +66-44224233 (R.O.); +66-44224400 (A.O.)

Abstract: This study investigated the impact of drying parameters on the quality of fingerroot (*Boesenbergia rotunda*) extract, focusing on phenolic compounds, flavonoids, and antioxidant activity. A Box–Behngen design was employed to evaluate the effects of maltodextrin concentration, inlet temperature, and outlet temperature on the extract's properties. The highest total phenolic content (18.96 μg of GAE/mg extract) and total flavonoid content (33.52 μg of GE/mg extract) were achieved using 20% maltodextrin, a 160 °C inlet temperature, and an 80 °C outlet temperature. Antioxidant activity, measured by DPPH and FRAP assays, was also influenced by drying parameters. Stepwise regression analysis revealed that maltodextrin concentration significantly affected all responses, while the inlet temperature had no significant effect. The outlet temperature significantly influenced FRAP activity. The developed mathematical models accurately predicted experimental values, validating the effectiveness of the RSM and Deep-Learning Machine. Optimal drying conditions for maximizing phenolic compounds were determined to be 20% maltodextrin, a 150 °C inlet temperature, and a 70 °C outlet temperature, resulting in TPC 15.33 μg of GAE/mg extract, TF 28.75 μg of GE/mg extract, IC_{50} value of 3.99 μg/mL, FRAP value at 4.44 μmoL Fe^{2+}/mg extract of phenolic content, and 18.96 μg of the GAE/mg extract. Similar conditions were found to be optimal for maximizing flavonoid content. These findings provide valuable insights for optimizing the drying process of fingerroot extract to preserve its bioactive compounds and enhance its potential applications.

Keywords: deep-learning machine; response-surface methodology; fingerroot; optimization; percentage yield; antioxidant activity

1. Introduction

Boesenbergia rotunda, a medicinal plant from the ginger family, has been extensively studied for its bioactive compounds and therapeutic properties. Researchers have employed various solvents, such as methanol, ethanol, water, and their combinations, along with extraction techniques like maceration, Soxhlet extraction, ultrasound-assisted extraction, and supercritical fluid extraction, to obtain these compounds [1–6]. The major bioactive compounds extracted include flavonoids (e.g., pinostrobin, pinocembrin, and alpinetin), chalcones (e.g., Panduratin A, Panduratin B, and Panduratin C), and essential oils (e.g., camphene, β-pinene, and borneol) [5,7,8]. These compounds have exhibited various bioactivities, such as antioxidant, anti-inflammatory, antimicrobial, and anticancer properties [9–11]. Optimization techniques, like response-surface methodology (RSM), have been employed to enhance the extraction yield and bioactivity [2,5], while purification methods, such as column chromatography, preparative HPLC, and crystallization, have been used to isolate and purify specific bioactive compounds from the crude extracts [7,8].

The extracts obtained from *Boesenbergia rotunda*, particularly those containing flavonoids, chalcones, and essential oils, have exhibited significant antioxidant properties in various studies. Cheng et al. [9] isolated and characterized antioxidant compounds from *B. rotunda* extracts, finding that the extracts contained substantial amounts of phenolic compounds and flavonoids, which contributed to their strong antioxidant activity. Morikawa et al. [11] investigated the structure–activity relationships of pimarane-type diterpenes from *B. rotunda*, and some of these diterpenes displayed notable antioxidant effects, suggesting their potential as natural antioxidant agents. Although studying a different plant species (*Alpinia zerumbet*), Elzaawely et al. [3] employed similar extraction methods and solvents as those used for *B. rotunda*, demonstrating the ability of these techniques to obtain antioxidant-rich extracts from related plants in the ginger family. These findings collectively indicate the significant antioxidant potential of the bioactive compounds extracted from *B. rotunda*, warranting further exploration of these extracts as natural sources of antioxidants for various applications.

Deep-learning and machine-learning techniques have been increasingly applied in various areas of biological sciences, including genomics and bioinformatics for genome sequencing and assembly, gene-expression analysis, protein structure prediction, and biomarker discovery [12,13]; proteomics and metabolomics for protein–protein interaction prediction, metabolic pathway analysis, and small molecule identification and characterization [14,15]; molecular modeling and drug discovery for the virtual screening of drug candidates, protein–ligand interactions and binding affinity predictions, and quantitative structure–activity relationship (QSAR) modeling [16,17]; biomedical imaging for image segmentation and analysis, disease diagnosis and classification, and microscopy image analysis [18,19]; and systems biology and computational biology for modeling biological networks and pathways, simulation of biological systems, and integration and analysis of multi-omics data [20,21]. Deep-learning architectures like convolutional neural networks (CNNs), recurrent neural networks (RNNs), and deep-belief networks have been successfully applied in these areas, often outperforming traditional machine-learning methods in tasks such as pattern recognition, image analysis, and sequence analysis [22,23]. While the provided references do not cover these applications, deep learning and machine learning are rapidly emerging fields in biological sciences, offering powerful tools for data analysis, modeling, and prediction in various domains.

Deep-learning models are built on artificial neural networks with multiple interconnected layers, enabling them to learn hierarchical representations of data and automatically extract increasingly complex features from raw inputs, making them effective for tasks like image recognition, natural language processing, and speech recognition [22,23]. The backpropagation algorithm, a supervised learning technique that calculates error gradients to adjust network parameters, is a fundamental component that allows for the efficient training of these multi-layer networks [24]. Nonlinear activation functions, such as the rectified linear unit (ReLU), introduce nonlinearity into the models, enhancing their ability to learn complex, nonlinear relationships within the data [25]. Additionally, deep learning leverages the principle of transfer learning, where pre-trained models can be fine-tuned on specific tasks, transferring knowledge gained from large datasets to related domains with limited data, a process known as feature transferability [26]. Furthermore, deep-learning models can scale to handle large datasets and benefit from parallel computing architectures like GPUs, accelerating the training process and enabling the development of increasingly complex and powerful models [23].

Artificial intelligence (AI) and machine learning (ML) have emerged as indispensable tools for optimizing food-drying processes. ML models, ranging from shallow- to deep-learning techniques, have effectively addressed challenges like uneven drying, nutrient degradation, and high energy consumption. By intelligently configuring these models and integrating them with real-time measurement systems, industries can achieve dynamic optimization and automated decision-making. Moreover, AI's ability to predict drying times and analyze energy usage patterns offers significant potential for resource efficiency. Despite these advancements, challenges such as data quality, integration complexity, and ethical considerations persist [27].

Deep learning (DL) offers significant advantages in food-drying optimization. DL models can accurately predict key parameters like drying time, energy consumption, and product quality, enabling efficient process control. By integrating DL with real-time measurements, drying parameters can be dynamically adjusted to maintain optimal conditions. Furthermore, DL-based algorithms can automate decision-making and reduce human errors, improving process efficiency. Additionally, DL can identify opportunities to optimize resource usage, such as minimizing energy consumption and waste, contributing to sustainable food production [28–31].

Deep learning, a subset of machine learning, has emerged as a promising tool for optimizing food-drying processes. These studies demonstrate the potential of DL to enhance various aspects of food drying, including modeling, prediction, and process control. By leveraging the power of AI, DL can address challenges such as energy efficiency, product quality, and operational optimization, leading to more sustainable and efficient food production [28–31].

This research aims to enhance the antioxidant properties of *Boesenbergia rotunda* extract through an optimized drying process. By employing Response Surface Methodology (RSM) and Dynamic Linear Models (DLM), we sought to identify the optimal drying conditions that maximize antioxidant activity while minimizing product degradation. This study contributes to the development of efficient and effective drying techniques for preserving bioactive compounds in natural products.

2. Materials and Methods

All chemicals of analytical grade, including 1,1-diphenyl-2-picryhydrazyl free radical (DPPH), 2,4,6-tri(2-pyridyl)-s-triazine (TPTZ), ferric, chloride-6-hydrate, ferrous, sulphate 7-hydrate, acetate buffer pH 4.6, gallic acid, Folin–Ciocalteu phenol reagent, anhydrous sodium carbonate (Na_2CO_3), catechin, sodium nitrite ($NaNO_2$), and aluminum chloride ($AlCl_3$), were obtained from Sigma–Aldrich Co. (St. Louis, MO, USA). Solvents such as ethanol were purchased from Mallinckrodt–Baker (Phillipsburg, NJ, USA).

2.1. Fingerroot Powder Preparation

Fingerroot powder was prepared from *Boesenbergia rotunda* specimens collected in Nakhon Ratchasima Province, Thailand, during November 2023. The collected finger roots were thoroughly rinsed to remove any adhering dirt or debris and subsequently subjected to a drying process at 60 °C for 24 h to remove moisture content. After drying, the fingerroot was ground into a fine powder using an Ultra Centrifugal Mill Model ZM-1000 (Retsch, Hann, Germany), and the resulting powder was sieved through a mesh size of 0.2 mm to obtain a uniform particle-size distribution. Finally, the sieved finger root powder was stored in vacuum-sealed packages at a temperature of −20 °C until further use, ensuring the maintenance of its quality and preventing potential degradation during storage. This standardized procedure facilitated the preparation of a homogeneous fingerroot powder from *Boesenbergia rotunda*, suitable for various applications, such as extraction of bioactive compounds or incorporation into formulations.

2.2. Fingerroot Extraction

The ethanol-extraction method for fingerroot (*Boesenbergia rotunda*) followed a similar procedure to the water extraction, except 57% ethanol was used as the solvent instead of water. One gram of finger root powder was accurately weighed and extracted three times with 10 mL of 57% ethanol. The combined extract was then adjusted to a final volume of 50 mL. Aliquots of 2 mL were taken from the adjusted extract and transferred into individual test tubes. Unlike water extraction, the ethanol extracts were dried using a vacuum dryer instead of freeze-drying. The dried extracts obtained from the vacuum drying process were stored at −20° C until further use or analysis [24].

2.3. Total Phenolic Contents

The total soluble phenolic constituents of the extracts were determined using the Folin–Ciocalteu method, with gallic acid employed as the standard reference compound [3,26]. In this procedure, 20 μL of the crude extract solution, standard gallic acid solution, or blank solvent were added to a 1.5 mL cuvette, followed by the addition of 1.58 mL of deionized water and 100 μL of the Folin–Ciocalteu reagent. The mixture was thoroughly mixed and incubated for 5 min at room temperature. After incubation, 300 μL of a 2% (*w*/*v*) sodium carbonate (Na_2CO_3) solution were introduced, and the mixture was allowed to stand at room temperature for 2 h to facilitate the development of the blue-colored complex resulting from the reduction of the Folin–Ciocalteu reagent by the phenolic compounds present in the sample. Subsequently, the absorbance of the solution was measured at a wavelength of 765 nm using a spectrophotometer. The total soluble phenolic content was quantified by establishing a calibration curve with known concentrations of the gallic acid standard, and the results were expressed as gallic acid equivalents (GAE) based on the linear regression equation obtained from the calibration curve [3,4,26].

2.4. Total Flavonoid

The total flavonoid contents of the extracts were determined using the aluminum chloride colorimetric method [4,31]. In this assay, 250 μL of the crude extract solution, standard solution, or blank were added to a 1.5 mL cuvette, followed by the addition of 1.25 mL of deionized (DI) water and 75 μL of 5% sodium nitrite ($NaNO_2$) solution. The mixture was incubated for 6 min at room temperature. After incubation, 150 μL of 10% aluminum chloride ($AlCl_3$) solution were introduced, followed by the sequential addition of 0.5 mL of 1 M sodium hydroxide (NaOH) and 275 μL of DI water. The resulting mixture was then incubated for an additional 5 min to allow the development of the pink-colored flavonoid–aluminum complex. Subsequently, the absorbance of the solution was measured at a wavelength of 510 nm using a spectrophotometer. The total flavonoid contents were quantified by establishing a calibration curve with known concentrations of the catechin standard, and the results were expressed as catechin equivalents based on the linear regression equation obtained from the calibration curve.

2.5. DPPH Assay

The DPPH (2,2-diphenyl-1-picrylhydrazyl) free-radical scavenging activity of Thai basil extracts (water, ethanol, and ethyl acetate), as well as the positive controls butylated hydroxytoluene (BHT) and ascorbic acid, were evaluated using a DU 800 Spectrophotometer (Beckman Coulter, Brea, CA, USA) according to the method described by Oonsivilai et al. [5]. An aliquot (100 μL) of the extract at varying concentrations or the positive control solutions prepared in methanol was mixed with 1.9 mL of a methanolic DPPH solution. The reaction mixtures were incubated in the dark for a specified duration, during which the DPPH radical was reduced by the antioxidant compounds present in the extracts or positive controls, resulting in a discoloration of the DPPH solution. The decrease in absorbance at a specific wavelength (typically 515–520 nm) was measured spectrophotometrically and used to calculate the percentage of DPPH radical scavenging activity. The IC_{50} values, representing the concentration of the extract or positive control required to scavenge 50% of the DPPH radicals, were determined by nonlinear regression analysis using SigmaPlot 9.1 software (Systat Software Inc., Chicago, IL, USA) [5].

2.6. FRAP Assay

The ferric-reducing antioxidant power (FRAP) assay was performed according to the method described by Oonsivilai et al. [5]. The FRAP reagent was prepared by combining an acetate buffer (pH 3.6), a 10 mM solution of 2,4,6-tripyridyl-s-triazine (TPTZ) in 40 mM hydrochloric acid, and a 20 mM ferric chloride solution in the ratio of 10:1:1 (*v*/*v*/*v*), respectively. An aliquot (50 μL) of the extract was added to 1.5 mL of the freshly prepared FRAP reagent, and the mixture was incubated for 4 min at room temperature to facilitate the reduction of the ferric–tripyridyltriazine complex to the ferrous form by the antioxidant compounds present in the extract. After the incubation period, the absorbance of the reaction mixture was measured at 593 nm using a spectrophotometer. A calibration curve was constructed using ferric sulfate solutions of known concentrations ranging from 100 to 2000 μM. The FRAP values of the extracts were calculated from the linear regression equation obtained from the calibration curve and expressed as millimoles of ferric equivalents per gram of dry plant material [5,26].

2.7. Spray Drying Condition

2.7.1. Experimental Design

Response Surface Methodology (RSM), a well-established optimization technique, was employed to identify the optimal drying conditions that maximize the extraction yield, total phenolic contents, total flavonoid contents, and antioxidant activities [31–33]. Three independent variables—maltodextrin concentration (X_1: 20, 25, and 30% *w*/*v*), inlet temperature (X_2: 140, 150, and 160 °C), and outlet temperature (X_3: 70, 80, and 90 °C)—were investigated for their influence on the dependent variable, extraction yield. A Box–Behnken design was utilized to randomize the experimental order, and the resulting data were analyzed using regression analysis [34,35]. The independent variables are in Table 1.

Table 1. Encoded and coded levels of independent variables used in the experimental design.

Symbols	Independent Variables	Coded Levels		
		−1	0	1
X_1	Maltrodextrin (% *w*/*v*))	20	25	30
X_2	Inlet Temperature (°C)	140	150	160
X_3	Outlet Temperature (°C)	70	50	90

2.7.2. Experimental Validation of the Optimal Conditions

A model-driven optimization approach was employed to maximize the extraction yield, total phenolic contents (TPC), total flavonoid contents (TFC), and antioxidant activities as measured by DPPH and FRAP assays. The validity of the model was subsequently assessed by performing three independent extraction experiments under the predicted optimal conditions, which deviated from those employed for model development.

2.7.3. Deep-Learning Machine Modeling

Deep-learning models are built on artificial neural networks with multiple interconnected layers, enabling them to learn hierarchical representations of data and automatically extract increasingly complex features from raw inputs, making them effective for tasks like image recognition, natural language processing, and speech recognition [22,23]. The backpropagation algorithm, a supervised learning technique that calculates error gradients to adjust network parameters, is a fundamental component allowing for efficient training of these multi-layer networks [35]. Nonlinear activation functions, such as the rectified linear unit (ReLU), introduce nonlinearity into the models, enhancing their ability to learn complex, nonlinear relationships within the data [36]. Additionally, deep learning leverages the principle of transfer learning, where pre-trained models can be fine-tuned on specific tasks, transferring knowledge gained from large datasets to related domains with limited data, a process known as feature transferability [37]. Furthermore, deep-learning models can scale to handle large datasets and benefit from parallel computing architectures like GPUs, accelerating the training process and enabling the development of increasingly complex and powerful models [23].

2.8. Statistical Analysis

All experiments were performed in triplicate, and mean values (on a dry basis) with standard deviations were reported. The experimental data were analyzed using an analysis of variance (ANOVA). SPSS® software Version 17 (SPSS Inc., Chicago, IL, USA) was used to perform all statistical calculations.

3. Results

3.1. Validation of the Experimental Design

This study investigated the influence of drying parameters on the quality of a fingerroot extract initially obtained using 57% ethanol. A factorial design was employed, analyzing the effects of maltodextrin concentration (20, 25, 30% w/v), inlet temperature (140, 150, 160 °C), and outlet temperature (70, 80, 90 °C) on the extract's content of phenolic compounds, flavonoids, and its free-radical scavenging activity.

The total phenolic contents in the dried fingerroot extract ranged from 5.63 to 18.96 μg of GAE/mg extract. The combination of 20% maltodextrin, 160 °C inlet temperature, and 80 °C outlet temperature yielded the highest phenolic content. Similarly, the total flavonoid contents varied between 5.55 and 33.52 μg of GE/mg extract, with the same drying conditions (20% maltodextrin, 160 °C inlet temperature, 80 °C outlet temperature) producing the most significant amount.

Free-radical scavenging activity, measured by the DPPH method, ranged from 0.67 to 4.22 μg/mL of extract. Interestingly, drying conditions with either 20% or 30% maltodextrin, inlet temperatures of 140 or 150 °C, and outlet temperatures of 70 or 90 °C resulted in the highest IC_{50} and FRAP values (1.28 to 4.829 μmol Fe^2/mg extract). However, the drying condition with 20% maltodextrin, 160 °C inlet temperature, and 80 °C outlet temperature exhibited the strongest FRAP value. For detailed results of all experimental conditions, please refer to Tables 2 and 3.

When taking the total amount of phenolic compounds obtained from dried fingerroot extracts, the relationship is found with the independent variables using stepwise regression analysis, which can display results in the form of regression analysis. Details are shown in Tables 4–6.

Tables 2 and 3 present the fingerroot experimental conditions with the predicted experimental values, and deep-learning machine results of TPC, TF, DPPH, and FRAP assays, as well as the extraction yield of the recovered bioactive compounds from fingerroot.

The dried extract yield ranges from 23.34 ± 0.68% (run 15, 30%, 140 °C, 70 °C) to 67.29 ± 0.46% (run 5, 30%, 150 °C, 70 °C). The TPC ranged from 5.63 μg of the GAE/mg extract (run 13, 25%, 160 °C, 90 °C) to 18.96 μg of the GAE/mg extract (run 9, 20%, 160 °C, 80 °C). The TF ranged from 5.55 μg of the GE/mg extract (run 13, 25%, 160 °C, 90 °C) to 33.52 μg of the GE/mg extract (run 9, 20%, 160 °C, 80 °C). The antioxidant activity evaluated by the DPPH assay ranged between IC50 4.22 μg/mL (run 1, 20%, 140 °C, 80 °C) and IC50 0.67 μg/mL (run 10 and 5, 30%, 150 °C, 70 °C). In what concerns the FRAP assay, results ranged between 1.28 μmol of the FSE/mg extract (run 6, 25%, 150 °C, 80 °C) and 4.83 μmol of the FSE/mg extract (run 9, 20%, 160 °C, 80 °C). Interestingly, the results obtained for the drying condition at maltodextrin 20%, inlet temperature 150 °C, and outlet temperature 70 °C, were similar for both antioxidant activity responses. In addition, the inlet temperature and outlet temperature did not affect the DPPH and FRAP responses. The highest DPPH response (1.24 μg/mL) was achieved at maltodextrin 20%, inlet temperature 150 °C, and outlet temperature 70 °C.

The experimental values for the four responses were similar ($p > 0.05$) to the respective predicted ones and DLM data, allowing the validation of the experimental design and demonstrating the effectiveness and adequacy of the RSM and Deep-Learning Machine (Tables 4–6).

Table 2. Experimental and predicted values of TPC and TF of *Boesenbergia rotunda* crude extract.

Independent Variables				Dependent Variables					
Point	SWE Condition			Y1, TPC (μg GAE/mg)			Y2, TF (μg GE/mg)		
Run	X_1 (M, % *w*/*v*)	X_2 (T (In), °C)	X_3 (T (Out), °C)	Exp	Pred	DLM	Exp	Pred	DLM
1	20	140	80	14.97 ± 4.24 ^{defg}	14.89	14.92	27.74 ± 3.95 ^{e}	25.86	27.55
2	30	140	80	12.17 ± 0.81 ^{bcde}	10.83	12.15	15.02 ± 2.14 ^{c}	19.79	15.00
3	25	160	70	16.77 ± 6.22 ^{efg}	12.45	16.77	11.63 ± 0.79 ^{abc}	10.20	11.52
4	20	150	70	8.86 ± 0.40 ^{abc}	15.34	8.81	13.63 ± 5.95 ^{bc}	19.83	13.53
5	30	150	70	7.05 ± 1.07 ^{ab}	11.27	6.97	7.27 ± 1.37 ^{ab}	6.82	7.05
6	25	150	80	10.31 ± 1.85 ^{abcd}	12.01	11.55	12.3 ± 3.79 ^{abc}	10.01	9.87
7	25	150	80	13.78 ± 2.94 ^{cdef}	12.01	11.55	8.94 ± 2.62 ^{abc}	10.01	9.87
8	30	150	90	10.54 ± 0.46 ^{abcd}	8.68	10.52	10.06 ± 1.97 ^{abc}	3.86	10.03
9	20	160	80	18.96 ± 2.62 ^{g}	13.19	18.60	33.52 ± 9.21 ^{e}	28.75	33.23
10	25	140	90	11.97 ± 2.67 ^{bcde}	11.57	11.96	6.63 ± 1.04 ^{ab}	8.06	6.56
11	20	150	90	11.96 ± 0.68 ^{bcde}	12.75	11.97	13.19 ± 3.32 ^{bc}	13.64	13.03
12	30	160	80	8.72 ± 2.16 ^{abc}	9.13	8.71	10.16 ± 1.44 ^{abc}	12.04	10.02
13	25	160	90	5.63 ± 1.60 ^{a}	9.87	5.55	5.55 ± 0.20 ^{a}	9.87	5.57
14	25	150	80	10.7 ± 3.39 ^{abcd}	12.01	11.55	8.8 ± 3.38 ^{abc}	10.01	9.87
15	25	140	70	17.77 ± 2.47 ^{fg}	14.15	17.75	21.21 ± 0.78 ^{d}	16.89	20.96

Exp: Experimental values, performed in a random order and expressed as the average of triplicate determinations from different experiments ($n = 3$). Pred: Predicted values, based on BBD evaluation. DLM: Deep-Learning Machine. $^{a–g}$ Same letter show no significant difference at $p < 0.05$, different letter shows significant difference ($p < 0.05$) in the same column.

Table 3. Experimental and predicted values of DPPH, FRAP, and yield of *Boesenbergia rotunda* crude extract obtained by BBD.

Independent Variables				Dependent Variables								
Point	SWE Condition			Y3, DPPH (μg/mL)			Y4, FRAP (μmol Fe^{2+}/mg)			Y5, Yield (%)		
Run	X_1 (M, % *w/v*)	X_2 (T (In), °C)	X_3 (T (Out), °C)	Exp [a]	Pred [b]	DLM [c]	Exp [a]	Pred [b]	DLM [c]	Exp [a]	Pred [b]	DLM [c]
1	20	140	80	4.22 ± 0.12 [d]	3.83	4.10	4.278 ± 0.77 [fg]	4.30	4.281	36.4 ± 0.96 [c]	35.22	36.42
2	30	140	80	2.95 ± 1.11 [bcd]	1.98	2.93	3.216 ± 0.24 [de]	3.60	3.220	62.72 ± 2.35 [hi]	59.90	62.80
3	25	160	70	1.95 ± 0.41 [abc]	2.73	1.93	3.219 ± 0.30 [de]	3.21	3.221	52.41 ± 1.90 [e]	51.87	52.53
4	20	150	70	3.21 ± 0.40 [cd]	3.96	3.20	3.608 ± 1.04 [ef]	4.00	3.610	45.17 ± 1.69 [d]	35.67	45.20
5	30	150	70	0.67 ± 0.88 [a]	2.11	0.70	2.911 ± 0.15 [cde]	2.94	2.911	67.29 ± 0.46 [j]	60.34	66.64
6	25	150	80	2.02 ± 0.64 [abc]	2.60	3.02	1.28 ± 0.26 [a]	1.84	1.840	57.23 ± 2.09 [fg]	51.42	56.86
7	25	150	80	3.06 ± 0.38 [cd]	2.60	3.02	2.161 ± 0.14 [bc]	1.84	1.840	54.94 ± 1.65 [ef]	51.42	56.86
8	30	150	90	1.44 ± 0.40 [ab]	1.23	1.45	1.893 ± 0.22 [ab]	1.50	1.890	60.93 ± 2.72 [gh]	67.17	60.93
9	20	160	80	4.14 ± 0.92 [d]	3.22	4.05	4.829 ± 0.75 [g]	4.44	4.776	32.21 ± 2.07 [b]	42.95	32.25
10	25	140	90	0.67 ± 0.25 [a]	2.46	0.68	2.637 ± 0.52 [bcd]	2.65	2.639	53.7 ± 1.15 [ef]	50.98	53.70
11	20	150	90	3.03 ± 1.22 [cd]	3.09	3.00	3.247 ± 0.27 [de]	3.22	3.251	44.36 ± 2.11 [d]	42.50	44.41
12	30	160	80	2.13 ± 1.52 [abc]	1.37	2.11	2.38 ± 0.14 [bcd]	2.36	2.380	65.89 ± 1.63 [ij]	67.62	66.00
13	25	160	90	1.31 ± 0.68 [a]	1.85	1.32	2.177 ± 0.30 [bc]	2.59	2.181	56.55 ± 2.21 [f]	58.70	56.60
14	25	150	80	4.05 ± 0.56 [d]	2.60	3.02	2.076 ± 0.53 [abc]	1.84	1.840	58.19 ± 0.68 [fg]	51.42	56.86
15	25	140	70	4.13 ± 1.17 [d]	3.34	4.02	4.666 ± 0.42 [g]	4.25	4.672	23.34 ± 0.68 [a]	44.14	23.46

[a] Experimental values, performed in a random order and expressed as the average of triplicate determinations from different experiments ($n = 3$). [b] Predicted values, based on BBD evaluation. [c] Deep-Learning Machine. [a–j] Same letter show no significant difference at $p < 0.05$, different letter shows significant difference ($p < 0.05$) in the same column.

Table 4. Model summary and analysis of variance (ANOVA) of TPC and TF *Boesenbergia rotunda* crude extract.

Source	Sum of Squares	df	Mean Square	F-Value	*p*-Value		Source	Sum of Squares	df	Mean Square	F-Value	*p*-Value	
Model	52.26	3	17.42	1.22	0.3473	not significant	Model	705.63	9	78.40	2.19	0.2012	not significant
A-X_1	33.09	1	33.09	2.32	0.1556		A-X_1	259.58	1	259.58	7.24	0.0432	
B-X_2	5.78	1	5.78	0.4059	0.5371		B-X_2	11.86	1	11.86	0.3309	0.5900	
C-X_3	13.39	1	13.39	0.9404	0.3530		C-X_3	41.91	1	41.91	1.17	0.3289	
							AB	28.30	1	28.30	0.7898	0.4149	
							AC	2.61	1	2.61	4.87	0.7981	
							BC	18.06	1	18.06	2.74	0.5095	
							A^2	119.53	1	119.53	1.88	0.1274	
							B^2	128.84	1	128.84	1.64	0.1164	
							C^2	80.37	1	80.37	2.39	0.1945	
Residual	156.62	11	14.24				Residual	179.18	5	35.84			
Lack of Fit	149.40	9	16.60	4.59	0.1915	not significant	Lack of Fit	171.32	3	57.11	14.54	0.0650	not significant
Pure Error	7.23	2	3.61				Pure Error	7.85	2	3.93			
Cor Total	208.88	14					Cor Total	884.81	14				
				Std.Dev = 3.77 R-Squared = 0.2502 Mean = 12.01 R-Squared = 0.0457 C.V. % = 3.42 Adeq Precision = 3.4153							Std.Dev = 5.99 R-Squared = 0.7975 Mean = 13.71 R-Squared = 0.4330 C.V. % = 43.66 Adeq Precision = 5.0921		

Table 5. Model summary and analysis of variance (ANOVA) of DPPH and FRAP *Boesenbergia rotunda* crude extract.

Source	Sum of Squares	df	Mean Square	F-Value	*p*-Value		Source	Sum of Squares	df	Mean Square	F-Value	*p*-Value	
Model	9.15	3	3.05	2.66	0.1002	not significant	Model	13.78	9	1.53	5.37	0.0394	significant
A-X_1	6.86	1	6.86	5.98	0.0325		A-X_1	3.87	1	3.87	13.56	0.0143	
B-X_2	0.7442	1	0.7442	0.6486	0.4377		B-X_2	0.6006	1	0.6006	2.11	0.2064	
C-X_3	1.54	1	1.54	1.34	0.2712		C-X_3	2.48	1	2.48	8.68	0.0320	
							AB	0.4809	1	0.4809	1.69	0.2507	
							AC	0.1079	1	0.1079	0.3785	0.5653	
							BC	0.2435	1	0.2435	0.8542	0.3978	
							A^2	2.29	1	2.29	8.05	0.0364	
							B^2	4.06	1	4.06	14.23	0.0130	
							C^2	0.3049	1	0.3049	1.07	0.3485	
Residual	12.62	11	1.15				Residual	1.43	5	0.2851			
Lack of Fit	10.56	9	1.17	1.14	0.5516	not significant	Lack of Fit	0.9532	3	0.3177	1.35	0.4532	not significant
Pure Error	2.06	2	1.03				Pure Error	0.4723	2	0.2362			
Cor Total	21.77	14					Cor Total	15.21	14				
			Std.Dev = 1.07 R-Squared = 0.4202 Mean = 2.60 R-Squared = 0.2621 C.V. % = 41.22 Adeq Precision = 4.9353								Std.Dev = 0.5339 R-Squared = 0.9063 Mean = 2.97 R-Squared = 0.7376 C.V. % = 17.97 Adeq Precision = 6.7545		

Table 6. Model summary and analysis of variance (ANOVA) of %yield *Boesenbergia rotunda* crude extract.

Source	Sum of Squares	df	Mean Square	F-Value	*p*-Value	
Model	1430.18	3	476.73	6.20	0.0101	significant
A-X_1	1217.46	1	1217.46	15.84	0.0022	
B-X_2	119.35	1	119.35	1.55	0.2387	
C-X_3	93.37	1	93.37	1.21	0.2940	
Residual	845.68	11	76.88			
Lack of Fit	840.10	9	93.34	33.48	0.0293	significant
Pure Error	5.58	2	2.79			
Cor Total	2275.86	14				
				Std.Dev = 8.77 R-Squared = 0.6384 Mean = 51.42 R-Squared = 0.5271 C.V. % = 17.05 Adeq Precision = 7.1552		

3.2. *Analysis of Response Surface*

The independent variable X_1 (maltodextrin concentration, % *w*/*v*) showed a significant effect ($p < 0.05$) on TF, DPPH, FRAP, and yield response, whereas the independent variable X_2 (Inlet temperature, °C) showed no significant effect ($p > 0.05$) on all responses. In addition, the X_3 (Outlet temperature, °C) showed a significant effect ($p < 0.05$) on FRAP response. Furthermore, the quadratic term for X_1 and X_2 showed a significant effect on Y4 response (FRAP). The R^2 is useful for checking the model fitness, being in this study classified as strong for FRAP response (R^2 adjust was relatively close to 1) (Tables 4–6). Considering all response-surface models, the "lack of fit" was significant ($p < 0.05$) for yield response but not significant ($p > 0.05$) for TPC, TF, DPPH, and FRAP response.

3.2.1. Fitting Model

The parameters of the equation were determined through linear and multiple regression analysis of the experimental results.

$$\text{Total phenolic contents} = 45.27942 - 0.406750X_1 - 0.085X_2 - 0.129375X_3$$

$$\text{Total flavonoid contents} = 1335.61375 - 5.83042X_1 - 18.213\,X_2 + 3.64454\,X_3 - 0.0532\,X_1X_2 + 0.01615\,X_1X_3 + 0.02125\,X_2X_3 + 0.227583\,X_1{}^2 + 0.059071\,XX_2{}^2 - 0.046654\,X_3{}^2$$

$$\text{DPPH} = 15.31492 - 0.18525X_1 - 0.0305X_2 - 0.043875X_3$$

$$\text{FRAP} = 284.89475 - 0.412750\,X_1 - 3.19655\,X_2 - 0.803425\,X_3 - 0.006935\,X_1X_2 - 0.003285\,X_1X_3 + 0.002467\,X_2X_3 + 0.031535\,X_1{}^2 + 0.010484\,X_2{}^2 - 0.002874\,X_3{}^2$$

$$\text{Yield} = -95.52675 + 2.46725X_1 + 0.386250X_2 + 0.341625X_3$$

The RSM model was used to generate 3D contour plots in order to represent graphically the relationship between independent (maltodextrin concentration, inlet temperature, outlet temperature) and dependent variables (TPC, TF, DPPH, FRAP, Yield). As shown in Figures 1–5.

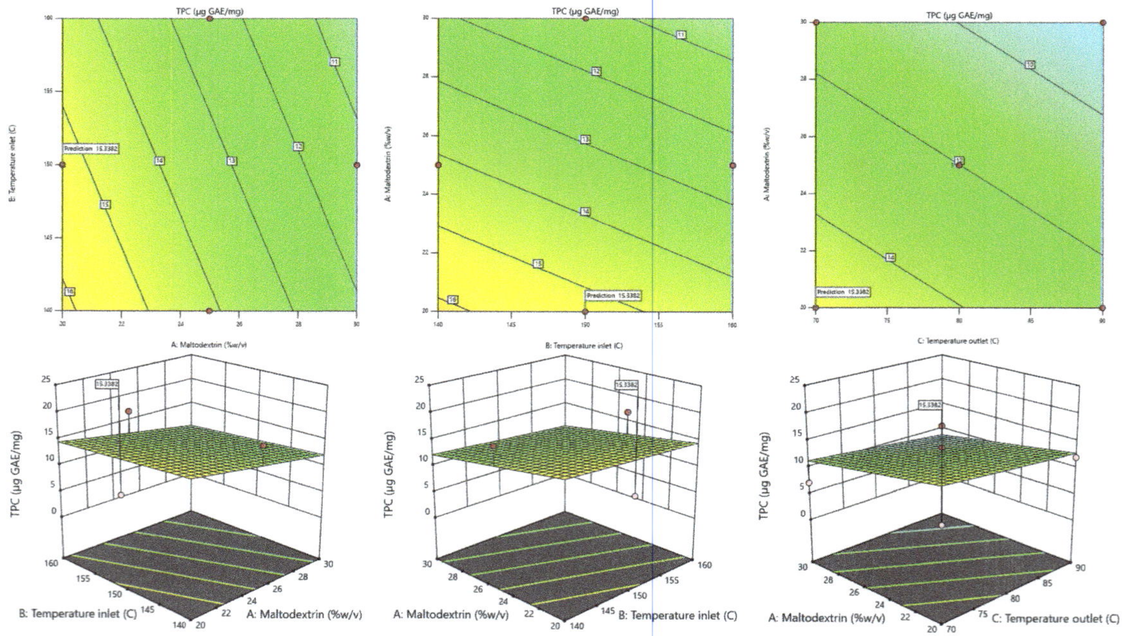

Figure 1. Response-surface plots depicting the interactive effects of total phenolic contents.

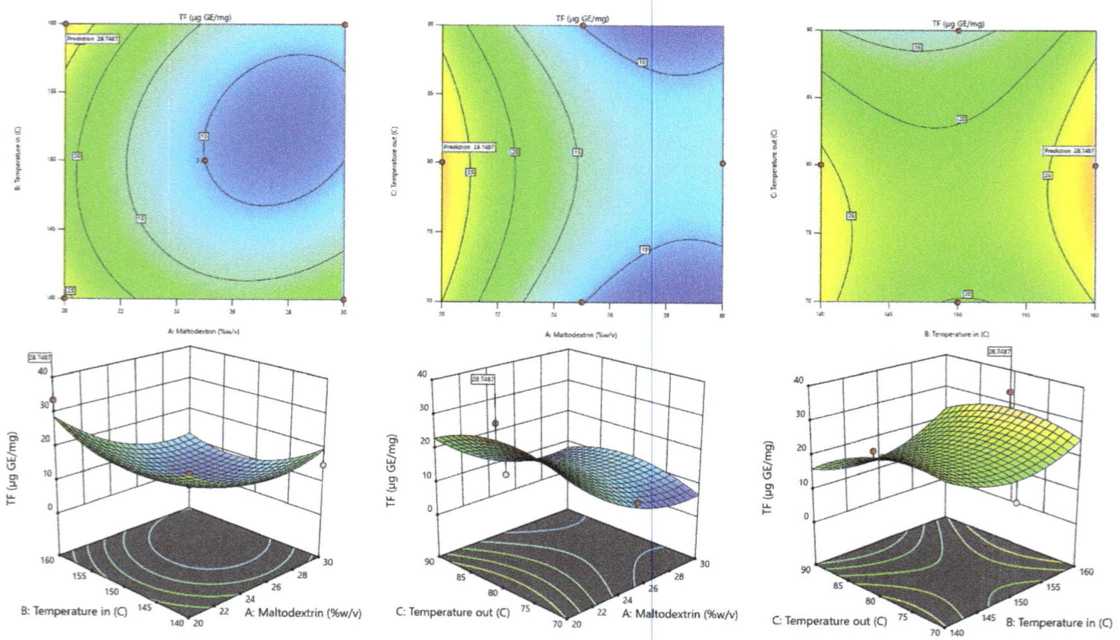

Figure 2. Response-surface plots depicting the interactive effects of total flavonoid contents.

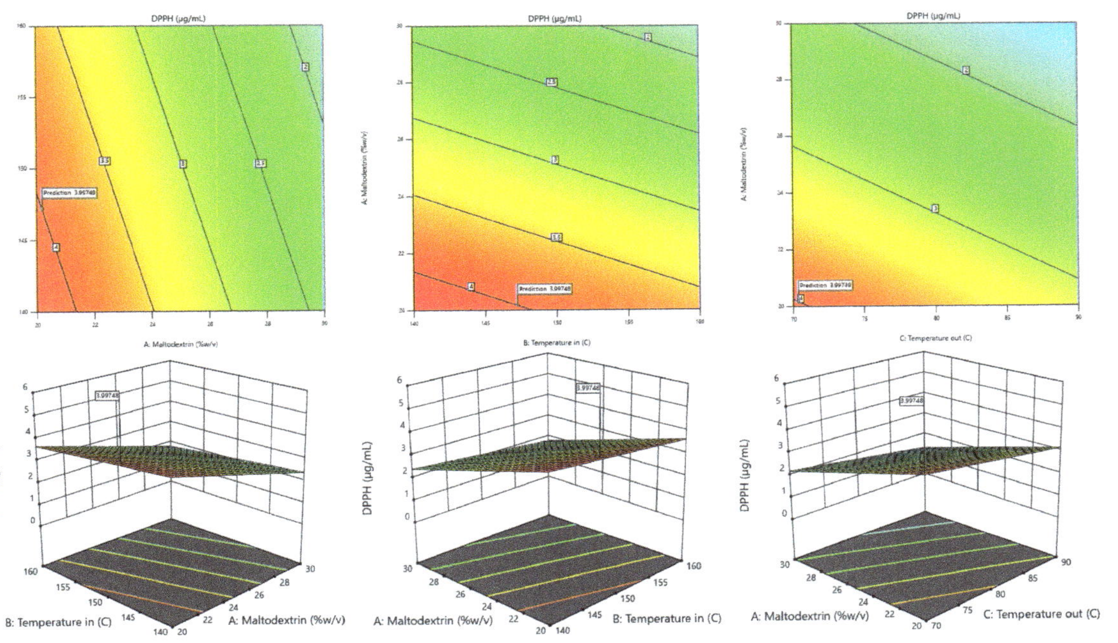

Figure 3. Response-surface plots depicting the interactive effects of IC_{50} value.

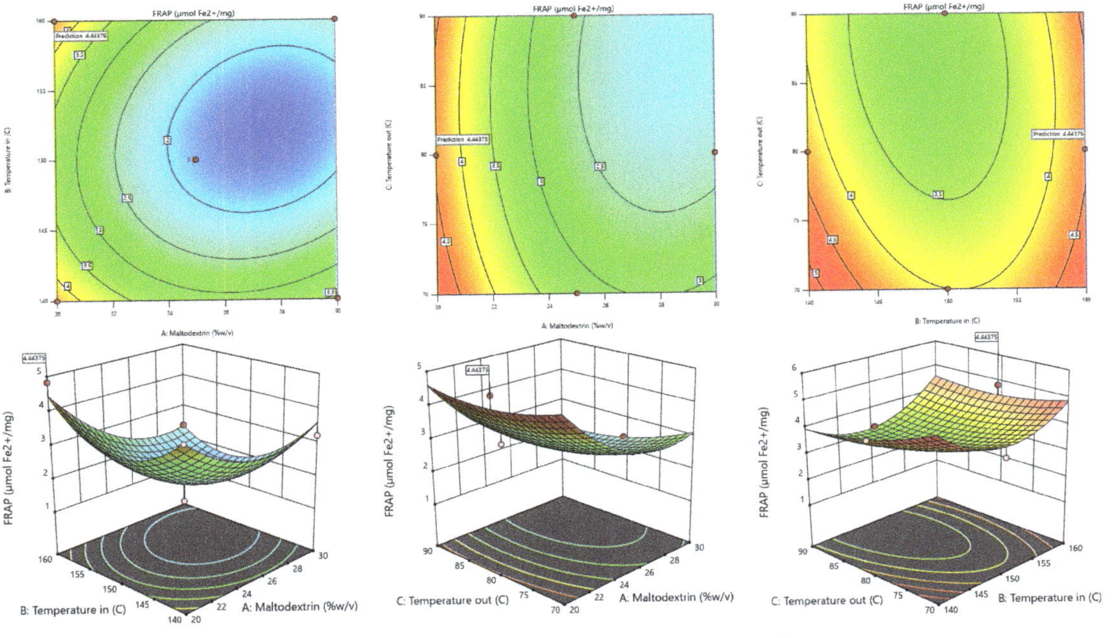

Figure 4. Response-surface plots depicting the interactive effects of FRAP value.

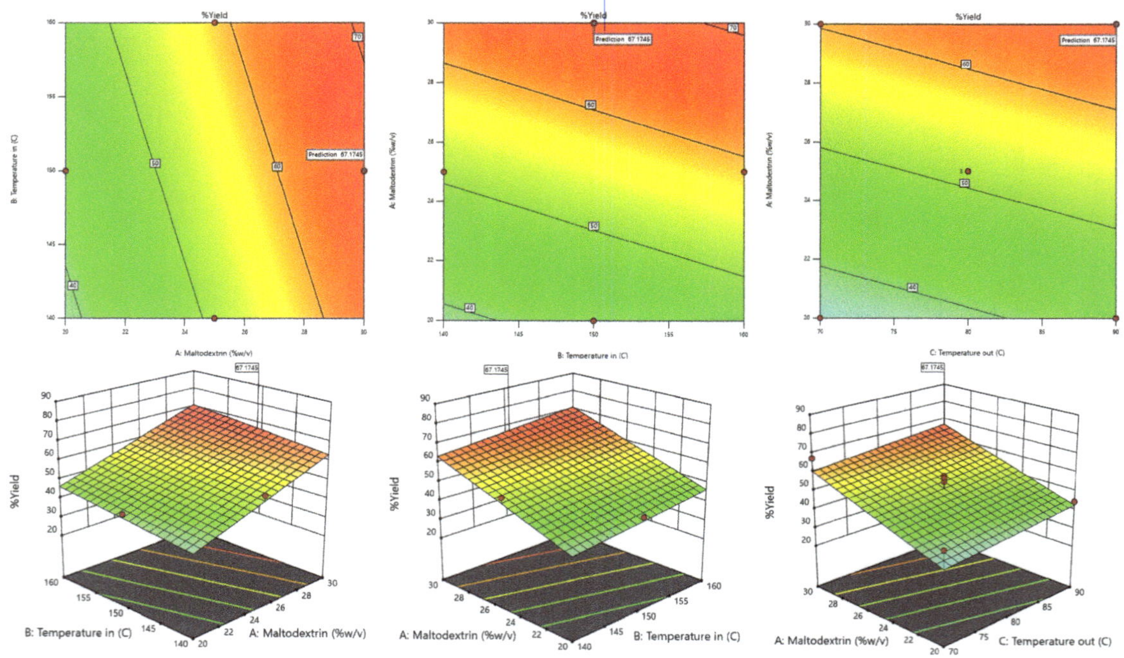

Figure 5. Response-surface plots depicting the interactive effects of yield percentage.

3.2.2. Total Phenolic Contents

When taking the TPC value obtained from dried fingerroot extract, the relationship with the independent variables is found by using Stepwise regression analysis, which can display results in the form of regression analysis. Details are shown in Table 4.

The table showing the results of the regression analysis determines the second-order mathematical model. This can be used to create a mathematical model or mathematical equation for predicting the dependent variable (response) by taking the values of the factors obtained from the analysis of the coefficients of the regression equation of the said dependent variable and writing them in the form The linear model equation is shown in Equation (1), which has an R^2 value of 0.2502.

$$Y1 = 45.27942 - 0.406750X_1 - 0.085X_2 - 0.129375X_3 \quad (1)$$

The reduced second-order mathematical model in Equation (1) is used as a function to determine the appropriate conditions for drying using a linear model solution, which helps to find the optimum condition that provides the highest amount of total phenolic contents. The results of the experiment show that the drying conditions were at a maltodextrin concentration of 20% w/v, an inlet temperature of 160 °C, and an outlet temperature of 80 °C, which results in the highest amount of phenolic compound When compared with the values calculated from model predictions at the same condition, it was found that the number of phenolic compounds was equal to 18.96 µg GAE/mg of dried fingerroot extract. When calculating the most suitable conditions for drying to obtain the highest number of phenolic compounds, a prediction by the model found that a maltodextrin concentration of 20% w/v, an inlet temperature of 150 °C, and an outlet temperature of 70 °C resulted in the number of phenolic compounds equal to 15.34 µg GAE/mg of dried fingerroot extract.

3.2.3. Total Flavonoid Contents

When taking the TF value obtained from dried fingerroot extract, the relationship with the independent variables was found using stepwise regression analysis, which can display results in the form of regression analysis. Details are shown in Table 4.

The results of the regression analysis can determine the second-order mathematical model, which can be used to create a mathematical model or mathematical equation for predicting the dependent variable (response) by taking the values of the factors obtained from the analysis of the coefficients of the regression equation of the said dependent variable and writing them in the form. The quadratic model equation is shown in Equation (2), which has an R^2 value of 0.7975.

$$Y1 = 1335.61375 - 5.83042X_1 - 18.213\,X_2 + 3.64454\,X_3 - 0.0532\,X_1X_2 + 0.01615\,X_1X_3 + 0.02125\,X_2X_3 + 0.227583\,{X_1}^2 + 0.059071\,{X_2}^2 - 0.046654\,{X_3}^2 \quad (2)$$

The reduced second-order mathematical model in Equation (2) is used as a function to determine the appropriate conditions for drying. A non-linear model solution can be used to find the optimum condition that provides the highest amount of total flavonoid contents. The results of the experiment show that the drying conditions were at a maltodextrin concentration of 20% *w/v*, an inlet temperature of 160 °C, and an outlet temperature of 80 °C, resulting in the highest number of flavonoid compounds. When compared with the values calculated from model predictions under the same conditions, the number of flavonoid compounds was found to be 33.52 µg GE/mg of dried fingerroot extract. When calculating the optimum drying conditions that resulted in the highest flavonoid content from a prediction by the model, it was found that the drying conditions at a maltodextrin concentration of 20% *w/v*, an inlet temperature of 160 °C, and an outlet temperature of 80 °C in the number of flavonoid compounds equaled to 28.75 µg GE/mg of dried fingerroot extract.

3.2.4. DPPH Assay

When taking the IC_{50} value obtained from dried fingerroot extract, the relationship with the independent variables was found using stepwise regression analysis, which can display results in the form of regression analysis. Details are shown in Table 5.

Based on the table showing the results of the regression analysis, the second-order mathematical model can be determined and used to create a mathematical model or mathematical equation for predicting the dependent variable (response) by taking the values of the factors obtained from the analysis of the coefficients of the regression equation of the said dependent variable and writing them in the form. The linear model equation is shown in Equation (3), which has an R^2 value of 0.4202.

$$Y1 = 15.31492 - 0.18525X_1 - 0.0305X_2 - 0.043875X_3 \quad (3)$$

The reduced second-order mathematical model in Equation (3) is used as a function to determine the appropriate conditions for drying using a linear model solution and find the optimum condition with the highest IC_{50} value. The results of the experiment show that the drying conditions using maltodextrin concentrations of 20 % *w/v*, inlet temperatures of 140, and an outlet temperature of 80 °C result in the highest IC_{50} value. When compared with the values calculated from model predictions at the same conditions, it was found that the IC_{50} value was equal to 4.22 µg/mg of dried fingerroot extract, which, when calculating the most suitable conditions for drying, resulted in the highest IC_{50} value from the model prediction. It was found that the drying condition at a maltodextrin concentration of 20.191% *w/v*, an inlet temperature of 147.227 °C, and an outlet temperature of 70.352 °C resulted in an IC_{50} value of 3.99748 µg/mg of dried fingerroot extract.

3.2.5. FRAP Assay

When taking the FRAP value obtained from dried fingerroot extract, the relationship with the independent variables was found using stepwise regression analysis, which can display results in the form of regression analysis. Details are shown in Table 5.

Based on the table showing the results of the regression analysis, the second-order mathematical model can be determined and used to create a mathematical model or mathematical equation for predicting the dependent variable (response) by taking the values of the factors obtained from the analysis of the coefficients of the regression equation of the said dependent variable and writing them in the form The quadratic model equation is shown in Equation (4), which has an R^2 value of 0.9063.

$$Y1 = 284.89475 - 0.412750\,X_1 - 3.19655\,X_2 - 0.803425\,X_3 - 0.006935\,X_1X_2 - 0.003285\,X_1X_3 + 0.002467\,X_2X_3 + 0.031535\,X_1{}^2 + 0.010484\,X_2{}^2 - 0.002874\,X_3{}^2 \quad (4)$$

The reduced second-order mathematical model in Equation (4) is used as a function to determine the appropriate conditions for drying. A non-linear model solution is used to find the optimum condition with the highest FRAP value. The results of the experiment showed that the drying conditions were at a maltodextrin concentration of 20% *w*/*v*, an inlet temperature of 160 °C, and an outlet temperature of 80 °C, resulting in the highest FRAP value. When compared with the values calculated from model predictions at the same conditions, it was found that the FRAP value was equal to 4.829 µmol Fe^{2+}/mg dried fingerroot extract when calculating the most appropriate conditions for drying, resulting in the highest FRAP value from the model prediction. It was found that drying conditions with a maltodextrin concentration of 20% *w*/*v*, an inlet temperature of 160 °C, and an outlet temperature of 80 °C resulted in a FRAP value of 4.44 µmol Fe^{2+}/mg of substance dried fingerroot extract.

3.2.6. Yield Percentage and Moisture Contents

The moisture content of the dried fingerroot and fingerroot extract was determined to be 6.49 ± 0.13% and 5.73 ± 0.14%, respectively. When taking the yield percentage obtained from dried fingerroot extract, the relationship with the independent variables can be found using stepwise regression analysis, which can display results in the form of regression analysis. Details are shown in Table 6.

From the table showing the results of the regression analysis, the second-order mathematical model can be used and determined to create a mathematical model or mathematical equation for predicting the dependent variable (response) by taking the values of the factors obtained from the analysis of the coefficients of the regression equation of the said dependent variable and writing them in the form The linear model equation is shown in Equation (5), which has an R^2 value of 0.9063.

$$Y1 = -95.52675 + 2.46725X_1 + 0.386250X_2 + 0.341625X_3 \quad (5)$$

The reduced second-order mathematical model in Equation (5) is used as a function to determine the appropriate conditions for drying using a linear model solution, and the optimum condition with the highest percentage yield is found. The results of the experiment showed that the drying conditions were at a maltodextrin concentration of 30% *w*/*v*, an inlet temperature of 160 °C, and an outlet temperature of 80 °C in the highest percentage yield. When compared with the values calculated from model predictions at the same conditions, it was found that the percentage yield was equal to 65.89 when calculating the most appropriate conditions for drying, resulting in the highest percentage yield from the model prediction. It was found that drying conditions with a maltodextrin concentration of 30% *w*/*v*, an inlet temperature of 150 °C, and an outlet temperature of 90 °C resulted in a percentage yield of 67.18.

3.3. Data Acquisition and Preprocessing for Deep-Learning Machines

Optimizing extraction conditions within a laboratory setting can be a cumbersome and time-consuming endeavor, particularly when multiple variables are involved, each possessing various levels. The Box–Behnken design serves as a prevalent strategy to streamline this process by minimizing the required number of experiments. However, recent advancements in deep-learning machines (DLMs) have demonstrated their superiority over Response Surface Methodology (RSM) in achieving superior optimization outcomes (Figures 6–10).

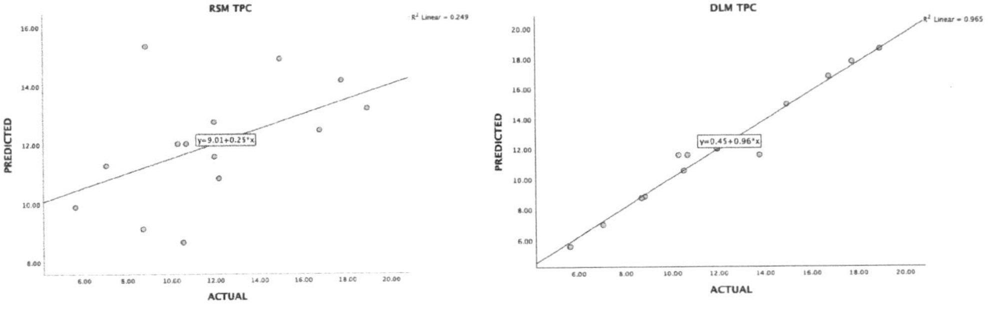

Figure 6. A plot of predicted and experimental values for total phenolic contents from *Boesenbergia rotunda* crude extract using RSM and DLM.

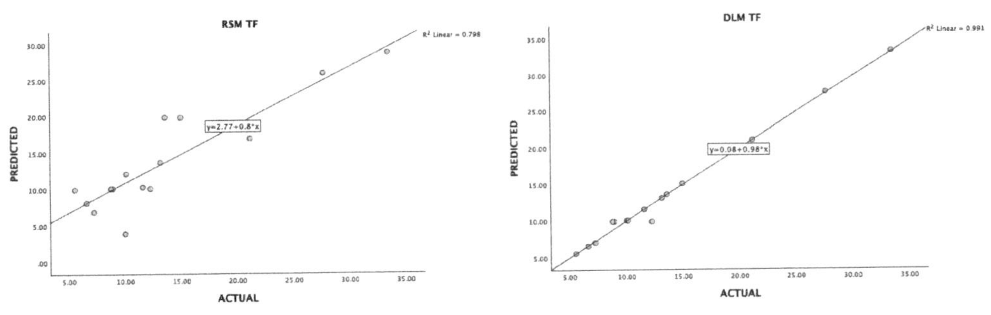

Figure 7. A plot of predicted and experimental values for total flavanoid contents from *Boesenbergia rotunda* crude extract using RSM and DLM.

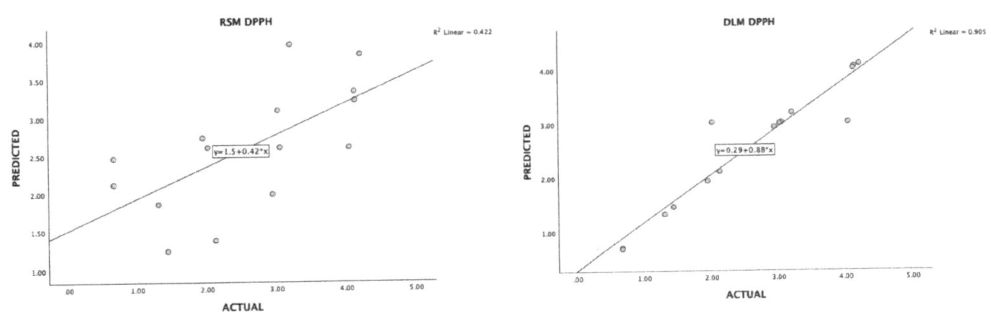

Figure 8. A plot of predicted and experimental values for DPPH radical scavenging activity from *Boesenbergia rotunda* crude extract using RSM and DLM.

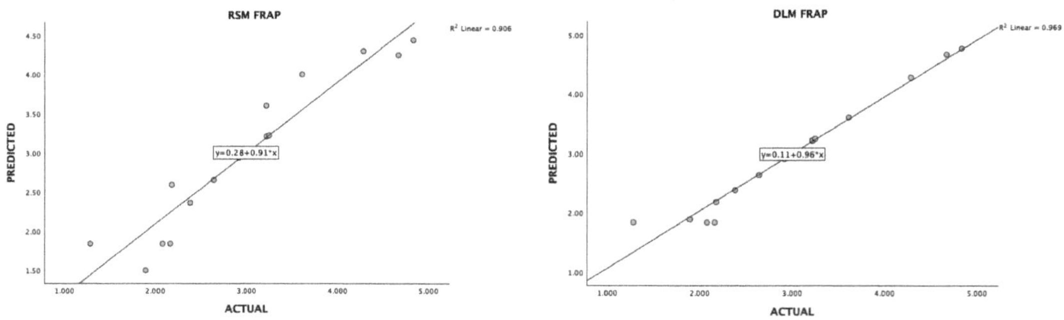

Figure 9. A plot of predicted and experimental values for FRAP activity from *Boesenbergia rotunda* crude extract using RSM and DLM.

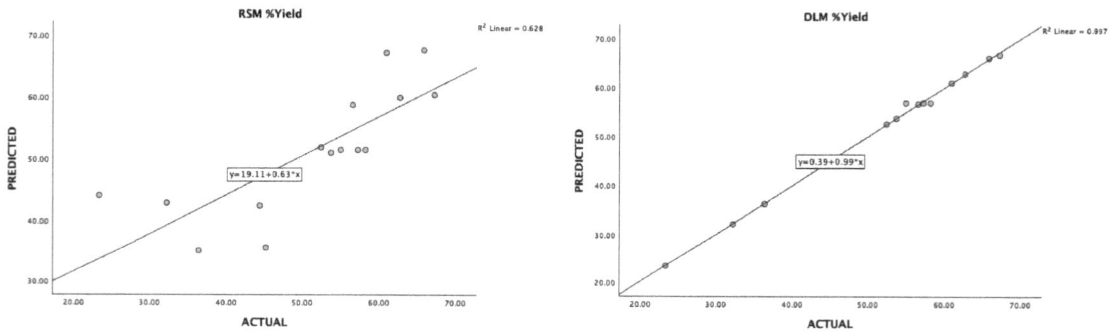

Figure 10. A plot of predicted and experimental values for yield percentage from *Boesenbergia rotunda* crude extract using RSM and DLM.

4. Discussion

Building upon research on fingerroot (Alpinia galanga), this study investigates two crucial stages: extraction and drying. The first study focused on optimizing the microwave-assisted extraction (MAE) process to maximize the total phenolic and flavonoid contents in fingerroot extracts. Here, water extraction yielded significantly higher phenolic content (up to 18.96 µg of GAE/mg extract) compared to ethanol extraction, aligning with findings by [38,39]. Additionally, optimized MAE conditions (20% (*w*/*v*) maltodextrin, microwave 500 W, 160 °C inlet temperature, 80 °C outlet temperature) resulted in the highest overall phenolic contents.

From previous research, this study explores the impact of drying parameters on a pre-extracted fingerroot extract obtained using 57% ethanol. Interestingly, a combination of 20% maltodextrin, a 160 °C inlet temperature, and an 80 °C outlet temperature during the drying process resulted in the highest levels of preserved phenolic and flavonoid compounds. However, other drying conditions (variations in maltodextrin concentration and inlet and outlet temperatures) yielded extracts with the strongest free-radical scavenging activity (FRAP values). This observation underscores the intricate relationship between drying parameters and the retention of bioactive compounds within the fingerroot extract. These findings align with the ongoing efforts to identify factors influencing the potential health benefits of fingerroot, as previously explored by Kanjanasirirat et al. [38].

This research investigates a critical aspect of fingerroot preservation—the drying process—with the aim of maximizing its health benefits. While this research lies on optimizing drying parameters (such as maltodextrin concentration and inlet/outlet temperature) to retain the highest levels of beneficial compounds (phenolics, flavonoids) and free-radical scavenging activity within pre-extracted fingerroot extracts, a complementary study by

Fahrudin et al. [40] explores the impact of different drying methods (freeze drying, oven drying, sun drying) on the overall quality of fingerroot powder. Notably, [40] investigate aspects such as moisture content, color, solubility, and morphology, which provide a broader perspective on the influence of drying on fingerroot beyond the specific bioactive compounds targeted in the present study. This comprehensive approach underscores the importance of considering various drying techniques and their multifaceted effects on fingerroot quality.

This research investigates the optimal drying conditions for preserving the valuable bioactive compounds in fingerroot extract (likely fingerroot) using a spray-drying technique. The study identified that a combination of 20% maltodextrin, 160 °C inlet temperature, and 80 °C outlet temperature yielded the highest levels of phenolics, flavonoids, and FRAP value, indicating strong overall antioxidant activity. Interestingly, a broader range of drying conditions (20–30% maltodextrin, 140–150 °C inlet temperature, 70–80 °C outlet temperature) resulted in the highest IC50 value, suggesting maximized free-radical scavenging activity. Furthermore, the use of response-surface methodology allowed for slight adjustments in drying parameters to target specific bioactive compounds.

While the reference itself describes the current study, the DPPH assay method used to measure IC50 values likely originates from the work of [41]. Other references like [41–44], although not directly cited here, offer valuable insights into spray-drying optimization for various materials. By comparing their findings on factors like carrier material (maltodextrin) and temperature with this study on fingerroot extract, future research can establish broader trends in spray drying-optimization for bioactive compound preservation.

Overall, this study effectively utilizes spray drying to optimize fingerroot extract production by identifying drying conditions that maximize the retention of valuable bioactive compounds and antioxidant activity. This approach aligns with existing research and paves the way for further exploration of spray-drying parameters for different materials.

Researchers are exploring Deep Learning (DL) for drying process optimization. In Zhang et al. [45], a Deep-Learning Machine (DLM) was applied to identify drying conditions that maximize the yield and bioactivity of extracted fingerroot compounds. In this study, researchers focused on developing a Deep-Learning Model Predictive Control (DL-MPC) system to optimize drying efficiency in a complex paddy drying system. While this fingerroot experiment aimed to improve product quality, the paddy drying research prioritized reducing computational cost and achieving faster control compared to traditional methods. Both studies leverage deep learning for drying process optimization but in different ways. The fingerroot experiment employed a Deep-Learning Machine (DLM) to predict optimal drying conditions based on factors like temperature and maltodextrin concentration. The good agreement between predicted and actual results validates the DLM's effectiveness for this specific application. Meanwhile, the paddy drying research proposes a Deep-Learning Model Predictive Control (DL-MPC) system. This system uses deep-learning models to forecast future drying behavior and optimize control decisions, like adjusting air temperature, in real time. Notably, the research emphasizes the significant reduction in computational time compared to traditional methods, making DL-MPC suitable for online control in industrial drying settings. Both studies achieved positive outcomes using deep learning for drying process optimization. The fingerroot experiment identified ideal drying conditions (20% maltodextrin, 150 °C inlet temperature, 70 °C outlet temperature) that maximized extract yield and antioxidant activity. This success highlights the potential of Deep-Learning Machines (DLMs) for optimizing fingerroot extraction processes. In the paddy drying research, researchers implemented a Deep-Learning Model Predictive Control (DL-MPC) system for a complex multistage drying system. This system achieved significant improvements in computational speed compared to traditional methods. Field tests further confirmed the effectiveness of the DL-MPC controller in maintaining consistent moisture content in the paddy and achieving smoother control over the drying process compared to manual adjustments. Despite both studies utilizing deep learning for drying process optimization, their focus and goals differed. The fingerroot experiment aimed to maximize the yield and bioactivity (TPC, TF, DPPH, FRAP) of extracted compounds from a specific botanical material—fingerroot. Here, the DLM identified optimal drying conditions based on factors like temperature and maltodextrin concentration. In contrast, the paddy drying research dealt with optimizing efficiency in a complex multistage drying system for a staple crop—paddy rice. Their primary goal was to reduce computational cost and improve control speed compared to traditional methods. Here, the Deep-Learning Model Predictive Control (DL-MPC) system focused on optimizing drying efficiency (reducing drying time and energy consumption) while ensuring the paddy rice maintains the desired moisture content.

Traditional drying methods, such as oven drying, spray drying, and freeze drying, often struggle to provide precise control over drying conditions, leading to the potential degradation of bioactive compounds and suboptimal product quality. Empirical methods, relying on trial-and-error experiments, can be time-consuming and may not offer a systematic understanding of the complex relationships between process variables, hindering the optimization of drying processes.

Advanced optimization techniques offer promising solutions for complex drying processes. Artificial Neural Networks (ANNs) can model nonlinear relationships but require substantial data and are often difficult to interpret. Genetic Algorithms (GAs) explore a wide range of solutions but can be computationally intensive and may get stuck in local optima. Response-Surface Methodology (RSM) effectively models nonlinear processes but is limited to a smaller number of variables.

RSM-DLM offers several advantages over traditional methods for drying process optimization. By combining the strengths of RSM and DLM, RSM-DLM provides a more comprehensive approach to modeling and optimizing dynamic processes. Additionally, DLM's ability to incorporate uncertainty and variability makes the model more robust to real-world conditions. The flexibility of RSM-DLM allows it to handle a variety of process variables and constraints. Comparative studies, if available, can further demonstrate the effectiveness of RSM-DLM compared to other methods in optimizing drying processes.

5. Conclusions

By studying the appropriate drying conditions of dried fingerroot extract, it was found that factors affecting physical and chemical properties, with a maltodextrin concentration level of 30% (*w*/*v*), will result in the highest amount of dried fingerroot extract. A maltodextrin concentration level of 20% (*w*/*v*) will make the dried fingerroot extract have the most yellow color, and an outlet temperature of 90 °C will result in the moisture content, while the amount of free water makes the value minimal. In terms of pH, alkalinity and solubility were not significantly different ($p > 0.05$).

The study of the most suitable conditions for drying fingerroot extract found that drying with a spray dryer obtained the highest amount of active ingredients. Also, the response-surface methodology was used to find the most suitable conditions for drying the fingerroot extract. The results from the research can be summarized as follows: The drying conditions that resulted in the highest number of phenolic compounds were a maltodextrin concentration of 20% (*w*/*v*), an inlet temperature of 160 °C, and an outlet temperature of 80 °C, which corresponded to the number of flavonoid compounds and the FRAP value.

Meanwhile, the drying conditions using maltodextrin concentrations of 20 and 30% (*w*/*v*), inlet temperatures of 140 and 150 °C, and outlet temperatures of 70 and 80 °C resulted in the IC_{50} value. Also, a prediction by mathematical model found that drying conditions using a maltodextrin concentration of 20% (*w*/*v*), an inlet temperature of 160 °C, and an outlet temperature of 70 °C resulted in the highest number of phenolic compounds.

Meanwhile, the drying condition using a maltodextrin concentration of 20% (*w*/*v*), an inlet temperature of 160 °C, and an outlet temperature of 78.9 °C resulted in the highest flavonoid compound content. The drying conditions are most suitable for the antioxidant capacity. Based on the prediction using the model, the details are as follows. The drying condition using a maltodextrin concentration of 30% *w*/*v*, an inlet temperature of 160 °C, and an outlet temperature of 70 °C resulted in the highest IC_{50} value, while the drying condition at the concentration level of maltodextrin 20% w/v, an inlet temperature of 140 °C, and an outlet temperature of 70 °C resulted in the highest FRAP value.

When comparing the stability results of important substances in dried fingerroot extracts by drying with a spray dryer, it was found that the number of important substances in the dried fingerroot extract decreased, while the yield of the dried fingerroot extract increased. Compared to drying with a vacuum oven and freeze dryer, the physical characteristics of dried fingerroot extracts dried with a vacuum oven and freeze dryer are lumpy, hard, and insoluble in water but soluble in ethanol. The physical characteristics of the white fingerroot extract that are dried with a spray dryer will be in the form of a powder that is soluble in water. Therefore, it is suitable for application in functional food products.

Author Contributions: P.P.-l.-o.; writing-review and editing, K.P.; formal analysis, data curation, writing-original draft preparation, P.B.; formal analysis, data curation, J.S.; writing-review and editing, J.O.-A.; writing-review and editing, R.O.; conceptualization, methodology, validation, writing-review and editing, A.O.; conceptualization, methodology, investigation, resources, software, writing-review and editing. All authors have read and agreed to the published version of the manuscript.

Funding: This research has received funding support from the Suranaree University of Technology Research and Development Fund.

Data Availability Statement: The original contributions presented in the study are included in the article, further inquiries can be directed to the corresponding authors.

Acknowledgments: The authors would like to thank Suranaree University of Technology for providing laboratory facilities.

Conflicts of Interest: The authors hereby declare that there are no conflicts of interest in this article.

References

1. Chahyadi, A.; Hartati, R.; Wirasutisna, K.R.; Elfahri, Y. *Boesenbergia pandurata* Roxb., an Indonesian medicinal plant: Phytochemistry, biological activity, plant biotechnology. *Procedia Chem.* **2014**, *13*, 13–37. [CrossRef]
2. Atun, S.; Handayani, S.; Rakhmawati, A. Potential bioactive compounds isolated from Boesenbergia rotunda as antioxidant and antimicrobial agaents. *Phramacognosy J.* **2018**, *10*, 513–518. [CrossRef]
3. Elzaawely, A.A.; Xuan, T.D.; Koyama, H.; Tawata, S. Antioxidant activity and contents of essential oil and phenolic compounds in flowers and seeds of *Alpinia zerumbet* (Pers.) B.L. Burtt & R.M. Sm. *Food Chem.* **2017**, *104*, 1648–1653. [CrossRef]
4. Sahak, M.K.A.; Mohamed, A.M.; Hashim, N.H.; Haron, D.E. Biological Activities of Triterpenoids from the Rhizomes of *Boesenbergia rotunda* (L.) Mansf. *Rec. Nat. Prod.* **2016**, *10*, 160–169.
5. Tan, L.T.H.; Lee, L.H.; Yin, W.F.; Chan, C.K.; Kadir, H.A.; Chan, K.G.; Goh, B.H. Combining analytic hierarchy process and statistical analysis for optimization of ultrasonic-assisted extraction of polyphenols from starfruit (*Averrhoa carambola* L.) residues. *Molecules* **2019**, *24*, 1949.
6. Yusuf, N.A.; Shukor, A.F.A.; Jemaon, N. Optimization of extraction conditions and analysis of bioactive compounds from *Boesenbergia rotunda* rhizomes. *J. Trop. Agric. Food Sci.* **2015**, *43*, 225–235.
7. Jaipong, J.; Yusufoglu, H.S.; Nualkaew, N. Flavonoid enriched fraction from *Boesenbergia rotunda* and isolation of panduratin A. *Biochem. Biophys. Rep.* **2018**, *16*, 22–28. [CrossRef]
8. Jeenapongsa, R.; Yoovathaworn, K.; Srityanyong, S.; Pongprayoon, U.; Sraisook, K. Anti-inflammatory activity of ((1′S)-a-Substituted-b-Methyl-4′-Hydroxypheylpropanoic Acid) Isolated from Boesenbergia pandurata in Human Synovial Fibroblast Cell Line. *J. Pharmacol. Sci.* **2003**, *93*, 89–97.
9. Cheng, X.R.; Liao, W.W.; Xiao, C.H.; Wang, Z.H.; Dai, W.; Wang, Z.Z. Isolation and characterization of antioxidants from Boesenbergia rotunda. *Med. Chem. Res.* **2015**, *24*, 4210–4217.
10. Mahattanadul, R.R.; Chulikhit, Y.; Srisawat, T.; Malaivijitnond, S.; Eiam-Ong, S. In vitro study on the anticancer potential of fingerroot extract (*Boesenbergia rotunda*). *Foods* **2019**, *8*, 622. [CrossRef]
11. Morikawa, T.; Funakoshi, K.; Ninomiya, K.; Yasuda, D.; Miyagawa, K.; Kamura, T.; Yoshikawa, M.; Muraoka, O. Structure-diversity enhancing activity relationships of pimarane-type diterpenes and their conformational behaviors. *J. Nat. Prod.* **2008**, *71*, 1375–1385.
12. Ching, T.; Himmelstein, D.S.; Beaulieu-Jones, B.K.; Kalinin, A.A.; Do, B.T.; Way, G.P.; Greene, C.S. Opportunities and obstacles for deep learning in biology and medicine. *J. R. Soc. Interface* **2018**, *15*, 20170387. [CrossRef] [PubMed]
13. Eraslan, G.; Avsec, Ž.; Gagneur, J.; Klau, G.W. Deep learning: New computational modelling techniques for genomics. *Nat. Rev. Genet.* **2019**, *20*, 389–403. [CrossRef] [PubMed]

14. Görgülü, M.T.; Görgülü, Ö.Y.; Sevilgen, F.E. Deep Learning Applications in Metabolomics and Proteomics. In *Deep Learning in Bioinformatics*; Springer: Cham, Switzerland, 2022; pp. 285–303.
15. Ma, J.; Sheridan, R.P.; Liaw, A.; Dahl, G.E.; Svetnik, V. Deep neural nets as a method for quantitative structure-activity relationships. *J. Chem. Inf. Model.* **2018**, *55*, 263–274. [CrossRef]
16. Chen, H.; Engkvist, O.; Wang, Y.; Olivecrona, M.; Blaschke, T. The rise of deep learning in drug discovery. *Drug Discov. Today* **2018**, *23*, 1241–1250. [CrossRef] [PubMed]
17. Jumper, J.; Evans, R.; Pritzel, A.; Green, T.; Figurnov, M.; Ronneberger, O.; Hassabis, D. Highly accurate protein structure prediction with AlphaFold. *Nature* **2021**, *596*, 583–589. [CrossRef]
18. Litjens, G.; Kooi, T.; Bejnordi, B.E.; Setio, A.A.A.; Ciompi, F.; Ghafoorian, M.; Sánchez, C.I. A survey on deep learning in medical image analysis. *Med. Image Anal.* **2017**, *42*, 60–88. [CrossRef]
19. Ronneberger, O.; Fischer, P.; Brox, T. U-net: Convolutional networks for biomedical image segmentation. In *International Conference on Medical Image Computing and Computer-Assisted Intervention*; Springer: Cham, Switzerland, 2015; pp. 234–241.
20. Karr, J.R.; Sanghvi, J.C.; Macklin, D.N.; Gutschow, M.V.; Jacobs, J.M.; Bolival, B.; Covert, M.W. A whole-cell computational model predicts phenotype from genotype. *Cell* **2012**, *150*, 389–401. [CrossRef]
21. Oughtred, R.; Stark, C.; Breitkreutz, B.J.; Rust, J.; Boucher, L.; Chang, C.; Tyers, M. The BioGRID interaction database: 2019 update. *Nucleic Acids Res.* **2019**, *47*, D529–D541. [CrossRef]
22. LeCun, Y.; Bengio, Y.; Hinton, G. Deep learning. *Nature* **2015**, *521*, 436–444. [CrossRef]
23. Schmidhuber, J. Deep learning in neural networks: An overview. *Neural Netw.* **2015**, *61*, 85–117. [CrossRef]
24. Taokaenchan, N.; Areseesom, P.; Aresesom, K.; Maifaey, V.; Kongtanajaruanun, R.; Sateinperakul, S. Comparison of Different Extraction Methods and Solvent Ratios on Yield, Content of Total Phenolic and Flavonoid in *Kaempferia parviflora* Wallich. ex Baker. *J. Agric. Prod.* **2022**, *4*, 77–90. Available online: https://li01.tci-thaijo.org/index.php/japmju/article/view/252339/173854 (accessed on 1 May 2024).
25. Prasongdee, P.; Posridee, K.; Oonsivilai, A.; Oonsivilai, R. A Culinary and Medicinal Gem: Exploring the Phytochemical and Functional Properties of Thai Basil. *Foods* **2024**, *13*, 632. [CrossRef] [PubMed]
26. Samruan, W.; Gasaluck, P.; Oonsivilai, R. Total Phenolic and Flavonoid Contents of Soybean Fermentation by *Bacillus subtilis* SB-MYP-1. *Adv. Mater. Res.* **2014**, *931*, 1587–1591. [CrossRef]
27. Ashtiani, S.-H.M.; Martynenko, A. Toward intelligent food drying: Integrating artificial intelligence into drying systems. *Dry. Technol.* **2024**, *42*, 1240–1269. [CrossRef]
28. Khan, M.I.; Sablani, S.S.; Joardder, M.U.H.; Karim, M.A. Application of machine learning-based approach in food drying: Opportunities and challenges. *Dry. Technol.* **2022**, *40*, 1051–1067. [CrossRef]
29. Martynenko, A.; Misra, N.N. Machine learning in drying. *Dry. Technol.* **2020**, *38*, 596–609. [CrossRef]
30. Sun, Q.; Zhang, M.; Mujumder, A.S. Recent developments of artificial intelligence in drying of fresh food: A review. *Crit. Rev. Food Sci. Nutr.* **2019**, *59*, 2258–2275. [CrossRef] [PubMed]
31. Oonsivilai, A.; Janjamraj, N.; Oonsivilai, R. Optimized harmonic for H-bridges connection converters using particle swarm optimization. *Int. Rev. Electr. Eng.* **2013**, *8*, 528–538.
32. Oonsivilai, A.; Oonsivilai, R. Parameter estimation of frequency response twin-screw food extrusion process using genetic algorithms. *WSEAS Trans. Syst.* **2008**, *7*, 1207–1217.
33. Pao-la-or, P.; Marungsri, B.; Chirinang, P.; Posridee, K.; Oonsivilai, R.; Oonsivilai, A. Boosting *Purnica granatum* L. Seed Oil Yield: An Adaptive Neuro-Fuzzy Interference System Fuels SC-CO_2 Extraction Breakthrough. *Foods* **2024**, *13*, 161. [CrossRef] [PubMed]
34. Pao-la-or, P.; Marungsri, B.; Posridee, K.; Oonsivilai, R.; Oonsivilai, A. Supercritical CO_2 Extraction of Seed Oil from *Psophocarpus tetragonolobus* (L.) DC.: Optimization of Operating Conditions through Response Surface Methodology and Probabilistic Neural Network. *Processes* **2023**, *11*, 1949. [CrossRef]
35. Rumelhart, D.E.; Hinton, G.E.; Williams, R.J. Learning representations by back-propagating errors. *Nature* **1986**, *323*, 533–536. [CrossRef]
36. Nair, V.; Hinton, G.E. Rectified linear units improve restricted Boltzmann machines. In Proceedings of the 27th International Conference on Machine Learning (ICML-10), Haifa, Israel, 21–24 June 2010; pp. 807–814.
37. Yosinski, J.; Clune, J.; Bengio, Y.; Lipson, H. How transferable are features in deep neural networks. In Proceedings of the 27th International Conference on Neural Information Processing Systems, Montreal, QC, Canada, 8–13 December 2014; Volume 2, pp. 3320–3328.
38. Kanjanasirirat, P.; Suksatu, A.; Manopwisedjaroen, S.; Munyoo, B.; Tuchinda, P.; Jearawuttanakul, K.; Thitithanyanont, A. High-content screening of Thai medicinal plants reveals *Boesenbergia rotunda* extract and its component Panduratin A as anti-SARS-CoV-2 agents. *Sci. Rep.* **2020**, *10*, 19963. [CrossRef] [PubMed]
39. La-aedon, K.; Patitungkho, S. Effect of Organic Growing Media on Growth, total phenolic content and antioxidants activity of *Boesenbergia pandurate*. *Rajamangala Univ. Technol. Tawan-Ok Res. J.* **2022**, *15*, 1–11.
40. Fahrudin, F.I.; Sulaiman, R.; Sukaryadi, Y. Effect of Drying Methods on physicochemical characteristics of *Boesenbergia rotunda* (L.) Mansf. Powder. *Int. J. Adv. Sci. Technol.* **2020**, *29*, 3952–3962.
41. Mishra, K.; Ojha, H.; Chaudhury, N.K. Estimation of antiradical properties of antioxidants using DPPH assay: A critical review and results. *Food Chem.* **2012**, *130*, 1036–1043. [CrossRef]

42. Sahin-Nadeem, H.; Adnan, A.; Dincer, C.; Torun, M.; Topuz, A. Influence of inlet air temperature and carrier material on the production of instant soluble sage (*Salvia fruticosa* Miller) by spray drying. *LWT* **2013**, *52*, 31–38. [CrossRef]
43. Akther, S.; Jothi, J.S.; Badsha, M.R.; Alim, M.A. Drying methods effect on bioactive compounds, phenolic profile, and antioxidant capacity of mango powder. *J. King Saud Univ. Sci.* **2023**, *35*, 102370. [CrossRef]
44. Çalışkan, G.; Dirim, K. The effect of different drying processes and the amounts of maltodextrin addition on the powder properties of sumac extract powders. *Powder Technol.* **2016**, *287*, 308–314. [CrossRef]
45. Zhang, Y.; Fang, Z.; Li, C.; Li, C. Deep-Learning-Based Model Predictive Control of an Industrial-Scale Multistate Counter-Flow Paddy Drying Process. *Foods* **2024**, *13*, 43. [CrossRef] [PubMed]

foods

MDPI

Article

Integrating Image Analysis and Machine Learning for Moisture Prediction and Appearance Quality Evaluation: A Case Study of Kiwifruit Drying Pretreatment

Shuai Yu [1], Haoran Zheng [1], David I. Wilson [2,*], Wei Yu [1] and Brent R. Young [1]

1 Department of Chemical & Materials Engineering, University of Auckland, Auckland 1010, New Zealand; syu981@aucklanduni.ac.nz (S.Y.)

2 Electrical and Electronic Engineering Department, Auckland University of Technology, Auckland 1010, New Zealand

* Correspondence: david.wilson@aut.ac.nz

Abstract: The appearance of dried fruit clearly influences the consumer's perception of the quality of the product but is a subtle and nuanced characteristic that is difficult to quantitatively measure, especially online. This paper describes a method that combines several simple strategies to assess a suitable surrogate for the elusive quality using imaging, combined with multivariate statistics and machine learning. With such a convenient tool, this study also shows how one can vary the pretreatments and drying conditions to optimize the resultant product quality. Specifically, an image batch processing method was developed to extract color (hue, saturation, and value) and morphological (area, perimeter, and compactness) features. The accuracy of this method was verified using data from a case study experiment on the pretreatment of hot-air-dried kiwifruit slices. Based on the extracted image features, partial least squares and random forest models were developed to satisfactorily predict the moisture ratio (MR) during drying process. The MR of kiwifruit slices during drying could be accurately predicted from changes in appearance without using any weighing device. This study also explored determining the optimal drying strategy based on appearance quality using principal component analysis. Optimal drying was achieved at 60 °C with 4 mm thick slices under ultrasonic pretreatment. For the 70 °C, 6 mm sample groups, citric acid showed decent performance.

Keywords: kiwifruit drying; pretreatment; image processing; HSV color space; principal component analysis; partial least squares; random forest

1. Introduction

Over the past several years, the consumption of fruit has risen due to the increasing awareness of their nutritional properties and appreciable content of essential vitamins, minerals, and other nutrients beneficial to human health. As fresh fruit is highly perishable, extending its shelf life has emerged as a significant challenge [1]. Drying, or dehydration, is one of the oldest and most common processing technologies that, by reducing the moisture content, extends the shelf life by inhibiting microbial proliferation and diminishing the rate of deteriorative chemical reactions [2]. Furthermore, dried fruit has lower transportation and storage costs due to reductions in weight and volume [3–5].

Currently, hot air drying is the most applied method for drying processes in the food industry. Despite its lower drying efficiency compared to that of other new drying methods like freeze drying, the cost advantage is more important to producers. However, the deleterious effects of hot air drying on the physical, structural, chemical, organoleptic, and nutritional properties of foods can lead to a significant diminishment in consumer acceptability [2]. The shape, color, taste, odor, and nutritional composition of food products are all important indicators that help drive sales and profits [6]. Notably, when the appearance attributes of dried products deviate substantially from their fresh counterparts, they may fail to elicit the desired consumer appeal and purchasing intent. Consequently, preserving the external physical qualities is of paramount importance to render processed fruits attractive and acceptable to consumers. In practical production settings, most producers rely on subjective human visual evaluation or conventional measurement instruments to assess the appearance quality of the products. Although techniques employing instruments such as Lab* colorimeters or spectrophotometers provide accurate quantification of color, they are limited in scope as they assess only a partial area of the sample, rendering them impractical for the simultaneous evaluation of an entire batch of fruit [7]. The exigency of the continuous monitoring of appearance quality attributes in food processing necessitates the development and implementation of rapid, nondestructive testing methodologies.

Computer vision systems (CVSs) are considered efficient for quality control. Numerous investigations have documented the prospective applicability of CVSs in monitoring alterations in the visual characteristics of various dehydrated food products, including apples [8], ginseng [9], blueberries [10], bananas [11], and peas [12]. However, previous studies have primarily focused on using the CIELAB color space to evaluate color changes [13,14]. Few studies have attempted to use the HSV color space. Quevedo et al. [15] reported that CIELAB and HSV color spaces perform well regarding food color monitoring as they effectively represent the human eyes' perception of color. Basak et al. [16] employed color features obtained from three color spaces (RGB, HSV, and HSL) in conjunction with a machine learning model to predict the total soluble solids and pH value of strawberry fruits. Their findings indicated that the HSV-based support vector machine regression (SVM-R) model exhibited the highest predictive accuracy. Pardede et al. [17] employed four color spaces, namely, RGB, HSV, HSL, and Lab*, in conjunction with a support vector machine (SVM) algorithm to assess fruit ripeness. Their results demonstrated that the HSV color feature exhibited the highest accuracy in determining the stage of fruit ripeness. Consequently, we investigated the potential of the HSV color space as a viable approach for monitoring and characterizing the color alterations occurring during food drying processes.

In addition to color characteristics, most previous studies also calculated shrinkage by comparing area retractions [10]. The shrinkage phenomenon occurring in foods during the drying process is irregular and anisotropic, rendering it challenging to accurately characterize the shrinkage behavior solely based on the changes in the projected area. The shape factors of 2D images can be considered more accurate quality indicators for describing irregular shrinkage. Only a few studies have focused on morphological changes during drying. Ding et al. [18] investigated the effect of morphological differences on the stacking density of milk powder. Nasri and Belhamri [19] considered shape factors' influence on potato-case solar drying in drying kinetics. In general, the development and implementation of an image batch processing method, integrated into food drying equipment, holds significant potential for a wide range of industrial and research applications.

Kiwifruit (*Actinidia deliciosa* cv. Hayward) plays an essential role in New Zealand's agricultural products and is exported to over 60 countries worldwide each year [20]. It is rich in nutritional elements, especially ascorbic acid, thus being attractive to consumers [13,21]. Improving the quality of the appearance of dried fruit can be achieved not only by optimizing the processing conditions but also by adding pretreatment steps. Some pretreatment methods for food before dehydration have been used, such as chemical methods (hyperosmotic, alkali, sulfite, and acid), gas treatments (sulfur dioxide, carbon dioxide, and ozone), thermal blanching (hot water, steam, super-heated steam, and ohmic and microwave heating), and nonthermal processes (ultrasound, freezing, pulsed electric field, and high hydrostatic pressure) [22–24]. However, sulfite pretreatment of food products is discouraged because its residues pose health risks when consumed [25]. Consequently, chemical pretreatment options, which have no documented health risks but are readily and widely available, such as citric acid and sodium chloride treatment, have been suggested [22]. In the food industry, hot water blanching is widely employed as a pretreatment [26]. Dandamrongrak et al. [27] also explored the effect of four pretreatments (blanching, freezing, chilling, as well as combined blanching and freezing) on the drying of bananas and found that the blanched sample was most preferred in terms of color. However, hot water blanching may have other adverse effects, such as on texture and flavor. In addition, ultrasonic pretreatment is a method being investigated as a likely alternative to conventional pretreatment. It has been shown to improve drying speed by altering plant tissue microstructure and product quality by reducing drying time [28,29]. With ultrasound, the internal structure of the food tissue is greatly altered, facilitating the transfer of water from the inside to the outside [30,31]. Pan et al. [32] confirmed that ascorbic acid and citric acid improved the color of banana crisps in freeze drying. Limited investigations have been conducted on the impact of nonthermal pretreatment methods on the drying process of kiwifruit. Nadian et al. [33] confirmed that acid ascorbic could prevent the browning of kiwifruit during the initial drying stage. Investigating the optimal pretreatment methodologies for drying kiwifruit presents an intriguing research avenue worthy of further exploration.

This study was segmented into three parts. Most notably, we developed an image batch processing method premised on the HSV color space, which facilitates the extraction of both color and morphological features. The accuracy of feature extraction was validated through a case study encompassing drying experiments on pretreated kiwifruit samples. In addition, we constructed moisture ratio prediction models by employing continuous image features extracted within a rolling time window (in half-hourly intervals) as input variables, utilizing partial least squares (PLS) regression and random forest (RF) regression. Ultimately, the optimal pretreatment method and drying conditions were determined by employing an unsupervised learning approach, principal component analysis (PCA), with the image features extracted at the end of the drying process serving as the input data.

The main contributions of this study are as follows:

(1) We developed of an automatic appearance feature extraction method based on the HSV color space. This method exhibits substantial potential for integration into processing equipment, thereby enabling the real-time monitoring of food appearance during actual production processes in the future.
(2) We proposed an alternative approach with monitoring moisture loss by providing a means to assess image characteristics through more convenient analytical techniques.
(3) Based on the image feature data at the end of drying for each set of experimental samples, we used this measurement to quickly optimize the drying process by establishing good operating conditions.

2. Materials and Methods

2.1. Image Feature Extraction Batch Processing Steps

In this study, the images were processed using an algorithm written in the Python environment (version 3.8). To extract and quantify pertinent features from images for integration into machine learning models, we devised a feature extraction batch processing method. Figure 1 illustrates the complete workflow of our proposed image feature extraction steps.

In the dried kiwifruit pretreatment experiment, our aim was to obtain continuous image feature data during the drying process, including color and morphology. We chose the HSV color space to quantify the color change in kiwifruit. Subsequently, based on the significant color difference between the kiwifruit slices and the background, we segmented the prospect and background by setting a threshold range for the HSV color space. Subsequently, we used the 'Mask' function in the OpenCV library to create a mask, setting the pixel values within the specified color range to white and the rest to black. Edge detection was performed based on 'cv2.Canny' function.

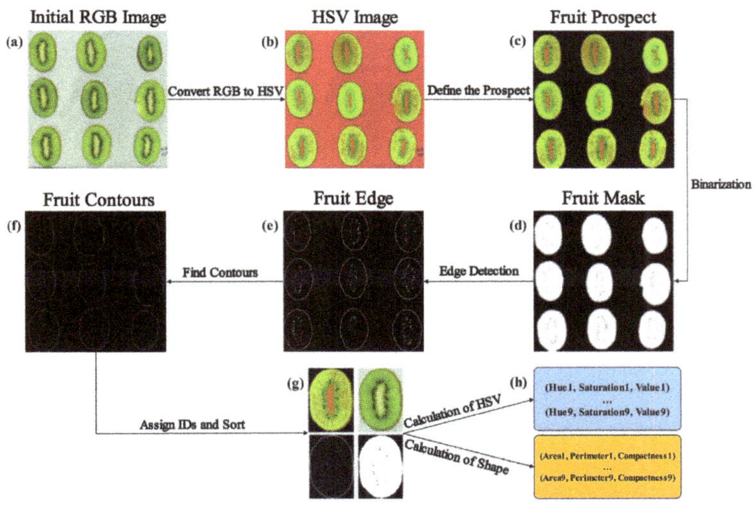

Figure 1. The steps of feature extraction from kiwifruit slice samples: (**a**) the original RGB image; (**b**) RGB image converted to HSV; (**c**) HSV-based background removal; (**d**) mask image of samples; (**e**) edge detection results; (**f**) contour extraction results; (**g**) sorted each slice according to their respective spatial interrelationships; (**h**) quantification of colorimetric and morphological attributes.

To quantify the evolution of their appearance characteristics throughout the drying process, it was imperative to establish a systematic correspondence between the same slices across different time points. This necessitated the assignment of unique identifiers (IDs) to each slice, thereby enabling the tracking of individual slices across a sequence of images. The underlying rationale for this approach was predicated on the observation that the relative positions of the kiwifruit slices remain largely invariant during the drying process. Consequently, we proposed the following methodological steps to facilitate the accurate matching and tracking of individual slices:

Step 1: Ascertain the coordinate values of the centroidal points for all bounded contour regions present within the image.

Step 2: Partition the bounded contour regions into three distinct subgroups based on their respective y-coordinate values.

Step 3: Sort each subgroup in ascending order according to the respective x-coordinate values of the constituent bounded contour regions.

Step 4: Assign unique identifiers (IDs) to each bounded contour regions.

Upon assigning the IDs, all images acquired during a given desiccation experiment were treated as a temporally ordered sequential dataset. The culminating step entailed the extraction of features for each individual kiwifruit section.

2.1.1. Color Feature Extraction

A conical space model can be used to describe HSV color space. Hue (*H*) refers to the color or is equivalent to the wavelength of the transmission or reflection of the object. Traditionally, the value for hue is represented around a circle with a value spanning 0 to 360 degrees. Saturation (*S*) represents how close the color is to the spectral color and is approximately equivalent to spectral purity. The value range is 0–100%. The larger the value, the more saturated the color. Value (*V*) refers to the brightness of the color, usually in the range of 0–100% (black to white) [34]. In Python's OpenCV library, *H* ranges from 0 to 180, and *S* and *V* range from 0 to 255. The thresholds for HSV were *H* (35–75), *S* (50–255), *V* (50–255), which were determined by iterative testing to ensure that kiwifruit contours could be accurately extracted throughout the drying process.

For RGB to HSV, the first step is scaling the RGB values to between 0 and 1. Then, the RGB image is converted to HSV with the following equation:

$$H = \begin{cases} 0,\ if\ Max = Min \\ \left(\frac{G-B}{Max-Min}\right) \times 60 + 0,\ if\ Max = R \\ \left(\frac{B-R}{Max-Min}\right) \times 60 + 120,\ if\ Max = G \\ \left(\frac{R-G}{Max-Min}\right) \times 60 + 240,\ if\ Max = B \end{cases} \quad (1)$$

$$S = \begin{cases} 0,\ if\ Max = 0 \\ \left(\frac{Max-Min}{Max}\right),\ others \end{cases} \quad (2)$$

$$V = Max\ (R, G, B) \quad (3)$$

where Max = max (R, G, B); Min = min (R, G, B).

2.1.2. Morphology Feature Extraction

In this study, three shape factors (area, perimeter and compactness) were considered. Compactness is used to describe the shape characteristics during drying. When the fruit is perfectly round, its compactness reaches a maximum value of 1. As the shape of the fruit becomes more noncircular, the compactness tends to 0. The formulae for the three morphological features are as follows:

$$Area = \frac{Pixel\ of\ Area_t}{Pixel\ of\ Area_{t0}} \quad (4)$$

where $Area_t$ and $Area_{t0}$ are the current fruit area fruit and fruit area before drying by counting the number of pixels;

$$Perimeter = \frac{Pixel\ of\ Perimeter_t}{Pixel\ of\ Perimeter_{t0}} \quad (5)$$

where $Preimeter_t$ and $Preimeter_{t0}$ are the current fruit perimeter fruit area before drying by counting the number of pixels.

$$Compactness = \frac{4\pi \times Area_t}{{Perimeter_t}^2} \quad (6)$$

2.2. *Drying Experiment of Case Study*

2.2.1. Sample Preparation

In the present study, kiwifruits (*Actinidia deliciosa* cv. Hayward) were purchased from a local supermarket in Auckland, New Zealand. When purchasing, samples were chosen that were similar in color and size, moderately hard, and had no obvious external defects. All the samples were stored in a refrigerator at 4.0 °C ± 0.5 °C to delay oxidation and deterioration caused by changes in characteristics. Before the drying experimentation, the samples were transferred from the refrigerator and kept for 3 h at room temperature. After that, fresh kiwifruits were washed to remove foreign matter and cut into the target thickness using a stainless-steel precision slicer, which could ensure a consistent thickness of the kiwifruit slices. To determine the initial moisture content of the kiwifruit slices, approximately 50 g of kiwi sample was placed in an oven and dried at 105 °C until a constant weight was achieved. This procedure was repeated three times, and the average value of the measurements was calculated. The initial moisture content of the sample in this study was 81% ± 0.07% on a wet basis.

2.2.2. Experiment Design

The experiment employed a full factorial design, with two different desiccation temperatures (60 °C, 70 °C) and two distinct kiwifruit slice thicknesses (4 mm, 6 mm) as the treatment factors. Additionally, three pretreatment methods were implemented: immersion in a citric acid solution (CA), immersion in a sodium chloride solution (SCS), and ultrasonic pretreatment (ULT), along with a nonpretreated control group (Control). The specific experimental conditions are delineated in Table 1. To ensure the reliability and validity of the obtained results, each treatment combination was replicated in triplicate.

Table 1. Full factorial experimental design for kiwifruit.

Group Name	Drying Temperature (°C)	Slice Thickness (mm)	Pretreatment
60C4Control	60	4	None
70C4Control	70	4	
60C6Control	60	6	
70C6Control	70	6	
60C4CA	60	4	Citric acid solution
70C4CA	70	4	
60C6CA	60	6	
70C6CA	70	6	
60C4SCS	60	4	Sodium chloride solution
70C4SCS	70	4	
60C6SCS	60	6	
70C6SCS	70	6	
60C4ULT	60	4	Ultrasound
70C4ULT	70	4	
60C6ULT	60	6	
70C6ULT	70	6	

2.2.3. Pretreatment Processes

(i) Citric Acid (CA)

For citric acid solution pretreatment, a 10 g/L solution of citric acid and water was used [31]. After weighing with a digital balance (Model: BL6001, Giorgio Bormac s.r.l, Carpi, Italy) and photographing, the sliced fruit samples were immersed in the solution for 10 min to ensure complete coverage. The fruit samples and solution had a weight ratio of 1 g/20 mL. The pretreated samples were taken out of the solution and blotted dry with a paper towel. After pretreatment, the samples were weighed and photographed again.

(ii) Sodium Chloride Solution (SCS)

The hypertonic solution pretreatment method used sodium chloride and water mixed with a ratio of 5 g/L to form the solution. The sliced fruit samples were soaked in the prepared solution at 0.1 g/mL for 30 min in a container to ensure complete coverage. The pretreated samples were then removed from the container and dried on a surface with a paper towel.

(iii) Ultrasound (ULT)

According to a previous study [35], the ultrasonic pretreatment method involved using an ultrasonic bath (Model: XUBA3, Grant instruments Ltd., Royston, United Kingdom) with a volume of 2.5 L and a fixed frequency of 44 KHz. The sliced fruit samples were immersed and treated with ultrasound in the machine for 30 min to ensure complete coverage. The surface was dried using a paper towel before sending to a tray dryer.

2.2.4. Drying Process and Image Acquisition

The tray dryer used in this study was a UOP8 MKII (Armfield Ltd., Ringwood, UK). The drying air temperature could be easily controlled through a PID controller built into the dryer by a software (version UOP8 MKII Armsoft v.306), which varied the electrical supply to the heating element. The dryer was switched on and preheated for 30 min to reach the target temperature before drying, and the air velocity was set at 1.5 m/s. At the initiation of the drying process, the relative humidity of the hot air employed was approximately 22%.

As drying is an energy-consuming process, the time taken to reach the target moisture content is also a vital indicator of the suitability of the drying conditions. This experiment used five hours (300 min) as a reference time to monitor whether the samples had dried to a safe moisture level (moisture ratio = 0.05 ± 0.01) [33].

The image acquisition system consisted of a digital camera (Model: D300S, Nikon, Tokyo, Japan), lightbox with a 5500K planar LED array light source, and a personal computer (PC). The camera was securely mounted in a fixed position directly above the lightbox. During image capture, the sample specimens were precisely positioned beneath the camera's lens for optimal imaging. A total of nine kiwifruit slices were placed on each tray in a 3 × 3 pattern, as shown in Figure 2. The camera was controlled by a PC connected to it, utilizing Camera Control Pro 2 software (version 2.37.1) for camera management and image acquisition. To ensure optimal image exposure, the camera settings were adjusted to an aperture value of f/5.6, an exposure time of 1/125 s, and an ISO sensitivity of 200. The acquired image had a resolution of 4288 × 2848 pixels. Kiwifruit sample weights were recorded half-hourly for the calculation of moisture ratio, and photographs were taken simultaneously for further image analysis. After recording, the samples were quickly returned to the drying chamber to continue drying. The data acquisition process for the entire drying experiment is depicted in Figure 2.

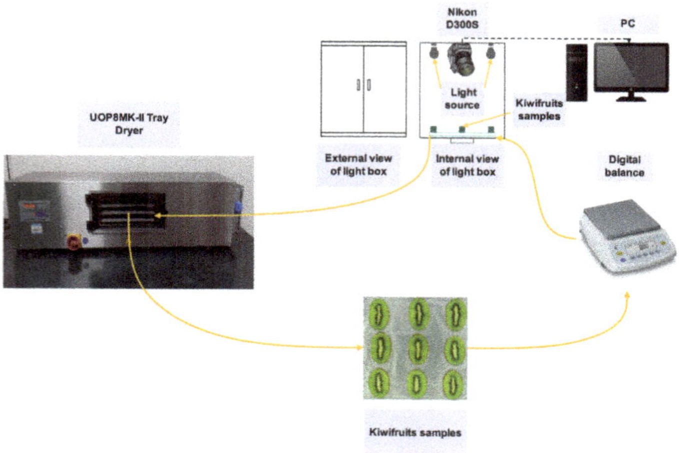

Figure 2. The diagram of experimental setup for image acquisition.

2.2.5. Calculation of Moisture Ratio

The moisture content (M_t) (g water g^{-1} dry matter) of kiwifruit slices was calculated as

$$M_t = \frac{W_t - W_d}{W_d} \tag{7}$$

where W_t and W_d in grams (g) are the current weight and the dried weight of sample, respectively. The moisture ratio (MR) was calculated as

$$MR = \frac{M_t - M_e}{M_0 - M_e} \tag{8}$$

where M_t and M_0 are the equilibrium moisture content and the initial moisture content, respectively. However, M_e can be neglected for products having a high initial moisture content. Therefore, Equation (8) was simplified as follows:

$$MR \sim \frac{M_t}{M_0} \tag{9}$$

2.3. Image-Feature-Based Moisture Ratio Prediction

2.3.1. The Case Study Dataset for Regression

Integrating the color and morphological characteristics with the parameters derived from the drying experiments facilitated the prediction of the moisture ratio of the fruit slices, as delineated in Table 2.

Table 2. The dataset structure.

Variable	Name	Range of Values
X_1	Hue	0 to 180
X_2	Saturation	0 to 255
X_3	Value	0 to 255
X_4	Area	0 to 1
X_5	Perimeter	0 to 1
X_6	Compactness	0 to 1
Y	Moisture Ratio	0 to 1

2.3.2. Partial Least Squares (PLS) Regression

Partial least squares (PLS) is often used as the primary regression technique for multivariate data to express the relationship between independent and dependent variables [36]. PLS regression also has been used in the food industry. For example, Sahin and Demir [37] applied PLS to the determination of the antioxidant capacities of fruit juices. Sun et al. [38] developed a flavor prediction model for garlic using PLS.

In this study, PLS regression models were developed to predict the moisture ratio of kiwifruit samples during the drying process by using color and morphological features. K-fold cross-validation is a model validation technique that assesses the generalizability of a machine learning model to an independent dataset [39]. In the context of this study, the dataset was divided into $k = 5$ folds, and model performance was assessed using established evaluation metrics, with each fold alternately serving as the test set.

The model evaluation metrics for regression tasks include the multiple correlation coefficient (R^2), the cross-validated R-squared (Q^2), and the Mean Squared Error (MSE), which can be calculated by Equations (10)–(12), respectively:

$$R^2 = 1 - \frac{SS_{res}}{SS_{tot}} \tag{10}$$

where SS_{res} is the residual sum of squares; SS_{tot} is the total sum of squares. The closer R^2 is to 1, the better the model fits the data.

$$Q^2 = 1 - \frac{PRESS}{SS_{tot}} \tag{11}$$

where $PRESS$ means predictive residual sum of squares;

$$MSE = \frac{1}{n}\sum_{i=1}^{n}\left(y_i - \hat{y}_i\right)^2 \tag{12}$$

where y_i is the actual value, $\hat{y}_i$ is the predicted value, and n is the number of samples.

2.3.3. Random Forest (RF) Regression

Random forest is an ensemble technique capable of performing both regression and classification tasks using multiple decision trees and a method called bootstrap aggregating [40]. The fundamental idea is to combine multiple decision trees to determine the final output rather than relying on a single decision tree. Random forests generate a large number of classification trees. New objects are classified by inputting vectors into each tree in the forest. Each tree assigns a classification, a process known as "voting", and the classification with the highest number of votes is selected.

In this study, random forest regression was used to make moisture ratio predictions for each kiwifruit slice during the drying process. We exploited the facilities of the scikit-learn library in the Python environment to train a random forest model comprising 20 decision trees. The model evaluation metrics for regression tasks includes mean absolute error (MAE), root mean squared error ($RMSE$), and multiple correlation coefficient (R^2). The equation for R^2 is the same as above, and the other two are calculated as follows:

$$RMSE = \sqrt{\frac{1}{n}\sum_{i=1}^{n}\left(y_i - \hat{y}_i\right)^2} \tag{13}$$

where y_i is the actual value, $\hat{y}_i$ is the predicted value, and n is the number of samples.

$$MAE = \frac{1}{n}\sum_{i=1}^{n}\left|y_i - \hat{y}_i\right| \tag{14}$$

Feature importance refers to the relative significance of each input variable in determining the prediction of a machine learning model. This concept is crucial for model interpretation, as it helps identify the most critical variable.

The dataset and cross-validation methods used for training the random forest algorithm are consistent with those in Section 2.3.2.

2.4. *Optimal Drying Strategy Determination by Principal Component Analysis*

Principal component analysis (PCA) can compress high-dimensional index data into a low-dimensional space while retaining most of the critical information [41]. For drying processing, the new low-dimensional principal component space allowed for better observation of the differences and clustering among samples, which could guide the selection of optimal drying conditions.

In this study, the results obtained by using true values in the PCA were inaccurate due to the initial variations in the color and shape factors of the fruit slice samples. To minimize the impact of initial variation, all data were analyzed using the relative change before and after drying. The appearance features of the fresh fruits were determined as a reference group. After standardizing the color and morphological features data, PCA was employed to obtain the similarities in dried fruits under different drying conditions and pretreatment methods to determine the optimal drying strategy.

3. Results and Discussion

3.1. *Visualization and Accuracy of Feature Extraction Batch Processing Method*

Accurate image segmentation and edge detection are crucial for reliable image feature extraction, as inaccuracies can lead to inconsistent and erroneous morphological data. As illustrated in Figure 3a,b, our method accurately identified the correct contours at all stages of kiwifruit drying in most instances. The method accurately extracted the contours between the fruit flesh and the background. The edges are clear and coherent, effectively delineating the shape features of the kiwifruit slices.

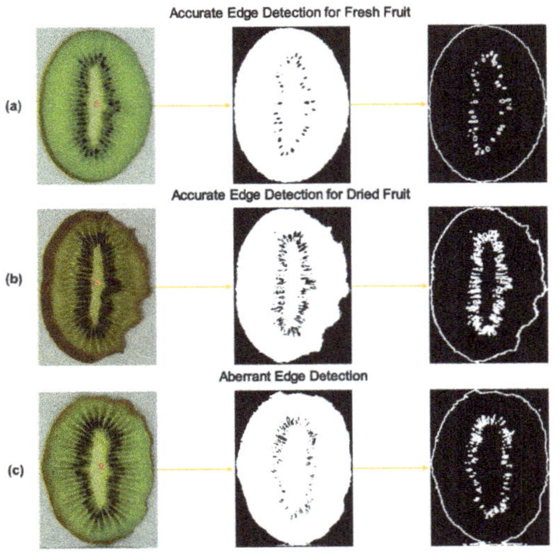

Figure 3. The diagrams of image edge detection for some samples: (**a**) accurate edge detection for fresh sample; (**b**) accurate edge detection for dried sample; (**c**) an example of an abnormal result.

However, there were exceptions where the image processing encountered problems with contour extraction accuracy, resulting in unsuccessful data ID matching and some instances exhibiting areas outside the expected boundary range. As depicted in Figure 3c, the lower edge of the kiwifruit slice was not segmented correctly, and a portion of the background was incorrectly incorporated as part of the slice. The main reason for this issue was that the color of the bottom part of the kiwifruit slice was very similar to the color of the tray, lacking a significant color difference.

To address the problem that appears in Figure 3c, improvements can be considered from the following aspects. Firstly, using a tray with a greater color difference or increasing the distance between the kiwifruit slice and the tray can be explored. In the future, we can also consider introducing more intelligent segmentation algorithms, such as methods that combine region growing, texture analysis, or deep-learning-based segmentation algorithms. By comprehensively utilizing various information such as color, texture, and spatial relationships, these algorithms can improve the segmentation performance in low-contrast regions.

After the image processing phase, a total of 2700 groups of features were identified. Following the exclusion of these abnormal data, a subset of 2459 groups of valid features was retained for the training of the machine learning model.

The accuracy of the proposed method exceeds 90%, indicating its robust performance. While this method has the potential for application in various food processing contexts, the HSV threshold would need to be adjusted to accommodate specific conditions. However, this conclusion warrants further verification through comprehensive testing. In future applications, the method could be integrated into existing drying equipment to facilitate online dynamic drying appearance prediction and quality control.

3.2. Prediction of the Moisture Ratio during the Drying Process

3.2.1. Partial Least Squares Model

A partial least squares (PLS) model was developed to predict the change in moisture ratio (MR) of kiwifruit samples throughout the drying process, utilizing appearance features and drying conditions as variables.

Figure 4a delineates the variation in R^2 and Q^2 values concomitant with the increase in the number of PLS components. The R^2 value is indicative of a model's fitting accuracy on the training dataset, whereas the Q^2 value elucidates a model's capacity for generalization, as inferred from cross-validation procedures. Notably, as the count of PLS components escalates to five, the Q^2 value culminates at a peak of 0.85, concurrently with the R^2 value reaching 0.87. Complementarily, Figure 4b presents a comparative analysis of the predicted versus actual MR for all the dried kiwifruit samples. The model exhibits a mean squared error (*MSE*) of 0.017. These findings substantiate the efficacy of the PLS model in accurately predicting the moisture ratio of kiwifruit during the drying process.

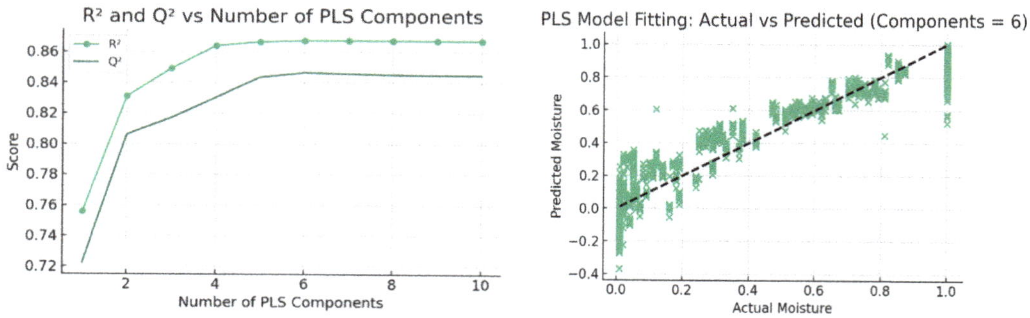

(a) (b)

Figure 4. (**a**) R^2 and Q^2 versus the number of PLS components for the PLS model; (**b**) predicted MR versus actual MR during drying for kiwifruit samples.

3.2.2. Random Forest Model

In this study, the monitoring of weight alterations and the image capture of kiwifruit samples were undertaken on a half-hourly basis. A random forest regression model was developed, aiming to predict the moisture ratio utilizing HSV color and morphological features. The model's accuracy is depicted in Figure 5a; the disparity between the model's predictions and the actual values is remarkably minimal.

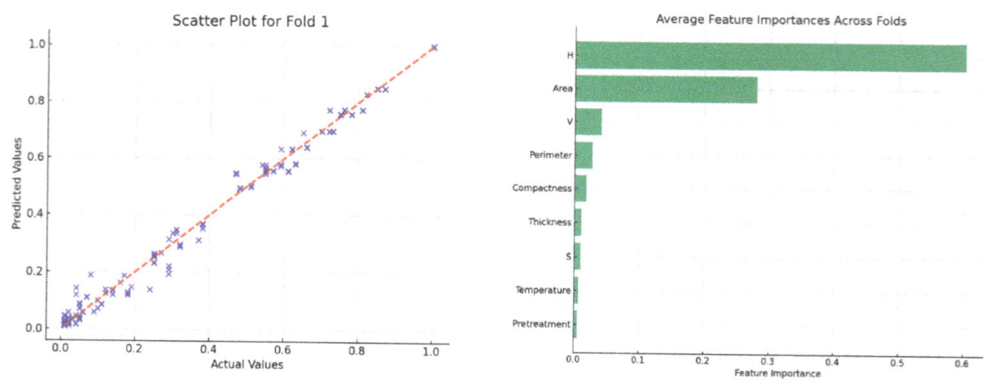

(a) (b)

Figure 5. (**a**) Predicted MR versus actual MR during drying for fold one; (**b**) feature importance of random forest regression.

Further, a detailed analysis of the significance of each feature within the model was executed, as delineated in Figure 5b. This analysis highlighted 'Hue' as the most influential factor, with its importance score exceeding 0.6. Following closely was 'Area', with a significance score around 0.3. Other attributes such as 'Value', 'Perimeter', and 'Compactness' also emerged as factors of substantial importance. These empirical results are in harmony with our initial hypotheses. A strong correlation was observed between the color shift and area reduction in kiwifruit slices during the drying process with the variations in the moisture ratio.

The model evaluation metrics in Table 3 show that the model demonstrates exceptional performance across all folds. The mean R^2 of this model is very close to 1 (0.9923), indicating that the model explains a substantial proportion of the variance in the data. Both the *MAE* (0.0211) and *RMSE* (0.0312) are very small.

Table 3. Model evaluation by random forest regression.

Fold	R^2	*MAE*	*RMSE*
1	0.9939	0.0185	0.0288
2	0.9920	0.0214	0.0317
3	0.9930	0.0205	0.0303
4	0.9920	0.0204	0.0307
5	0.9907	0.0242	0.0345
Mean	0.9923	0.0211	0.0312
Standard deviation	0.0011	0.0018	0.0019

This suggests that the image features for predicting the moisture ratio during the drying process in kiwifruit slices are highly effective.

To summarize, both the PLS and RF models presented above were relatively accurate in predicting the moisture ratio of kiwifruit samples during drying. Meanwhile, the prediction results of the models demonstrate the applicability of using color and morphological features to predict the moisture ratio of kiwifruit slices in real time. In the food industry, the confirmation of the drying endpoint is mainly performed manually based on experience. The use of these two types of models in this study may help with constructing soft sensors in the future. There is the potential to predict whether drying is complete by monitoring appearance changes during the drying process.

3.3. *Effect of Different Pretreatments on Appearance (Color and Morphology) Changes*

The moisture ratio (MR) of all samples was calculated using Equation (9), and it was found that all 60 °C, 6 mm samples did not reach the target MR after 300 min of drying. As mentioned above, a drying time of 300 min was used as a reference to represent the maximum acceptable drying time, with an MR equal to 0.05 representing the end of drying in this study. Therefore, the appearance change in the samples (60 °C, 6 mm) is not discussed as the drying was not completed. Section 3.3.1 discusses the similarity of the appearance of the 4 mm slice thickness samples at different drying temperatures (60 °C and 70 °C) to the fresh kiwifruit slices. Section 3.3.2 discusses the similarity of the appearance of samples with different slice thicknesses (4 mm and 6 mm) compared to fresh kiwifruit slices after drying at 70 °C. The PCA data were normalized. Thus, the coordinate system in the PCA figure and the Euclidean distance are dimensionless.

3.3.1. Appearance Similarity Comparison by PCA for 4 mm Samples

The PCA scores and loadings plots for the 4 mm kiwifruits samples are shown in Figure 6a,b, respectively. The variance explained by PC1, PC2, and PC3 was 48.77%, 23.76%, and 17.21%, respectively, cumulatively accounting for 89.74% of the total variance. As shown in Figure 6b, regarding the appearance parameters, PC1 was positively correlated with hue, area, and value, while only negatively correlated with compactness. Figure 6a depicts a pronounced divergence of all dried sample groups from the fresh fruit (labeled as 'Fresh'), signifying substantial alterations in appearance, attributable to the drying process.

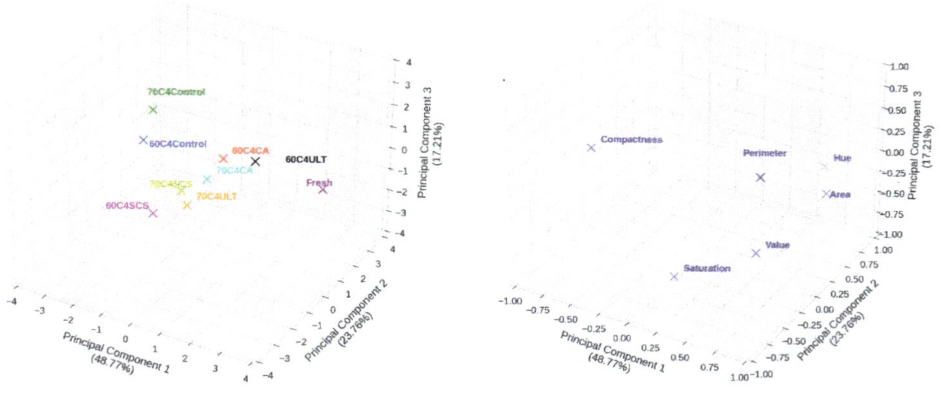

(a) (b)

Figure 6. (**a**) PCA score plot of the first three principal components was used to explore the degree of variation in the appearance for different pretreatments for 4 mm samples; (**b**) PCA loading plot of the first three principal components allowing the determination of the relationships between different appearance variables.

Table 4 presents the Euclidean distances between each sample point and the fresh fruit reference point. Notably, the points labeled 60C4ULT and 60C4CA are positioned proximate to those of the fresh fruit, suggesting that the application of citric acid and ultrasonic pretreatments at a drying temperature of 60 °C and a slice thickness of 4 mm potentially preserves green color effectively. In addition, the 70C4SCS and 70C4ULT groups are located close to each other in Figure 6a and moderately deviate from the fresh fruit point. This means they had similar changes in appearance at the end of drying.

Table 4. Euclidean distance to fresh fruit for 4 mm thick samples.

Number	Name	Distance to Fresh
1	60C4ULT	3.44
2	70C4CA	3.81
3	70C4SCS	4.46
4	70C4ULT	4.68
5	60C4CA	5.32
6	60C4SCS	5.51
7	60C4Control	5.54
8	70C4Control	5.66

The control samples, irrespective of being dried at 60 °C or 70 °C, are distinctly departed from the fresh fruit in Figure 6a, indicating that regardless of the pretreatment used, it is beneficial to inhibit color and morphological changes during the process of drying kiwifruit.

A comparison of ultrasonic, citric acid, and unpretreated samples with fresh fruit is shown in Figure 7. The results of human eye perception are consistent with the data, showing that ultrasonic-pretreated samples best retained the green and had less morphological modification. In contrast, nonpretreated samples browned heavily after drying and showed a large degree of shrinkage.

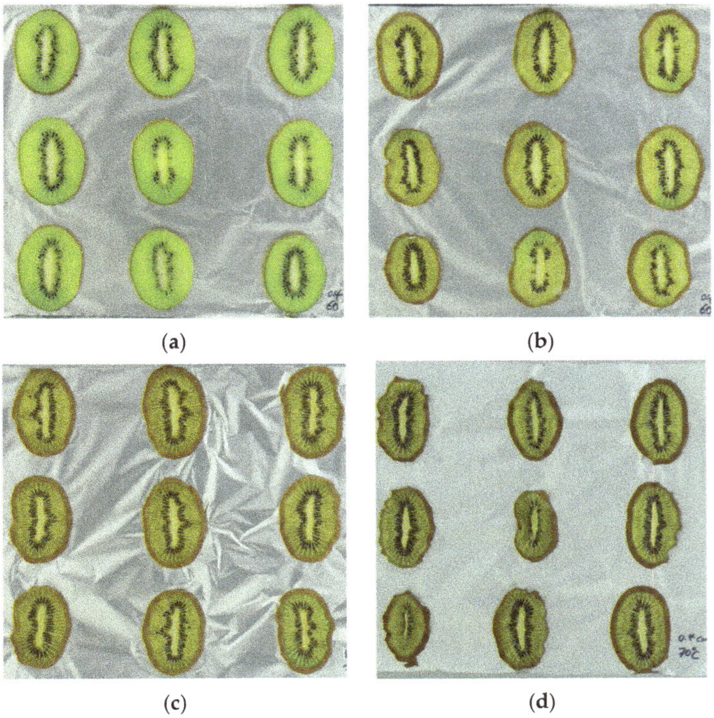

Figure 7. (**a**) Fresh fruit before drying; (**b**) 60C4ULT after drying; (**c**) 70C4CA after drying; (**d**) 70C4Control after drying.

In general, an analysis integrating the data from Figure 6 and Table 4 suggested that for the 4 mm sliced kiwifruit samples, ultrasonic pretreatment was optimal at a drying temperature of 60 °C. In contrast, citric acid pretreatment yielded favorable results at an elevated temperature of 70 °C. Furthermore, all selected pretreatment methods in this study mitigated appearance-quality deterioration to varying extents compared to the samples without pretreatment.

3.3.2. Appearance Similarity Comparison by PCA for 70 °C Samples

Figure 8a,b show the PCA scores and loading plots for the 70 °C drying temperature of kiwifruit samples, respectively. PC1, PC2, and PC3 accounted for 47.01%, 22.88%, and 16.40% of the variance, cumulatively constituting 86.29%. Figure 8b shows that PC1 was positively correlated with all parameters except compactness, while PC2 had positive correlations with all parameters except saturation.

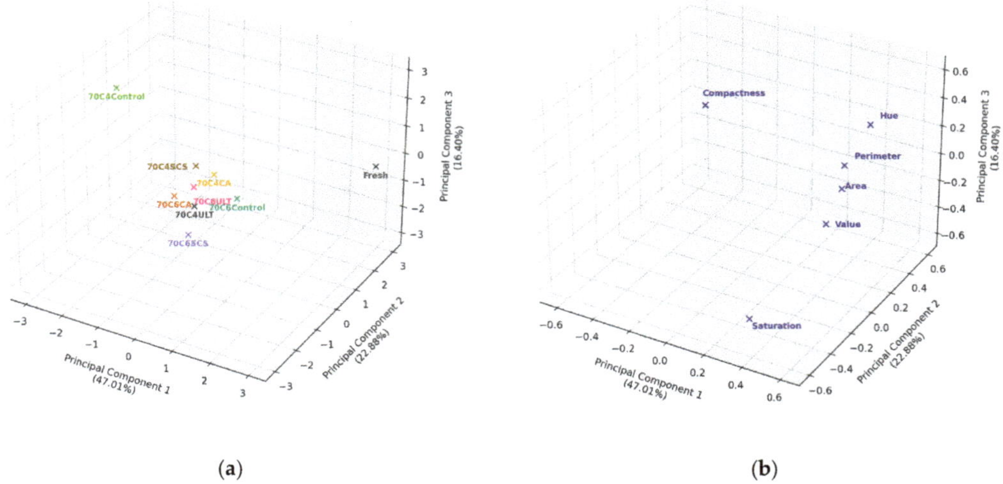

Figure 8. (**a**) PCA score plot of the first three principal components was used to explore the degree of variation in the appearance for different pretreatments for 70 °C samples; (**b**) PCA loading plot of the first three principal components allowing the determination of the relationships between different appearance variables.

It was observed that samples with the same pretreatment differed in appearance at the 70 °C drying temperature for different slice thicknesses. This discrepancy is likely attributable to the extended drying duration necessary for the 6 mm samples. This protracted drying period presumably intensified the degree of browning and shrinkage in the fruits, thereby engendering appearance differences among the sliced samples. As drying progresses, the moisture content of the fruit gradually decreases, leading to an acceleration of the Maillard reaction. As the Maillard reaction accelerates, interactions between sugars and amino acids at higher temperatures lead to the formation of more brown compounds [4]. However, the score plot shows the position of the individual samples in the principal component space, and, as can be seen in Figure 8a, most of the sample groups are relatively clustered in position between each other, which means that there was not much difference in the appearance features between the samples, except for the 70C4Control sample.

As in the previous section, the Euclidean distances of each sample from the fresh fruit were computed, as delineated in Table 5. The 70C4Control group exhibited the most pronounced appearance deviation from the fresh fruit, possibly due to the rapid drying of thinner slices at high temperatures, which precipitated significant pigment deterioration and shrinkage. The 70C4CA samples demonstrated the shortest Euclidean distance (3.84) from the reference point (Fresh). Remarkably, the unpretreated 6 mm samples (70C6Control) subjected to drying at 70 °C did not undergo severe appearance deterioration. This may have been because thicker samples are more likely to produce a porous hard outer case at higher temperatures, thereby retarding shape change [33].

Table 5. Euclidean distance to fresh fruit for 70 °C samples.

Number	Name	Distance to Fresh
1	70C4CA	3.84
2	70C6Control	4.41
3	70C6ULT	4.47
4	70C4ULT	4.51
5	70C4SCS	4.52
6	70C6SCS	4.85
7	70C6CA	4.89
8	70C4Control	6.63

As shown in Figure 9, when the drying temperature was raised to 70 °C, the human eye perceived that only the citric-acid-pretreated samples (b) retained a little green color and did not undergo severe morphology changes.

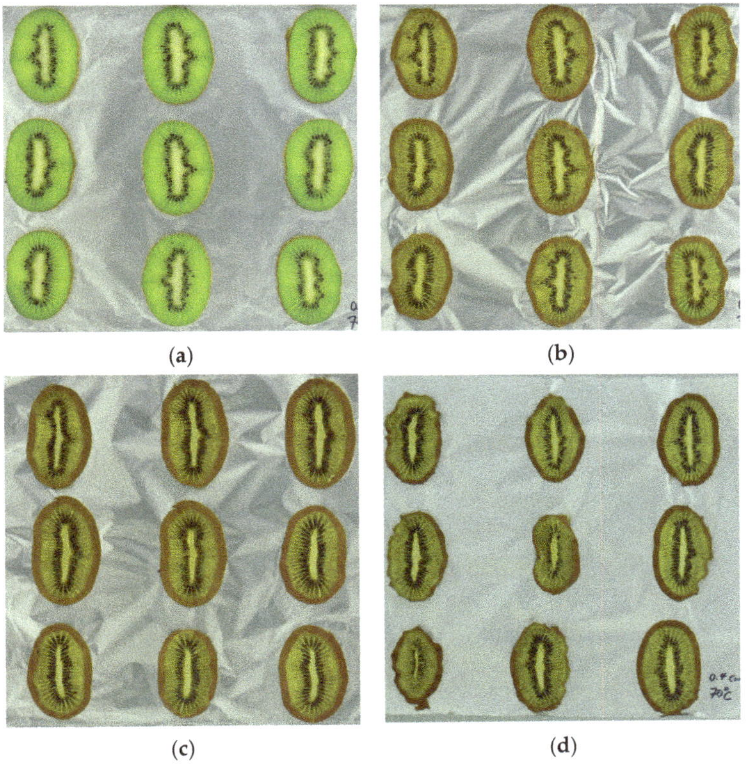

Figure 9. (**a**) Fresh fruit before drying; (**b**) 70C4CA after drying; (**c**) 70C6Control after drying; (**d**) 70C4Control after drying.

In summary, at a drying temperature of 70 °C, citric acid appeared to be the most efficacious pretreatment for the 4 mm slices. However, for thicker slices, the various pretreatments did not exhibit notably effective performance.

4. Conclusions

In this study, an image batch processing method was developed, facilitating the extraction of color and morphological features with an accuracy exceeding 90% for kiwifruit slices. This result indicates the potential to integrate such method into smart drying equipment and help optimize the drying process.

Furthermore, partial least squares (PLS) and random forest (RF) models were developed using color and morphological features to predict the moisture ratio. While the RF model exhibited superior predictive capability, PLS models, due to their greater explainability, might be more suitable for real-world industrial applications. The satisfactory prediction outcomes suggest vision-based soft sensors have potential application in monitoring moisture content in the food industry without relying on weighing equipment.

Principal component analysis (PCA) was employed to compare the color and morphological differences among the samples. Specifically, for kiwifruit slices with a 4 mm thickness that underwent ultrasonic pretreatment, minimal appearance changes were observed at 60 °C. However, when the drying temperature increased to 70 °C, the slices pretreated with citric acid closely resembled fresh fruit.

Overall, through the integration of image processing and machine learning techniques, the proposed approach presents an efficient and accurate method for monitoring the fruit drying process. However, this study is a case study of kiwifruit slice drying, and the general applicability of the real-time image feature extraction method to different types of food will be verified in future research. The application of these methodologies in dynamic fruit drying systems is anticipated to enhance the productivity and effectiveness of fruit drying operations.

Author Contributions: Conceptualization, W.Y. and D.I.W.; methodology, S.Y. and H.Z.; software, S.Y., H.Z. and D.I.W.; validation, S.Y. and B.R.Y.; formal analysis, S.Y.; investigation, S.Y.; resources, S.Y.; data curation, S.Y.; writing—original draft preparation, S.Y.; writing—review and editing, S.Y., B.R.Y., D.I.W., H.Z. and W.Y.; visualization, S.Y. and H.Z.; supervision, W.Y. and B.R.Y.; project administration, W.Y. and B.R.Y. All authors have read and agreed to the published version of the manuscript.

Funding: This research received no external funding.

Institutional Review Board Statement: Not applicable.

Informed Consent Statement: Not applicable.

Data Availability Statement: The data presented in this study are available on request from the corresponding author due to the data are part of an ongoing study.

Acknowledgments: The authors wish to acknowledge the China Scholarship Council for the support of study.

Conflicts of Interest: The authors declare no conflicts of interest.

References

1. Ozgen, F.; Celik, N. Evaluation of Design Parameters on Drying of Kiwi Fruit. *Appl. Sci.* **2018**, *9*, 10. [CrossRef]
2. Nadian, M.H.; Rafiee, S.; Aghbashlo, M.; Hosseinpour, S.; Mohtasebi, S.S. Continuous Real-Time Monitoring and Neural Network Modeling of Apple Slices Color Changes during Hot Air Drying. *Food Bioprod. Process.* **2015**, *94*, 263–274. [CrossRef]
3. Liu, C.; Liu, W.; Lu, X.; Chen, W.; Yang, J.; Zheng, L. Potential of Multispectral Imaging for Real-Time Determination of Colour Change and Moisture Distribution in Carrot Slices during Hot Air Dehydration. *Food Chem.* **2016**, *195*, 110–116. [CrossRef] [PubMed]
4. Raponi, F.; Moscetti, R.; Monarca, D.; Colantoni, A.; Massantini, R. Monitoring and Optimization of the Process of Drying Fruits and Vegetables Using Computer Vision: A Review. *Sustainability* **2017**, *9*, 2009. [CrossRef]
5. Kaleta, A.; Górnicki, K.; Winiczenko, R.; Chojnacka, A. Evaluation of Drying Models of Apple (Var. Ligol) Dried in a Fluidized Bed Dryer. *Energy Convers. Manag.* **2013**, *67*, 179–185. [CrossRef]
6. Hosseinpour, S.; Rafiee, S.; Aghbashlo, M.; Mohtasebi, S.S. Computer Vision System (CVS) for In-Line Monitoring of Visual Texture Kinetics During Shrimp (*Penaeus* Spp.) Drying. *Dry. Technol.* **2015**, *33*, 238–254. [CrossRef]
7. Zielinska, M.; Markowski, M. Color Characteristics of Carrots: Effect of Drying and Rehydration. *Int. J. Food Prop.* **2012**, *15*, 450–466. [CrossRef]
8. Aghilinategh, N.; Rafiee, S.; Hosseinpour, S.; Omid, M.; Mohtasebi, S.S. Real-Time Color Change Monitoring of Apple Slices Using Image Processing during Intermittent Microwave Convective Drying. *Food Sci. Technol. Int.* **2016**, *22*, 634–646. [CrossRef]

9. Martynenko, A.I. Porosity Evaluation of Ginseng Roots from Real-Time Imaging and Mass Measurements. *Food Bioprocess Technol* **2011**, *4*, 417–428. [CrossRef]
10. Chen, Y.; Martynenko, A. Computer Vision for Real-Time Measurements of Shrinkage and Color Changes in Blueberry Convective Drying. *Dry. Technol.* **2013**, *31*, 1114–1123. [CrossRef]
11. Seyedabadi, E.; Khojastehpour, M.; Abbaspour-Fard, M.H. Online Measuring of Quality Changes of Banana Slabs during Convective Drying. *Eng. Agric. Environ. Food* **2019**, *12*, 111–117. [CrossRef]
12. Iheonye, A.; Gariepy, Y.; Raghavan, V. Computer Vision for Real-Time Monitoring of Shrinkage for Peas Dried in a Fluidized Bed Dryer. *Dry. Technol.* **2020**, *38*, 130–146. [CrossRef]
13. Xu, R.; Zhou, X.; Wang, S. Comparative Analyses of Three Pretreatments on Color of Kiwifruits during Hot Air Drying. *Int. J. Agric. Biol. Eng.* **2020**, *13*, 228–234. [CrossRef]
14. Horuz, E.; Bozkurt, H.; Karataş, H.; Maskan, M. Effects of Hybrid (Microwave-Convectional) and Convectional Drying on Drying Kinetics, Total Phenolics, Antioxidant Capacity, Vitamin C, Color and Rehydration Capacity of Sour Cherries. *Food Chem.* **2017**, *230*, 295–305. [CrossRef] [PubMed]
15. Quevedo, R.A.; Aguilera, J.M.; Pedreschi, F. Color of Salmon Fillets By Computer Vision and Sensory Panel. *Food Bioprocess Technol* **2010**, *3*, 637–643. [CrossRef]
16. Basak, J.K.; Madhavi, B.G.K.; Paudel, B.; Kim, N.E.; Kim, H.T. Prediction of Total Soluble Solids and pH of Strawberry Fruits Using RGB, HSV and HSL Colour Spaces and Machine Learning Models. *Foods* **2022**, *11*, 2086. [CrossRef] [PubMed]
17. Pardede, J.; Husada, M.G.; Hermana, A.N.; Rumapea, S.A. Fruit Ripeness Based on RGB, HSV, HSL, L*a*b* Color Feature Using SVM. In Proceedings of the 2019 International Conference of Computer Science and Information Technology (ICoSNIKOM), Medan, Indonesia, 28–29 November 2019; IEEE: Piscataway, NJ, USA, 2019; pp. 1–5. [CrossRef]
18. Ding, H.; Li, B.; Boiarkina, I.; Wilson, D.I.; Yu, W.; Young, B.R. Effects of Morphology on the Bulk Density of Instant Whole Milk Powder. *Foods* **2020**, *9*, 1024. [CrossRef] [PubMed]
19. Nasri, M.Y.; Belhamri, A. Effects of the Climatic Conditions and the Shape on the Drying Kinetics, Application to Solar Drying of Potato-Case of Maghreb's Region. *J. Clean. Prod.* **2018**, *183*, 1241–1251. [CrossRef]
20. Veeregowda, P.M.; Jeffery, P.B.; Johnston, J.W.; East, A.R. A Survey of Retail Conditions in the Kiwifruit Supply Chains of India and Singapore. *N. Z. J. Crop Hortic. Sci.* **2022**, *50*, 274–285. [CrossRef]
21. Zhou, X.; Li, R.; Lyng, J.G.; Wang, S. Dielectric Properties of Kiwifruit Associated with a Combined Radio Frequency Vacuum and Osmotic Drying. *J. Food Eng.* **2018**, *239*, 72–82. [CrossRef]
22. Deng, L.-Z.; Mujumdar, A.S.; Zhang, Q.; Yang, X.-H.; Wang, J.; Zheng, Z.-A.; Gao, Z.-J.; Xiao, H.-W. Chemical and Physical Pretreatments of Fruits and Vegetables: Effects on Drying Characteristics and Quality Attributes—A Comprehensive Review. *Crit. Rev. Food Sci. Nutr.* **2019**, *59*, 1408–1432. [CrossRef] [PubMed]
23. Rodrigues, S.; Fernandes, F.A.N. Use of Ultrasound as Pretreatment for Dehydration of Melons. *Dry. Technol.* **2007**, *25*, 1791–1796. [CrossRef]
24. Miranda, G.; Berna, À.; Salazar, D.; Mulet, A. Sulphur Dioxide Evolution during Dried Apricot Storage. *LWT—Food Sci. Technol.* **2009**, *42*, 531–533. [CrossRef]
25. Kamiloglu, S.; Toydemir, G.; Boyacioglu, D.; Beekwilder, J.; Hall, R.D.; Capanoglu, E. A Review on the Effect of Drying on Antioxidant Potential of Fruits and Vegetables. *Crit. Rev. Food Sci. Nutr.* **2016**, *56*, S110–S129. [CrossRef] [PubMed]
26. Guida, V.; Ferrari, G.; Pataro, G.; Chambery, A.; Di Maro, A.; Parente, A. The Effects of Ohmic and Conventional Blanching on the Nutritional, Bioactive Compounds and Quality Parameters of Artichoke Heads. *LWT—Food Sci. Technol.* **2013**, *53*, 569–579. [CrossRef]
27. Dandamrongrak, R.; Mason, R.; Young, G. The Effect of Pretreatments on the Drying Rate and Quality of Dried Bananas. *Int J Food Sci Tech* **2003**, *38*, 877–882. [CrossRef]
28. Tao, Y.; Wang, P.; Wang, Y.; Kadam, S.U.; Han, Y.; Wang, J.; Zhou, J. Power Ultrasound as a Pretreatment to Convective Drying of Mulberry (*Morus alba* L.) Leaves: Impact on Drying Kinetics and Selected Quality Properties. *Ultrason. Sonochemistry* **2016**, *31*, 310–318. [CrossRef] [PubMed]
29. Chemat, F.; Zill-e-Huma; Khan, M.K. Applications of Ultrasound in Food Technology: Processing, Preservation and Extraction. *Ultrason. Sonochemistry* **2011**, *18*, 813–835. [CrossRef] [PubMed]
30. Bozkir, H.; Rayman Ergün, A.; Tekgül, Y.; Baysal, T. Ultrasound as Pretreatment for Drying Garlic Slices in Microwave and Convective Dryer. *Food Sci. Biotechnol.* **2019**, *28*, 347–354. [CrossRef]
31. Tepe, T.K.; Kadakal, Ç. Determination of Drying Characteristics, Rehydration Properties, and Shrinkage Ratio of Convective Dried Melon Slice with Some Pretreatments. *Food Process. Preserv.* **2022**, *46*, e16544. [CrossRef]
32. Pan, Z.; Shih, C.; McHugh, T.H.; Hirschberg, E. Study of Banana Dehydration Using Sequential Infrared Radiation Heating and Freeze-Drying. *LWT—Food Sci. Technol.* **2008**, *41*, 1944–1951. [CrossRef]
33. Nadian, M.H.; Abbaspour-Fard, M.H.; Sadrnia, H.; Golzarian, M.R.; Tabasizadeh, M. Optimal Pretreatment Determination of Kiwifruit Drying via Online Monitoring: Pretreatment Determination of Kiwifruit Drying. *J. Sci. Food Agric.* **2016**, *96*, 4785–4796. [CrossRef]
34. Wu, D.; Sun, D.-W. Colour Measurements by Computer Vision for Food Quality Control—A Review. *Trends Food Sci. Technol.* **2013**, *29*, 5–20. [CrossRef]

35. Fernandes, F.A.N.; Rodrigues, S. Ultrasound as Pre-Treatment for Drying of Fruits: Dehydration of Banana. *J. Food Eng.* **2007**, *82*, 261–267. [CrossRef]
36. Wold, S.; Sjöström, M.; Eriksson, L. PLS-Regression: A Basic Tool of Chemometrics. *Chemom. Intell. Lab. Syst.* **2001**, *58*, 109–130. [CrossRef]
37. Sahin, S.; Demir, C. Determination of Antioxidant Properties of Fruit Juice by Partial Least Squares and Principal Component Regression. *Int. J. Food Prop.* **2016**, *19*, 1455–1464. [CrossRef]
38. Sun, Y.; Zhang, M.; Ju, R.; Mujumdar, A. Novel Nondestructive NMR Method Aided by Artificial Neural Network for Monitoring the Flavor Changes of Garlic by Drying. *Dry. Technol.* **2021**, *39*, 1184–1195. [CrossRef]
39. Stone, M. Cross-Validatory Choice and Assessment of Statistical Predictions. *J. R. Stat. Society. Ser. B (Methodol.)* **1974**, *36*, 111–147. [CrossRef]
40. Breiman, L. Random forests. *Mach. Learn.* **2001**, *45*, 5–32. [CrossRef]
41. Bro, R.; Smilde, A.K. Principal Component Analysis. *Anal. Methods* **2014**, *6*, 2812–2831. [CrossRef]

foods

MDPI

Article

Dynamic Nondestructive Detection Models of Apple Quality in Critical Harvest Period Based on Near-Infrared Spectroscopy and Intelligent Algorithms

Zhiming Guo [1,2,*], Xuan Chen [1], Yiyin Zhang [1], Chanjun Sun [1], Heera Jayan [1], Usman Majeed [2], Nicholas J. Watson [3] and Xiaobo Zou [1,2]

1 China Light Industry Key Laboratory of Food Intelligent Detection & Processing, School of Food and Biological Engineering, Jiangsu University, Zhenjiang 212013, China; cx18756041020@163.com (X.C.); zyysnb123@gmail.com (Y.Z.); chanjun.sun@ujs.edu.cn (C.S.); heerajayan93@outlook.com (H.J.); zou_xiaobo@ujs.edu.cn (X.Z.)

2 International Joint Research Laboratory of Intelligent Agriculture and Agri-Products Processing, Jiangsu University, Zhenjiang 212013, China; majeedusman55@gmail.com

3 School of Food Science and Nutrition, University of Leeds, Leeds LS2 9JT, UK; n.j.watson@leeds.ac.uk

* Correspondence: guozhiming@ujs.edu.cn

Abstract: Apples are usually bagged during the growing process, which can effectively improve the quality. Establishing an in situ nondestructive testing model for in-tree apples is very important for fruit companies in selecting raw apple materials for valuation. Low-maturity apples and high-maturity apples were acquired separately by a handheld tester for the internal quality assessment of apples developed by our group, and the effects of the two maturity levels on the soluble solids content (SSC) detection of apples were compared. Four feature selection algorithms, like ant colony optimization (ACO), were used to reduce the spectral complexity and improve the apple SSC detection accuracy. The comparison showed that the diffuse reflectance spectra of high-maturity apples better reflected the internal SSC information of the apples. The diffuse reflectance spectra of the high-maturity apples combined with the ACO algorithm achieved the best results for SSC prediction, with a prediction correlation coefficient (Rp) of 0.88, a root mean square error of prediction (RMSEP) of 0.5678 °Brix, and a residual prediction deviation (RPD) value of 2.466. Additionally, the fruit maturity was predicted using PLS-LDA based on color data, achieveing accuracies of 99.03% and 99.35% for low- and high-maturity fruits, respectively. These results suggest that in-tree apple in situ detection has great potential to enable improved robustness and accuracy in modeling apple quality.

Keywords: NIR spectroscopy; apple; dynamic detection; machine learning; maturity

1. Introduction

In recent years, apple, as a highly favored fruit, has shown continuous growth in both global production and consumption [1]. Apples are bagged during growth to prevent and remove fruit rust and surface blemishes, promote coloration, reduce pesticide residues, and improve storability [2]. Apples are removed from their protective bags about 14–20 days before the harvest period, which is very important for surface coloring and sugar conversion. It is also important for fruit harvesting companies to select apple materials for valorization and harvesting. However, current studies are still limited to the static quality inspection of fruits, while in-depth studies on the dynamic changes in apple quality during the harvesting period are relatively limited [3]. In fact, apples undergo a series of physiological and chemical changes after harvest, such as respiration and sugar metabolism, processes that directly affect their taste and storage life [4]. Therefore, establishing a dynamic nondestructive testing model for apple quality during harvesting is of great practical significance in realizing the comprehensive monitoring of fruit production, storage, and marketing.

A commonly used technique to achieve this goal is near-infrared (NIR) spectroscopy, because of its efficient and nondestructive nature [5–7]. With the development of chemometric methods and the availability of instruments, NIR spectroscopy has been widely used for the determination of fruit SSC [8]. Research in the field of fruit detection using NIR spectroscopy mainly focuses on establishing spectral signal models and developing corresponding processing methods. In terms of model building, intelligent algorithms such as principal component analysis (PCA), partial least squares (PLS), and support vector machines (SVM) have been realized to construct static models for the nondestructive testing of fruit quality [9–12]. Numerous researchers have conducted studies indicating that the fruit quality may change significantly due to biological variability, which affects light propagation and light–substance interactions. The predictive power of the modeling may be reduced as a result [13]. Peirs et al. [14] found that the orchards, seasons, and varieties have a great influence on the spectral variation of apples.

The near-infrared spectra of reflectance modes are more readily available and have relatively high intensity levels [15]. Several studies have been conducted to measure SSC online using reflectance modes, such as those focusing on blueberries [16], oranges [17], and apples [18], suggesting that the reflectance mode is suitable for the accurate determination of fruit's internal qualities. Therefore, in this study, the diffuse reflectance mode was used to detect the diffuse reflectance spectra of apples online. Fruit ripening and senescence are complex physiological as well as biochemical processes. The physicochemical compositions of fruits vary greatly with the maturity, which may affect the robustness of SSC and hardness calibration models [19]. The fruit maturity indicators are the soluble solids, titratable acid, and color. Shao et al. [20] investigated the relationship between soluble solids and the near-infrared hyperspectral data of medium-ripe and fully ripened winter jujubes. Yu et al. [21] constructed a quantitative maturity model for Kurrer balsam pears, choosing the hardness, soluble solids content, and titratable acid to quantify the maturity of the pears. Pourdarbani et al. [22] classified "Fuji" apples into four stages of maturity based on color and spectral data. Fruits at the freshly bagged and uncolored stage exhibit lower maturity, and those harvested when they reach the harvest standard demontrate higher maturity. Zhang et al. [23] collected spectral data on apples to evaluate their maturity levels. On the other hand, DeLong et al. [24] used a handheld chlorophyll meter to obtain chlorophyll content data on apples, and the developed model was used to succesfully assess the harvest maturity of "Minneiska" apples. Therefore, the nondestructive testing of dynamic change processes in new application scenarios is necessary to investigate the effect of ripeness on apple quality detection during the harvesting period. The goal is to establish a quality detection model that is insensitive to changes in apple ripeness while on the tree.

The purpose of this study is to model the dynamic nondestructive detection of apples on trees during the harvesting period after the removal of apple bags. The technical approach is depicted in Figure 1. The specific objectives are to (1) compare and analyze the change rules of the apple surface coloring and brix on the tree during the harvesting period and establish a mechanism linking near-infrared spectroscopy and different maturity levels; (2) analyze and compare the optimization results of reflectance spectral feature algorithms for two maturity levels of apples before and after the harvesting period, to provide a recommended solution for the dynamic quality inspection of apples; and (3) study the effect of color parameters (L*, a*, b*) on the modeling of the dynamic change process. A classification model was built with PLS-LDA.

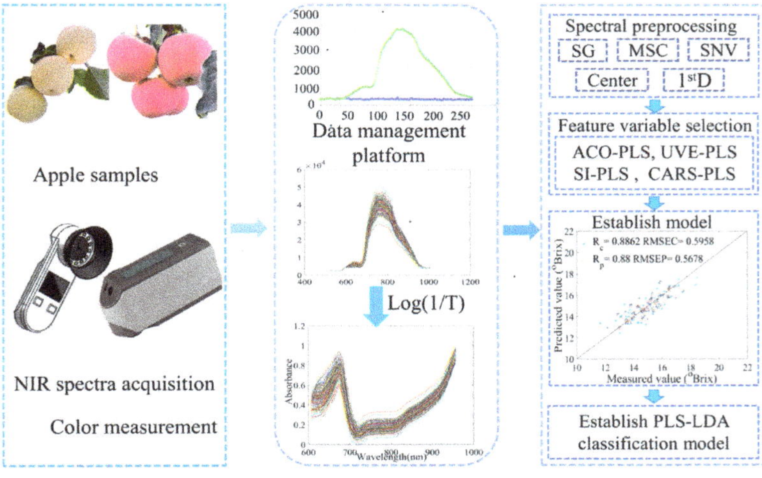

Figure 1. Detecting apple SSC and constructing a dynamic quality detection model for the apple harvesting period using a handheld tester.

2. Materials and Methods

2.1. Sample Collection

The apples used in this experiment were produced in an orchard at an 80°16′ eastern longitude and 41°18′ northern latitude in the Aksu region, Xinjiang. The Fuji variety apples were similar to those available in the market and represented the predominant cultivar in the area. The color of the sun-exposed and shaded sides of the apples was determined when the apples were first bagged, and the spectra of the apples were detected in situ on the tree using a prototype handheld monitoring terminal developed by our group. The apples were then collected to determine their physicochemical information. Then, 14 days later, when the harvest standard was reached, as identified by a professional harvester, samples that were relatively the same size as those of the low-maturity apples and had no scars on their surfaces were detected in situ on the tree. The harvest dates and numbers of samples at different maturity stages are shown in Table 1. The sample surfaces were wiped with damp gauze and allowed to dry naturally. Then, the samples were numbered one by one and the spectral and physicochemical information of the samples was collected.

Table 1. Harvest date and number of samples at different maturity stages.

Date	Samples	Maturity
4 October 2023	100	LM
5 October 2023	100	LM
28 October 2023	80	HM
29 October 2023	80	HM

2.2. Color Measurement

The color of the samples was measured with a spectrophotometer (CM-2300d, Konica Minolta, Osaka, Japan). In the CIE L*a*b* color space, the color is estimated by three parameters: L* denotes the luminance from black (0) to white (100); a* denotes the green–red trend, with negative values toward green (−) and positive values toward red (+); and b* denotes the blue–yellow trend, with negative values toward blue (−) and positive values toward yellow (+) [25–27]. The spectrophotometer was first calibrated with a standard white calibration plate CM-A145. Then, two separate equatorial measurements were taken of the sunward and dorsal surfaces of the sample. The color parameters of the sunward and dorsal sides of the samples were averaged over the two measurements.

2.3. Measurement of Apple SSC

The SSC measurement procedure was as follows. After removing the peel, a 10 × 10 × 10 mm tissue block was removed at the equator of the apple with a sampler. The SSC of the apples was determined by squeezing the juice from the removed tissue blocks wrapped in gauze. A portable refractometer (PAL-1, Atago, Tokyo, Japan) with automatic temperature compensation was used. The determination of the suspended solids in the apples was carried out according to the agriculture industry standard (NY/T2637-2014). The refractometer was first zeroed with pure water and then the squeezed juice was dropped onto the test lens. The refractometer prism was cleaned with pure water at the end of each test. The average of the two measurements was considered as the SSC value for each sample.

2.4. Prototype of Handheld Detection Terminal

The internal quality of a sample can be detected using the apple internal quality handheld tester. As shown in Figure 2, the apple internal quality handheld tester and an Internet of Things (IoT) cloud data system comprise the detector. When testing, the apple internal quality handheld tester can acquire the reflectance spectrum information of the apple. At the same time, the data are transferred to the IoT cloud data system via a communication module. In the Internet of Things (IoT) cloud data system, the model of the inspection is used for calculation and can predict the inspection results. Data can also be queried, downloaded, and statistically analyzed in the detection database.

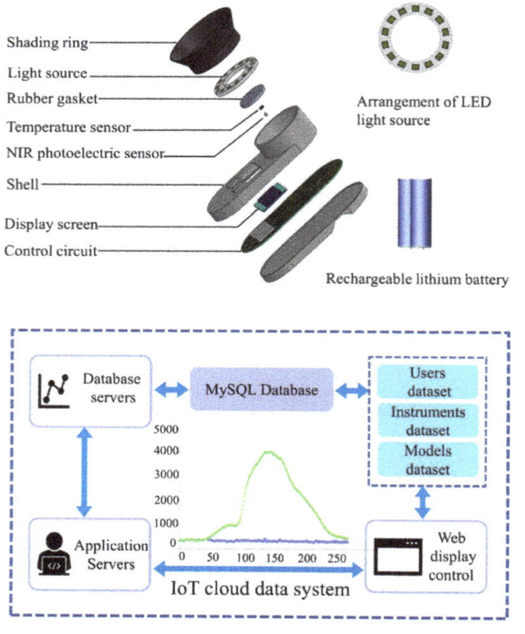

Figure 2. Structure of apple internal quality handheld tester and IoT cloud data testing system.

The hardware of the handheld inspection terminal of the test system mainly consists of a light-emitting diode (LED) light source, a near-infrared photoelectric sensor, a light-shielding ring, a rubber gasket, a temperature sensor, a casing, a rechargeable lithium battery, a control circuit, a display, and so on. The LED point light source is used to irradiate the tested apple. The rubber gasket protects the apple from mechanical damage. In addition, the shading ring excludes stray light other than diffuse reflected light, fully ensuring that it is not affected by ambient light during outdoor inspection. During the detection process, the light illuminates the apple and then passes through a series of internal transmissions and is finally received by the visible near-infrared photoelectric sensor. After the control circuit processes the received signal, it is then transmitted to the cloud server through the 4G/5G module. The final detection results can be calculated by using the cloud model.

In addition, the apple internal quality handheld detector measures 118 × 50 × 38 mm and weighs about 0.12 kg. It has the advantages of a small size, high integration, low energy consumption, and ease of secondary development, making it easy to carry and more convenient for apple quality assessment. Compared with traditional portable and handheld NIR detection devices in the market, it has a greatly simplified design structure and significantly reduced cost. In conclusion, the advantages of this design are low power consumption, a low cost, and ultra-portability. It can be used by fruit companies, farmers, and apple producers for a wide range of applications.

2.5. NIR Spectra Acquisition

The diffuse reflectance spectra of the samples were collected using the apple internal quality handheld detector, as described in the previous section. After starting the spectral acquisition software, as optimization was performed many times before this experiment, the average number of times was set to 5, the exposure time was 75 ms, the sampling frequency was 10 Hz, and the pulse-width modulation (PWM) value was 60. The apple was attached via the equatorial position to the rubber washer, and, after pressing the detection button, the miniature light source began to flash and the spectral data appeared in the spectral acquisition software. The spectral acquisition was completed upon clicking the save button. Two measurements were taken at the equator of each sample on the sun-exposed side and the shaded side. Our operations were conducted on the surfaces of the apples, eliminating the need for peeling. In order to resolve the testing errors caused by dark noise, a total reflection reference spectrum was acquired using a PTFE reflector before the apple spectrum acquisition. In addition, dark noise spectra were acquired simultaneously with the apple spectra and the diffuse reflectance spectral intensity was converted into the relative absorbance lg(1/R).

$$R = \frac{I_\lambda - D_\lambda}{R_\lambda - D_\lambda}$$

where R is the corrected spectrum; I_λ is the original apple spectrum collected; D_λ is the dark noise spectrum; and R_λ is the total reflection reference spectrum.

2.6. Data Processing Algorithms

2.6.1. Spectral Preprocessing

In the process of acquiring apples' diffuse reflectance spectra, the spectral detection performance may be affected by the influence of instrument errors, manual manipulation, and variations in the apple samples. To eliminate these effects, spectral preprocessing was performed. Savitzky–Golay smoothing (SG) was first applied for noise reduction, and then attempts were made to further process the apple spectra using the standard normal variable transform (SNV), multiple scattering correction (MSC), normalization (center), first-order differential (1^{st}D), and further processing [28].

2.6.2. PLS Model

The partial least squares method is the most classical of the chemometric methods for visible/near-infrared spectral analysis. The partial least squares algorithm first decomposed the apple spectral matrix and the SSC matrix and obtained the score matrix and the loading matrix from these two original matrices [29,30]. In turn, a one-way linear regression relationship between the score matrix of the spectral variable and the score matrix of the SSC variable was established.

2.6.3. Feature Variable Selection

The composition of food is usually complex, with a large number of second and third overtones in the visible/near-infrared region. Interactions between overtone bands result in spectra containing a large amount of irrelevant information. These factors affect the accuracy and robustness of spectral prediction models, and the speed of the detection model may be reduced as a result [31,32]. The four feature variable selection algorithms used in this experiment were the synergetic interval (SI), the continuous projection algorithm (UVE), competitive adaptive reweighted sampling (CARS), and the ant colony optimization algorithm (ACO). The SI algorithm divides the entire spectrum into different subintervals and then combines these different subintervals to find a combination that gives the model the highest accuracy [33–35]. To simplify the model, the interval with the smallest RMSEC is chosen during construction. UVE excludes variables considered "useless" from the dataset through thresholds such as variance and the correlation coefficient. By excluding irrelevant variables, redundant and irrelevant information in the data can be effectively reduced, thus helping to improve the performance of the model. CARS evaluates and ranks the importance of each feature and obtains a subset of optimal feature variables. Compared with some other traditional feature selection methods, the CARS algorithm handles the interrelationships between features better, improving the accuracy and stability of feature selection [36]. ACO-PLS is a powerful variable selection algorithm used to solve regression problems. It mimics the behavior of ants in mapping food channels, capturing the target portion of the spectrum from the spectral data. Firstly, it initializes the pheromone vector and other related parameters. Then, it initiates the ants and uses the Monte Carlo roulette wheel spinning method to randomly select a variable from the set until the number of variables reaches the maximum. After completing the variable selection in stages, it establishes a partial least squares model and outputs the root mean square error. All four algorithms are widely used in fruit spectral feature selection.

2.6.4. Model Performance Evaluation

Feature variable selection and PLS regression were performed using MATLAB R2017b (MathWorks, Natick, MA, USA). Then, the parameters of the PLS model were imported into the software developed by our group and five preprocessing methods were used to predict the SSC of the samples with different maturity levels. The correlation coefficient of calibration (Rc), root mean square error of calibration (RMSEC), correlation coefficient of prediction (Rp), root mean square error of prediction (RMSEP), and residual prediction deviation (RPD) were used to compare and evaluate the performance of the model [37]. The degree of linear correlation between the predicted SSC values and true SSC values is indicated by Rc and Rp. RMSEC and RMSEP indicate the deviation between the predicted SSC and true SSC. Higher Rc and Rp values and lower RMSEC and RMSEP values indicate the excellent performance of the established PLS model. The RPD value is the standard deviation divided by the RMSEP. An RPD value >1.8 indicates good model stability and an RPD value > 2 indicates that the predictive accuracy of the model is excellent.

2.6.5. PLS-LDA Classification Model

Partial least squares–linear discrimination analysis (PLS-LDA) is a multivariate analysis technique mainly used to deal with classification and discrimination problems; it combines the ideas of partial least squares regression (PLS) and linear discriminant analysis (LDA) [38]. PLS-LDA is a supervised discriminant analysis method where the dataset contains both an independent variable (X) and a dependent variable (Y), with the ability to model high-dimensional and high-correlation samples that capture the maximum variance of X associated with Y in the space.

3. Results

3.1. Apple Quality at Different Maturity Stages

3.1.1. Comparison of Color Parameters

The pigments in the skin of red apples consist mainly of chlorophyll, carotenoids, and anthocyanins. Among them, chlorophyll contributes to the green color, carotenoids mainly exhibit a yellow color, and anthocyanins are responsible for the red color of the fruit. The three major classes of pigments exhibit their own dynamic formation patterns during fruit development. Chlorophyll and carotenoids mainly form during the early stage of fruit growth and development, while the formation of pericarp anthocyanins initiates gradually as the fruit ripens, often serving as a marker for the onset of fruit ripening. As shown in Table 2, as the harvest period progresses, the maturity of apples gradually increases. Correspondingly, the L* value gradually decreases, the a* value increases, and the b* value gradually decreases, culminating in the gradual formation of a red color in apples. The a*/b* ratio and the a* value are directly correlated to the anthocyanin content.

Table 2. Measured color change statistics of apple samples.

Parameter	Maturity	Samples	Range	Mean
L*	LM	200	39.6–84.38	78.45 ± 4.29
	HM	160	26.41–82.26	68.69 ± 8.00
a*	LM	200	−5.16–27.87	2.46 ± 5.77
	HM	160	−2.41–35.51	13.84 ± 8.89
b*	LM	200	16.31–88.90	27.77 ± 6.16
	HM	160	15.15–40.60	26.60 ± 4.87

3.1.2. SSC Measurement Statistics

The SSC ranges of the low- and high-maturity apples were 9.3–17.0 °Brix and 11.7–19.4 °Brix, respectively. The mean SSC of the low- and high-maturity apples was 13.4 ± 1.2 °Brix and 15.1 ± 1.4 °Brix, respectively. After removing the outliers using Monte Carlo outlier detection, accurate predictions were obtained, as shown in Table 3. The apple samples were randomly divided into calibration and prediction datasets, with a ratio of 3:2. Calibration and prediction sets are independent of each other and do not overlap. On the other hand, outliers in normal samples can have a significant impact on the intelligent algorithms building the model, leading to a decrease in the accuracy of the model predictions.

Table 3. Statistics of apple SSC values measured for spectral acquisition methods.

Maturity	Dataset	Samples	Range (°Brix)	Mean (°Brix)
LM	Total	200	9.3–17.0	13.4 ± 1.2
	Calibration	120	9.3–17.2	13.2 ± 1.3
	Prediction	80	9.6–16.8	13.2 ± 1.0
HM	Total	160	11.7–19.4	15.1 ± 1.4
	Calibration	96	11.7–18.5	14.8 ± 1.3
	Prediction	64	12.2–19.4	15.0 ± 1.4

3.2. Spectral Analysis and Pretreatment

The raw spectra were converted into reflectance spectra based on white and dark references. Figure 3 shows the reflectance spectra of the low- and high-maturity apples, respectively. The principal component analysis of the raw spectra of the low- and high-maturity apples is shown in Figure 3c. As can be seen from the figure, the calculated scores of the first three principal components account for 99.57% of the spectral information. There is less overlap between the spectra of the low- and high-maturity apples, indicating that the difference in the reflectance spectra of the samples collected at the two maturity levels is more significant. The effective spectral range was chosen to be 608–958 nm to avoid noise at low energy. The wider absorption range showed significant variation in the peel pigmentation content among apples at different maturity. Both the low- and high-maturity apples' reflectance spectra had a chlorophyll-related peak at 675 nm. The trough at 710 nm may contain important information related to fruit maturity [39], which may affect the SSC values of the apples. The absorption peak near 750 nm may be the overtone of the O-H band in water. The absorption peak at 820 nm is associated with the N-H overtone [40].

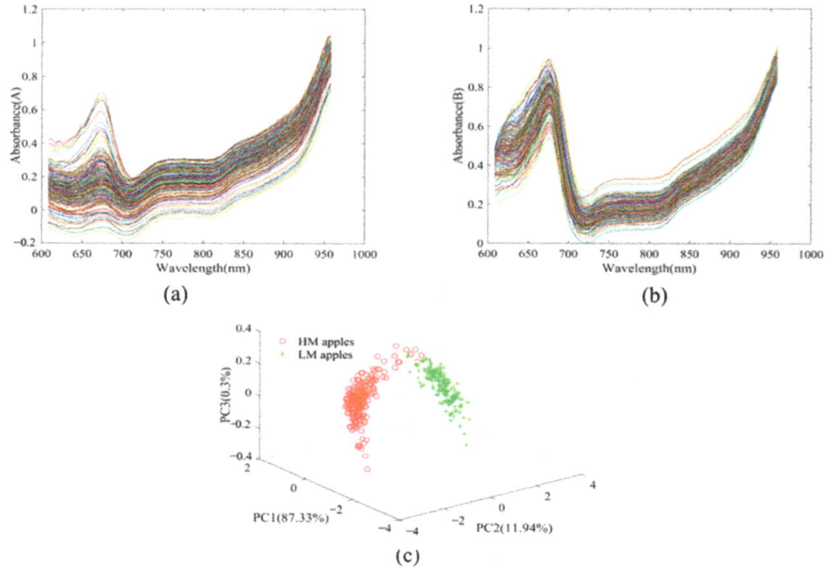

Figure 3. Spectral analysis of apple samples. (**a**) Spectrum of LM apples, (**b**) spectrum of HM apples, (**c**) PCA scatter plot of LM apples (green +) and HM apples (red ○).

As shown in Table 4, the best performance was seen for the reflectance spectra of highly mature apples combined with the SG-SNV preprocessing model, with Rp, RMSEP, and RPD values of 0.8350, 0.7876 °Brix, and 1.778. Therefore, SG-SNV was chosen as the preprocessing method before feature selection for the reflectance spectra. In the low-maturity apples' reflectance spectra, the SG-SNV method did not have as high an RPD value as SG-1stD, but it was also chosen for preprocessing in order to control the variables to obtain a good model of the reflectance spectra.

Table 4. SSC prediction results of PLS LM apples and HM apples with different spectral preprocessing methods.

Maturity	Preprocessing	Samples	Rc	RMSEC	Rp	RMSEP	RPD
LM	SG	200	0.7703	0.7551	0.7916	0.7493	1.601
	SG-SNV		0.7649	0.7639	0.7919	0.7414	1.619
	SG-MSC		0.7753	0.7462	0.7743	0.7676	1.563
	SG-Center		0.7703	0.7551	0.7916	0.7493	1.601
	SG-1stD		0.7957	0.7163	0.7977	0.7219	1.662
HM	SG	160	0.7988	0.7954	0.7801	0.8220	1.703
	SG-SNV		0.8356	0.7932	0.8350	0.7876	1.778
	SG-MSC		0.8343	0.7961	0.8294	0.8022	1.745
	SG-Center		0.8456	0.7709	0.8206	0.8186	1.710
	SG-1stD		0.8439	0.7747	0.8029	0.8538	1.640

3.3. Spectral Models Based on Feature Selection

3.3.1. SI-PLS Model Results

The entire spectrum was decomposed into 10 intervals and combined with four subintervals to build the PLS model with accurate prediction. The best prediction results were obtained when the combination had the lowest interval RMSEP. For low-maturity apples, the best combination of selected subintervals was [2 4 7 10], corresponding to 644–678, 716–750, 824–856, and 926–958 nm. For high-maturity apples, the best spectral subintervals were [5 8 9 10], corresponding to 752–786, 858–890, 892–924, and 926–956 nm. The SI-PLS model predictions are shown in Figures 4b and 5b. The RPD values of the low- and high-maturity apples' spectral models were 1.729 and 1.733. The spectral model of the low-maturity apples showed less accuracy compared to that of the high-maturity apples.

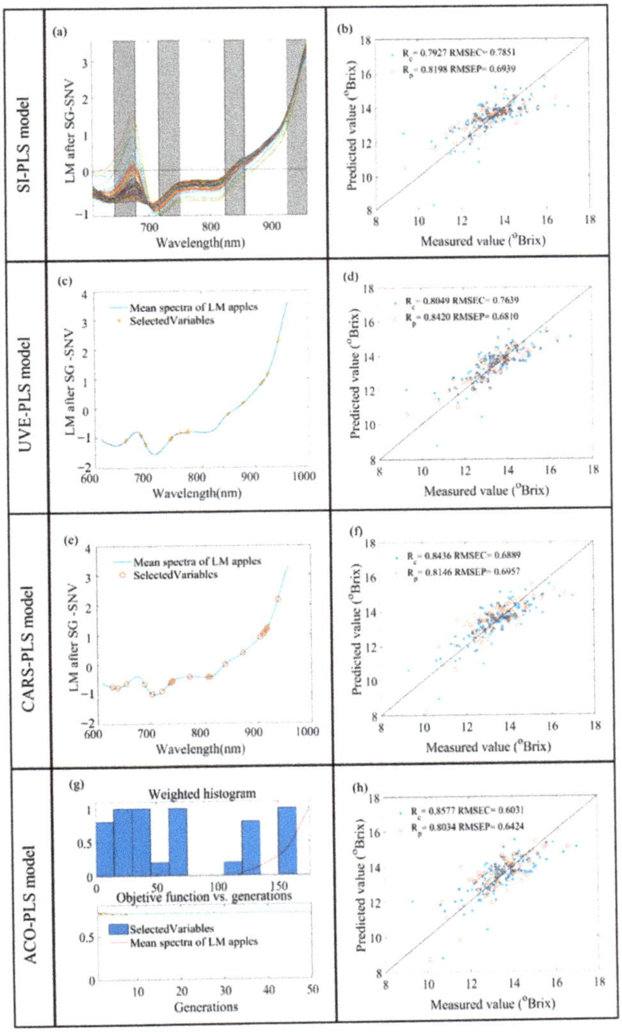

Figure 4. Results of SI, UVE, CARS, and ACO variable selection for LM apples (**a**,**c**,**e**,**g**). The PLS model built by the four variable selection algorithms demonstrates the relevance of the model (**b**,**d**,**f**,**h**) using the distributions of the calibration set (*) and the prediction set (○).

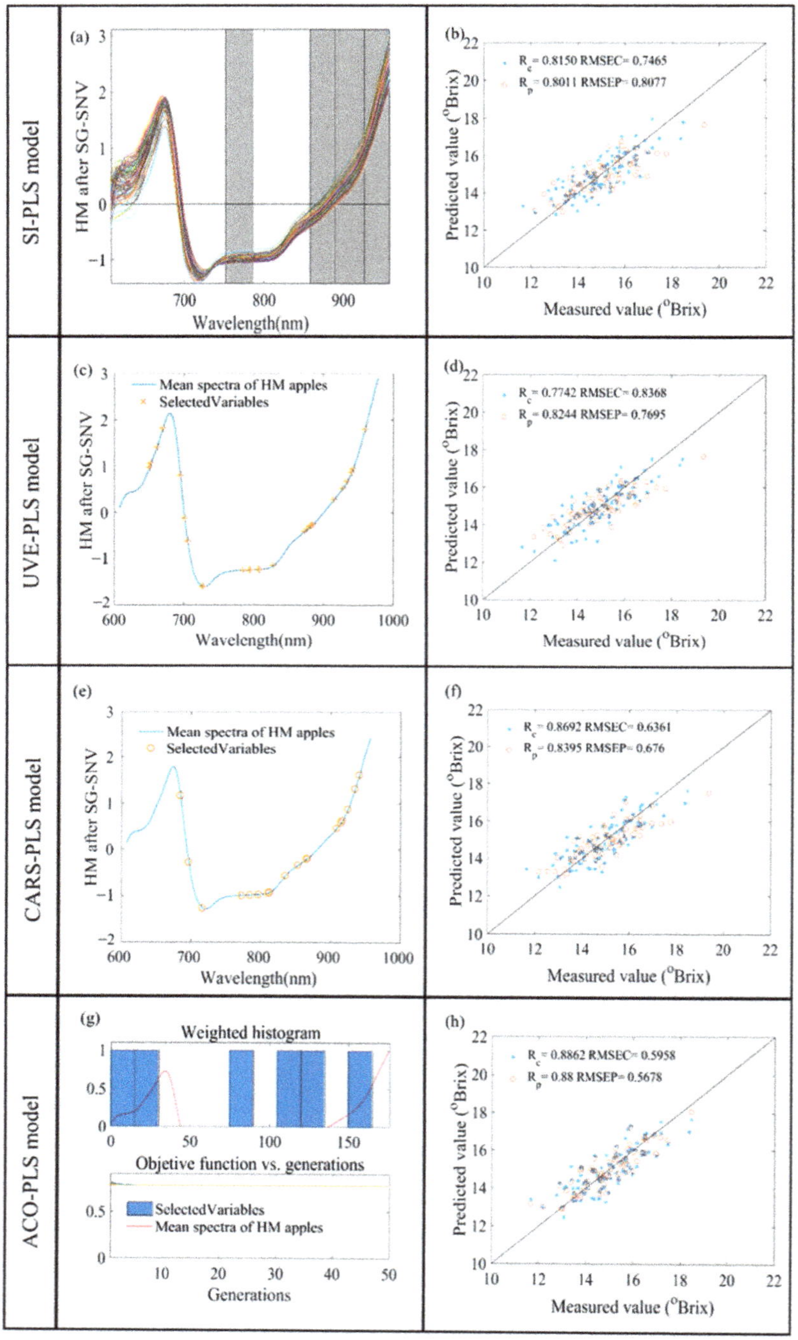

Figure 5. Results of SI, UVE, CARS, and ACO variable selection for HM apples (**a**,**c**,**e**,**g**). The PLS model built by the four variable selection algorithms demonstrates the relevance of the model (**b**,**d**,**f**,**h**) using the distributions of the calibration set (*) and the prediction set (○).

3.3.2. UVE-PLS Model Results

During UVE, the spectral matrix can be cross-validated to eliminate invalid wavelength variables. UVE measures the characteristic wavelength band via controlling the coefficient of variation. Fourteen and twenty-eight wavelength variables were selected in the reflectance spectra of low- and high-maturity apples, respectively. The selected wavelength variables are shown in Figures 4c and 5c. The Rp and RMSEP of the low-maturity apples' reflectance spectral model were 0.8420 and 0.6810 °Brix, respectively. The RPD values of the low-maturity apples' and high-maturity apples' reflectance spectral models were 1.762 and 1.819, respectively. An improvement in the accuracy of the reflectance spectral models of the apples for both maturity levels was realized compared to the full-spectrum model.

3.3.3. CARS-PLS Model Results

Fifty Monte Carlo samplings were set up and the optimal model was determined using five-fold cross-validation based on the RMSEC minimum. The number of samples was gradually increased, the number of selected variables was gradually decreased, and the speed of both the coarse and fine selection processes varied from fast to slow. When the sampling was performed 26 times and 4 times, respectively, the RMSEC value was the smallest, and then it increased gradually. The established models for Rc, RMSEC, Rp, and RMSEP are shown in Figures 4f and 5f. The RPD values of the reflectance spectral models for low- and high-maturity apples were 1.725 and 2.071. However, the RPD value of the reflectance spectral model for high-maturity apples was >2.

3.3.4. ACO-PLS Model Results

The characteristic wavelength points were extracted from the full spectrum using the ACO algorithm. The control parameters of the ACO optimization algorithm were verified by several experiments, as follows: the initial population size was 100, the maximum number of iterations was 50, the maximum number of loops was 10, the probability threshold of variable selection was 0.3, and the significance factor Q was 0.01. The pheromone attenuation coefficient p was 0.65, which reliably ensured the comprehensiveness of the transmitted information and the convergence speed of the algorithm. The selected wavelength variables are shown in Figures 4g and 5g, and the established models for Rc, RMSEC, Rp, and RMSEP are shown in Figures 4h and 5h. The RPD values of the reflectance spectral models for low- and high-maturity apples were 1.868 and 2.466. ACO-PLS achieved the best results in the reflectance spectral modeling of apples at both maturity levels.

3.4. Comparisons of Different PLS Models

Figures 4 and 5 show the modeling results of the low- and high-maturity apples' reflectance spectra processed through the SI, UVE, CARS, and ACO algorithms. In the collection of the low-maturity apples' reflectance spectra, the Rp values were all less than 0.85 and the RPD values were all less than 2. Meanwhile, in the high-maturity apples' reflectance spectra, the RP values were greater than 0.80; the RPD values were greater than 1.8 after processing with the UVE, CARS, and ACO algorithms; and the RPD values were greater than 2 after processing with the CARS and ACO algorithms. These suggest that high-maturity apples can be predicted more accurately with respect to the internal SSC values. The performance of the reflectance model for low-maturity apples, in descending order, was ACO-PLS > UVE-PLS > SI-PLS > CARS-PLS > PLS. The performance of the reflectance model for high-maturity apples, in descending order, was ACO-PLS > CARS-PLS > UVE-PLS > SI-PLS > PLS. The ACO-PLS model with reflectance spectra had the best prediction ability for high-maturity apples' SSC content. Meanwhile, the reflectance spectral model for high-maturity apples had better performance overall.

3.5. PLS-LDA Classification Model

To investigate the effect of color on maturity, the color space coordinates (L*, a*, b*) were used as input variables, and then a PLS-LDA classification model was built by combining these variables with the SSC. The SSC array was set to −1 for samples with low maturity and 1 for samples with high maturity. The classification accuracy of the training set is presented in Figure 6. The accuracy of the training and prediction sets was 99.03% and 99.35%, respectively. Overall, the PLS-LDA model was effective in analyzing the apple color during ripening.

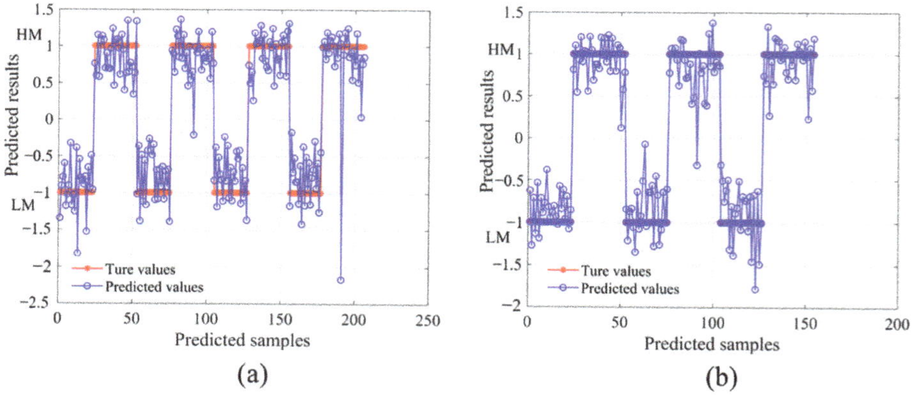

Figure 6. PLS-LDA modeling results with color space coordinates (L*, a*, b*) as input variables, true values (*), and predicted values (○). Comparison of prediction results of training set (**a**) with 99.03% accuracy and comparison of prediction results of prediction set (**b**) with 99.53% accuracy.

The diffuse reflectance spectra corresponding to the classified color coordinates (L*, a*, b*) were preprocessed with SG-SNV and then subjected to PLS modeling. The classified low-maturity apples had more acccurate values, with Rp = 0.8139, RMSEP = 0.7013, and RPD = 1.711, compared to the PLS model. After classification, high-maturity apples, with Rp = 0.8379, RMSEP = 0.7760, and RPD = 1.804, showed slight improvements in the model predictions. The results obtained via the LDA method showed reliability, and it could be applied to different classification methods supervised by discriminant analysis.

4. Discussion

The SSC was obtained for two maturity levels of apples by in situ testing on trees. Diffuse reflectance spectra were collected with a handheld inspection system developed by our group for the internal quality assessment of apples, aiming to model the dynamic in situ inspection of apples on trees for the period from bag removal to harvest. Due to the differences in the light transmission properties and cellular structures of apples, the original spectral trends of the apples at different maturity levels were generally similar, but the spectral intensities varied greatly from 650 nm to 800 nm.

Comparing the prediction results of the spectra of the two maturity groups of apples, the SSC prediction accuracy for high-maturity apples was higher than that for low-maturity apples among the various feature extraction methods. There are many reasons for the differences between the modeling results of low- and high-maturity apples. The signals collected by the spectrometer in reflectance mode arise from the apple surface, and prediction models developed by applying NIR spectroscopy for the non-invasive detection of the fruit SSC are often influenced by biological variability factors. Fruits with different physiological stages, varieties, growing regions, or seasons undergo changes in physical properties and chemical composition, which lead to differences in the fruit's cellular structure and optical properties.

The apple samples were randomly divided into mutually independent calibration and prediction sets. Among the five pretreatment methods (SG, SNV, MSC, center, and 1^{st}D), SG and SNV performed better. This was due to the fact that these two methods can eliminate the effects of the solid particle size, surface scattering, and light path variations from the spectra. Among the four feature wavelength extraction methods, the SI algorithm performed the worst, which may have been due to the fact that the optimal subintervals selected were not within the effective spectral range and did not effectively represent the SSC features. Compared to the CARS algorithm, the UVE algorithm removes SSC-independent variables from the spectral information. Most of the characteristic wavelength variables selected by the CARS algorithm were related to the internal chemical properties of the apples, seeking to improve the performance of the model, but potentially valid information was also removed. Compared to the other three spectral feature selection algorithms, the ACO algorithm achieved the best optimization results, and the highest accuracy for SSC prediction model was achieved using the high-maturity apple reflectance spectral model built by the ACO algorithm. This may have been due to the ability of the ACO algorithm to screen out highly competitive combinations of feature wavelengths. The ACO algorithm adaptively adjusted the search behavior through pheromone updating and showed strong adaptability to environmental changes, enabling it to consistently find a better solution. Meanwhile, the best results were achieved by using diffuse reflectance spectroscopy combined with the ACO algorithm to predict the SSC of apples at high maturity, with Rp, RMSEP, and RPD values of 0.88, 0.5678 °Brix, and 2.466, respectively. The PLS-LDA model was more effective regarding the apple color against ripeness, with accuracy of 99.03% for the training set model and 99.35% for the prediction set model. The post-spectral color classification showed great potential in predicting the apples' SSC.

In addition, the model has the potential to assess the external color and internal quality for the grading of apples for customer satisfaction. These tested and graded apples can better satisfy customers' purchasing needs and improve the efficiency and accuracy of apple valuation and harvesting.

5. Conclusions

This study acquired the diffuse reflectance spectra of low-maturity and high-maturity apple samples using a handheld inspection system for the internal quality assessment of apples developed by our group. Four feature wavelength selection algorithms, namely SI, UVE, CARS, and ACO, were employed for feature wavelength selection. The established ACO-PLS models for high-maturity apples achieved the best results in this study. The model's Rp, RMSEP, and RPD values were 0.88, 0.5678 °Brix, and 2.466, respectively. In order to consider the influence of color factors on the predictive ability of the model for low- versus high-ripeness apples during the apple harvest period, a PLS-LDA classification model for apple ripeness based on color was developed. The accuracy of the training set model was 99.03% and the accuracy of the prediction set model was 99.35%. The results showed that the new application scenario of dynamic inspection has great potential to achieve improved robustness and accuracy in modeling apple quality during the harvest period. This method can be expanded to other varieties of apples in the future to verify the generalization and applicability of the model.

Author Contributions: Z.G.: supervision, project administration, funding acquisition, conceptualization. X.C.: writing—original draft, MATLAB R2017b (MathWorks, Natick, MA, USA), investigation, formal analysis. Y.Z.: MATLAB R2017b (MathWorks, Natick, MA, USA), investigation. C.S.: visualization. H.J.: visualization, methodology. U.M.: writing—review and editing, formal analysis. N.J.W.: writing—review and editing. X.Z.: supervision, resources. All authors have read and agreed to the published version of the manuscript.

Funding: This research was funded by the National Key R&D Program of China (2023YFE0107100), Key R&D Project of Jiangsu Province (BE2022363), China Scholarship Council (202208690011), Natural Science Foundation of Jiangsu Province (BK20220524), Jiangsu Agriculture Science and Technology

Innovation Fund (CX(22)3069), Senior Talent Program of Jiangsu University (22JDG015), and Jiangsu Provincial Double-Innovation Doctor Program.

Institutional Review Board Statement: Not applicable.

Informed Consent Statement: Not applicable.

Data Availability Statement: The original contributions presented in the study are included in the article, further inquiries can be directed to the corresponding author.

Acknowledgments: The authors would like to thank Nongfu Spring Co., Ltd. (Hangzhou, China), for supporting the comprehensive apple demonstration base and providing the apple sample materials.

Conflicts of Interest: The authors declare no conflicts of interest.

References

1. Guo, Z.; Zhang, Y.; Wang, J.; Liu, Y.; Jayan, H.; El-Seedi, H.R.; Zou, X. Detection model transfer of apple soluble solids content based on NIR spectroscopy and deep learning. *Comput. Electron. Agric.* **2023**, *212*, 108127. [CrossRef]
2. Wang, F.; Zhao, Y.; Zheng, J.; Qi, K.; Fan, Y.; Yuan, X.; Ruan, H. Estimation model of soluble solids content in bagged and non-bagged apple fruits based on spectral data. *Comput. Electron. Agric.* **2021**, *191*, 106492. [CrossRef]
3. Chang, H.; Wu, Q.; Tian, H.; Yan, J.; Luo, X.; Xu, H. Non-Destructive Identification of Internal Watercore in Apples Based on Online Vis/NIR Spectroscopy. *Trans. ASABE* **2020**, *63*, 13844. [CrossRef]
4. Zhang, M.; Zhang, B.; Li, H.; Shen, M.; Tian, S.; Zhang, H.; Ren, X.; Xing, L.; Zhao, J. Determination of bagged 'Fuji' apple maturity by visible and near-infrared spectroscopy combined with a machine learning algorithm. *Infrared Phys.Technol.* **2020**, *111*, 103529. [CrossRef]
5. Rady, A.; Watson, N.J. Detection and quantification of peanut contamination in garlic powder using NIR sensors and machine learning. *J. Food Compos. Anal.* **2022**, *114*, 104820. [CrossRef]
6. Bai, Y.; Fang, Y.; Zhang, B.; Fan, S. Model robustness in estimation of blueberry SSC using NIRS. *Comput. Electron. Agric.* **2022**, *198*, 107073. [CrossRef]
7. Shalev, N.; Ghermandi, A.; Tchernov, D.; Shemesh, E.; Israel, A.; Brook, A. NIR spectroscopy and artificial neural network for seaweed protein content assessment in-situ. *Comput. Electron. Agric.* **2022**, *201*, 107304. [CrossRef]
8. Ireri, D.; Belal, E.; Okinda, C.; Makange, N.; Ji, C. A computer vision system for defect discrimination and grading in tomatoes using machine learning and image processing. *Artif. Intell. Agric.* **2019**, *2*, 28–37. [CrossRef]
9. Ding, F.; Zuo, C.; García-Martín, J.; Ge, Y.; Tu, K.; Peng, J.; Xiao, H.; Lan, W.; Pan, L. Non-invasive prediction of mango quality using near-infrared spectroscopy: Assessment on spectral interferences of different packaging materials. *J. Food Eng.* **2023**, *357*, 111653. [CrossRef]
10. Esquerre, C.A.; Achata, E.M.; García-Vaquero, M.; Zhang, Z.; Tiwari, B.K.; O'Donnell, C.P. Use of an NIR MEMS spectrophotometer and visible/NIR hyperspectral imaging systems to predict quality parameters of treated ground peppercorns. *LWT-Food Sci. Technol.* **2020**, *131*, 109761. [CrossRef]
11. Li, X.; Cai, M.; Li, M.; Wei, X.; Liu, Z.; Wang, J.; Jia, K.; Han, Y. Combining VIS-NIR and NIR hyperspectral imaging techniques with a data fusion strategy for the rapid qualitative evaluation of multiple qualities in chicken. *Food Control.* **2023**, *145*, 109416. [CrossRef]
12. Luo, X.; Ye, Z.; Xu, H.; Zhang, D.; Bai, S.; Ying, Y. Robustness improvement of NIR-based determination of soluble solids in apple fruit by local calibration. *Postharvest Biol. Technol.* **2018**, *139*, 82–90. [CrossRef]
13. Zhang, B.; Dai, D.; Huang, J.; Zhou, J.; Gui, Q.; Dai, F. Influence of physical and biological variability and solution methods in fruit and vegetable quality nondestructive inspection by using imaging and near-infrared spectroscopy techniques: A review. *Crit. Rev. Food Sci. Nutr.* **2018**, *58*, 2099–2118. [CrossRef]
14. Peirs, A.; Tirry, J.; Verlinden, B.; Darius, P.; Nicolaï, B.M. Effect of biological variability on the robustness of NIR models for soluble solids content of apples. *Postharvest Biol. Technol.* **2003**, *28*, 269–280. [CrossRef]
15. Bobelyn, E.; Serban, A.-S.; Nicu, M.; Lammertyn, J.; Nicolai, B.M.; Saeys, W. Postharvest quality of apple predicted by NIR-spectroscopy: Study of the effect of biological variability on spectra and model performance. *Postharvest Biol. Technol.* **2010**, *55*, 133–143. [CrossRef]
16. Hu, M.; Dong, Q.; Liu, B.; Opara, U.; Chen, L. Estimating blueberry mechanical properties based on random frog selected hyperspectral data. *Postharvest Biol. Technol.* **2015**, *106*, 1–10. [CrossRef]
17. Liu, Y.; Zhou, Y.; Pan, Y. Online quantitative analysis of soluble solids content in navel oranges using visible-near infrared spectroscopy and variable selection methods. *J. Innov. Opt. Health Sci.* **2014**, *7*, 1350065. [CrossRef]
18. Li, Y.; Peng, Y.; Li, Y.; Yin, T.; Wang, B. Optimization of Online Soluble Solids Content Detection Models for Apple Whole Fruit with Different Mode Spectra Combined with Spectral Correction and Model Fusion. *Foods* **2024**, *13*, 1037. [CrossRef]
19. Wang, F.; Zhao, C.; Yang, H.; Jiang, H.; Li, L.; Yang, G. Non-destructive and in-site estimation of apple quality and maturity by hyperspectral imaging. *Comput. Electron. Agric.* **2022**, *195*, 106843. [CrossRef]

20. Shao, Y.; Ji, S.; Xuan, G.; Wang, K.; Xu, L.Q.; Shao, J. Soluble solids content monitoring and shelf life analysis of winter jujube at different maturity stages by Vis-NIR hyperspectral imaging. *Postharvest Biol. Technol.* **2024**, *210*, 112773. [CrossRef]
21. Yu, S.; Tang, Y.; Lan, H.; Li, X.; Zhang, H.; Zeng, Y.; Niu, H.; Jin, X.; Liu, Y. Construction method of quantitative evaluation model for the maturity of Korla fragrant pear. *Int. J. Agric. Biol. Eng.* **2022**, *15*, 243–250. [CrossRef]
22. Pourdarbani, R.; Sabzi, R.; Kalantari, D.; Karimzadeh, R.; Ilbeygi, E.; Arribas, J.I. Automatic non-destructive video estimation of maturation levels in Fuji apple (*Malus Malus pumila*) fruit in orchard based on colour (Vis) and spectral (NIR) data. *Biosyst. Eng.* **2020**, *195*, 136–151. [CrossRef]
23. Zhang, B.; Zhang, M.; Shen, M.; Li, H.; Zhang, Z.; Zhang, H.; Zhou, Z.; Ren, X.; Ding, Y.; Xing, L.; et al. Quality monitoring method for apples of different maturity under long-term cold storage. *Infrared Phys. Technol.* **2021**, *112*, 103580. [CrossRef]
24. DeLong, J.; Harrison, P.; Harkness, L. An optimal harvest maturity model for 'Minneiska' apple fruit based on the delta-absorbance meter. *J. Hortic. Sci. Biotechnol.* **2020**, *95*, 637–644. [CrossRef]
25. Porras, I.; Brotons, J.M.; Conesa, A.; Castañer, R.; Pérez-Tornero, O.; Manera, F.J. Quality and fruit colour change in Verna lemon. *J. Appl. Bot. Food Qual.* **2015**, *88*, 215–221.
26. Yang, N.; Chang, K.; Tang, J.; Xu, L.; He, Y.; Huang, R.; Yu, J. Detection method of rice blast based on 4D light field refocusing depth information fusion. *Comput. Electron. Agric.* **2023**, *205*, 107614. [CrossRef]
27. Li, W.; Liu, Z.; Wang, H.; Zheng, Y.; Zhou, Y.; Duan, L.; Tang, Y.; Jiang, Y.; Li, X.; Jiang, Y. Harvest maturity stage affects watercore dissipation and postharvest quality deterioration of watercore 'Fuji' apples. *Postharvest Biol. Technol.* **2024**, *210*, 112736. [CrossRef]
28. Jiang, H.; Wang, J.; Chen, Q. Comparison of wavelength selected methods for improving of prediction performance of PLS model to determine aflatoxin B1 (AFB1) in wheat samples during storage. *Microchem. J.* **2021**, *170*, 106642. [CrossRef]
29. Abu-Khalaf, N.; Hmidat, M. Visible/Near Infrared (VIS/NIR) spectroscopy as an optical sensor for evaluating olive oil quality. *Comput. Electron. Agric.* **2020**, *173*, 105445. [CrossRef]
30. Wang, J.; Guo, Z.; Zou, C.; Jiang, S.; El-Seedi, H.R.; Zou, X. General model of multi-quality detection for apple from different origins by Vis/NIR transmittance spectroscopy. *J. Food Meas. Charact.* **2022**, *16*, 2582–2595. [CrossRef]
31. Guo, Z.; Wang, M.; Agyekum, A.A.; Wu, J.; Chen, Q.; Zuo, M.; Zou, X. Quantitative detection of apple watercore and soluble solids content by near infrared transmittance spectroscopy. *J. Food Eng.* **2020**, *279*, 109955. [CrossRef]
32. Guo, Z.; Zhai, L.; Zou, Y.; Sun, C.; Heera, J.; El-Seedi, H.R.; Jiang, S.; Cai, J.; Zou, X. Comparative study of Vis/NIR reflectance and transmittance method for on-line detection of strawberry SSC. *Comput. Electron. Agric.* **2024**, *218*, 108844. [CrossRef]
33. Liu, S.; Rong, Y.; Chen, Q.; Ouyang, Q. Colorimetric sensor array combined with chemometric methods for the assessment of aroma produced during the drying of tencha. *Food Chem.* **2024**, *432*, 137190. [CrossRef] [PubMed]
34. Soares, S.F.C.; Gomes, A.A.; Galvao, A.R.; Araujo, M.C.U.; Galvao, R.K.H. The successive projections algorithm. *Trends Anal. Chem.* **2013**, *42*, 84–98. [CrossRef]
35. Liu, L.; Zareef, M.; Wang, Z.; Li, H.; Chen, Q.; Ouyang, Q. Monitoring chlorophyll changes during Tencha processing using portable near-infrared spectroscopy. *Food Chem.* **2023**, *412*, 0308–8146. [CrossRef] [PubMed]
36. Zhang, M.; Shen, M.; Li, H.; Zhang, B.; Zhang, Z.; Quan, P.; Zhao, J. Modification of the effect of maturity variation on nondestructive detection of apple quality based on the compensation model. *Spectrochim. Acta Part A Mol. Biomol. Spectrosc.* **2022**, *267*, 120598. [CrossRef] [PubMed]
37. Zhang, Y.; Zareef, M.; Rong, Y.; Lin, H.; Chen, Q.; Ouyang, Q. Application of colorimetric sensor array coupled with chemometric methods for monitoring the freshness of snakehead fillets. *Food Chem.* **2024**, *439*, 138172. [CrossRef] [PubMed]
38. Carlos, H.; Celito, C.; Adriano, A.; Marco, F. Green analytical methodology for grape juice classification using FTIR spectroscopy combined with chemometrics. *Talanta Open* **2022**, *6*, 2666–8319.
39. Manuela, Z.-S.; Ingo, T.; Bernd, H. An approach to non-destructive apple fruit chlorophyll determination. *Postharvest Biol. Technol.* **2002**, *25*, 123–133.
40. Bahareh, J.; Saeid, M.; Ezzedin, M. Reflectance Vis/NIR spectroscopy for nondestructive taste characterization of Valencia oranges. *Comput. Electron. Agric.* **2012**, *85*, 64–69.

foods

MDPI

Article

Disambiguity and Alignment: An Effective Multi-Modal Alignment Method for Cross-Modal Recipe Retrieval

Zhuoyang Zou , Xinghui Zhu , Qinying Zhu, Hongyan Zhang and Lei Zhu *

College of Information and Intelligence, Hunan Agricultural University, Changsha 410128, China; zzy@stu.hunau.edu.cn (Z.Z.); zhuxh@hunau.edu.cn (X.Z.); zhuqy@stu.hunau.edu.cn (Q.Z.); hongyan_zhang@hunau.edu.cn (H.Z.)

* Correspondence: leizhu@hunau.edu.cn

Abstract: As a prominent topic in food computing, cross-modal recipe retrieval has garnered substantial attention. However, the semantic alignment across food images and recipes cannot be further enhanced due to the lack of intra-modal alignment in existing solutions. Additionally, a critical issue named food image ambiguity is overlooked, which disrupts the convergence of models. To these ends, we propose a novel Multi-Modal Alignment Method for Cross-Modal Recipe Retrieval (MMACMR). To consider inter-modal and intra-modal alignment together, this method measures the ambiguous food image similarity under the guidance of their corresponding recipes. Additionally, we enhance recipe semantic representation learning by involving a cross-attention module between ingredients and instructions, which is effective in supporting food image similarity measurement. We conduct experiments on the challenging public dataset Recipe1M; as a result, our method outperforms several state-of-the-art methods in commonly used evaluation criteria.

Keywords: cross-modal recipe retrieval; multi-modal alignment; food image ambiguity; deep learning

1. Introduction

With rising awareness of health and sustainability, issues such as food safety [1,2] and nutrition [3] have gained unprecedented attention. Food computing [4–8] plays a crucial role in promoting healthier lifestyles, mitigating food waste, and enhancing both the quality and safety of food products. Cross-modal recipe retrieval [9,10] is one of the hot topics in food computing, leveraging artificial intelligence (AI) [11,12] which aims to retrieve the corresponding recipes by queries of food images or vice versa. In this task, food images depict finished dishes, while recipes comprise text encompassing three key components: a title, a list of ingredients, and detailed instructions outlining the cooking process.

The principal challenge in cross-modal recipe retrieval lies in mitigating the inherent heterogeneity between two distinct modalities: the recipes and the food images. To solve this challenging task, numerous studies have delved into additional interactions between the two modalities. For instance, Refs. [13–15] tried to learn the consistent feature distribution of food images and recipe texts. Refs. [16–19] boosted the interaction between two modalities through cross-modal attention. Ref. [20] employed a joint transformer encoder to promote alignment. Due to the complexity of image–recipe pairs, many existing studies focused on exploiting the latent semantic information within a modality. As typical studies, refs. [21–25] aimed to focus on the crucial term within recipes, while others [26–28] attempted to capture the salient objects or regions from food images to improve the cross-modal similarity measurement. Due to the complexity of the textual structure in recipes, other researchers [29–33] investigated the interaction among the title, ingredients, and instructions to excavate important semantics. Furthermore, some studies introduced diverse augmentation mechanisms to enhance cross-modal feature representations. For example, Refs. [34–38] employed various Generative Adversarial Networks (GANs) to reconstruct information from food images and recipes to bridge the heterogeneity gap across modalities, while refs. [39,40] leveraged multilingual translation to enrich the recipe information. Crowdsourcing strategy is also used to construct program representations of recipes [41]. Thanks to the flourishing development of visual language pre-training recently, some pioneers [42–45] have further embedded complex semantic relationship information into common feature subspace by leveraging the pre-trained Contrastive Language–Image Pre-Training model (CLIP).

Despite the significant progress made so far, there is still room for further improvement in semantic distribution alignment across food images and recipes. To be specific, the prevailing efforts [18,29,30] concentrate on exploring inter-modal semantic alignment using conventional metric learning strategies, such as triplet loss. As shown in Figure 1a, the conventional metric learning strategy is devoted to reducing the distance between positive image–recipe pairs (the circles and squares with the same color) and enlarging the distance between negative samples (the circles and the gray squares) and is proficient in learning similarity relations within each image–recipe pair. However, semantic relations exist not only within each image–recipe pair, but extensively between different pairs. For example, the two image–recipe pairs in Figure 1a belong to the same food (chilli sauce), indicating strong semantic relations (highlighted by red lines) between the two food images as well as the two recipes. The conventional metric learning method (e.g., triplet loss), however, fails to capture this relation information. To be sure, there are lots of image pairs belonging to the same food in practical time. This situation indicates that effectively enhancing intra-modal semantic alignment is significant for improving recipe retrieval performance.

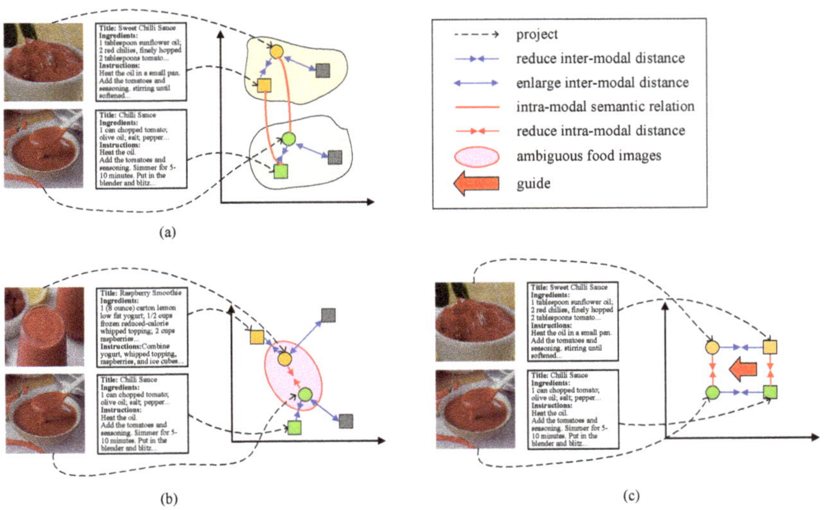

Figure 1. The demonstration of multi-modal alignment schemes for cross-modal recipe retrieval. (**a**) The prevailing learning strategy that ignores intra-modal alignment. (**b**) The food image ambiguity issue. (**c**) Our solution (negative samples are omitted). Circles represent images, and squares represent recipes. Shapes of the same color indicate positive pairs, while gray shapes indicate negative samples.

For this purpose, a straightforward method applied in lots of cross-modal retrieval tasks [46–48] is to utilize metric learning or contrastive learning strategy within each modality. However, there is a non-trivial issue, i.e., *food image ambiguity*, in cross-modal recipe retrieval that has not been considered. Specifically, foods that look similar may be made from quite different materials and via different preparation methods. Thus, these similar food images correspond to significantly distinct recipes. For example, in Figure 1b, the top food image is a cup of raspberry smoothie, and the bottom one is a bowl of chilli sauce. Theses two foods are visually similar to each other, yet they are crafted from distinct ingredients and have undergone quite different instructions. Unfortunately, existing methods embed their semantics from the two modalities to the common subspace independently. Due to the resemblance in appearance, these two images will be close in the common space, while their corresponding recipes will not. This leads to a dilemma; embeddings that have a large similarity (small distance) in the visual modality may have a small similarity (large distance) in the text modality. As a result, the two modalities are hard to align with each other, and the models are difficult to converge, which reduces the accuracy of retrieval. Stumped by this stand-out drawback, we observe that recipes are the more reliable modality. In other words, foods prepared using similar recipes will have similar visual appearances. Therefore, this study aims to answer the following two questions:

- **Q1:** How can we measure the similarity between ambiguous food images guided by their corresponding recipes?
- **Q2:** How can we further improve the fine-grained semantic alignment between ingredients and instructions within each recipe to support food image similarity measurement?

To this end, we propose a novel cross-modal recipe retrieval method called the Multi-Modal Alignment Method for Cross-Modal Recipe Retrieval (**MMACMR**). To answer **Q1**, we design a novel strategy, the Multi-Modal Disambiguity and Alignment strategy (**MDA** for short), which calculates the intra-modal similarity of recipes and guides the distances between corresponding images. As shown in Figure 1c, the green square is a recipe (chilli sauce) similar to the one shown by the orange square (sweet chilli sauce). Our MDA strategy attempts to pull them close and guide the distance between the green and orange circles (their corresponding images). For **Q2**, considering ingredients play a significant role within instructions, we introduce sentence-level cross-attention to focus on important ingredients in the instructions and further enhance the representations of recipes. In a nutshell, this work is a pioneering effort to further narrow cross-modal heterogeneity between food images and recipes by considering both multi-modal (inter-modal and intra-modal) alignment while mitigating the impact of food image ambiguity.

To sum up, the main contributions of this article are four fold:

- We propose a novel framework called MMACMR which addresses the problem of ambiguous food images in cross-modal recipe retrieval;
- We introduce a novel deep learning strategy named MDA which promotes the alignment of two modalities without adding new parameters;
- We enhance the representation of recipes by focusing on important ingredients within instructions at the sentence level;
- We conduct extensive experiments on the challenging dataset Recipe1M. The results demonstrate that the proposed technique outperforms several state-of-the-art methods.

The remainder of this article is organized as follows. The technical details and specific learning process are outlined in Section 2, the experimental particulars are discussed in Section 3, and we conclude the paper in Section 4.

2. Method

In this section, we first present the notations involved in this paper and provide the problem formulation for cross-modal recipe retrieval in Section 2.1. Then, we elaborate on the technique details of our method MMACMR, including the models in Section 2.2, the strategy in Section 2.3, and the algorithm in Section 2.4.

2.1. Notations and Problem Formulations

2.1.1. Notations

Without loss of generality, we denote sets as uppercase, handwritten, bold letters (e.g., $\mathcal{D}$) and matrices as uppercase letters (e.g., $\mathbf{W}$). The i-th row of $\mathbf{W}$ is denoted by $\mathbf{W}_i$, and the element found in the j-th column of i-th row in $\mathbf{W}$ is denoted as $\mathbf{W}_{ij}$. We represent the transpose of a matrix $\mathbf{W}$ as $\mathbf{W}^\top$. Notation $\|\cdot\|_2$ denotes the L2 norm of a matrix. We use $softmax(\cdot)$ to represent the softmax function. To ease reading, we summarize the frequently used notations in Table 1.

Table 1. Summary of notations.

Notation	Definition
$\mathcal{D}$	A cross-modal recipe dataset
$\mathbf{X}_i^v$	The food image of the i-th pair
$\mathbf{X}_i^r$	The recipe of the i-th pair
$\mathbf{X}_i^{tit}$	The title of the recipe $\mathbf{X}_i^r$
$\mathbf{X}_i^{ing}$	The ingredients of the recipe $\mathbf{X}_i^r$
$\mathbf{X}_i^{ins}$	The instructions of the recipe $\mathbf{X}_i^r$
$\mathbf{E}_i^{tit}$	The embedding of the title in a recipe $\mathbf{X}_i^r$
$\mathbf{E}_i^{ing}$	The embedding of the ingredients in a recipe $\mathbf{X}_i^r$
$\mathbf{E}_i^{ins}$	The embedding of the instructions in a recipe $\mathbf{X}_i^r$
$\mathbf{R}$	The recipe embedding
$\mathbf{V}$	The food image embedding
f^r	The recipe encoder
f^v	The image encoder
$\boldsymbol{\theta}^r$	The parameters of recipe encoder
$\boldsymbol{\theta}^v$	The parameters of image encoder
$\mathcal{L}_{tri}$	The N-pairs triplet loss function
$\mathcal{L}_{RGI}$	The RGI loss function

2.1.2. Problem Formulations

Let $\mathcal{D} = \{\mathbf{X}_i^v, \mathbf{X}_i^r\}_{i=1}^n$ denote a cross-modal recipe dataset comprising n image–recipe pairs, where $\mathbf{X}_i^v$ and $\mathbf{X}_i^r = \langle \mathbf{X}_i^{tit}, \mathbf{X}_i^{ing}, \mathbf{X}_i^{ins} \rangle$ represent the food image and recipe of the i-th pair, respectively. $\mathbf{X}_i^{tit}$, $\mathbf{X}_i^{ing}$, and $\mathbf{X}_i^{ins}$ denote the title, list of ingredients, and list of instructions of the recipe, respectively. Note that each title comprises a single sentence, while both ingredients and instructions consist of several sentences. Given a recipe $\mathbf{X}_i^r$ as a query, cross-modal recipe retrieval aims to search for the most similar food image $\mathbf{X}_i^v$ from this dataset $\mathcal{D}$, or vise versa. To enhance consistent feature distribution alignment across food images and recipes, we attempt to optimize an improved recipe encoder $\mathbf{R} = f^r(\mathbf{X}^r; \boldsymbol{\theta}^r)$ and an image encoder $\mathbf{V} = f^r(\mathbf{X}^v; \boldsymbol{\theta}^v)$ under the guidance of a novel learning strategy dubbed MDA. This strategy integrates two losses: an N-pairs triplet loss $\mathcal{L}_{tri}$ to focus on inter-modal semantic alignment and an RGI loss $\mathcal{L}_{RGI}$ to focus on semantic consistency within the same modality. By considering both inter-modal and intra-modal alignment, this approach effectively avoids the harmful effects of food image ambiguity. Therefore, the objective function is formulated as follows:

$$(\hat{\boldsymbol{\theta}}^v, \hat{\boldsymbol{\theta}}^r) = \arg\min_{\boldsymbol{\theta}^v, \boldsymbol{\theta}^r} (\mathcal{L}_{tri} + \lambda \mathcal{L}_{RGI}), \tag{1}$$

where $\boldsymbol{\theta}^v$ and $\boldsymbol{\theta}^r$ are two learnable parameter vectors for image and recipe encoders, and λ is a pre-defined balance parameter.

2.2. Framework Overview

An overview of our method MMACMR is depicted in Figure 2. Following prevailing solutions [29,49], the backbone of MMACMR comprises an image encoder $f^v(\cdot; \boldsymbol{\theta}^v)$ and a recipe encoder $f^r(\cdot; \boldsymbol{\theta}^r)$ which project food images and recipes into a common feature subspace. In this subspace, the cross-modal features can be aligned effectively so that the similarity between images and recipes can be measured with accuracy. Below, we provide details of them.

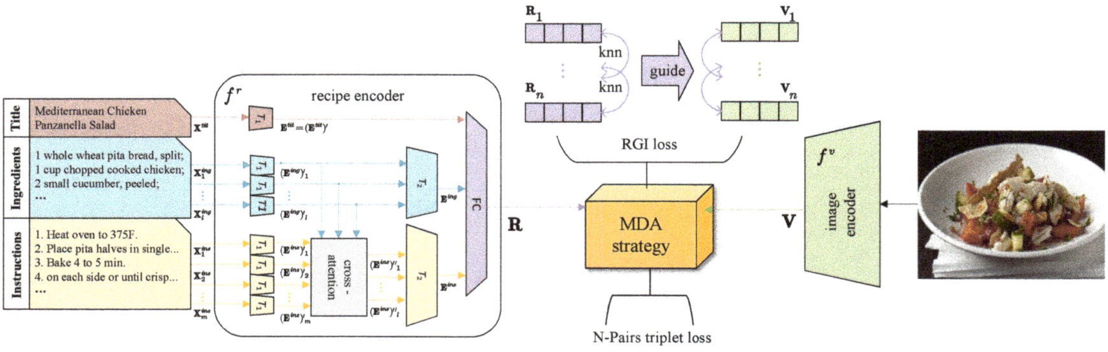

Figure 2. The framework of MMACMR, which comprises two branches of modality encoder, f^r for recipe texts and f^v for food images, along with the MDA strategy.

2.2.1. Image Encoder

To fully capture the global semantic relations between fine-grained features in the content of each food image, we adopt the base-size model of Vision Transformer (ViT-B) [50] as the image encoder $f^v(\cdot;\boldsymbol{\theta}^v)$. It is initialized with the weights pre-trained on ImageNet [51] and fine-tuned on the cross-modal recipe dataset. Given a food image $\mathbf{X}_i^v$, the embedding of $\mathbf{X}_i^v$ is denoted as $\mathbf{V}_i = f^v(\mathbf{X}_i^v;\boldsymbol{\theta}^v)$.

2.2.2. Improved Recipe Encoder

To focus on the consistent fine-grained semantics between ingredients and instructions, we improve the hierarchical transformer-based recipe encoder [29]. This encoder consists of two levels of transformers, denoted as T_1 and T_2, with identical architectures. The first level encodes the title $\mathbf{X}_i^{tit}$, ingredients $\mathbf{X}_i^{ing}$, and instructions $\mathbf{X}_i^{ins}$ at word level and then outputs their sentence-level embeddings, while the second level encoder receives the sentence-level embeddings of ingredients and instructions and produces component-level embeddings. Such a widely adopted recipe embedding scheme unfortunately overlooks a fundamental yet crucial rule in a recipe; the instructions are steps tailored to the ingredients, with the ingredients playing a determining role in shaping the instructions to some extent. To obey this rule, we plug a cross-attention module for instructions between the two transformers for two purposes: (1) to focus on the salient ingredient and (2) to highlight semantic relationships between ingredients and instructions.

Specifically, given a recipe set $\{\mathbf{X}_i^r\}_{i=1}^n$, as shown in Figure 2, the first level module T_1 receives the word-level tokens of the three components separately and outputs the average embeddings of every sentence of every components, denoted as $((\mathbf{E}_i^{tit})', (\mathbf{E}_i^{ing})', (\mathbf{E}_i^{ins})') = T_1(\mathbf{X}_i^{tit}, \mathbf{X}_i^{ing}, \mathbf{X}_i^{ins})$, where $(\mathbf{E}_i^{ing})' = \{(\mathbf{E}_i^{ing})_k'\}_{k=1}^l$, $(\mathbf{E}_i^{ins})' = \{(\mathbf{E}_i^{ins})_k'\}_{k=1}^m$. To highlight the effect of ingredients to instructions at sentence level and enhance the semantic relationship learning, cross-attention is carried out between $(\mathbf{E}_i^{ing})'$ and $(\mathbf{E}_i^{ins})$. Firstly, within a recipe, we construct an affinity matrix $\mathbf{W}$ as an attention map:

$$\mathbf{W} = softmax(\frac{((\mathbf{E}_i^{ing})'\mathbf{W}^{ing})((\mathbf{E}_i^{ins})'\mathbf{W}^{ins})^\top}{d}), \tag{2}$$

where $(\mathbf{E}_i^{ing})' \in R^{l\times d}$, $(\mathbf{E}_i^{ins})' \in R^{m\times d}$, and d is the dimension of each ingredient and each instruction. $\mathbf{W}^{ing} \in R^{d\times d}$ and $\mathbf{W}^{ins} \in R^{d\times d}$ are learnable weight matrices. Each element $\mathbf{W}_{jk}$ means the normalized correlation between the j-th ingredient and the k-th instruction. Thereby, the embedding of instructions can be enhanced by focusing on the consistent semantics between instructions and ingredients as follows:

$$(\mathbf{E}_i^{ins})'' = \mathbf{W}(\mathbf{E}_i^{ins})'. \tag{3}$$

After the second-level processing, we obtain the component-level features of title, ingredients, and instructions: $\mathbf{E}_i^{tit}$, $\mathbf{E}_i^{ing} = T_2(\mathbf{E}_i^{ing})'$ and $\mathbf{E}_i^{ins} = T_2(\mathbf{E}_i^{ins})''$. Finally, these three component embeddings are concatenated and fed into a linear layer; thus, we obtain the final recipe feature, $\mathbf{R}_i = FC([\mathbf{E}_i^{tit}; \mathbf{E}_i^{ing}; \mathbf{E}_i^{ins}]; \boldsymbol{\theta}^l)$, where $FC(\cdot; \boldsymbol{\theta}^l)$ is a linear layer, $\boldsymbol{\theta}^l$ are the parameters of it, and symbol $[\cdot; \cdot; \cdot]$ denotes the concatenation operation.

2.3. Multi-Modal Disambiguity and Alignment

To enhance the consistent feature distribution alignment across food images and recipes, we extend the prevailing learning scheme (only inter-modal metric learning, e.g., triplet loss) by considering both inter- and intra-modal alignment. To do so, we employ N-pairs triplet loss to realize inter-modal alignment within each batch, while we propose a novel RGI loss to steer the model towards capturing intra-modal consistent semantics effectively by preventing the misrecognition of ambiguous food images.

2.3.1. Inter-Modal Alignment: N-Pairs Triplet Loss

Given an anchor food image $\mathbf{V}_i$, a positive recipe $\mathbf{R}_i^+$, and a negative recipe $\mathbf{R}_j^-$, where $i \neq j$, the N-pairs triplet loss for visual modality can be defined as follows:

$$\mathcal{L}_{tri}^{v} = \sum_{|V|}[(\mathbf{V}_i, \mathbf{R}_i^+, \mathbf{R}_j^-) = (S(\mathbf{V}_i, \mathbf{R}_j^-) - S(\mathbf{V}_i, \mathbf{R}_i^+) + m)]_+, \tag{4}$$

where $[\cdot]_+ = max(0, \cdot)$, $S(\cdot)$ is the similarity function (we use cosine similarity here), $|V|$ is the number of the image sample in the batch, and m is a pre-defined margin (we set $m = 0.3$ in this work). Similarly, the N-pairs triplet loss for text modality can be written in the same way. Consequently, we formulate the whole N-pairs triplet loss as follows:

$$\mathcal{L}_{tri} = \mathcal{L}_{tri}^{v} + \mathcal{L}_{tri}^{r}. \tag{5}$$

2.3.2. Intra-Modal Alignment with Disambiguity: RGI Loss

As discussed above, N-pairs triplet loss is a satisfactory scheme for reducing heterogeneity between images and recipes. Using it within each modality, however, is far from a suitable intra-modal alignment solution due to the disturbance of food image ambiguity. Nor is this all; the prevailing recipe retrieval approaches [18,29] only consider cross-modal similarity measurement, which narrows the distance between anchor and positive samples while enlarging the distances between the anchor and negative samples. Such a limitation, on the one hand, leads to a discrepancy between the two modalities, making it difficult for model convergence. On the other hand, it is easy to match one of the ambiguous images, resulting in low retrieval performance.

Fortunately, recipes, or, more rigorously, text, are the more reliable modality owing to their ability to abstract semantic expression word by word. Thus, inspired by [52], we design a novel learning strategy termed RGI loss which chooses the similarity relations between recipes as guidance to determine the relations between corresponding food images. Specifically, if we assume that $< \mathbf{R}_i, \mathbf{R}_j >$ is a recipe pair in a batch, we aim to preserve the similarity relation for it and project this relation to the corresponding image pair $< \mathbf{V}_i, \mathbf{V}_j >$. Given a recipe $\mathbf{R}_i$, we first rank other recipes in this batch by the similarity to $\mathbf{R}_i$ using the K-nearest neighbors (KNN) algorithm [53]. From the ranked recipes, we select the nearest neighbor as the positive sample $\mathbf{R}_j^+$ and a randomly selected recipe that is not among the top 10 neighbors as the negative recipe $\mathbf{R}_k^-$, $i \neq j \neq k$. Inspired by the angular loss [54], our RGI loss for text modality is defined as follows:

$$\mathcal{L}_{RGI}^{r} = \left[\left\| \mathbf{R}_i - \mathbf{R}_j^+ \right\|_2^2 - 4\tan^2\alpha \left\| \mathbf{R}_k^- - C_i \right\|_2^2 \right]_+, \tag{6}$$

$$C_i = \frac{\mathbf{R}_i + \mathbf{R}_j^+}{2}, \tag{7}$$

where $\tan^2\alpha = 1$ is a pre-defined upper bound. For the visual modality, we no longer compute the KNN for images, while we adopt the rank of the neighbors of corresponding recipes directly. The RGI loss for the visual modality is defined in the same way:

$$\mathcal{L}^{v}_{RGI} = \left[\left\| \mathbf{V}_i - \mathbf{V}_j^{+} \right\|_2^2 - 4\tan^2\alpha \left\| \mathbf{V}_k^{-} - C_i \right\|_2^2 \right]_+, \tag{8}$$

$$C_i = \frac{\mathbf{V}_i + \mathbf{V}_j^{+}}{2}, \tag{9}$$

where $\tan^2\alpha = 1$ is a pre-defined upper bound. Note that the indices of the visual modality are the same as the text modality. Thus, the entire RGI loss is formulated as follows:

$$\mathcal{L}_{RGI} = \lambda_1 \mathcal{L}^{r}_{GRI} + \lambda_2 \mathcal{L}^{v}_{RGI}, \tag{10}$$

where λ_1 and λ_2 are hyper-parameters for adjusting the relation projection.

2.3.3. Total Loss

Finally, the total loss can be written as follows:

$$\mathcal{L}_{total} = \mathcal{L}_{tri} + \lambda \mathcal{L}_{RGI}, \tag{11}$$

where λ is a balance hyper-parameter for adjusting the performance of the two loss functions.

2.4. Optimization

Our method undergoes end-to-end optimization. The optimization procedure is outlined in Algorithm 1.

Algorithm 1 Optimization procedure of MMACMR

Input: cross-modal recipe dataset $\mathcal{D} = \{\mathbf{X}_i^v, \mathbf{X}_j^r\}_{i,j=1}^n$, number of epoch T.
Output: parameters $\boldsymbol{\theta}^v$, $\boldsymbol{\theta}^r$ of modality encoders.

```
Initialize θ^v, θ^r;
for t = 1 to T do
  repeat
    Compute embeddings V and R;
    for i = 1 to n do
      Calculate Equation (5)
      Rank the recipes neighbors via KNN algorithm;
      Rank the images neighbors follow recipes;
      Calculate Equation (10);
    end for
    Update the parameters θ^v, θ^r by Equation (11) via gradient descent algorithm.
  until Convergence
end for
```

3. Experiments and Discussion

This section presents extensive experiments conducted to assess our method's performance. We begin by introducing the experiment settings, followed by a detailed discussion of the experimental results.

3.1. Experiment Settings

3.1.1. Dataset

We implement experiments on the Recipe1M [9] dataset, which is by far the largest public multi-modal recipe dataset available. Recipe1M comprises over 1 million cooking recipe texts and 800 K food images which are collected from more than 24 popular cooking websites. We adhere to the official splits for data, with 238,399 image–recipe pairs allocated for training, 51,119 pairs for validation, and 51,303 pairs for testing.

3.1.2. Baselines

We benchmark our approach against the state-of-the-art baselines below:

- CCA [9] stands for Canonical Correlation Analysis, a classical statistical method used to learn a joint embedding space;
- JE [9] was the first to conduct the cross-modal recipe retrieval task on the Recipe1M dataset. It uses a joint encoder and a classifier to learn the information from food images and recipes;
- AdaMin [10] combines the retrieval loss and classifies the loss to improve the robustness of models and proposes a novel strategy to mine the significant triplets;
- R2GAN [35] promotes the modality alignment by employing a GAN mechanism equipped with two discriminators and one generator;
- MCEN [14] bridges the semantic gap between the two modalities using stochastic latent variable models;
- SN [16] employs three attention mechanisms on three components of recipes to capture the relationship between sentences;
- SCAN [13] introduces semantic consistency loss to regularize the representations of images and recipes;
- HF-ICMA [20] exploits the global and local similarity between the two modalities by considering inter- and intra-modal fusion;
- SEJE [22] constructs a two-phase feature framework and divides the processes of data pre-processing and model training to extract additional semantic information;
- M-SIA [17] argues that multiple aspects in recipes are related to multiple regions in food images and leverages multi-head attention to bridge them;
- X-MRS [39] augments recipe representations by utilizing multilingual translation;
- LCWF-GI [31] employs latent weight factors to fuse the three components of recipes by considering their complex interaction;
- H-T [29] captures the latent semantic information in recipes by applying self-supervised loss to push components sourced from the same close recipe;
- LMF-CSF [30] introduces a low-rank fusion strategy to combine the components in recipes and generate superior representations.

3.1.3. Evaluation Criteria

Similar to the majority of previous studies [9,29,44], we sample 1 K and 10 K image–recipe pairs from the test partition and assess the retrieval performance for image-to-recipe and and recipe-to-image tasks using median rank (MedR) and recall rate at top k (R@k). Among these metrics, MedR represents the median index of the retrieved samples for each query, measuring the ability of models to understand the semantic correlation between two modalities and the accuracy of retrieval. A lower MedR value indicates better performance. R@k indicates that the percentage of the ground truth index is among the first k retrieved samples, which is also known as sensitivity or the true positive rate, measuring the ability of models to correctly identify all relevant instances. A higher R@k value indicates better performance. Here, we evaluate the top 1 (R@1), top 5 (R@5), and top 10 (R@10). By using these two metrics, we can evaluate the comprehensive performance of the models. Every evaluation is repeated 10 times, and the mean results are returned.

3.1.4. Implementation Details

In line with prior research [49], we use food images with a depth of three channels in the RGB color space. All the images in our experiments are resized to 256 pixels in their shorter dimension and then cropped to 224×224 pixels. The image encoder utilizes a pre-trained ViT-based model, yielding an output size of 1024. Regarding recipes, sentences in three components are truncated to a maximum length of 15, and every ingredients or instructions list has a maximum of 20 sentences. Each transformer in the hierarchical transformer recipe encoder comprises two layers, and each layer has four attention heads. Every component in the recipes is embedded as 512 dimensions, and the final embedding of a recipe is output as 1024 dimensions. The model is trained utilizing the Adam optimizer, the batch size is set as 128, and the learning rate is $\eta = 10^{-4}$. The balance parameters $\lambda_1 = 0.09$, $\lambda_2 = 0.1$, and $\lambda = 0.01$.

3.1.5. Experimental Environment

Our experiments are conducted using Python 3.7 with the PyTorch 1.31.1 framework. We utilize a deep learning workstation equipped with an Intel(R) Core i9-12900K 3.9 GHz processor, 128 GB of RAM, 1 TB SSD, and 2 TB HDD storage. The workstation runs on the Ubuntu-22.04.1 operating system and is powered by two NVIDIA GeForce RTX 3090Ti GPUs (NVIDIA, Palo Alto, CA, USA).

3.2. Comparison with State-of-the-Art Methods

We compare the performance of our method with the baselines mentioned above. The results are reported in Table 2. It is easy to see that MMACMR is superior to the best results of existing works using all the metrics. Concretely, our method achieves a 3.3, 1.1, 0.6 R{1, 5, 10} improvement for image to recipe and a 3.7, 1.2, 0.7 R{1, 5, 10} improvement for recipe to image in the 1 K size compared to the SOTA method LMF-CSF [30] and achieves a 3.5, 3.1, 2.7 R{1, 5, 10} improvement for image to recipe and a 4.0, 3.1, 2.8 R{1, 5, 10} improvement for recipe to image in the 10 K size compared to the SOTA method LMF-CSF [30]. In addition, the MedR of our method in the 10 K size dataset decreases to 2.1 for image to recipe and 2.2 for recipe to image compared to 3.0 in LMF-CSF [30]. These results demonstrate the effectiveness of our MMACMR. In other words, our approach to addressing the questions mentioned above is effective for cross-modal recipe retrieval.

Table 2. Comparison with SOTA methods. MedR(↓) and R@k(↑) in 1 K and 10 K size. The best results are marked in bold font.

	Methods	Image to Recipe				Recipe to Image			
		MedR	**R@1**	**R@5**	**R@10**	**MedR**	**R@1**	**R@5**	**R@10**
1 K	CCA [9]	15.7	14.0	32.0	43.0	24.8	9.0	24.0	35.0
	JE [9]	5.2	24.0	51.0	65.0	5.1	25.0	52.0	65.0
	AdaMin [10]	2.0	39.8	69.0	77.4	2.0	40.2	68.1	78.7
	R2GAN [35]	2.0	39.1	71.0	81.7	2.0	40.6	72.6	83.3
	MCEN [14]	2.0	48.2	75.8	83.6	1.9	48.4	76.1	83.7
	ACME [34]	1.0	51.8	80.2	87.5	1.0	52.8	80.2	87.6
	SN [16]	1.0	52.7	81.7	88.9	1.0	54.1	81.8	88.9
	SCAN [13]	1.0	54.0	81.7	88.8	1.0	54.9	81.9	89.0
	HF-ICMA [20]	1.0	55.1	86.7	92.4	1.0	56.8	87.5	93.0
	SEJE [22]	1.0	58.1	85.8	92.2	1.0	58.5	86.2	92.3
	M-SIA [17]	1.0	59.3	86.3	92.6	1.0	59.8	86.7	92.8
	X-MRS [39]	1.0	64.0	88.3	92.6	1.0	63.9	87.6	92.6
	H-T [29]	1.0	60.0	87.6	92.9	1.0	60.3	87.6	93.2
	LCWF-GI [31]	1.0	59.4	86.8	92.5	1.0	60.1	86.7	92.7
	H-T(ViT) [29]	1.0	64.2	89.1	93.4	1.0	64.5	89.3	93.8
	LMF-CSF [30]	1.0	65.8	89.7	94.3	1.0	65.5	89.4	94.3
	Ours	**1.0**	**69.1**	**90.8**	**94.9**	**1.0**	**69.2**	**90.6**	**95.0**

Table 2. *Cont.*

	Methods	Image to Recipe				Recipe to Image			
		MedR	R@1	R@5	R@10	MedR	R@1	R@5	R@10
10 K	JE [9]	41.9	-	-	-	39.2	-	-	-
	AdaMin [10]	13.2	14.9	35.3	45.2	12.2	14.8	34.6	46.1
	R2GAN [35]	13.9	13.5	33.5	44.9	12.6	14.2	35.0	46.8
	MCEN [14]	7.2	20.3	43.3	54.4	6.6	21.4	44.3	55.2
	ACME [34]	6.7	22.9	46.8	57.9	6.0	24.4	47.9	59.0
	SN [16]	7.0	22.1	45.9	56.9	7.0	23.4	47.3	57.9
	SCAN [13]	5.9	23.7	49.3	60.6	5.1	25.3	50.6	61.6
	HF-ICMA [20]	5.0	24.0	51.6	65.4	4.2	25.6	54.8	67.3
	SEJE [22]	4.2	26.9	54.0	65.6	4.0	27.2	54.4	66.1
	M-SIA [17]	4.0	29.2	55.0	66.2	4.0	30.3	55.6	66.5
	X-MRS [39]	3.0	32.9	60.6	71.2	3.0	33.0	60.4	70.7
	H-T [29]	4.0	27.9	56.4	68.1	4.0	28.3	56.5	68.1
	LCWF-GI [31]	4.0	27.9	56.0	67.8	4.0	28.6	55.8	67.5
	H-T(ViT) [29]	3.0	33.5	62.1	72.8	3.0	33.7	62.2	72.7
	LMF-CSF [30]	3.0	34.6	62.7	73.2	3.0	34.3	62.5	72.8
	Ours	**2.1**	**38.1**	**65.8**	**75.9**	**2.2**	**38.3**	**65.6**	**75.6**

3.3. Scalability Analysis

In order to investigate the scalability of our method, we conduct experiments on datasets larger than 10 K in size. As shown in Figure 3, the MedR results of MMACMR are consistently lower than those of all other methods across all dataset sizes. In addition, it can be seen that, with the increase in test size, the performance gap between our method and others also widens. We argue that, on the one hand, the enhancement of recipe embedding promotes the alignment between the two modalities. On the other hand, as the dataset size increases, so does the number of ambiguous food images, leading to a higher probability of matching incorrect recipes. By effectively addressing this issue, our method demonstrates improved robustness and scalability as the dataset size enlarges.

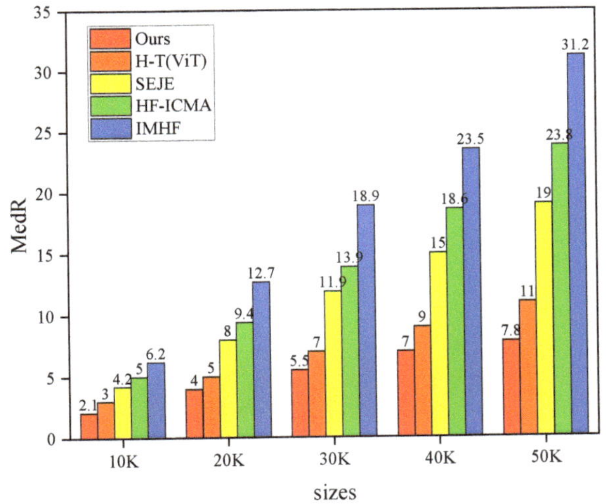

Figure 3. Scalability analysis. The abscissa represents the dataset size ranging from 10 K to 50 K, while the ordinate represents the MedR value.

3.4. Ablation Studies

In this subsection, we conduct ablation experiments to assess the contribution of each part of our model to the overall performance. Table 3 reports the image-to-recipe retrieval results of different parts of MMACMR in 1 K and 10 K test size. In Table 3, Base is the baseline framework consisting of the food image encoder (ViT-B) and the original hierarchical transformer recipe encoder coupled with the N-pairs triplet loss. IR means introducation of the improved recipe encoder, and $\mathcal{L}_{RGI}$ is our RGI loss. A $\surd$ symbol under the columns Base, IR, and $\mathcal{L}_{RGI}$ indicates the use of that part. On the right, we list the MedR, R@1, R@5, and R@10 results for the image-to-recipe and recipe-to image tasks. We first evaluate the Base framework, then introduce the improved recipe encoder and RGI loss separately. Finally, we combine all three parts. It can be observed that the addition of both IR and $\mathcal{L}_{RGI}$ boosts the baseline model. This indicates that the solutions we propose to address the questions mentioned above are effective. When employing all subassemblies, we achieve the best performance, further validating the effectiveness of each element in our approach. Note that the method without IR obtains the same scores as the full method in R@5 and R@10 for image to recipe, and R@5 for recipe to image, for the 10 K size dataset. Additionally, it achieves better performance in MedR for recipe to image in 10 K size. Therefore, we attribute the main contribution to the MDA strategy.

Table 3. Ablation study. MedR (↓) and R@k (↑) in 1 K and 10 K size. The best results are marked in bold font. A $\surd$ symbol indicates that the corresponding part in this column is being used.

	Base	I R	$\mathcal{L}_{RGI}$	Image to Recipe				Recipe to Image			
				MedR	R@1	R@5	R@10	MedR	R@1	R@5	R@10
1 K	√			1.0	58.3	86.2	91.8	1.0	59.6	86.1	92.2
	√	√		1.0	67.2	90.0	94.5	1.0	67.6	90.0	94.5
	√		√	1.0	68.6	90.5	94.7	1.0	68.2	90.3	94.7
	√	√	√	**1.0**	**69.1**	**90.8**	**94.9**	**1.0**	**69.2**	**90.6**	**95.0**
10 K	√			4.1	26.8	54.7	66.5	4.0	37.5	55.1	66.8
	√	√		3.0	35.9	64.5	74.7	3.0	36.6	64.7	74.9
	√		√	2.2	37.7	65.8	75.9	**2.0**	38.0	65.6	75.4
	√	√	√	**2.1**	**38.1**	**65.8**	**75.9**	2.2	**38.3**	**65.6**	**75.6**

3.5. Qualitative Results

3.5.1. Qualitative Results on Image-to-Recipe Retrieval

To more intuitively analyze the representative results of MMACMR in image-to-recipe retrieval, we select four food images as queries to retrieve the recipes from the test set using our method and the SOTA method H-T (ViT) [29]. As shown in Figure 4, from left to right, the queries are "Chickpeas and Spinach with Smoky Paprika", "Blue Ribbon Apple Crumb Pie", "Apricot Nectar Cake", and "Sweet and Spicy Grilled Pork Tenderloin". In the first two samples, the categories of food are relatively easy to distinguish; both of these methods retrieve approximate recipes. However, in the first example, H-T (ViT) [29] does not retrieve the main ingredient, apricot nectar, while our method successfully retrieves it. The same situation occurs in the second example, where H-T (ViT) [29] retrieves a recipe whose corresponding image is similar to the query image but it misrecognizes the pork tenderloin as chicken thighs. In contrast, MMACMR retrieves the ground truth recipe. We attribute this to our MDA strategy, which can better address the problem of ambiguous food images and recognize the ingredients correctly. In the third example, H-T (ViT) [29] identifies some beans and vegetable leaf in the image but misclassifies their types, and the retrieved entire recipe deviates significantly from the ground truth. In the last example, the food image is difficult to recognize by human eye. H-T (ViT) [29] retrieves a recipe whose corresponding image has a similar color to the query (actually, it is a shortcut for models to classify objects which have not been seen before). However, our method retrieves the correct recipe even though the query image is ambiguous. We believe this is because MMACMR can reduce the distances between images with similar recipes, allowing the correct sample to be retrieved even when the query is hard to distinguish.

Figure 4. Examples of image-to-recipe retrieval results for the 10 K test set. The first row contains the query images, the second row shows the recipes retrieved using our method (all of which are the ground truth recipes; therefore, the first row is their corresponding food images), the third row displays the recipes retrieved using H-T (ViT) [29], and the last row presents the corresponding food images of the recipes from the third row. The key ingredients not retrieved by H-T [29] but retrieved by our method are highlighted in red.

3.5.2. Qualitative Results on Recipe-to-Image Retrieval

We also conduct experiments to visualize the results of recipe-to-image retrieval for the 1 K test set, which are presented in Figure 5. From top to bottom, the query recipes are titled "Fruit Salad", "Italian Beef Roast", and "Pesto Salmon", followed by the top five retrieved images using our method and the SOTA method H-T (ViT) [29]. In the first example, both methods retrieve five food images of fruit salad, but our method retrieves the ground truth as the top one, while H-T (ViT) [29] retrieves it in the top three. In the second example, the two methods retrieve the correct image in the top two. However, all the food images MMACMR retrieves are roast beef, while the third and fifth retrieved images of H-T (ViT) [29] do not match the recipe query. In the last example, our method retrieves the ground truth image as the top one, while H-T (ViT) [29] fails to retrieve the correct food image. At the same time, the second image retrieved by MMACMR is similar to the correct one, while the first and third images retrieved by H-T (ViT) [29] deviate significantly from the ground truth. We attribute these achievements to the capability of our method to address the problem of ambiguous food images, allowing MMACMR to retrieve images that have similar recipes.

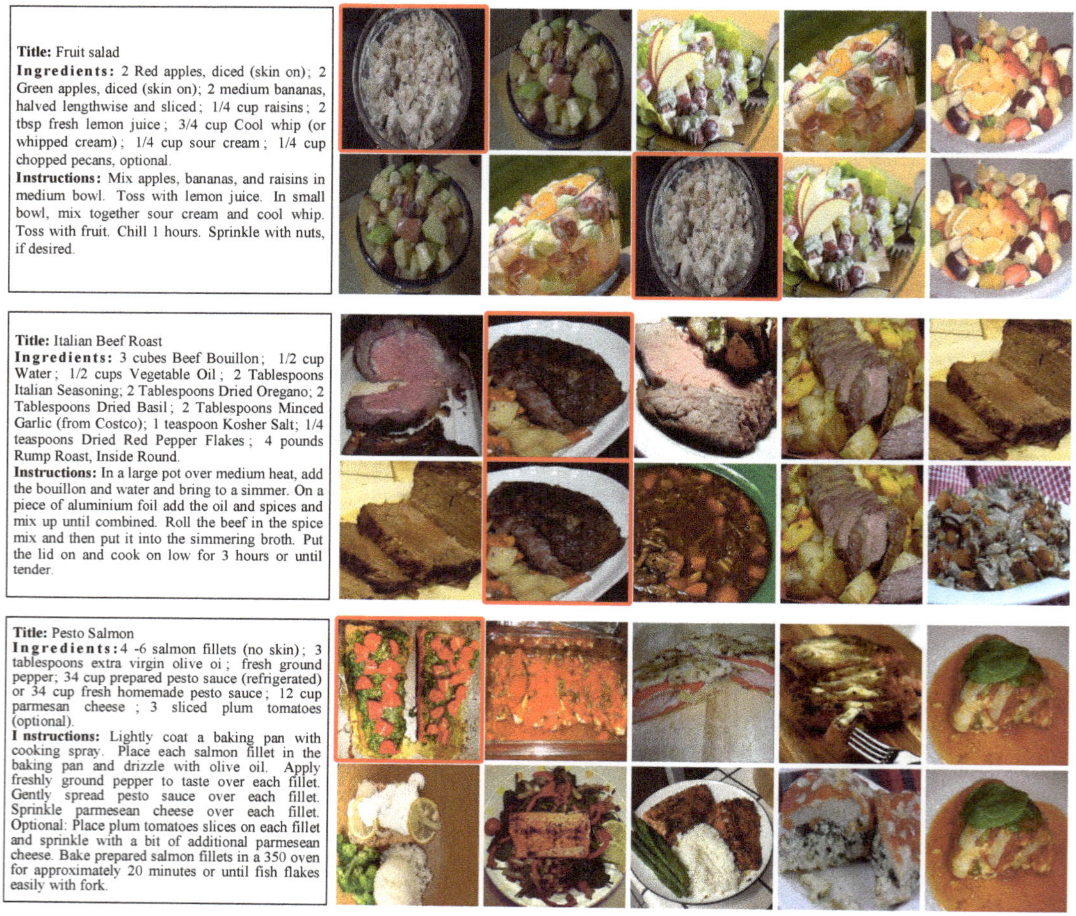

Figure 5. Examples of recipe-to-image retrieval results for the 10 K test set. The left side shows the query recipes, while the right side displays the top 5 retrieved images using our method or H-T (ViT) [29]. The ground truth images are marked by a red box.

4. Conclusions

In this paper, we propose a novel cross-modal recipe retrieval method named MMACMR which addresses the problem of ambiguous food images in retrieval using a novel training strategy, MDA, that guides the similarity within food images by recipe. Additionally, we improve the recipe encoder to ensure the precision of recipe embeddings. We conduct extensive experiments on the challenging public dataset Recipe1M, and the experimental results demonstrate the effectiveness of our method. Given the necessity of analyzing vast numbers of food data, our method could offer significant practical value in the food industry by enhancing user convenience and efficiency.

However, due to the complexity of recipe texts, some information representing the dish preparation program is still not captured by our method. In the future, we aim to focus on the fine-grained information in recipes using visual language pre-training models.

Author Contributions: Methodology, Z.Z.; Software, Z.Z.; Validation, Q.Z.; Investigation, Z.Z.; Resources, X.Z. and H.Z.; Data curation, Q.Z.; Writing—original draft, Z.Z.; Writing—review & editing, L.Z.; Supervision, X.Z., H.Z. and L.Z.; Project administration, X.Z.; Funding acquisition, X.Z. and L.Z. All authors have read and agreed to the published version of the manuscript

Funding: This work was supported in part by the National Natural Science Foundation of China (62202163), the Natural Science Foundation of Hunan Province (2022JJ40190), the Scientific Research Project of Hunan Provincial Department of Education (22A0145), the Key Research and Development Program of Hunan Province (2020NK2033), the Hunan Provincial Department of Education Scientific Research Outstanding Youth Project (21B0200), and the Hunan Provincial Natural Science Foundation Youth Fund Project (2023JJ40333).

Data Availability Statement: The original contributions presented in the study are included in the article, further inquiries can be directed to the corresponding author.

Acknowledgments: The authors would like to thank the *Foods* journal for providing this opportunity to submit the manuscript.

Conflicts of Interest: The authors declare no conflicts of interest. The funders had no role in the design of the study; in the collection, analyses, or interpretation of data.

Abbreviations

The following abbreviations are used in this manuscript:

MMACMR	Multi-Modal Alignment Method for Cross-Modal Recipe Retrieval
MDA	Multi-Modal Disambiguity and Alignment
RGI	Recipe Guide Image
AI	Artificial intelligence
LSTM	Long Short-Term Memory
GANs	Generative Adversarial Networks
ViT	Vision Transformer
CLIP	Contrastive Language–Image Pre-training
KNN	K-Nearest Neighbors
SOTA	State Of The Art
MedR	Median Rank
SSD	Solid-State Disk
RAM	Random-Access Memory
HDD	Hard Disk Drive
CCA	Canonical Correlation Analysis

References

1. Guo, Z.; Jayan, H. Fast Nondestructive Detection Technology and Equipment for Food Quality and Safety. *Foods* **2023**, *12*, 3744. [CrossRef] [PubMed]
2. Guo, Z.; Wu, X.; Jayan, H.; Yin, L.; Xue, S.; El-Seedi, H.R.; Zou, X. Recent developments and applications of surface enhanced Raman scattering spectroscopy in safety detection of fruits and vegetables. *Food Chem.* **2023**, *434* , 137469. [CrossRef] [PubMed]
3. Thames, Q.; Karpur, A.; Norris, W.; Xia, F.; Panait, L.; Weyand, T.; Sim, J. Nutrition5k: Towards automatic nutritional understanding of generic food. In Proceedings of the IEEE/CVF Conference on Computer Vision and Pattern Recognition, Nashville, TN, USA, 20–25 June 2021; pp. 8903–8911.
4. Min, W.; Wang, Z.; Liu, Y.; Luo, M.; Kang, L.; Wei, X.; Wei, X.; Jiang, S. Large scale visual food recognition. *IEEE Trans. Pattern Anal. Mach. Intell.* **2023**, *45*, 9932–9949. [CrossRef] [PubMed]
5. Min, W.; Wang, Z.; Yang, J.; Liu, C.; Jiang, S. Vision-based fruit recognition via multi-scale attention CNN. *Comput. Electron. Agric.* **2023**, *210*, 107911. [CrossRef]
6. Min, W.; Liu, C.; Xu, L.; Jiang, S. Applications of knowledge graphs for food science and industry. *Patterns* **2022**, *3*, 100484. [CrossRef] [PubMed]
7. Wang, W.; Min, W.; Li, T.; Dong, X.; Li, H.; Jiang, S. A review on vision-based analysis for automatic dietary assessment. *Trends Food Sci. Technol.* **2022**, *122*, 223–237. [CrossRef]
8. Liu, Y.; Min, W.; Jiang, S.; Rui, Y. Convolution-Enhanced Bi-Branch Adaptive Transformer with Cross-Task Interaction for Food Category and Ingredient Recognition. *IEEE Trans. Image Process.* **2024**, *33*, 2572–2586. [CrossRef] [PubMed]

9. Salvador, A.; Hynes, N.; Aytar, Y.; Marin, J.; Ofli, F.; Weber, I.; Torralba, A. Learning cross-modal embeddings for cooking recipes and food images. In Proceedings of the IEEE Conference on Computer Vision and Pattern Recognition, Honolulu, HI, USA, 21–26 July 2017; pp. 3020–3028.
10. Carvalho, M.; Cadène, R.; Picard, D.; Soulier, L.; Thome, N.; Cord, M. Cross-modal retrieval in the cooking context: Learning semantic text-image embeddings. In Proceedings of the 41st International ACM SIGIR Conference on Research & Development in Information Retrieval, Ann Arbor, MI, USA, 8–12 July 2018; pp. 35–44.
11. Min, W.; Zhou, P.; Xu, L.; Liu, T.; Li, T.; Huang, M.; Jin, Y.; Yi, Y.; Wen, M.; Jiang, S.; Jain, R. From Plate to Production: Artificial Intelligence in Modern Consumer-Driven Food Systems. *arXiv* **2023**, arXiv:2311.02400.
12. Guo, Z.; Zhang, Y.; Wang, J.; Liu, Y.; Jayan, H.; El-Seedi, H.R.; Alzamora, S.M.; Gómez, P.L.; Zou, X. Detection model transfer of apple soluble solids content based on NIR spectroscopy and deep learning. *Comput. Electron. Agric.* **2023**, *212*, 108127. [CrossRef]
13. Wang, H.; Sahoo, D.; Liu, C.; Shu, K.; Achananuparp, P.; Lim, E.P.; Hoi, S.C. Cross-modal food retrieval: Learning a joint embedding of food images and recipes with semantic consistency and attention mechanism. *IEEE Trans. Multimed.* **2021**, *24*, 2515–2525. [CrossRef]
14. Fu, H.; Wu, R.; Liu, C.; Sun, J. Mcen: Bridging cross-modal gap between cooking recipes and dish images with latent variable model. In Proceedings of the IEEE/CVF Conference on Computer Vision and Pattern Recognition, Seattle, WA, USA, 13–19 June 2020; pp. 14570–14580.
15. Chen, Y.; Zhou, D.; Li, L.; Han, J.M. Multimodal encoders for food-oriented cross-modal retrieval. In Proceedings of the Web and Big Data: 5th International Joint Conference, APWeb-WAIM 2021, Guangzhou, China, 23–25 August 2021; Proceedings, Part II 5; Springer International Publishing: Cham, Switzerland, 2021; pp. 253–266.
16. Zan, Z.; Li, L.; Liu, J.; Zhou, D. Sentence-based and noise-robust cross-modal retrieval on cooking recipes and food images. In Proceedings of the 2020 International Conference on Multimedia Retrieval, Dublin, Ireland, 8–11 June 2020; pp. 117–125.
17. Li, L.; Li, M.; Zan, Z.; Xie, Q.; Liu, J. Multi-subspace implicit alignment for cross-modal retrieval on cooking recipes and food images. In Proceedings of the 30th ACM International Conference on Information & Knowledge Management, Virtual Event, 1–5 November 2021; pp. 3211–3215.
18. Shukor, M.; Couairon, G.; Grechka, A.; Cord, M. Transformer decoders with multimodal regularization for cross-modal food retrieval. In Proceedings of the IEEE/CVF Conference on Computer Vision and Pattern Recognition, New Orleans, LA, USA, 18–24 June 2022; pp. 4567–4578.
19. Li, L.; Hu, C.; Zhang, H.; Maradapu Vera Venkata sai, A. Cross-modal Image-Recipe Retrieval via Multimodal Fusion. In Proceedings of the 5th ACM International Conference on Multimedia in Asia, Taiwan, China, 6–8 December 2023; pp. 1–7.
20. Li, J.; Sun, J.; Xu, X.; Yu, W.; Shen, F. Cross-modal image-recipe retrieval via intra-and inter-modality hybrid fusion. In Proceedings of the 2021 International Conference on Multimedia Retrieval, Taipei, Taiwan, 21–24 August 2021; pp. 173–182.
21. Chen, J.J.; Ngo, C.W.; Feng, F.L.; Chua, T.S. Deep understanding of cooking procedure for cross-modal recipe retrieval. In Proceedings of the 26th ACM International Conference on Multimedia, Seoul, Republic of Korea, 22–26 October 2018; pp. 1020–1028.
22. Xie, Z.; Liu, L.; Wu, Y.; Zhong, L.; Li, L. Learning text-image joint embedding for efficient cross-modal retrieval with deep feature engineering. *ACM Trans. Inf. Syst. (TOIS)* **2021**, *40*, 1–27. [CrossRef]
23. Xie, Z.; Liu, L.; Li, L.; Zhong, L. Efficient Deep Feature Calibration for Cross-Modal Joint Embedding Learning. In Proceedings of the 2021 International Conference on Multimodal Interaction, Montréal, QC, Canada, 18–22 October 2021; pp. 43–51.
24. Xie, Z.; Liu, L.; Li, L.; Zhong, L. Learning joint embedding with modality alignments for cross-modal retrieval of recipes and food images. In Proceedings of the 30th ACM International Conference on Information & Knowledge Management, Virtual Event, 1–5 November 2021; pp. 2221–2230.
25. Xie, Z.; Liu, L.; Wu, Y.; Li, L.; Zhong, L. Learning tfidf enhanced joint embedding for recipe-image cross-modal retrieval service. *IEEE Trans. Serv. Comput.* **2021**, *15*, 3304–3316. [CrossRef]
26. Cao, D.; Chu, J.; Zhu, N.; Nie, L. Cross-modal recipe retrieval via parallel-and cross-attention networks learning. *Knowl.-Based Syst.* **2020**, *193*, 105428. [CrossRef]
27. Li, J.; Xu, X.; Yu, W.; Shen, F.; Cao, Z.; Zuo, K.; Shen, H.T. Hybrid fusion with intra-and cross-modality attention for image-recipe retrieval. In Proceedings of the 44th International ACM SIGIR Conference on Research and Development in Information Retrieval, Virtual Event, 11–15 July 2021; pp. 244–254.
28. Xie, Z.; Li, L.; Zhong, L.; Liu, J.; Liu, L. Cross-Modal Retrieval between Event-Dense Text and Image. In Proceedings of the 2022 International Conference on Multimedia Retrieval, Newark, NJ, USA, 27–30 June 2022; pp. 229–238.
29. Salvador, A.; Gundogdu, E.; Bazzani, L.; Donoser, M. Revamping cross-modal recipe retrieval with hierarchical transformers and self-supervised learning. In Proceedings of the IEEE/CVF Conference on Computer Vision and Pattern Recognition, Nashville, TN, USA, 20–25 June 2021; pp. 15475–15484.
30. Zhao, W.; Zhou, D.; Cao, B.; Zhang, K.; Chen, J. Efficient low-rank multi-component fusion with component-specific factors in image-recipe retrieval. *Multimed. Tools Appl.* **2024**, *83*, 3601–3619. [CrossRef]
31. Zhao, W.; Zhou, D.; Cao, B.; Liang, W.; Sukhija, N. Exploring latent weight factors and global information for food-oriented cross-modal retrieval. *Connect. Sci.* **2023**, *35*, 2233714. [CrossRef]
32. Wahed, M.; Zhou, X.; Yu, T.; Lourentzou, I. Fine-Grained Alignment for Cross-Modal Recipe Retrieval. In Proceedings of the IEEE/CVF Winter Conference on Applications of Computer Vision, Waikoloa, HI, USA, 3–8 January 2024; pp. 5584–5593.

33. Wang, H.; Lin, G.; Hoi, S.C.; Miao, C. Learning structural representations for recipe generation and food retrieval. *IEEE Trans. Pattern Anal. Mach. Intell.* **2022**, *45*, 3363–3377. [CrossRef]
34. Wang, H.; Sahoo, D.; Liu, C.; Lim, E.P.; Hoi, S.C. Learning cross-modal embeddings with adversarial networks for cooking recipes and food images. In Proceedings of the IEEE/CVF Conference on Computer Vision and Pattern Recognition, Long Beach, CA, USA, 15–20 June 2019; pp. 11572–11581.
35. Zhu, B.; Ngo, C.W.; Chen, J.; Hao, Y. R2gan: Cross-modal recipe retrieval with generative adversarial network. In Proceedings of the IEEE/CVF Conference on Computer Vision and Pattern Recognition, Long Beach, CA, USA, 15–20 June 2019; pp. 11477–11486.
36. Sugiyama, Y.; Yanai, K. Cross-modal recipe embeddings by disentangling recipe contents and dish styles. In Proceedings of the 29th ACM International Conference on Multimedia, Virtual Event, 20–24 October 2021; pp. 2501–2509.
37. Wang, H.; Lin, G.; Hoi, S.; Miao, C. Paired cross-modal data augmentation for fine-grained image-to-text retrieval. In Proceedings of the 30th ACM International Conference on Multimedia, Lisboa, Portugal, 10–14 October 2022; pp. 5517–5526.
38. Yang, J.; Chen, J.; Yanai, K. Transformer-Based Cross-Modal Recipe Embeddings with Large Batch Training. In Proceedings of the International Conference on Multimedia Modeling, Bergen, Norway, 9–12 January 2023; Springer: Cham, Switzerland, 2023; pp. 471–482.
39. Guerrero, R.; Pham, H.X.; Pavlovic, V. Cross-modal retrieval and synthesis (x-mrs): Closing the modality gap in shared subspace learning. In Proceedings of the 29th ACM International Conference on Multimedia, Virtual Event, 20–24 October 2021; pp. 3192–3201.
40. Zhu, B.; Ngo, C.W.; Chen, J.; Chan, W.K. Cross-lingual adaptation for recipe retrieval with mixup. In Proceedings of the 2022 International Conference on Multimedia Retrieval, Newark, NJ, USA, 27–30 June 2022; pp. 258–267.
41. Papadopoulos, D.P.; Mora, E.; Chepurko, N.; Huang, K.W.; Ofli, F.; Torralba, A. Learning program representations for food images and cooking recipes. In Proceedings of the IEEE/CVF Conference on Computer Vision and Pattern Recognition, New Orleans, LA, USA, 18–24 June 2022; pp. 16559–16569.
42. Huang, X.; Liu, J.; Zhang, Z.; Xie, Y. Improving Cross-Modal Recipe Retrieval with Component-Aware Prompted CLIP Embedding. In Proceedings of the 31st ACM International Conference on Multimedia, Ottawa, ON, Canada, 29 October–3 November 2023; pp. 529–537.
43. Sun, J.; Li, J. PBLF: Prompt Based Learning Framework for Cross-Modal Recipe Retrieval. In Proceedings of the International Symposium on Artificial Intelligence and Robotics, Shanghai, China, 21–23 October 2022; Springer: Singapore, 2022; pp. 388–402.
44. Shukor, M.; Thome, N.; Cord, M. Vision and Structured-Language Pretraining for Cross-Modal Food Retrieval. *arXiv* **2022**, arXiv:2212.04267.
45. Voutharoja, B.P.; Wang, P.; Wang, L.; Guan, V. MALM: Mask Augmentation based Local Matching for Food-Recipe Retrieval. *arXiv* **2023**, arXiv:2305.11327.
46. Zhang, C.; Song, J.; Zhu, X.; Zhu, L.; Zhang, S. Hcmsl: Hybrid cross-modal similarity learning for cross-modal retrieval. *ACM Trans. Multimed. Comput. Commun. Appl. (Tomm)* **2021**, *17*, 1–22. [CrossRef]
47. Zhu, L.; Zhang, C.; Song, J.; Liu, L.; Zhang, S.; Li, Y. Multi-graph based hierarchical semantic fusion for cross-modal representation. In Proceedings of the 2021 IEEE International Conference on Multimedia and Expo (ICME), Shenzhen, China, 5–9 July 2021; pp. 1–6.
48. Yi, Z.; Zhu, X.; Wu, R.; Zou, Z.; Liu, Y.; Zhu, L. Multi-Label Weighted Contrastive Cross-Modal Hashing. *Appl. Sci.* **2023**, *14*, 93. [CrossRef]
49. Zou, Z.; Zhu, X.; Zhu, Q.; Liu, Y.; Zhu, L. CREAMY: Cross-Modal Recipe Retrieval by Avoiding Matching Imperfectly. *IEEE Access* **2024**, *12*, 33283–33295. [CrossRef]
50. Dosovitskiy, A.; Beyer, L.; Kolesnikov, A.; Weissenborn, D.; Zhai, X.; Unterthiner, T.; Dehghani, M.; Minderer, M.; Heigold, G.; Gelly, S.; et al. An image is worth 16x16 words: Transformers for image recognition at scale. *arXiv* **2020**, arXiv:2010.11929.
51. Deng, J.; Dong, W.; Socher, R.; Li, L.J.; Li, K.; Fei-Fei, L. Imagenet: A large-scale hierarchical image database. In Proceedings of the 2009 IEEE Conference on Computer Vision and Pattern Recognition, Miami, FL, USA, 20–25 June 2009; pp. 248–255.
52. Thomas, C.; Kovashka, A. Preserving semantic neighborhoods for robust cross-modal retrieval. In Proceedings of the Computer Vision–ECCV 2020: 16th European Conference, Glasgow, UK, 23–28 August 2020; Proceedings, Part XVIII 16; Springer International Publishing: Cham, Switzerland, 2020; pp. 317–335.
53. Guo, G.; Wang, H.; Bell, D.; Bi, Y.; Greer, K. KNN model-based approach in classification. In Proceedings of the On The Move to Meaningful Internet Systems 2003: CoopIS, DOA, and ODBASE: OTM Confederated International Conferences, CoopIS, DOA, and ODBASE 2003, Catania, Sicily, Italy, 3–7 November 2003; Springer: Berlin/Heidelberg, Germany, 2003; pp. 986–996.
54. Wang, J.; Zhou, F.; Wen, S.; Liu, X.; Lin, Y. Deep metric learning with angular loss. In Proceedings of the IEEE International Conference on Computer Vision, Venice, Italy, 22–29 October 2017; pp. 2593–2601.

foods

MDPI

Article

Exploration of Convective and Infrared Drying Effect on Image Texture Parameters of 'Mejhoul' and 'Boufeggous' Date Palm Fruit Using Machine Learning Models

Younes Noutfia and Ewa Ropelewska *

Fruit and Vegetable Storage and Processing Department, The National Institute of Horticultural Research, Konstytucji 3 Maja 1/3, 96-100 Skierniewice, Poland; younes.noutfia@inhort.pl
* Correspondence: ewa.ropelewska@inhort.pl

Abstract: Date palm (*Phoenix dactylifera* L.) fruit samples belonging to the 'Mejhoul' and 'Boufeggous' cultivars were harvested at the Tamar stage and used in our experiments. Before scanning, date samples were dried using convective drying at 60 °C and infrared drying at 60 °C with a frequency of 50 Hz, and then they were scanned. The scanning trials were performed for two hundred date palm fruit in fresh, convective-dried, and infrared-dried forms of each cultivar using a flatbed scanner. The image-texture parameters of date fruit were extracted from images converted to individual color channels in RGB, Lab, XYZ, and UVS color models. The models to classify fresh and dried samples were developed based on selected image textures using machine learning algorithms belonging to the groups of Bayes, Trees, Lazy, Functions, and Meta. For both the 'Mejhoul' and 'Boufeggous' cultivars, models built using Random Forest from the group of Trees turned out to be accurate and successful. The average classification accuracy for fresh, convective-dried, and infrared-dried 'Mejhoul' reached 99.33%, whereas fresh, convective-dried, and infrared-dried samples of 'Boufeggous' were distinguished with an average accuracy of 94.33%. In the case of both cultivars and each model, the higher correctness of discrimination was between fresh and infrared-dried samples, whereas the highest number of misclassified cases occurred between fresh and convective-dried fruit. Thus, the developed procedure may be considered an innovative approach to the non-destructive assessment of drying impact on the external quality characteristics of date palm fruit.

Keywords: *Phoenix dactylifera* L.; image features; flatbed scanner; artificial intelligence; classification models

1. Introduction

Date palm (*Phoenix dactylifera* L.) is a very important fruit crop grown in many regions of the world, especially in hot and dry areas, and it is considered a promising tree for the irrigated dry zones of developing nations [1,2]. For many nations, date fruit is an essential subsistence food [1] and an important component of a healthy diet due to its functional properties as well as its high sugar content, flavonoids, anthocyanins, phenols, vitamins, minerals, and carotenoids [3]. Hence, the nutritional composition of date fruit differs according to several factors and can range from 44 to 88% for total sugar, 2.3–5.6% for protein, 6.4–11.5% for fibers, 0.2–0.5% for fat, and 0.2–0.5% for oil. Thus, both edible parts of date fruit and seeds can be used in food products as a rich source of biochemical and functional compounds such as antioxidants and fibers [1], especially at the Tamar stage.

At this complete stage of maturity, the fruit usually has a reduced moisture content of less than 20–25%, and it is considered appropriate for consumption in many regions and for storage under specific and mastered conditions [4,5]. However, some physiological and physical disorders can occur at this stage, mainly related to skin separation and sugar spots. These visual defects negatively impact the visual quality of the fruit and its market value [1].

Therefore, drying can provide shelf-life extension [6] and postpone the perishability and degeneration of date fruit.

Drying as a pretreatment, in association with efficient storage availability, ensures high-quality date fruit all year round in markets and in places where they are not produced [7]. By removing water, transport efficiency is increased by reducing the product volume as well as packaging costs. Besides this logistical improvement, moisture level, fruit sugars, and other biochemical compounds are more concentrated, allowing for more protection against deterioration factors, microorganism proliferation and growth, undesired chemical reactions, and high risk of deterioration during storage [8], leading to preservation of quality properties and shelf life extension. Moreover, date fruit may be dried using various methods, such as open sun drying, solar drying, hot air oven drying, vacuum drying, microwave drying, microwave vacuum drying, drum drying, or freeze drying [9].

The sun drying of date fruit has been the most common way throughout history. However, the process is challenging because of the inability to control drying parameters that may fluctuate drastically depending on climatic conditions. Furthermore, the sun drying process requires a long time while date fruit is under uncontrolled conditions of illumination and airflow, in addition to a possible source of contamination with insects, sand particles, soil, and dust. Thus, sun drying is not considered an effective method to apply for date fruit in the perspective of quality preservation and amelioration prior to further postharvest operations (e.g., storage). Consequently, other drying methods are more suitable for maintaining and improving the internal and external qualities of date fruit: homogenous skin color; adequate skin hardness; and high preservation of nutrient compounds [10–13]. One of the most common and effective methods is convective drying, which ensures a reduction in moisture content and quality preservation [14]. Also, infrared drying can be more beneficial compared with conventional drying due to rapid processing, shorter heating times, chemical compound preservation, and reduced risk of flavor loss [15]. Artificial intelligence and image analysis can be considered innovative and alternative approaches to assess the quality of date fruit before and after drying instead of tedious, time-consuming, and destructive analyses [5]. In doing so, machine vision enables the quantitative analysis of the quality of the qualitative criteria of the sample. Machine learning (ML) models can be successful in the classification of different samples of date fruit based on selected image parameters using several algorithms and models (e.g., traditional ML algorithms) [16]. Furthermore, image textures and determined geometric parameters using image analysis can be useful in the objective characterization of date fruit [17].

Considering the previous elements, the objective of this study was to compare the effect of convective and infrared drying on the quality of 'Mejhoul' and 'Boufeggous' date palm fruit in terms of external appearance. The assessment of the fruit quality was performed objectively and non-destructively using innovative artificial intelligence models developed based on selected image texture parameters.

2. Materials and Methods

2.1. Materials

Two well-known commercial cultivars of Moroccan date palm fruit, 'Mejhoul' and 'Boufeggous', were used in these experiments. Fruit samples were harvested at the Tamar stage in November 2023 in an orchard located at Henabou-Erfoud in southeastern Morocco (31°26′10″ N, 4°13′58″ W) and stored in cardboard boxes in a cold room at 2–4 °C until experiment starting. Before and after drying experiments, two hundred fruits of 'Mejhoul' and 'Boufeggous' without any visual defects were subjected to imaging.

2.2. Date Fruit Drying

Date fruit belonging to 'Mejhoul' and 'Boufeggous' cultivars were dried using a CONVECO semi-industrial dryer (CONVECO Sp. z o.o., Glinianka, Poland) equipped with a power supply cabinet, control cabinet, heat recovery (recuperation) units, and a ventilation duct system. For both convective and infrared drying, two technological repetitions were carried out, and the same temperature of 60 °C was used. For the infrared drying, the waving range was at a frequency of 10 min at a radiation distance of 170 mm with a radiation power of 50 Hz.

2.2.1. Convective Drying

For each technological repetition, date fruit samples were spread out in one layer into trays placed at different levels of the drying chamber rack (Figure 1). Each tray contained approximately 3–4 kg of 'Mejhoul' or 'Boufeggous' to provide a total amount of dried material of approximately 25–26 kg per cultivar and per technological repetition. This process was performed through hot air at a temperature of 60 °C, and the airflow side (right or left) was changed every 10 min. This process was carried out until reaching the desired weight loss and estimated water content (achieved in 240 min).

Figure 1. Illustration of convective drying of 'Mejhoul' and 'Boufeggous'.

2.2.2. Infrared Drying

The surfaces of the date samples were heated using lamps (radiant heaters) located above trays with date fruit, with a distance (between lamps and fruit samples) of 17 cm (Figure 2). The infrared drying was performed at parameters of 60 °C and 50 Hz. The whole process lasted 150 min, and the date fruit was side-changed in the middle of the process so that the lower side of the fruit was on top and heated consequently by the lamps. The changing side of airflow (right or left) was every 10 min. As for convective drying, infrared drying was performed in two technological repetitions, and the airflow side (right or left) was changed every 10 min.

Figure 2. Illustration of infrared drying of 'Mejhoul'.

2.3. Image Analysis

The fresh, convective-dried, and infrared-dried date fruit were scanned using the Epson Perfection flatbed scanner (Epson, Suwa, Nagano, Japan) on a white background covered by a box and saved in TIFF format. Twenty fruits were in one image, and the scanning was performed for two hundred date palm fruit samples of 'Mejhoul' and 'Boufeggous' in fresh, convective-dried, and infrared-dried forms. Before image processing, the white background was changed to black to facilitate image segmentation and separation of fruit from the background, and images were saved in BMP format. Image processing, including image segmentation, ROI (region of interest) determination, and texture extraction, was performed using MaZda 4.7 software (Łódź University of Technology, Institute of Electronics, Łódź, Poland) [18–20]. The date fruit images were converted to individual color channels R, G, B, L, a, b, X, Y, Z, U, V, and S. Exemplary original color images and images in selected color channels of fresh, convective-dried, and infrared-dried date fruit of 'Mejhoul' and 'Boufeggous' are presented respectively in Figures 3 and 4. The dried fruit samples were of better quality in terms of appearance and characterized by shinier skin, especially in the case of infrared-dried fruit. The image segmentation was performed based on the brightness threshold, and the lighter date fruit samples were separated from the black background. Each whole date fruit was treated as an ROI, and for each ROI, 2160 image textures were computed based on the run-length matrix, co-occurrence matrix, histogram, gradient map, and autoregressive model.

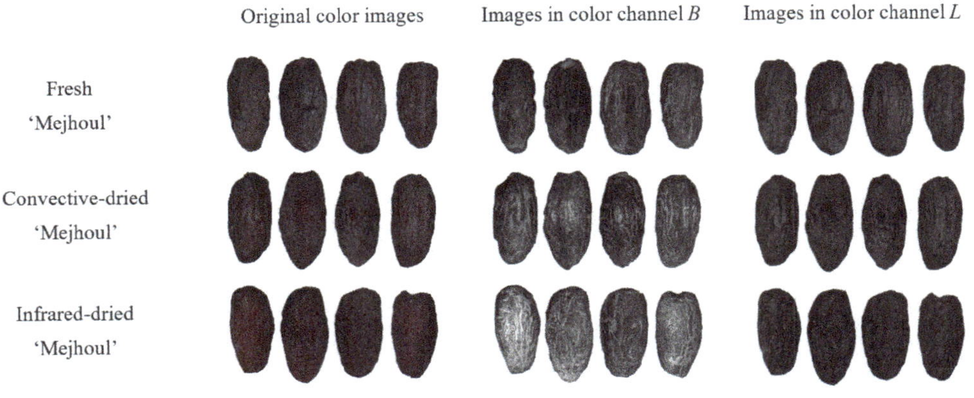

Figure 3. Exemplary original color images and images in selected color channels of fresh, convective-dried (CD), and infrared-dried (ID) date fruit 'Mejhoul'. B refers to blue, and L refers to lightness.

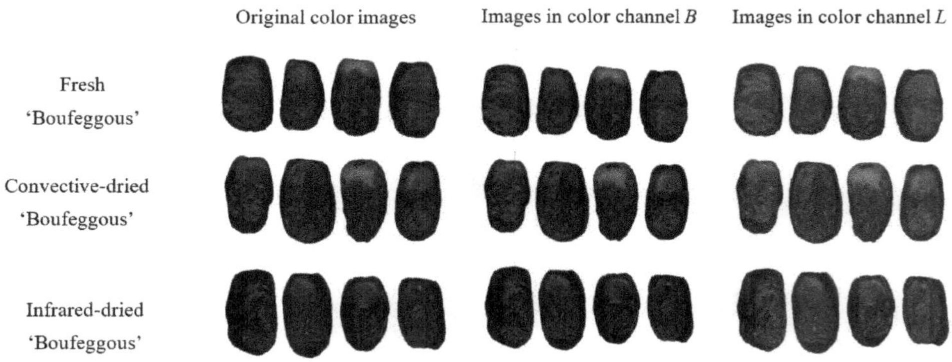

Figure 4. Exemplary original color images and images in selected color channels of fresh, convective-dried, and infrared-dried date fruit 'Boufeggous'. *B* refers to blue, and *L* refers to lightness.

2.4. Mean Comparison of Selected Image Textures

Graphs of mean values of selected image textures and analysis of mean comparison were carried out using STATISTICA 3.1 (Dell Inc., Tulsa, OK, USA, StatSoft Polska, Kraków, Poland) software. The normality of the distribution was checked using Shapiro–Wilk, Lilliefors, and Kolmogorov–Smirnov tests. The means were compared using Tukey's test at a significance level of $p < 0.05$.

2.5. Machine Learning Models for Distinguishing Fresh, Convective, and Infrared-Dried Date Fruit

The classification models were built based on selected image texture parameters of fresh and dried date fruit using WEKA 3.9 machine learning software (Machine Learning Group, University of Waikato, Hamilton, New Zealand) [21–23]. The image textures with the highest discriminative power were selected using the Best First and the correlation-based feature selection subset evaluator. The models for distinguishing fresh, convective, and infrared-dried date fruit were built using a 10-fold cross-validation mode by dividing the dataset into 10 parts and considering nine parts as the training sets and one part as the test set. This process was repeated 10 times for different training/test sets, and the result was the average of 10 estimates. The machine learning algorithms belonging to the groups of Bayes, Trees, Lazy, Functions, and Meta were tested to select algorithms that provided highly accurate results. The confusion matrices with accuracies for individual classes, average accuracies, and the values of True Positive Rate (TPR), False Positive Rate (FPR), Precision, Recall, F-Measure, Matthews Correlation Coefficient (MCC), Receiver Operating Characteristic Area (ROC Area), and Precision–Recall Area (PRC Area) (Equations (1)–(8)) [24–27] were determined.

$$\text{Accuracy} = \frac{(\text{TP} + \text{TN})}{\text{TP} + \text{TN} + \text{FN} + \text{FP}} \tag{1}$$

$$\text{TPR} = \text{Recall} = \frac{\text{TP}}{\text{TP} + \text{FN}} \tag{2}$$

$$\text{FPR} = \frac{\text{FP}}{\text{FP} + \text{TN}} \tag{3}$$

$$\text{Precision} = \frac{\text{TP}}{\text{TP} + \text{FP}} \tag{4}$$

$$\text{F} - \text{Measure} = \frac{2\text{TP}}{2\text{TP} + \text{FP} + \text{FN}} \tag{5}$$

$$\text{MCC} = \frac{(\text{TP} * \text{TN} - \text{FP} * \text{FN})}{\sqrt{((\text{TP} + \text{FP})(\text{TP} + \text{FN})(\text{TN} + \text{FP})(\text{TN} + \text{FN}))}} \tag{6}$$

$$\text{ROC Area} = \text{Area Under TPR vs. FPR Curve} \tag{7}$$

$$\text{PRC Area} = \text{Area Under Precision vs. Recall Curve} \tag{8}$$

TP: True Positive; TN: True Negative; FP: False Positive; FN: False Negative

3. Results and Discussion

3.1. 'Mejhoul': Discrimination between Fresh, Convective, and Infrared Dried Fruit Based on Machine Learning Models and Selected Image Texture Parameters

The selected image texture parameters of fresh, convective-dried, and infrared-dried 'Mejhoul' date fruit samples were compared, and graphs of mean values are shown in Table 1. In the case of RHMean, when *R* was color channel *R* (red) from the RGB color space, and HMean was histogram's mean, there was no significant difference ($p > 0.05$) between fresh and convective-dried fruit samples, and these samples were included in one homogenous group. For other analyzed image textures, such as GHMean, BHMean, LHMean, aHMean, bHMean, XHMean, YHMean, and ZHMean (the first letter means the color channel as follows: *G* (green) and *B* (blue) from the RGB color space; *L* (lightness component from black to white), *a* (red or green), and *b* (yellow or blue) from the Lab color space; and *X* (component with color information), *Y* (lightness), and *Z* (component with color information) from the XYZ color space, fresh, convective and infrared samples were significantly different.

Table 1. Mean values of selected image texture parameters of fresh, convective-dried, and infrared-dried 'Mejhoul' date fruit.

Image Texture Parameter	Fresh 'Mejhoul'	Convective-Dried 'Mejhoul'	Infrared-Dried 'Mejhoul'
RHMean	70.35 a	70.08 ab	69.44 b
GHMean	59.65 a	57.44 b	55.54 c
BHMean	55.51 a	53.45 b	51.60 c
LHMean	91.94 a	90.28 b	88.77 c
aHMean	130.02 a	130.79 b	131.24 c
bHMean	130.46 a	130.73 b	130.91 b
XHMean	12.74 a	12.41 b	11.85 c
YHMean	12.58 a	12.08 b	11.40 c
ZHMean	10.56 a	10.07 b	9.30 c

Note. In the same row, means with different letters are significantly different ($p < 0.05$) according to Tukey HSD test.

Models developed based on selected texture parameters extracted from images in different color channels *R*, *G*, *B*, *L*, *a*, *b*, *X*, *Y*, *Z*, *U*, *V*, and *S* allowed for distinguishing fresh and dried 'Mejhoul' date fruit with very high correctness. The accuracy average of classification of fresh, convective-dried, and infrared-dried samples reached 99.33% for a model built using Random Forest (Table 2). However, slightly lower accuracies were obtained for models developed using IBk (99.25%), Multilayer Perceptron (98.67%), Logit Boost and Bayes Net (98.42%), and PART (97.83%). The higher differences occurred between fresh and infrared-dried 'Mejhoul' date fruit, whereas the greatest misclassification was observed between fresh and convective-dried fruit.

Other performance metrics, such as True Positive Rate (TPR), False Positive Rate (FPR), Precision, Recall, F-Measure, Matthews Correlation Coefficient (MCC), Receiver Operating Characteristic Area (ROC Area), and Precision–Recall Area (PRC Area) confirmed the highest correctness of the classification of infrared-dried 'Mejhoul' date fruit (Table 3). The greatest differentiation of infrared-dried fruit in terms of selected image textures from fresh and convective-dried samples was particularly visible in the case of the machine learning model built using Multilayer Perceptron. The values of TPR, Precision, Recall, F-Measure, MCC, ROC Area, and PRC Area were equal to 1.000, and FPR was 0.000. It confirmed that there was no mixing of cases between infrared-dried and other classes. In the case of each model, the lowest values of TPR, Recall, F-Measure, MCC, and ROC Area and the highest FPR were obtained for convective-dried 'Mejhoul' fruit. These results confirmed that convective-dried date fruit was classified with the lowest correctness.

Table 2. Distinguishing accuracies of fresh, convective, and infrared-dried 'Mejhoul' date fruit using machine learning models.

Algorithm	Predicted Class (Number of Cases)			Actual Class	Average Accuracy (%)
	Fresh 'Mejhoul'	Convective-Dried 'Mejhoul'	Infrared-Dried 'Mejhoul'		
trees	400	0	0	Fresh 'Mejhoul'	
Random	6	393	1	Convective-dried 'Mejhoul'	99.33
Forest	0	1	399	Infrared-dried 'Mejhoul'	
rules	399	1	0	Fresh 'Mejhoul'	
PART	23	376	1	Convective-dried 'Mejhoul'	97.83
	0	1	399	Infrared-dried 'Mejhoul'	
meta	394	6	0	Fresh 'Mejhoul'	
Logit	12	387	1	Convective-dried 'Mejhoul'	98.42
Boost	0	0	400	Infrared-dried 'Mejhoul'	
lazy	400	0	0	Fresh 'Mejhoul'	
IBk	8	391	1	Convective-dried 'Mejhoul'	99.25
	0	0	400	Infrared-dried 'Mejhoul'	
functions	397	3	0	Fresh 'Mejhoul'	
Multilayer	13	387	0	Convective-dried 'Mejhoul'	98.67
Perceptron	0	0	400	Infrared-dried 'Mejhoul'	
Bayes	391	9	0	Fresh 'Mejhoul'	
Bayes	9	390	1	Convective-dried 'Mejhoul'	98.42
Net	0	0	400	Infrared-dried 'Mejhoul'	

Table 3. Performance metrics of distinguishing fresh and dried date fruit 'Mejhoul' using machine learning models developed based on texture parameters.

Algorithm	Class	TPR	FPR	Precision	Recall	F-Measure	MCC	ROC Area	PRC Area
trees Random Forest	Fresh 'Mejhoul'	1.000	0.008	0.985	1.000	0.993	0.989	1.000	1.000
	Convective-dried 'Mejhoul'	0.983	0.001	0.997	0.983	0.990	0.985	0.998	0.998
	Infrared-dried 'Mejhoul'	0.998	0.001	0.998	0.998	0.998	0.996	1.000	1.000
rules PART	Fresh 'Mejhoul'	0.998	0.029	0.945	0.998	0.971	0.956	0.985	0.939
	Convective-dried 'Mejhoul'	0.940	0.003	0.995	0.940	0.967	0.951	0.966	0.970
	Infrared-dried 'Mejhoul'	0.998	0.001	0.998	0.998	0.998	0.996	0.998	0.994
meta Logit Boost	Fresh 'Mejhoul'	0.985	0.015	0.970	0.985	0.978	0.966	0.999	0.998
	Convective-dried 'Mejhoul'	0.968	0.008	0.985	0.968	0.976	0.964	0.996	0.996
	Infrared-dried 'Mejhoul'	1.000	0.001	0.998	1.000	0.999	0.998	0.999	0.998
lazy IBk	Fresh 'Mejhoul'	1.000	0.010	0.980	1.000	0.990	0.985	0.991	0.966
	Convective-dried 'Mejhoul'	0.978	0.000	1.000	0.978	0.989	0.983	0.981	0.985
	Infrared-dried 'Mejhoul'	1.000	0.001	0.998	1.000	0.999	0.998	0.999	0.998

Table 3. *Cont.*

Algorithm	Class	TPR	FPR	Precision	Recall	F-Measure	MCC	ROC Area	PRC Area
functions Multilayer Perceptron	Fresh 'Mejhoul'	0.993	0.016	0.968	0.993	0.980	0.970	0.996	0.990
	Convective-dried 'Mejhoul'	0.968	0.004	0.992	0.968	0.980	0.970	0.992	0.974
	Infrared-dried 'Mejhoul'	1.000	0.000	1.000	1.000	1.000	1.000	1.000	1.000
Bayes Bayes Net	Fresh date fruit	0.978	0.011	0.978	0.978	0.978	0.966	0.999	0.998
	Convective-dried date fruit	0.975	0.011	0.977	0.975	0.976	0.964	0.997	0.996
	Infrared-dried date fruit	1.000	0.001	0.998	1.000	0.999	0.998	0.999	0.998

TPR—True Positive Rate; FPR—False Positive Rate; MCC—Matthews Correlation Coefficient; ROC Area—Receiver Operating Characteristic Area; PRC Area—Precision–Recall Area.

3.2. 'Boufeggous': Discrimination between Fresh, Convective, and Infrared Dried Fruit Based on Machine Learning Models and Selected Texture Parameters

In the case of the 'Boufeggous' cultivar, the mean comparison of selected image texture parameters of fresh, convective-dried, and infrared-dried samples revealed the greatest similarity between fresh and convective-dried fruit. Among the nine image texture features analyzed for fresh and convective-dried 'Boufeggous' samples, only RHMean, GHMean, LHMean, and XHMean were in the same homogenous group with no statistically significant difference. Graphs of texture mean values are presented in Table 4.

Table 4. Mean values of selected image texture parameters of fresh, convective-dried, and infrared-dried 'Boufeggous' date fruit.

Image Texture Parameter	Fresh 'Boufeggous'	Convective-Dried 'Boufeggous'	Infrared-Dried 'Boufeggous'
RHMean	65.32 a	65.25 a	63.70 b
GHMean	56.54 a	56.18 a	53.20 b
BHMean	49.58 a	51.53 b	48.22 c
LHMean	88.10 a	88.04 a	85.40 b
aHMean	129.06 a	129.58 b	130.07 c
bHMean	131.64 a	130.59 b	131.01 c
XHMean	11.56 a	11.47 a	10.81 b
YHMean	11.61 a	11.60 a	10.74 b
ZHMean	9.08 a	9.72 b	8.84 c

Note. In the same row, means with different letters are significantly different ($p < 0.05$) according to Tukey HSD test.

For this date palm cultivar, classification accuracies were slightly lower (Table 5) than 'Mejhoul'. The average accuracy of distinguishing fresh, convective-dried, and infrared-dried date fruit reached 94.33% for a model built using Random Forest, compared to 99.33% for 'Mejhoul'. Nevertheless, the lowest average accuracy for analyzed models was observed for the IBk algorithm, and it was equal to 91.00%.

As was reported for 'Mejhoul', fresh and infrared-dried 'Boufeggous' date fruit were distinguished with the highest correctness. Thus, the highest number of misclassified cases was between fresh and convective-dried fruit. In the case of the model developed using IBk, 57 cases belonging to the actual class of convective-dried 'Boufeggous' fruit were misclassified as fresh samples, and 47 cases of fresh fruit were incorrectly classified as convective-dried fruit.

Table 5. Distinguishing accuracies of fresh, convective, and infrared dried 'Boufeggous' date fruit using machine learning models.

Algorithm	Predicted Class (Number of Cases) Fresh 'Boufeggous'	Convective-Dried 'Boufeggous'	Infrared-Dried 'Boufeggous'	Actual Class	Average Accuracy (%)
trees	377	22	1	Fresh 'Boufeggous'	
Random	42	356	2	Convective-dried 'Boufeggous'	94.33
Forest	0	1	399	Infrared-dried 'Boufeggous'	
rules	361	38	1	Fresh 'Boufeggous'	
PART	48	348	4	Convective-dried 'Boufeggous'	92.25
	0	2	398	Infrared-dried 'Boufeggous'	
meta	367	31	2	Fresh 'Boufeggous'	
Logit	35	365	0	Convective-dried 'Boufeggous'	94.25
Boost	0	1	399	Infrared-dried 'Boufeggous'	
lazy	352	47	1	Fresh 'Boufeggous'	
IBk	57	342	1	Convective-dried 'Boufeggous'	91.00
	1	1	398	Infrared-dried 'Boufeggous'	
functions	370	26	4	Fresh 'Boufeggous'	
Multilayer	41	359	0	Convective-dried 'Boufeggous'	93.92
Perceptron	1	1	398	Infrared-dried 'Boufeggous'	
Bayes	373	26	1	Fresh 'Boufeggous'	
Bayes	43	357	0	Convective-dried 'Boufeggous'	93.67
Net	1	5	394	Infrared-dried 'Boufeggous'	

The classification values of TPR, FPR, Precision, Recall, F-Measure, MCC, ROC Area, and PRC Area of fresh, convective-dried, and infrared-dried 'Boufeggous' date fruit are presented in Table 6. The highest classification accuracy of infrared-dried date fruit was confirmed by the lowest FPR and the highest values of other metrics. The value of 1.000 was obtained in the case of ROC Area and PRC Area for a model built using Random Forest and ROC Area for a model developed using Bayes Net. The lowest FPR was found for a model built by Bayes Net.

Table 6. Performance metrics of distinguishing fresh and dried 'Boufeggous' date fruit using machine learning models developed based on texture parameters.

Algorithm	Class	TPR	FPR	Precision	Recall	F-Measure	MCC	ROC Area	PRC Area
trees	Fresh 'Boufeggous'	0.943	0.053	0.900	0.943	0.921	0.880	0.990	0.980
Random	Convective-dried 'Boufeggous'	0.890	0.029	0.939	0.890	0.914	0.873	0.990	0.980
Forest	Infrared-dried 'Boufeggous'	0.998	0.004	0.993	0.998	0.995	0.993	1.000	1.000
rules	Fresh 'Boufeggous'	0.903	0.060	0.883	0.903	0.892	0.838	0.961	0.938
PART	Convective-dried 'Boufeggous'	0.870	0.050	0.897	0.870	0.883	0.826	0.967	0.927
	Infrared-dried 'Boufeggous'	0.995	0.006	0.988	0.995	0.991	0.987	0.995	0.984
meta	Fresh 'Boufeggous'	0.918	0.044	0.913	0.918	0.915	0.873	0.988	0.975
Logit	Convective-dried 'Boufeggous'	0.913	0.040	0.919	0.913	0.916	0.874	0.987	0.978
Boost	Infrared-dried 'Boufeggous'	0.998	0.003	0.995	0.998	0.996	0.994	0.999	0.999
lazy	Fresh 'Boufeggous'	0.880	0.073	0.859	0.880	0.869	0.803	0.905	0.800
IBk	Convective-dried 'Boufeggous'	0.855	0.060	0.877	0.855	0.866	0.800	0.899	0.804
	Infrared-dried 'Boufeggous'	0.995	0.003	0.995	0.995	0.995	0.993	0.996	0.992
functions	Fresh 'Boufeggous'	0.925	0.053	0.898	0.925	0.911	0.866	0.973	0.948
Multilayer	Convective-dried 'Boufeggous'	0.898	0.034	0.930	0.898	0.913	0.872	0.979	0.963
Perceptron	Infrared-dried 'Boufeggous'	0.995	0.005	0.990	0.995	0.993	0.989	0.998	0.995
Bayes	Fresh 'Boufeggous'	0.933	0.055	0.894	0.933	0.913	0.869	0.982	0.964
Bayes	Convective-dried 'Boufeggous'	0.893	0.039	0.920	0.893	0.906	0.860	0.981	0.966
Net	Infrared-dried 'Boufeggous'	0.985	0.001	0.997	0.985	0.991	0.987	1.000	0.999

TPR—True Positive Rate; FPR—False Positive Rate; MCC—Matthews Correlation Coefficient; ROC Area—Receiver Operating Characteristic Area; PRC Area—Precision–Recall Area.

The above-mentioned tables showed an important trend in the variation in the above-mentioned texture parameters that can be ranged into two different groups: group with high values for ID drying (aHMean and bHMean); and group with lower values for ID drying (GHMean, BHMean, LHMean, XHMean, YHMean, and ZHMean). Nevertheless, both groups are characterized by convective drying texture features comprising fresh and infrared values. This trend may reflect significant texture variation between fresh, convective, and infrared dried 'Mejhoul' date fruit with high differences between ID and fresh compared to CD and fresh fruit. For 'Boufeggous', these findings are applied only for the case of aHMean and YHMean features.

Regarding developed models, Random forest was the most accurate and predicting algorithm for both 'Mejhoul' and 'Boufeggous' while PART and lazy IBk were respectively the lower predicting models for 'Mejhoul' and 'Boufeggous'. Such constatation was reported by [28] while comparing the effectiveness and accuracy of six algorithms used for orange (*Citrus sinensis*) classification. Among Logistic Regression, Naïve Bayes, Decision Trees, Neural Networks, K-nearest neighbors, and Random Forests models, authors found that Logistic Regression was more accurate and precise, with an accuracy of 91% and a precision score of 92%.

In another study related to cultivar date fruit classification, [29] employed algorithms developed on the basis of ANN and LR approaches to discriminate between seven date fruit cultivars. The lowest accuracies of 91 and 92.2% compared to the results of this paper were obtained for distinguishing color, shape, and pomological features extracted from images.

In relation to these findings and the scope of this study, the previous literature data considered mainly the application of destructive analysis (instead of non-destructive methods) to assess date fruit quality in terms of physicochemical properties under different experimental conditions and various postharvest treatments [30–33]. In this way, mathematical models were used in addition to "response surface methodology" to express the drying kinetics and behavior of date fruit [34–38] without considering machine learning approaches for the specific case of date fruit drying. Nevertheless, machine learning models were employed for other fruit species to determine and assess dried fruit quality. For example, Raihen and Akter [39] developed deep-learning models for the classification of dried grapefruit types. Sağlam and Çetin [40] applied machine learning models to estimate the drying characteristics of dried apple slices by employing artificial neural networks, k-nearest neighbors, random forests, etc. For the three apple cultivars, the highest correlation coefficients were about 0.98 for moisture ratio estimation using a Random Forest algorithm. The effectiveness and usefulness of machine learning models in rapid estimation of chemical properties of apples were also proven with five machine learning algorithms [41]. Also, machine learning algorithms were used to predict the sweetness of dried bananas using Linear Discriminant Analysis (LDA), K Nearest Neighbors (KNN), Decision Tree (CART), Random Forest (RF), and Support Vector Machines (SVM) models. Among these five algorithms, authors reported that RF and SVM allowed for a high prediction accuracy of 86% [42]. The above-mentioned studies stated the importance of using ML algorithms to predict and estimate the biochemical properties of dried fruit with high accuracy, and the results presented in this paper are comparable in terms of high accuracy. Our study presented in this paper revealed the usefulness of machine learning models in determining the external quality of dried date fruit. However, further research can be related to the estimation of the physicochemical properties of date fruit using image processing and artificial intelligence.

4. Conclusions

Through this work, the effect of mild convective and infrared drying techniques on the changes in the external quality of date fruit was revealed in a non-destructive and objective manner using a flatbed scanner. The developed machine learning models, using selected texture parameters from images in different color channels R, G, B, L, a, b, X, Y, Z, U, V, and S, classified accurately fresh, convective-dried, and infrared-dried 'Mejhoul' and

'Boufeggous' date palm fruit. Thus, samples were classified with high correctness of 99.33% and 94.33 for 'Mejhoul' and 'Boufeggous', respectively, using the Random Forest algorithm, which was more accurate than other machine learning models tested under this work. Furthermore, the usefulness of image features and models built using artificial intelligence was confirmed and proved the greater effect of infrared drying on the improvement in the external appearance of date fruit for both cultivars.

From this perspective, selected texture features from date fruit images can be used as inputs to develop more accurate models using traditional machine learning and deep-learning algorithms that link those externally extracted features with other internal quality parameters such as biochemical and nutritional compounds. Additionally, other innovative drying methods can be evaluated for other date fruit cultivars with a focus on the elucidation of machine learning approaches that added value to the total quality assessment.

Author Contributions: Conceptualization, E.R. and Y.N.; methodology, Y.N. and E.R.; software, E.R.; validation, E.R.; investigation, E.R. and Y.N.; writing—original draft preparation, E.R.; writing—review and editing, Y.N. and E.R.; project administration, Y.N.; funding acquisition, Y.N. All authors have read and agreed to the published version of the manuscript.

Funding: This research is part of project No. 2022/45/P/NZ9/03904, co-funded by the National Science Centre and the European Union's Horizon 2020 research and innovation program under the Marie Skłodowska-Curie grant agreement no. 945339. Project title, "A novel approach to the assessment of date fruit quality (*Phoenix dactylifera* L.) under different storage conditions, using innovative models based on image analyses and machine learning" (M-LEARN4DATE).

Institutional Review Board Statement: Not applicable.

Informed Consent Statement: Not applicable.

Data Availability Statement: The original contributions presented in the study are included in the article, further inquiries can be directed to the corresponding author.

Conflicts of Interest: The authors declare no conflicts of interest.

References

1. Alsmairat, N.; Othman, Y.; Ayad, J.; Al-Ajlouni, M.; Sawwan, J.; El-Assi, N. Anatomical Assessment of Skin Separation in Date Palm (*Phoenix dactylifera* L. var. Mejhoul) Fruit during Maturation and Ripening Stages. *Agriculture* **2023**, *13*, 38. [CrossRef]
2. Rakesh Reddy, S.V.; Singh, R.S.; Meena, R.; Berwal, M.K.; Sarolia, D.K.; Palpandian, P. Impact of Hot Water Pre-Treatments on the Drying Efficiency and Quality of Dates cv. Medjool. *Horticulturae* **2023**, *9*, 784. [CrossRef]
3. Uchoi, J.; Nikhumbhe, P.H.; Kumar, A.; Patidar, A.; Harish, G.D. Impact of inclined solar drier for dehydration quality in khadrawy dates during doka maturity stage at north western arid India. *Int. J. Curr. Microbiol. Appl. Sci.* **2020**, *9*, 119–125. [CrossRef]
4. Muñoz-Bas, C.; Muñoz-Tebar, N.; Candela-Salvador, L.; Pérez-Alvarez, J.A.; Lorenzo, J.M.; Viuda-Martos, M.; Fernández-López, J. Quality Characteristics of Fresh Date Palm Fruits of "Medjoul" and "Confitera" cv. from the Southeast of Spain (Elche Palm Grove). *Foods* **2023**, *12*, 2659. [CrossRef] [PubMed]
5. Noutfia, Y.; Ropelewska, E. What can artificial intelligence approaches bring to an improved and efficient harvesting and postharvest handling of date fruit (*Phoenix dactylifera* L.)? A review. *Postharvest Biol. Technol.* **2024**, *213*, 112926. [CrossRef]
6. Bassey, E.J.; Cheng, J.H.; Sun, D.W. Novel nonthermal and thermal pretreatments for enhancing drying performance and improving quality of fruits and vegetables. *Trends Food Sci. Technol.* **2021**, *112*, 137–148. [CrossRef]
7. Alsmairat, N.; Al-Qudah, T.; El-Assi, N.; Mehyar, G.; Gammoh, I.; Othman, Y.; Araj, S.; Al-Antary, T. Effect of drying process on physical and chemical properties of 'Medjool' date palm fruits. *Fresenius Environ. Bull.* **2019**, *28*, 1552–1559.
8. Özlem, E.; Yeliz, I. Modeling of drying processes of dates (*Phoenix*, *Arecaceae*) with oven or TGA and microbiological properties of fresh and dried dates. *Int. J. Fruit Sci.* **2020**, *20*, S1530–S1538.
9. Elwakeel, A.E.; Tantawy, A.A.; Alsebiey, M.M.; Elliby, A.K. The date fruit drying systems: Acritical over review. *Al-Azhar J. Agric. Eng.* **2022**, *3*, 26–36. [CrossRef]
10. Falade, K.O.; Abbo, E.S. Air-drying and rehydration characteristics of date palm (*Phoenix dactylifera* L.) fruits. *J. Food Eng.* **2007**, *79*, 724–730. [CrossRef]
11. Shahdadi, F.; Mirzaei, H.O.; Daraei Garmakhany, A. Study of phenolic compound and antioxidant activity of date fruit as a function of ripening stages and drying process. *J. Food Sci. Technol.* **2015**, *52*, 1814–1819. [CrossRef] [PubMed]
12. İzli, G. Total phenolics, antioxidant capacity, colour and drying characteristics of date fruit dried with different methods. *Food Sci. Technol.* **2016**, *37*, 139–147. [CrossRef]

13. Noutfia, Y.; Benali, A.; Alem, C.; Zegzouti, Y.F. Design of a solar dryer for small-farm level use and studying fig quality. *Acta Sci. Pol. Technol. Aliment.* **2018**, *17*, 359–365. [CrossRef] [PubMed]
14. Elghazali, M.N.; Tawfeuk, H.Z.; Gomaa, R.A.; Tantawy, A.A. Technological Studies on Aswan Dry Dates Products After Dehydration. *Assiut J. Agric. Sci.* **2020**, *51*, 32–49.
15. Tireki, S. Effective diffusivity determination of date (*Phoenix dactylifera* L.) leather in infrared drying: Effect of cooking time. *NOHU J. Eng. Sci.* **2023**, *12*, 1558–1565. [CrossRef]
16. Noutfia, Y.; Ropelewska, E. Innovative models built based on image textures using traditional machine learning algorithms for distinguishing different varieties of Moroccan date palm fruit (*Phoenix dactylifera* L.). *Agriculture* **2022**, *13*, 26. [CrossRef]
17. Noutfia, Y.; Ropelewska, E. Comprehensive characterization of date palm fruit 'Mejhoul' (*Phoenix dactylifera* L.) using image analysis and quality attribute measurements. *Agriculture* **2022**, *13*, 74. [CrossRef]
18. Szczypiński, P.M.; Strzelecki, M.; Materka, A. Mazda-a software for texture analysis. In Proceedings of the 2007 International Symposium on Information Technology Convergence (ISITC 2007), Jeonju, Republic of Korea, 23–24 November 2007; pp. 245–249.
19. Szczypiński, P.M.; Strzelecki, M.; Materka, A.; Klepaczko, A. MaZda—A software package for image texture analysis. *Comput. Methods Programs Biomed.* **2009**, *94*, 66–76. [CrossRef] [PubMed]
20. Strzelecki, M.; Szczypiński, P.; Materka, A.; Klepaczko, A. A software tool for automatic classification and segmentation of 2D/3D medical images. *Nucl. Instrum. Methods Phys. Res. Sect. A Accel. Spectrom. Detect. Assoc. Equip.* **2013**, *702*, 137–140. [CrossRef]
21. Witten, I.H.; Frank, E. *Data Mining: Practical Machine Learning Tools and Techniques*, 2nd ed.; Elsevier: San Francisco, CA, USA, 2005.
22. Bouckaert, R.R.; Frank, E.; Hall, M.; Kirkby, R.; Reutemann, P.; Seewald, A.; Scuse, D. *WEKA Manual for Version 3-9-1*; University of Waikato: Hamilton, New Zealand, 2016.
23. Frank, E.; Hall, M.A.; Witten, I.H. The WEKA Workbench. In *Data Mining: Practical Machine Learning Tools and Techniques*, 4th ed.; Online Appendix; Morgan Kaufmann: Burlington, MA, USA, 2016.
24. Ropelewska, E. Diversity of Plum Stones Based on Image Texture Parameters and Machine Learning Algorithms. *Agronomy* **2022**, *12*, 762. [CrossRef]
25. Sabanci, K.; Aslan, M.F.; Ropelewska, E.; Unlersen, M.F. A convolutional neural network-based comparative study for pepper seed classification: Analysis of selected deep features with support vector machine. *J. Food Process Eng.* **2022**, *45*, e13955. [CrossRef]
26. Cinar, I.; Koklu, M. Classification of Rice Varieties Using Artificial Intelligence Methods. *Int. J. Intell. Syst. Appl. Eng.* **2019**, *7*, 188–194. [CrossRef]
27. Ropelewska, E. Distinguishing lacto-fermented and fresh carrot slice images using the Multilayer Perceptron neural network and other machine learning algorithms from the groups of Functions, Meta, Trees, Lazy, Bayes and Rules. *Eur Food Res Technol* **2022**, *248*, 2421–2429. [CrossRef]
28. Vaishnav, D.; Rao, B.R. Comparison of machine learning algorithms and fruit classification using orange data mining tool. In Proceedings of the 2018 3rd International Conference on Inventive Computation Technologies (ICICT), Coimbatore, India, 15–16 November 2018; pp. 603–607.
29. Koklu, M.; Kursun, R.; Taspinar, Y.S.; Cinar, I. Classification of date fruits into genetic varieties using image analysis. *Math. Probl. Eng.* **2021**, *2021*, 4793293. [CrossRef]
30. Ramadan, B.R.; EL-Rify, M.N.A.; Abd El-Hamid, A.A.; Abd El-Majeed, M.H. Effect of some treatments on chemical composition and quality properties of Saidy date fruit (*Phoenix dactylifera* L.) during storage. *Assiut J. Agric. Sci.* **2016**, *47*, 107–124.
31. Tang, Z.X.; Shic, L.E.; Aleid, S.M. Date fruit: Chemical composition, nutritional and medicinal values, products. *J. Sci. Food Agric.* **2013**, *93*, 2351–2361. [CrossRef] [PubMed]
32. Juhaimi, F.A.; Ghafoor, K.; Özcan, M.M. Physical and chemical properties, antioxidant activity, total phenol and mineral profile of seeds of seven different date fruit (*Phoenix dactylifera* L.) varieties. *Int. J Food Sci. Nutr.* **2012**, *63*, 84–89. [CrossRef] [PubMed]
33. AL Juhaimi, F.; Ghafoor, K.; Özcan, M.M. Physicochemical properties and mineral contents of seven different date fruit (*Phoenix dactylifera* L.) varieties growing from Saudi Arabia. *Environ. Monit. Assess.* **2014**, *186*, 2165–2170. [CrossRef]
34. Hadibi, T.; Mennouche, D.; Boubekri, A.; Arıcı, M.; Wang, Y.; Li, M.; Hassanien, R.H.E.; Shirkole, S.S. Experimental investigation, performance analysis, and optimization of hot air convective drying of date fruits via response surface methodology. *Renew. Energy* **2024**, *226*, 120404. [CrossRef]
35. Sansaniwal, S.K.; Kumar, M.; Kumar Sahdev, R.; Bhutani, V.; Manchanda, H. Toward natural convection solar drying of date palm fruits (*Phoenix dactylifera* L.): An experimental study. *Environ. Prog. Sustain. Energy* **2022**, *41*, e13862. [CrossRef]
36. Al-Awaadh, A.M.; Hassan, B.H.; Ahmed, K.M.A. Hot Air Drying Characteristics of Sukkari Date (*Phoenix dactylifera* L.) and Effects of Drying Condition on Fruit Color and Texture. *Int. J. Food Eng.* **2015**, *11*, 421–434. [CrossRef]
37. Kumar, M.; Shimpy, M.; Sahdev, R.K.; Sansaniwal, S.K.; Bhutani, V.; Manchanda, H. Experimental forced convection greenhouse and indirect cabinet drying of date fruits: A comparative study. *J. Therm. Anal. Calorim.* **2023**, *148*, 5437–5454. [CrossRef]
38. Boubekri, A.; Benmoussa, H.; Mennouche, D. Solar drying kinetics of date palm fruits assuming a step-wise air temperature change. *J. Eng. Sci. Technol.* **2009**, *4*, 292–304.
39. Raihen, M.N.; Akter, S. Prediction modeling using deep learning for the classification of grape-type dried fruits. *Int. J. Math. Comput. Eng.* **2024**, *2*, 1–12. [CrossRef]
40. Sağlam, C.; Çetin, N. Machine learning algorithms to estimate drying characteristics of apples slices dried with different methods. *J. Food Process. Preserv.* **2022**, *46*, e16496. [CrossRef]

41. Çetin, N.; Sağlam, C. Rapid detection of total phenolics, antioxidant activity and ascorbic acid of dried apples by chemometric algorithms. *Food Biosci.* **2022**, *47*, 101670. [CrossRef]
42. Al-Mashhadany, S.A.; Hasan, H.A.; Al-Sammarraie, M.A.J. Using Machine Learning Algorithms to Predict the Sweetness of Bananas at Different Drying Times. *J. Ecol. Eng.* **2024**, *25*, 231–238. [CrossRef]

foods

MDPI

Article

Discrimination of New and Aged Seeds Based on On-Line Near-Infrared Spectroscopy Technology Combined with Machine Learning

Yanqiu Zhu [1], Shuxiang Fan [2], Min Zuo [3], Baohua Zhang [4], Qingzhen Zhu [1,*] and Jianlei Kong [3,*]

1 Key Laboratory for Theory and Technology of Intelligent Agricultural Machinery and Equipment of Jiangsu University, Zhenjiang 212013, China; 1000004953@ujs.edu.cn
2 College of Technology, Beijing Forestry University, Beijing 100083, China; fanshuxiang@outlook.com
3 National Engineering Research Center for Agri-Product Quality Traceability, Beijing Technology and Business University, Beijing 100048, China; zuomin@btbu.edu.cn
4 College of Artificial Intelligence, Nanjing Agricultural University, Nanjing 210095, China; bhzhang@njau.edu.cn
* Correspondence: qingzhen_zhu@ujs.edu.cn (Q.Z.); kongjianlei@btbu.edu.cn (J.K.)

Abstract: The harvest year of maize seeds has a significant impact on seed vitality and maize yield. Therefore, it is vital to identify new seeds. In this study, an on-line near-infrared (NIR) spectra collection device (899–1715 nm) was designed and employed for distinguishing maize seeds harvested in different years. Compared with least squares support vector machine (LS-SVM), k-nearest neighbor (KNN), and extreme learning machine (ELM), the partial least squares discriminant analysis (PLS-DA) model has the optimal recognition performance for maize seed harvest years. Six different preprocessing methods, including Savitzky–Golay smoothing (SGS), standard normal variate transformation (SNV), multiplicative scatter correction (MSC), Savitzky–Golay 1 derivative (SG-D1), Savitzky–Golay 2 derivative (SG-D2), and normalization (Norm), were used to improve the quality of the spectra. The Monte Carlo cross-validation uninformative variable elimination (MC-UVE), competitive adaptive reweighted sampling (CARS), bootstrapping soft shrinkage (BOSS), successive projections algorithm (SPA), and their combinations were used to obtain effective wavelengths and decrease spectral dimensionality. The MC-UVE-BOSS-PLS-DA model achieved the classification with an accuracy of 88.75% using 93 features based on Norm preprocessed spectral data. This study showed that the self-designed NIR collection system could be used to identify the harvested years of maize seed.

Keywords: maize seeds; on-line assessment; NIR spectral; effective wavelength selection

1. Introduction

Maize (*Zea mays* L.) is one of the indispensable food crops for human beings and is extensively cultivated worldwide. Maize is a vital raw material for the chemical and medical and health industries, as well as an essential feed source for animal aquaculture and husbandry [1,2]. Moreover, maize has various biological activities [3]. The quality of seed is crucial to agricultural production. High-quality maize seeds not only increase yields and ensure the consistency of plant growth, but are also conducive to using drones to conduct emasculation, pesticide spraying, and other operations [4]. The quality of seed can be determined by its physicochemical properties or germination ability. The nutrition of the maize seeds will be lost with the extension of the storage time, resulting in a low germination rate and weak seedlings [5]. The newly harvested seeds have high vigor, presenting a high germination rate. Therefore, the seedlings of high-quality seeds tend to be strong and healthy.

The traditional methods for inspecting the freshness of maize seeds include manual observation and chemical analysis [6]. The former requires observing the glossiness of the seed surface, while the latter requires soaking the maize seeds in red ink solution and examining the color of the embryo. However, these methods have disadvantages such as being time-consuming and susceptible to subjective influence, making them unsuitable for rapid maize seed assessment. The promotion of single seed sowing technology in China has put forward higher requirements for seed quality. In order to provide high-quality seeds for cultivation, it is necessary to develop a non-destructive, accurate, and rapid method for identifying aged maize seeds before seeding. Therefore, the technique of the on-line assessment of maize seed, which is used for evaluating and sorting maize seed according to quality at high speed, has enjoyed a great deal of attention from international machinery companies and researchers.

With the storage time extension, the internal protein, starch, fat, and cellulose components in seeds will change, resulting in the decline of the gemination rate. These substances are rich in hydrogen-containing groups such as C-H, O-H, and N-H [7], which can be characterized and identified by spectral technologies. Hyperspectral imaging technology has been widely used in seed quality assessment [8–10]. Ambrose et al. [11] used hyperspectral imaging technology to distinguish aged and normal corn seeds. The results showed that the classification accuracy for the calibration and prediction set were 97.6% and 95.6%, respectively. Zhou et al. [12] extracted visible and near-infrared spectral information from hyperspectral images of nine varieties of sweet corn seeds, which were used to build seven classifiers. The support vector machine SVM model developed using feature wavelengths selected by competitive adaptive reweighted sampling (CARS) obtained the optimal classification results with accuracies of 94.86% and 94.07% for germ and endosperm sides, respectively. Zhang et al. [13] applied the fused features of near-infrared hyperspectral imaging to establish classification models to identify the varieties of coated maize kernels and achieved a classification accuracy of over 90%. Wang et al. [6] identified maize seeds harvested in different years by using spectra from the endosperm side, embryo side, and both sides. Fan et al. [14] constructed a YOLOv7 model based on the spectral and image information of maize seeds, obtaining a classification accuracy of 99.7% for maize seed germination. The studies stated above indicated that hyperspectral imaging technology is a powerful tool for maize seed quality assessment. However, the time-consuming acquisition of hyperspectral images with high dimensionality and the high cost of spectral cameras pose great challenges for fast and real-time assessment, which has hindered the development of on-line inspection systems. Therefore, hyperspectral imaging technologies have mostly been used in lab conditions for fundamental research.

In comparison to hyperspectral imaging technology, the advantages of traditional NIR spectroscopy technology are convenience, low cost, high inspection efficiency, and multi-index measurement capability [15]. Recent improvements in miniaturization and mathematical tools has led to a wide application of NIR spectroscopy technology in the quality analysis of agricultural products, such as fruit and seeds [16]. In terms of fruit quality evaluation, NIR spectroscopy technology has been used for fruit sorting by detecting fruit sugar content and internal bruising in a commercial packing line with a speed of five or more fruits per second. For seed quality assessment, Wang et al. [17] used a self-designed NIR spectral collection device to acquire near-infrared spectra of normal maize seeds, artificially aged maize seeds, and heat-damaged maize seeds, which were utilized to establish a classification model using a partial least squares discriminant analysis (PLS-DA) algorithm. The classification accuracy of the prediction set was higher than 95%. Liu et al. [18] built a single-fiber spatially resolved device using visible/near-infrared spectroscopy to inspect maize seed vigor and achieved good prediction results. However, the studies stated above still involve the static assessment of maize seeds. Up to now, far too little attention has been paid to the on-line application of NIR spectroscopy for seed inspection. In a recent study, Wang et al. [19] applied self-designed on-line NIR spectroscopy equipment to evaluate insect-infested seeds and achieved a classification accuracy of 0.89 and 0.83 using full wavelengths or feature wavelengths, respectively. In addition, the protein content of maize seeds was predicted precisely, with an RPD of 2.08 and 2.11 for grain protein content and absolute grain protein content prediction, respectively. Similar results were reported for soybean seeds [20]. These previous studies demonstrated the enormous potential of NIR spectroscopy technology for the on-line assessment of maize seed quality.

Therefore, the aim of this study was to use NIR spectroscopy technology for the on-line discrimination of maize seeds harvested in different years. The specific objectives were (1) to acquire the near-infrared spectra of maize seeds using an on-line NIR device; (2) to establish classification models to identify the harvest year of maize seeds; and (3) to optimize the model using wavelength selection methods.

2. Materials and Methods

2.1. Preparation of Samples

JINGKE 968 is one of the high-quality varieties of maize seeds that has been widely planted in China. The JINGKE 968 maize seeds were harvested in 2018 and 2021 in the Gansu province of China. The seeds were sent to the lab and placed under room conditions with 60% relative humidity and a temperature of 24 °C. A total of 200 seeds, with 100 for each harvest year, were selected randomly to conduct standard germination tests at 25 °C following the guidelines of the International Seed Testing Association (ISTA) before the experiment. The normal seedlings were counted after 7 days, and the germination rate of the maize seeds harvested in 2018 and 2021 were 76% and 91%, respectively. A total of 400 maize seed samples without surface defects were then utilized in this study in January 2022; half of the samples were harvested in 2018, and the rest were harvested in 2021.

Table 1 shows the partition of the sample sets for establishing classification models. The training set consisted of 320 maize seeds that were selected randomly, including 160 samples harvested in 2018 and 160 samples harvested in 2021. The prediction set consisted of 80 samples, and the number of maize seeds harvested in 2018 and harvested in 2021 was 40, respectively. In addition, a label value (harvested in 2018 = −1 and harvested in 2021 = 1) was assigned to each sample.

Table 1. Class assignment and partition of sample sets.

Sample Class	No. of Samples	Training Set	Prediction Set	Assigned Class
Seeds harvested in 2018	200	160	40	−1
Seeds harvested in 2021	200	160	40	1

2.2. Spectral Data Acquisition

The NIR spectra were collected using an on-line spectra acquisition device (Ambrose, 2016) (Figure 1). The on-line NIR spectra acquisition device was composed of five parts: an optical fiber, a photoelectric sensor, a light source, a glass tube, and a spectrometer (Ocean optics, Orlando, FL, USA, NIR QUEST, spectra range of 899–1715 nm). The specific process of data collection is described as follows: a single seed triggers the photoelectric sensor as it slides down the tube, which then sends a signal to the spectrometer to collect a spectrum (Sraw) using a specific integration time. The spectrum is collected as the kernel travels down the tube and is displayed and stored in the host computer. Data collection is carried out through a preset integration time. The integration time in the preliminary experiment was set to 100 ms after many attempts.

Typically, the spectral data need to be corrected. In this experiment, data correction was conducted according to the formula below:

$$S_C = \frac{S_{raw}}{S_{white}} \tag{1}$$

where S_c was the corrected spectra, and S_{white} was the white reference acquired by turning on the light source together with no seed in the tube.

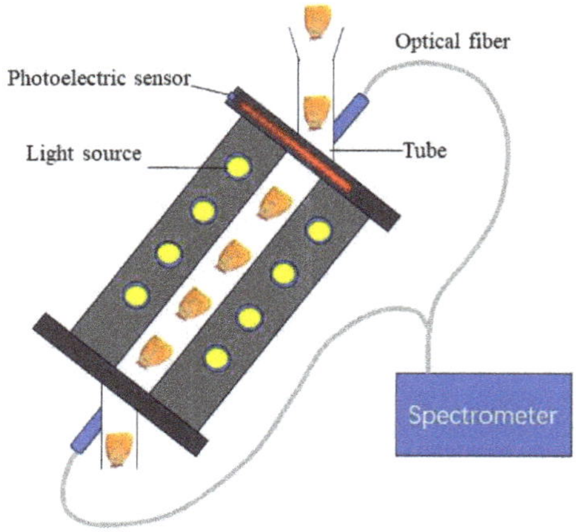

Figure 1. Schematic of the on-line spectra acquisition device.

2.3. Pretreatment of Spectra

The acquired spectral signal not only includes effective chemical information reflecting the samples' attributes, but also contains noise generated by the measurement environment and equipment. Thus, it is necessary to improve the quality of the raw spectral data before modeling [21]. In this study, six commonly used preprocessing methods were applied for data processing, including Savitzky–Golay smoothing (SGS), standard normal variate transformation (SNV), multiplicative scatter correction (MSC), Savitzky–Golay 1 derivative (SG-D1), Savitzky–Golay 2 derivative (SG-D2), and normalization (Norm). SGS can reduce noise by averaging several spectral points and retaining important features of the spectral curve [22]. The SNV and MSC can eliminate the influence of scattering from the sample surface on the spectral information. SG-D1 derivative with second-order filtering and a smoothing window of 11 points, and SG-D2 with a points second-order filtering smoothing window of 25 were used to identify the overlapping peaks of the original spectra and to reduce interference [23]. For the SGS method, a smaller window size could better preserve spectral features, but may not significantly decrease noise. In contrast, a larger window size is likely to blur significant spectral features during noise elimination. The smooth window was set to 9 after previous experience and many attempts. The Norm method normalizes the data using the Euclidean norm, and can reduce spectral differences caused by slight optical path variations [24]. Therefore, the performance of the above pretreatment methods was compared by evaluating the classification results of the built models.

2.4. Effective Wavelength Selection Algorithm

The spectral data, after preprocessing, contains a remarkable amount of uninformative and redundant spectral variables. In addition, there is multicollinearity between the informative variables. The effective wavelength selection method can reduce data dimensions and preserve important information, thereby improving model performance. In this study, four wavelength selection procedures including bootstrapping soft shrinkage (BOSS), CARS, Monte Carlo cross-validation uninformative variable elimination (MC-UVE), and the successive projections algorithm (SPA) were adopted. Moreover, the combination of MC-UVE-BOSS and MC-UVE-SPA algorithms has been proven to be a powerful tool and was consequently also used [25].

MC-UVE is a modified variable selection method proposed based on the MC (Monte Carlo) and UVE methods, and is commonly used to eliminate irrelevant information variables [26]. In the MC-UVE, the effectiveness of each variable was evaluated based on its stability defined as the ratio of the mean to the standard deviation of regression coefficients, which was calculated from PLS-DA models based on N subsets randomly sampled using the Monte Carlo approach. The specific details of the MC-UVE algorithm can be referred to in Cai's paper [27]. The performance of the MC-UVE method is superior to the original UVE method, which has been proven in some previous studies. However, the variables selected by MC-UVE are still redundant to a certain extent; it is therefore necessary to further extract effective variables by combining the approach with other algorithms.

BOSS is a novel variable selection method for reducing the collinearity of spectral data [28], which is derived from the idea of model population analysis and weighted bootstrap sampling. The weight of the variables is obtained by the absolute value of the regression coefficient of numerous PLS sub-models. The weights of variables were updated stepwise using the weighted bootstrap sampling method, resulting in a soft shrinkage of variables. The best variable set is then determined by the sub-model with minimum prediction error.

The CARS algorithm was proposed to eliminate the influence of uninformative variables and improve the performance of the established model [29]. In this algorithm, the absolute values of regression coefficients of the PLS model were used for evaluating the importance of each variable. This algorithm starts by developing a PLS-DA model with full wavelengths. The importance of each variable is evaluated by calculating the absolute values of regression coefficients at each iteration, followed by the elimination of variables using adaptive reweighted sampling and the exponential decreasing function. At each iteration, classification performance is evaluated using a subset of variables at each iteration; the subset with the lowest classification accuracy is determined as containing the best variables. For the CARS effective variables selection, the number of Monte Carlo sampling was set to 100, and 10-fold cross-validation was used.

SPA is mostly used to eliminate redundant information and reduce the collinearity of spectral data [30]. The SPA algorithm has two phases. First, one wavelength is selected, and its projection on the remaining wavelengths is calculated during each cycle. The wavelength with the maximum projection value is selected as the prospective effective wavelength. Then, the optimal wavelength can be selected based on the smallest RMSECV of the MLR model. Using SPA alone is usually not very effective, so it needs to be combined with other wavelength selection algorithms.

Combination wavelength selection algorithms were also considered in this study. Taking the example of MC-UVE-SPA, MC-UVE was firstly used to select a set of potential wavelengths, followed by further wavelength extraction using SPA. MC-UVE-BOSS is another combination method that uses MC-UVE and BOSS sequentially for selecting the effective wavelengths.

2.5. Model Construction

To gain a reliable and accurate classifier, four classification models, including least squares support vector machine (LS-SVM), partial least squares discriminant analysis (PLS-DA), k-nearest neighbor (KNN), and extreme learning machine (ELM), were built and compared.

LS-SVM is an evolutionary algorithm based on the standard SVM proposed by Suykens and Vandewalle [31], and is capable of quickly resolving linear and nonlinear multivariable analysis [32]. Compared with standard SVM, it maps input features into a high-dimensional space, reducing the complexity of calculation and obtaining the optimal solution by calculating the partial differentiation of each feature using a Lagrange multiplier. As a well-liked kernel function, the radial basis function (RBF) kernel function was chosen as it has advantages in handling nonlinear relationships between spectral data and the target category [33]. A 10-fold cross-validation coupled with grid-search method was employed to search the optimal parameter values of regularization parameter gamma (γ) and kernel function parameter sig2 ($\sigma 2$). PLS-DA is a supervised linear classifier based on PLS regression, which predicts the membership of the dataset by maximizing the covariance between the data matrix X (the maize spectral matrix) and the categorical Y matrix (sample labels). The predicted results obtained by PLS are continuous values, not strictly sample labels. Therefore, it is necessary to set a threshold to determine the prediction results of the model. The threshold was set as 0 in this study. In addition, a 10-fold cross validation was employed on the training dataset, followed by the determination of optimal latent variables according to the minimum classification accuracy. KNN is a statistical method for pattern recognition by tuning the hyperparameters, including distance metric and number of neighbors (K), which were optimized automatically to minimize 10-fold cross-validation loss [34]. ELM is a fast machine-learning algorithm proposed by Huang et al. [35], which shows high efficiency, rapidity, and good generalization performance in analyzing large-scale spectral data. The number of hidden neurons in ELM was increased from 10 to 150 with a step of 10, followed by the determination of the optimal hidden neurons according to the classification results.

2.6. Software

In this study, the spectral data analysis was performed in MATLAB 2018b (The Math Works, Natick, MA, USA) with the assistance of libpls toolbox [36]. Origin 2018 (Origin Lab Corporation, Northampton, MA, USA) was applied to construct figures.

3. Results and Discussion

3.1. Features of Spectra

The tested sample is irradiated with near-infrared radiation, with transmitted or reflected radiation measured by a spectrometer. The spectral characteristics change through scattering and absorption processes and can reflect the structure and content information of X-H groups (C–H, O–H and N–H) in the sample [37]. Figure 2a shows the raw average spectra (solid line) and deviation distribution (shadow region) of the maize seeds. The blue part represents the seeds harvested in 2021 and the red part represents the seeds harvested in 2018. It can be clearly seen that the spectra of new and aged maize seeds have similar intensity and trend. The prominent absorption peak and valley were located at around 1110 nm, 1200 nm, and 1300 nm. The spectral intensity peak and valley at about 1110 nm and 1200 nm might be relative to the second overtone of C-H [38], while the peak at around 1300 nm is possibly caused by the combination of the fundamental amide vibrations and the first overtone of Amide B [39]. It is difficult to directly distinguish seeds in different harvest years by only depending on spectral intensity. Therefore, it is necessary to establish an appropriate classification model using machine learning and chemometric methods.

3.2. Classification Results Using Full Spectra

Figure 2b–g represent the spectra pretreated by SGS, SNV, MSC, SG-D1, SG-D2, and Norm, respectively. It can be seen that SGS effectively reduced the noise of the spectral data, while MSC and SNV eliminated the influence of scattering. However, the SG-D1 and SG-D2 preprocessing methods did not improve the spectral quality. It is worth noting that the spectral curves of the seeds harvested in different years after Norm pretreatment showed more differences. However, there were still obvious overlaps between the two types of spectra.

In order to determine the optimal pretreatment method, the spectral data preprocessed by SGS, SNV, MSC, SG-D1, SG-D2, and Norm were used as inputs to establish PLS-DA, LS-SVM, KNN, and ELM models. The key parameters (γ, σ^2 for LS-SVM, LVs for PLS-DA, K for KNN, and hidden neurons for ELM) and classification results in the training set and the prediction set are summarized in Table 2.

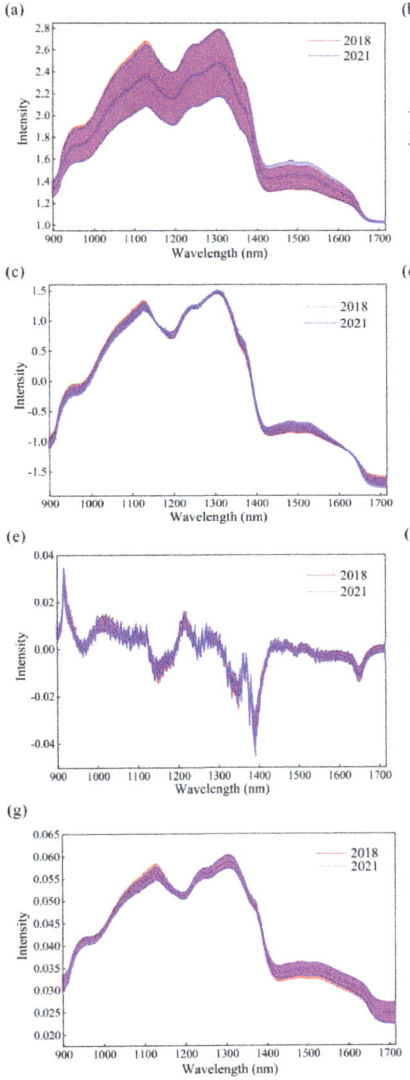

Figure 2. The average spectra (solid line) with their standard deviation (shadow region). (**a**) The original spectra curves; (**b–g**) stand for the spectra preprocessed by SGS, SNV, MSC, SG-D1, SG-D2, and Norm.

In comparison with the inspection results obtained by the raw spectra, the classification results yielded by the preprocessed spectra were improved to different degrees except for the model built with derivative spectra. Compared with other preprocessing methods, the spectra pretreated by SGS obtained the best results regardless of the classification methods. It can be seen that the result of the four classification models represented significant differences. Basically, PLS-DA obtained the best performance in inspecting seeds from different harvest years by comparing the four classification methods. When the number of LVs was equal to 7, 8, 8, 7, 5, and 10, the best discrimination results were obtained with spectral data preprocessed by SGS, SNV, MSC, SG-D1, SG-D2, and Norm preprocessing, respectively. The PLS-DA model based on spectra pretreated by Norm yielded relatively higher accuracy in discriminating between kernels from different harvest years, with classification accuracies of 96.56% and 91.25%, respectively for the training and prediction sets. Therefore PLS-DA will be used for further analysis. The above classification results indicated that near-infrared spectroscopy technology can achieve the on-line recognition of new and aged maize seeds. Seed quality properties, such as freshness, damage, and internal components, can be evaluated simultaneously through an on-line NIR device and multiple NIR calibration models, requiring high computing performance and shorter measurement time, especially when the seeds are inspected at a fast speed. In addition, the full spectra contain redundancy and multicollinearity variables among contiguous wavebands, thus causing time-consuming calibration processes and hindering the computing speed. Thus, it is necessary to select effective wavelengths to simplify calibration models.

Table 2. The discrimination results for seeds harvested years by different classification models and preprocessing methods using full wavelengths.

Models	Preprocessing Methods	(γ, σ2)	LVs	K	Hidden Neurons	Classification Accuracy of Training Set (%)			Classification Accuracy of Prediction Set (%)		
						2018	2021	Total	2018	2021	Total
LS-SVM	Raw	59,845.85; 14,662.89				100.00	99.38	99.69	82.50	87.50	85.00
	SNV	18,298.45; 11,9622.43				98.75	100.00	99.38	82.50	85.00	83.75
	MSC	1280.47; 43,116.50				95.00	95.63	95.31	85.00	87.50	86.25
	Norm	4062.37; 7016.89				100.00	100.00	100.00	92.50	87.50	90.00
	SGS	9440.09; 317.76				100.00	100.00	100.00	97.50	80.00	88.75
	S-G-1st	91.37; 4736.06				93.13%	92.50	92.81	77.50	80.00	78.75
	S-G-2nd	10,870.69; 52,589.62				97.50	97.50	97.50	77.50	67.50	72.50
PLSDA	Raw		9			98.13	97.50	97.81	87.50	85.00	86.25
	SNV		7			98.13	93.75	95.94	90.00	90.00	90.00
	MSC		8			98.75	97.50	98.13	90.00	85.00	87.50
	Norm		8			98.13	95.00	96.56	90.00	92.50	91.25
	SGS		7			86.88	87.50	87.19	90.00	85.00	87.50
	S-G-1st		5			96.88	98.13	97.50	70.00	75.00	72.50
	S-G-2nd		10			100.00	100.00	100.00	70.00	77.50	73.75
KNN	Raw			4		87.50	69.38	78.44	77.50	62.50	70.00
	SNV			10		73.75	69.38	71.56	65.00	67.50	66.25
	MSC			5		74.38	73.75	74.06	72.50	67.50	70.00
	Norm			8		93.75	92.50	93.13	90.00	85.00	87.50
	SGS			6		93.75	95.63	94.69	90.00	82.50	86.25
	S-G-1st			1		100.00	100.00	100.00	65.00	70.00	67.50
	S-G-2nd			5		73.13	75.00	74.06	67.50	57.50	62.50
ELM	Raw				140	85.13	83.70	84.41	73.40	68.75	71.08
	SNV				60	81.25	81.25	81.25	77.50	85.00	81.25
	MSC				100	86.25	82.50	84.38	82.50	80.00	81.25
	Norm				90	100.00	100.00	100.00	87.50	92.50	90.00
	SGS				110	98.13	99.38	98.75	92.50	82.50	87.50
	S-G-1st				50	80.00	77.50	78.75	77.50	82.50	80.00
	S-G-2nd				60	80.63	78.75	79.69	72.50	70.00	71.25

3.3. Classification Results Based on Effective Wavelengths

At first, four groups of effective wavelengths selected by MC-UVE, CARS, BOSS, and SPA were separately used as the inputs to establish PLS-DA models for identifying maize seeds harvested in different years. The discrimination results of the PLS-DA models built with effective wavelengths selected by different wavelength selection methods are shown in Table 3.

Table 3. The classification results for PLS-DA models using different wavelength selection methods.

Model	No. of Wavelengths	LVs	Classification Accuracy of Training Set (%)			Classification Accuracy of Prediction Set (%)		
			2018	2021	Total	2018	2021	Total
MC-UVE-PLS-DA	140	8	96.88	96.25	96.56	90.00	82.50	86.25
CARS-PLS-DA	48	7	93.13	92.50	92.81	82.50	80.00	81.25
BOSS-PLS-DA	107	7	96.88	95.00	95.94	80.00	77.50	78.75
SPA-PLS-DA	6	6	61.25	71.88	66.56	67.50	72.50	70.00
MC-UVE-SPA-PLS-DA	4	4	59.38	73.13	66.25	62.50	72.50	67.50
MC-UVE-BOSS-PLS-DA	93	8	95.63	94.38	95.00	85.00	92.50	88.75

When the numbers of LVs were equal to 8, 7, 7, and 6, the best classification results were acquired by MC-UVE-PLS-DA, CARS-PLS-DA, BOSS-PLS-DA, and SPA-PLS-DA, respectively, with classification accuracies of 86.25, 81.25, 78.75, and 70.00%. The classification accuracy of CARS-PLS-DA and BOSS-PLS-DA models on the prediction set were 81.25% and 78.75%, respectively. It can be clearly seen that SPA and MC-UVE-SPA could effectively compress data, with the selected number of wavelengths of 6 and 4, respectively, but the classification accuracy of the prediction set was smaller than 70%. The MC-UVE-PLS-DA model's performance was better than the CARS-PLS-DA model, with success rates of 90% and 82.5% for aged and new seeds, respectively. MC-UVE-PLS-DA obtained more promising classification results for seeds from different harvest years. However, the number of selected wavelengths was a little higher, and it can be reduced further by using other variable selection methods. MC-UVE-SPA-PLS- DA and MC-UVE-BOSS-PLS-DA models were then developed, respectively. The MC-UVE-BOSS-PLS-DA model obtained the optimal performance with an accuracy of 88.75% for the prediction set. Compared with the full-spectrum PLS-DA model, the MC-UVE-BOSS-PLS-DA model reduced the classification accuracy of prediction sets by 2.5%. However, the MC-UVE-BOSS-PLS-DA model only used 93 wavelengths (18% of full spectral data). It is worth noting that the MC-UVE-BOSS-PLS-DA model has a prediction accuracy of 92.5% for new seeds, which is the same accuracy as the full-spectrum PLS-DA model. The results demonstrated that the NIR spectra collected from the on-line acquisition device could establish a classification model for detecting maize seeds from different harvest years. The optimal models developed by 93 effective wavelengths were more suitable for on-line application.

3.4. Wavelength Selection Analysis of the Optimal Models

As discussed in Section 3.4, MC-UVE-BOSS, a combination of MC-UVE and BOSS, was an effective wavelength selection method in the identification of maize seeds from different harvest years. Figure 3 shows the wavelength selection process of the MC-UVE-BOSS method. In the process of MC-UVE, 320 samples in the training set were applied as the input for the MC-UVE algorithm for selecting effective wavelengths. The stability of each wavelength was calculated and is shown in Figure 3a. The absolute values of stability were sorted in a descending manner. A number (N) of the variables were then selected from the sorted stabilities as informative wavelengths, corresponding to the stability of the Nth variable as the cutoff value. As the N increased from 20 to 500 with a step of 20, a set of PLS models were developed and applied to the prediction set, followed by the calculation of RMSEP values. Figure 3b represents the RMSEP variation of the PLS calibration model with the increase in the number of variables. It can be seen that the lowest RMSEP value was acquired when the number of the selected wavelengths was 140, corresponding to the wavelength's stability outside the cut-off line (Figure 3a, red dotted line).

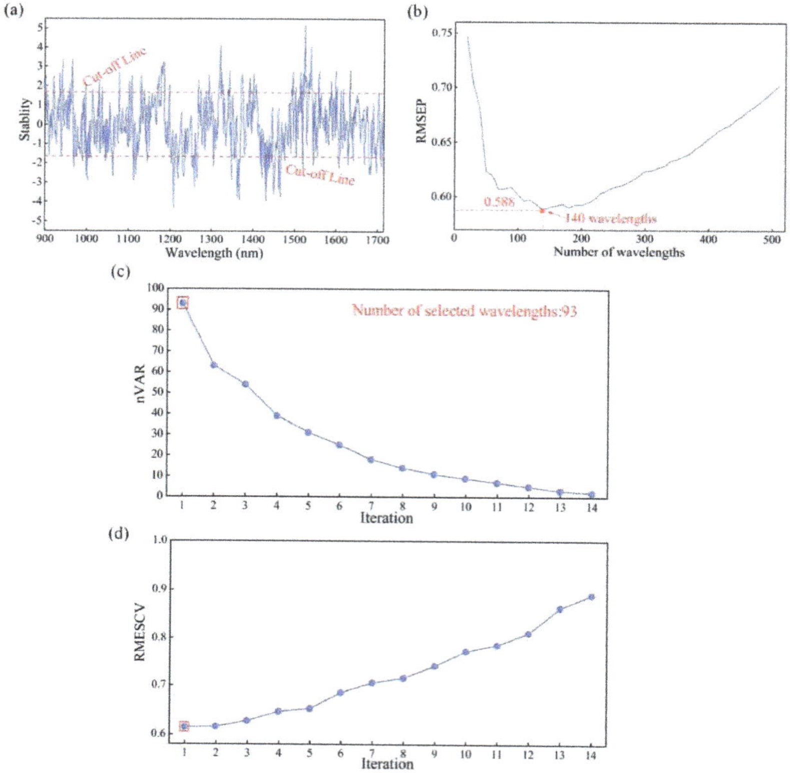

Figure 3. Feature wavelength selection for seeds harvested in different years by MC-UVE-BOSS from NIR spectra after pretreatment by Norm. (**a**) The stability of full wavelengths, (**b**) RMSEP variation with the number of selected wavelengths, (**c**) the variation in the number of variables (nVAR) and (**d**) RMSECV in sub-models in each iteration of BOSS.

After wavelength selection by the MC-UVE method, the BOSS method was used to further extract effective information. In the process of BOSS, 140 wavelengths selected by MC-UVE were used to generate a large number of subsets in variable space, which were used to build PLS sub-models. The sub-models with a smaller RMSECV were extracted (10%), obtaining the regression coefficients and then new weights for the variables. Based on the new weights, a weighted bootstrap sampling (WBS) method was used to generate a new subset. The number of selected variables decreased gradually and reached 1 after 14 iterations (Figure 3c). Meanwhile, the RMSECV in the sub-models increased during the iterations. It can be seen that the minimum RMSECV value was obtained at the first iteration, with the number of selected wavelengths of 93. The wavelengths selected by MC-UVE-BOSS were mainly located at 1000–1020, 1260–1300, 1420–1480, and 1500–1520 nm (Figure 4). The wavelengths near 1200 nm were associated with the attributes of the secondary stretching vibration of C–H bonds in starch, proteins, or lipids [40]. The selected wavelengths around 1400–1500 nm were the absorption wavelengths of the first overtone of O-H and N-H stretching, which were related to the water and protein contents of maize seeds [41]. The wavelengths near 1700 nm were relevant to the absorption wavelength of the first overtone of C-H, which was caused by the strong absorption of fatty acids contained in the maize kernels [42]. This indicates that the selected wavelengths contain rich information related to the nutritional composition of maize seeds.

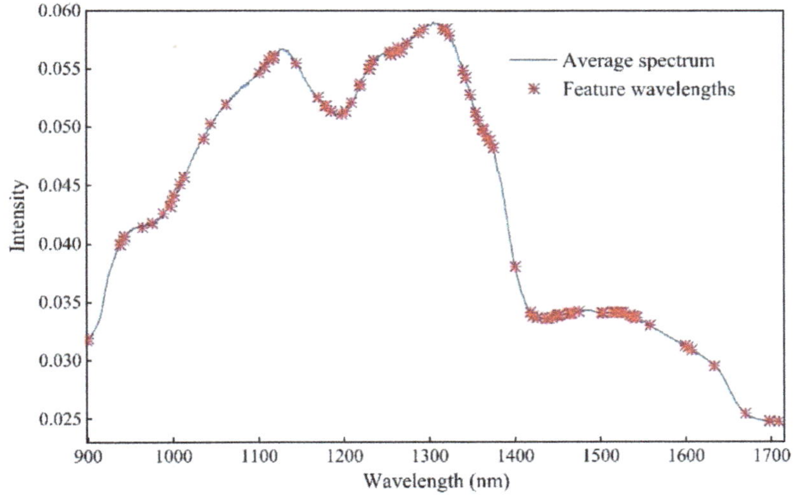

Figure 4. The distribution of feature wavelengths selected by MC-UVE-BOSS.

4. Conclusions

An on-line NIR device with a spectral range of 899–1715 nm was successfully utilized for the non-destructive and rapid classification of maize seeds harvested in different years. Four classic classification models, including PLS-DA, LS-SVM, KNN, and ELM, were established based on full spectra (512 wavelengths). The results indicated that the PLS-DA was more suitable for distinguishing maize seeds harvested in different years. The norm method has been determined as the optimal preprocessing method based on the PLS-DA model. In order to further improve the performance of the model, different wavelength selection methods were adopted. The classification accuracy of MC-UVE-BOSS-PLS-DA for the maize seed harvest year was 88.75%, which was slightly lower than the 91.25% achieved by the full-spectrum PLS-DA model. However, the number of wavelengths used by the MC-UVE-BOSS-PLSDA model was only about 1/6 of the full-spectrum PLS-DA model. Moreover, the classification accuracy of new seeds by the MC-UVE-BOSS-PLSDA model was equal to the full-spectrum PLS-DA model. In the next step, improvements in the classification accuracy will be studied by optimizing equipment and spectral data methods, thereby promoting the on-line application of this technology.

Author Contributions: Conceptualization, Q.Z. and J.K.; methodology, Y.Z. and S.F.; formal analysis, Y.Z.; investigation, Y.Z. and B.Z.; resources, M.Z. and J.K.; data curation, Q.Z. and S.F.; writing—original draft preparation, Y.Z.; writing—review and editing, Q.Z. and J.K. All authors have read and agreed to the published version of the manuscript.

Funding: This work was supported by the Natural Science Foundation of Jiangsu Province (Grant No. BK20230548), the National Natural Science Foundation of China (Grant No. 62277001), the Open Project of China Food Flavor and Nutrition Health Innovation Center (Grant No. CFC2023B-024), the Project of Faculty of Agricultural Equipment of Jiangsu University (Grant No. NGXB20240203), the Project of Beijing Municipal University Teacher Team Construction Support Plan (Grant No. BPHR 20220104), and the Jiangsu Postdoctoral Sustentation Fund (Grant No. 2020Z220).

Institutional Review Board Statement: Not applicable.

Data Availability Statement: The original contributions presented in the study are included in the article, further inquiries can be directed to the corresponding author.

Conflicts of Interest: The authors declare no conflicts of interest.

References

1. Guo, D.; Zhu, Q.; Huang, M.; Guo, Y.; Qin, J. Model updating for the classification of different varieties of maize seeds from different years by hyperspectral imaging coupled with a pre-labeling method. *Comput. Electron. Agric.* **2017**, *142*, 1–8. [CrossRef]
2. Wang, B.; Deng, J.; Jiang, H. Markov transition field combined with convolutional neural network improved the predictive performance of near-infrared spectroscopy models for determination of aflatoxin B1 in maize. *Foods* **2022**, *11*, 2210. [CrossRef] [PubMed]
3. Cui, Z.; Xie, J.; Wang, N.; Pan, L.; Song, T.; Zhang, Y.; Xu, X. NIR spectrometer for non-destruction measurement of oil contents in a corn seed. *Spectrosc. Spectr. Anal.* **2005**, *25*, 1807–1809.
4. Ahmad, F.; Adeel, M.; Qiu, B.; Ma, J.; Shoaib, M.; Shakoor, A.; Chandio, F.A. Sowing uniformity of bed-type pneumatic maize planter at various seedbed preparation levels and machine travel speeds. *Int. J. Agric. Biol. Eng.* **2021**, *14*, 165–171. [CrossRef]
5. Mcdonald, M.B. Seed deterioration: Physiology, repair and assessment. *Seed Sci. Technol.* **1999**, *27*, 177–237.
6. Wang, Z.; Huang, W.; Tian, X.; Long, Y.; Li, L.; Fan, S. Rapid and Non-destructive Classification of New and Aged Maize Seeds Using Hyperspectral Image and Chemometric Methods. *Front. Plant Sci.* **2022**, *13*, 849495. [CrossRef] [PubMed]
7. Zhang, J.; Fang, W.; Xu, C.; Xiong, A.; Zhang, M.; Goebel, R.; Bo, G. Current Optical Sensing Applications in Seeds Vigor Determination. *Agronomy* **2023**, *13*, 1167. [CrossRef]
8. Wang, Y.; Ren, Z.; Li, M.; Yuan, W.; Zhang, Z.; Ning, J. pH indicator-based sensor array in combination with hyperspectral imaging for intelligent evaluation of withering degree during processing of black tea. *Spectr. Acta Pt. A-Mol. Biomol. Spectr.* **2022**, *271*, 120959. [CrossRef]
9. Zhang, L.; An, D.; Wei, Y.; Liu, J.; Wu, J. Prediction of oil content in single maize kernel based on hyperspectral imaging and attention convolution neural network. *Food Chem.* **2022**, *395*, 133563. [CrossRef]
10. Sun, J.; Zhang, L.; Zhou, X.; Yao, K.; Tian, Y.; Nirere, A. A method of information fusion for identification of rice seed varieties based on hyperspectral imaging technology. *J. Food Process Eng.* **2021**, *44*, 13797. [CrossRef]
11. Ambrose, A.; Kandpal, L.M.; Kim, M.S.; Lee, W.H.; Cho, B.K. High speed measurement of corn seed viability using hyperspectral imaging. *Infrared Phys. Technol.* **2016**, *75*, 173–179. [CrossRef]
12. Zhou, Q.; Huang, W.; Fan, S.; Zhao, F.; Liang, D.; Tian, X. Non-destructive discrimination of the variety of sweet maize seeds based on hyperspectral image coupled with wavelength selection algorithm. *Infrared Phys. Technol.* **2020**, *109*, 103418. [CrossRef]
13. Zhang, C.; Zhao, Y.; Yan, T.; Bai, X.; Xiao, Q.; Gao, P.; Li, M.; Huang, W.; Bao, Y.; He, Y. Application of near-infrared hyperspectral imaging for variety identification of coated maize kernels with deep learning. *Infrared Phys. Technol.* **2020**, *111*, 103550. [CrossRef]
14. Fan, Y.; An, T.; Wang, Q.; Yang, G.; Huang, W.; Wang, Z.; Zhao, C.; Tian, X. Non-destructive detection of single-seed viability in maize using hyperspectral imaging technology and multi-scale 3D convolutional neural network. *Front. Plant Sci.* **2023**, *14*, 1248598. [CrossRef] [PubMed]
15. Rossi, G.B.; Lozano, V.A. Simultaneous determination of quality parameters in yerba mate (Ilex paraguariensis) samples by application of near-infrared (NIR) spectroscopy and partial least squares (PLS). *LWT-Food Sci. Technol.* **2020**, *126*, 109290. [CrossRef]
16. Zareef, M.; Chen, Q.; Hassan, M.M.; Arslan, M.; Hashim, M.M.; Ahmad, W.; Kutsanedzie, F.Y.H.; Agyekum, A.A. An Overview on the Applications of Typical Non-linear Algorithms Coupled with NIR Spectroscopy in Food Analysis. *Food Eng. Rev.* **2020**, *12*, 173–190.
17. Wang, Y.; Peng, Y.; Zhuang, Q.; Zhao, X. Feasibility analysis of NIR for detecting sweet corn seeds vigor. *J. Cereal Sci.* **2020**, *93*, 102977. [CrossRef]
18. Liu, W.; Luo, B.; Kang, K.; Xia, Y.; Zhang, H. Non-destructive detection of single corn seed vigor based on visible/near-infrared spatially resolved spectroscopy combined with chemometrics. *Spectr. Acta Part A-Mol. Biomol. Spectr.* **2024**, *312*, 124089. [CrossRef]
19. Wang, Z.; Huang, W.; Li, J.; Liu, S.; Fan, S. Assessment of protein content and insect infestation of maize seeds based on on-line near-infrared spectroscopy and machine learning. *Comput. Electron. Agric.* **2023**, *211*, 107969. [CrossRef]
20. Agelet, L.E.; Armstrong, P.R.; Clariana, I.R.; Hurburgh, C.R. Measurement of Single Soybean Seed Attributes by Near-Infrared Technologies. A Comparative Study. *J. Agric. Food Chem.* **2012**, *60*, 8314–8322. [CrossRef]
21. Rinnan, Å.; Berg, F.v.d.; Engelsen, S.B. Review of the most common pre-processing techniques for near-infrared spectra. *Trends Anal. Chem.* **2009**, *28*, 1201–1222. [CrossRef]
22. Pasquini, C. Near infrared spectroscopy: A mature analytical technique with new perspectives A review. *Anal. Chim. Acta* **2018**, *1026*, 8–36. [CrossRef]
23. Fan, S.; Li, J.; Xia, Y.; Tian, X.; Guo, Z.; Huang, W. Long-term evaluation of soluble solids content of apples with biological variability by using near-infrared spectroscopy and calibration transfer method. *Postharvest Biol. Technol.* **2019**, *151*, 79–87. [CrossRef]
24. Nicolaï, B.M.; Beullens, K.; Bobelyn, E.; Peirs, A.; Saeys, W.; Theron, K.I.; Lammertyn, J. Nondestructive measurement of fruit and vegetable quality by means of NIR spectroscopy: A review. *Postharvest Biol. Technol.* **2007**, *46*, 99–118.
25. Zhang, Y.; Wang, Z.; Tian, X.; Yang, X.; Cai, Z.; Li, J. Online analysis of watercore apples by considering different speeds and orientations based on Vis/NIR full-transmittance spectroscopy. *Infrared Phys. Technol.* **2022**, *122*, 104090. [CrossRef]
26. Cai, W.; Li, Y.; Shao, X. A variable selection method based on uninformative variable elimination for multivariate calibration of near-infrared spectra. *Chemom. Intell. Lab. Syst.* **2008**, *90*, 188–194. [CrossRef]

27. Li, J.B.; Huang, W.Q.; Chen, L.P.; Fan, S.X.; Zhang, B.H.; Guo, Z.M.; Zhao, C.J. Variable Selection in Visible and Near-Infrared Spectral Analysis for Noninvasive Determination of Soluble Solids Content of 'Ya' Pear. *Food Anal. Meth.* **2014**, *7*, 1891–1902. [CrossRef]
28. Deng, B.C.; Yun, Y.H.; Cao, D.S.; Yin, Y.L.; Wang, W.T.; Lu, H.M.; Luo, Q.Y.; Liang, Y.Z. A bootstrapping soft shrinkage approach for variable selection in chemical modeling. *Anal. Chim. Acta* **2016**, *908*, 63–74. [CrossRef]
29. Yun, Y.H.; Li, H.D.; Deng, B.C.; Cao, D.S. An overview of variable selection methods in multivariate analysis of near-infrared spectra. *TrAC Trends Anal. Chem.* **2019**, *113*, 102–115.
30. Sun, J.T.; Ma, B.X.; Dong, J.; Zhu, R.G.; Zhang, R.Y.; Jiang, W. Detection of internal qualities of hami melons using hyperspectral imaging technology based on variable selection algorithms. *J. Food Process Eng.* **2017**, *40*, 12496. [CrossRef]
31. Suykens, J.A.K.; Vandewalle, J. Least Squares Support Vector Machine Classifiers. *Neural Process. Lett.* **1999**, *9*, 293–300. [CrossRef]
32. Wu, D.; Shi, H.; Wang, S.; He, Y.; Bao, Y.; Liu, K. Rapid prediction of moisture content of dehydrated prawns using online hyperspectral imaging system. *Anal. Chim. Acta* **2012**, *726*, 57–66. [CrossRef] [PubMed]
33. Cheng, J.H.; Sun, D.W. Rapid and non-invasive detection of fish microbial spoilage by visible and near infrared hyperspectral imaging and multivariate analysis. *LWT-Food Sci. Technol.* **2015**, *62*, 1060–1068. [CrossRef]
34. Cai, Z.; Huang, W.; Wang, Q.; Li, J. Detection of early decayed oranges by structured-illumination reflectance imaging coupling with texture feature classification models. *Front. Plant Sci.* **2022**, *13*, 952942. [CrossRef] [PubMed]
35. Huang, G.B.; Zhu, Q.Y.; Siew, C.K. Extreme learning machine: Theory and applications. *Neurocomputing* **2006**, *70*, 489–501. [CrossRef]
36. Li, H.; Xu, Q.; Liang, Y. libPLS: An integrated library for partial least squares regression and linear discriminant analysis. *Chemometr. Intell. Lab. Syst.* **2018**, *176*, 34–43. [CrossRef]
37. Badaro, A.T.; Morimitsu, F.L.; Ferreira, A.R.; Clerici, M.T.P.S.; Barbin, D.F. Identification of fiber added to semolina by near infrared (NIR) spectral techniques. *Food Chem.* **2019**, *289*, 195–203. [CrossRef]
38. Marques, A.S.; Castro, J.N.F.; Costa, F.J.; Neto, R.M.; Lima, K.M.G. Near-infrared spectroscopy and variable selection techniques to discriminate Pseudomonas aeruginosa strains in clinical samples. *Microchem. J.* **2016**, *124*, 306–310. [CrossRef]
39. Wu, N.; Zhang, Y.; Na, R.; Mi, C.; Zhu, S.; He, Y.; Zhang, C. Variety identification of oat seeds using hyperspectral imaging: Investigating the representation ability of deep convolutional neural network. *RSC Adv.* **2019**, *9*, 12635–12644. [CrossRef]
40. Feng, L.; Zhu, S.; Zhang, C.; Bao, Y.; Feng, X.; He, Y. Identification of Maize Kernel Vigor under Different Accelerated Aging Times Using Hyperspectral Imaging. *Molecules* **2018**, *23*, 3078. [CrossRef]
41. Westad, F.; Schmidt, A.; Kermit, M. Incorporating chemical band-assignment in near infrared spectroscopy regression models. *J. Near Infrared Spec.* **2008**, *16*, 265–273. [CrossRef]
42. Wang, Y.; Peng, Y.; Qiao, X.; Zhuang, Q. Discriminant analysis and comparison of corn seed vigor based on multiband spectrum. *Comput. Electron. Agric.* **2021**, *190*, 106444. [CrossRef]

foods

Article

Modeling and Optimization of Triticale Wort Production Using an Artificial Neural Network and a Genetic Algorithm

Milana Pribić [1], Ilija Kamenko [2,*], Saša Despotović [3], Milan Mirosavljević [4] and Jelena Pejin [1]

1 Department of Biotechnology, Faculty of Technology Novi Sad, University of Novi Sad, Bulevar cara Lazara 1, 21000 Novi Sad, Serbia; milana.pribic@uns.ac.rs (M.P.); jpejin@uns.ac.rs (J.P.)
2 The Institute for Artificial Intelligence Research and Development of Serbia, Fruškogorska 1, 21000 Novi Sad, Serbia
3 Department of Food Technology and Biochemistry, Faculty of Agriculture, University of Belgrade, Nemanjina 6, 11080 Belgrade, Serbia; sdespot@agrif.bg.ac.rs
4 Institute of Field and Vegetable Crops, Maksima Gorkog 30, 21000 Novi Sad, Serbia; milan.mirosavljevic@nsseme.com
* Correspondence: ilija.kamenko@ivi.ac.rs

Abstract: Triticale grain, a wheat–rye hybrid, has been reported to comply very well with the requirements for modern brewing adjuncts. In this study, two triticale varieties, in both unmalted and malted forms, were investigated at various ratios in the grist, applying different mashing regimes and concentrations of the commercial enzyme Shearzyme® 500 L with the aim of evaluating their impact on wort production. In order to capture the complex relationships between the input (triticale ratio, enzyme ratio, mashing regime, and triticale variety) and output variables (wort extract content, wort viscosity, and free amino nitrogen (FAN) content in wort), the study aimed to implement the use of artificial neural networks (ANNs) to model the mashing process. Also, a genetic algorithm (GA) was integrated to minimize a specified multi-objective function, optimizing the mashing process represented by the ANN model. Among the solutions on the Pareto front, one notable set of solutions was found with objective function values of 0.0949, 0.0131, and 1.6812 for the three conflicting objectives, respectively. These values represent a trade-off that optimally balances the different aspects of the optimization problem. The optimized input variables had values of 23%, 9%, 1, and 3 for the respective input variables of triticale ratio, enzyme ratio, mashing regime, and triticale variety. The results derived from the ANN model, applying the GA-optimized input values, were 8.65% *w*/*w* for wort extract content, 1.52 mPa·s for wort viscosity, and 148.32 mg/L for FAN content in wort. Comparatively, the results conducted from the real laboratory mashing were 8.63% *w*/*w* for wort extract content, 1.51 mPa·s for wort viscosity, and 148.88 mg/L for FAN content in wort applying same input values. The presented data from the optimization process using the GA and the subsequent experimental verification on the real mashing process have demonstrated the practical applicability of the proposed approach which confirms the potential to enhance the quality and efficiency of triticale wort production.

Keywords: triticale; adjuncts; mashing; artificial neural network; genetic algorithm

1. Introduction

Brewing is one of the oldest food processes that began in the Middle East 10,000 years ago and currently is one of the leading food industries in the world. Over the last thirty years, there has been a remarkable increase in the demand for beers produced on a small scale and via a slower fermentation process, namely craft beers. One of the main characteristics of craft beers is the utilization of non-standard cereals [1–3].

Beer is typically brewed using raw ingredients including malt, yeast, hops, and water, which are transformed during consecutive and complex technological processes [3]. Traditionally, barley, in its malting form, is a common choice. The three main steps that can ensure biochemical changes occur within the grain during malting are steeping, germination, and kilning. The production of enzymes in the germinating cereal grain, which causes changes in the chemical constituents of barley in preparation for mashing, can be stated as the main aim of malting [4]. However, the enzymes developed during malting do have some limitations, as they can only work in certain environments, and their activity might be too low to properly degrade certain substances [5,6]. In comparison with endogenous enzymes, commercial enzymes are more active and stable at preferred mashing conditions. The addition of commercial enzymes can make the brewing process far more consistent and can create the opportunity to use cheaper or poorer quality raw materials [5]. With the above in mind, it can be assumed that today's brewing cannot continue without the utilization of commercial enzymes [7].

As the malting process is not economically efficient, brewers often seek various strategies for cost reduction. Accordingly, the use of barley malt substitutes, as a source of extract, can be part of the solution. These materials, which bring additional sources of carbohydrates and protein into the wort, are called adjuncts [1,8,9].

The main disadvantage in terms of adjunct processability is their lower levels of cytolytic, proteolytic, and amylolytic enzymes compared to barley malt. However, there are a few examples of deviations from this rule, as some adjuncts, among which is triticale, may provide further enzymatic activity [8,10]. Even in its unmalted form, triticale can produce high levels of α-amylase. Besides that, having a low starch gelatinization temperature (59–65 °C) brings an advantage in achieving efficient starch degradation, similar to barley malt [11,12].

Of all beer production stages, mashing is a crucial step, since it determines the composition of the wort, which has a major influence on yeast fermentation and final beer quality [13,14]. Grist ratios, mashing temperatures, as well as the time required for enzymes present in grains to convert specific substances, have to be adjusted to accomplish the optimal mashing process [14]. Since modern brewing practice requires modified mashing in terms of applied temperatures, owing to the increased amount of adjuncts in the grist, the standard Congress mashing method could be questioned [15].

In the past, food-related studies were conducted using more conventional approaches. However, nowadays, the development and adoption of new technologies are dramatically accelerated. This rapid technological change affects almost all areas of life, including the food industry [16].

The implementation of novel modeling and optimization methods in brewing holds the promise for enhancing economic efficiency without compromising the distinct sensory characteristics of the beer. For this reason, an artificial neural network (ANN) can be used. In recent years, the application of ANNs has become a useful tool to increase accuracy and time, and reduce costs in analytical methods. Therefore, ANN modeling has the potential to solve problems by bypassing traditional approaches [17,18].

The brewing process, as a bioprocess, can be illustrated by empiric mathematical models, due to variable interactions and complex biochemistry reactions, which require efficient the modeling methodologies. ANNs are distinguished by their nonlinear aspect; therefore, they can detect complex nonlinear correlations between dependent quality parameters and independent process parameters. The advantage of employing neural networks lies in the fact that without having exhaustive knowledge about a specific process, the network can learn from previous experiences to predict the system's behavior when some variables are modified. To make the selection of ANN topology more accurate, which depends on various features such as a number of hidden layers, number of neurons of hidden layers, and connections among neurons, using evolutionary algorithms can be a valuable asset. As such, GAs have been applied in studies of bioprocess optimization from experimental data using ANNs, to optimize their topologies and to deal with the optimization of multiple objective functions [17,18]. The concept of many-objective optimization has emerged as a powerful approach to efficiently explore the trade-offs between multiple objectives and identify a set of solutions that represent the best compromise solutions [19]. GAs mimic the biology of chromosome generation with operators such as mutation, crossover, and the selection criteria of the random population of individuals. By applying an objective function, each individual is evaluated based on a fitness criterion [20].

ANNs and GAs showed remarkable ability in modeling and optimizing various processes. Specifically, within the brewing industry, these artificial intelligence techniques were effectively employed to enhance fermentation processes [21]. A dynamic neural network was developed, comprising adapted neurons to account for process dynamics. This model accurately captured the intricate, non-linear nature of beer fermentation, enabling predictions of future process behaviors. Also, an ANN was used in beer classification to build a classification model based on the relevant features of beer [22]. As one of the most relevant components of beer taste, the control of acetic acid content is very important to maintain consistent beer quality. For that reason, an ANN and a support vector machine were applied to predict acetic acid content at the end of commercial-scale beer fermentation [23]. ANNs and GAs were not only applied in the brewing process, but also to optimize the further utilization of brewing bio-waste materials such as brewers' spent grain [20], where the biosynthesis of poly-β hydroxybutyrate from brewers' spent grain was modeled and optimized using an ANN–GA strategy. Besides that, better control of a brewery's waste water was achieved by developing a robust mathematical tool for performance prediction using an ANN [24].

Although ANNs and GAs have commonly been utilized in studying the final fermentation process in brewing, there remains a gap in research regarding their application in modeling and optimizing the production of wort. Therefore, this study aimed to develop a new approach to the mashing process, applying an ANN model that used the data obtained from a laboratory wort production of two triticale varieties—NS Paun and Odisej—in unmalted and malted forms. The conventional Congress mash protocol and a modified temperature variation regime, with and without the application of a commercial enzyme, were explored. Moreover, a multi-objective optimization approach based on a GA was developed to optimize the conditions needed to obtain the best wort quality.

2. Materials and Methods

2.1. Materials

Two winter triticale varieties—NS Paun and Odisej, both accepted in Serbia and the European Union—were examined in unmalted and malted forms. They were grown in experimental fields in Rimski Šančevi, Serbia. As a control sample, commercial barley malt type Pilsner, produced by malthouse Maltinex Soufflet Group, Bačka Palanka, Serbia, was used. Commercial enzyme Shearzyme® 500 L as xylanase (endo-1,4-; produced by *Aspergillus oryzae*) was kindly provided by Novozymes A/S (Bagsvaerd, Denmark). For free amino nitrogen (FAN) content determination, the following chemicals were purchased from Sigma-Aldrich: disodium hydrogen phosphate dodecahydrate, potassium dihydrogen phosphate, ninhydrin, fructose, potassium iodate, ethanol 96%, fructose, and glycine.

2.2. Methods

The flowchart in Figure 1 summarizes the overall process of mashing, modeling, and optimization of triticale wort production.

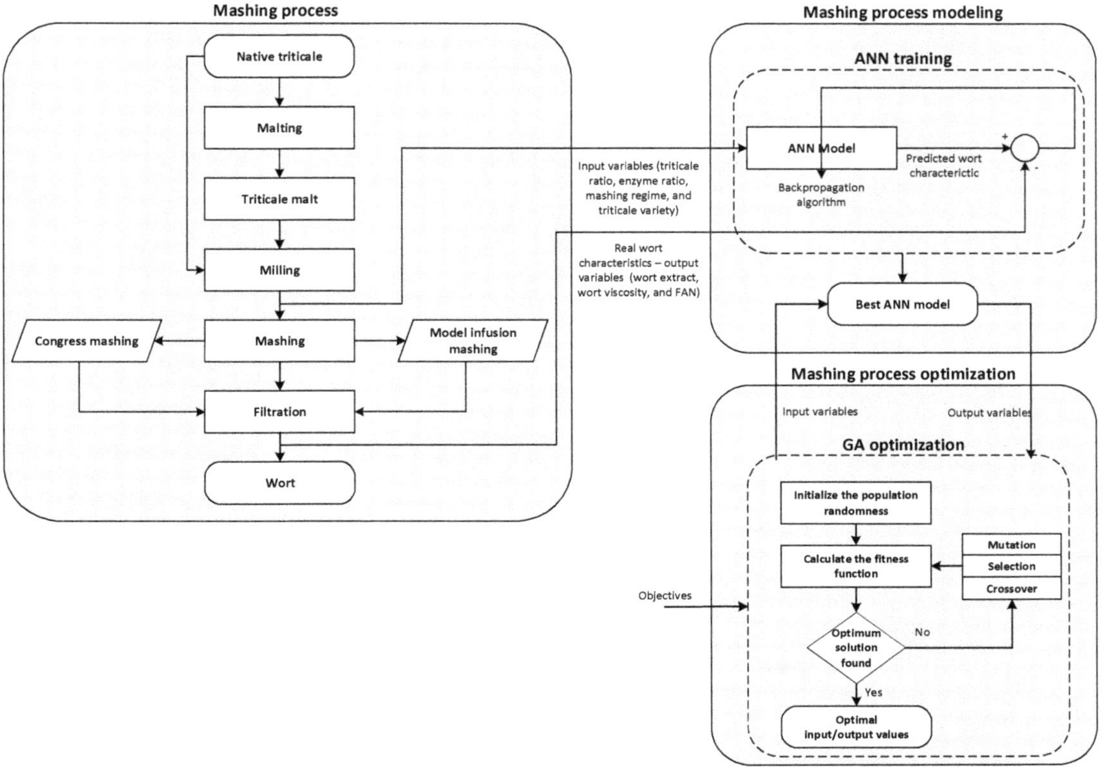

Figure 1. Flowchart of mashing, modeling, and optimization of triticale wort production.

2.2.1. Mashing Process

The mashing process consists of several successive steps, as can be seen in Figure 1. In this study, triticale was utilized in two forms—unmalted and malted cereal. In order to produce malt, a micro-malting pilot plant consisting of a steeping, germination, and kilning unit (Seeger, Germany, at the Faculty of Technology, Novi Sad, Serbia), with the following malting program: 3 days of steeping with a wet and dry cycle at 15 °C (1st day—6 h wet, 18 h dry; 2nd day—4 h wet, 20 h dry; and 3rd day—2 h wet, 22 h dry) was used. The germination phase lasted 4 days at 16 °C. After that, green malts were dried by applying the following kilning temperatures: 55 °C for 15 h, 72 °C for 5 h, 82 °C for 3 h. The rootlets generated during germination were eliminated from the malts by deculming.

To quantify the main wort quality indicators, cereal samples were finely milled in a laboratory disc mill, DLFU (Bühler GmbH, Braunschweig, Germany), at 0.2 mm. Mashes were prepared in mechanically stirred metal beakers in a mash bath (Lochner Labor Technik GmbH, Berching, Germany) using 50 g of grist in total mixed with 200 mL distilled water. After the mashing process, to obtain extracted liquid wort, mashes were filtered.

In this study, mashing processes were performed as described in Table 1.

Table 1. Mashing processes applied in the wort production.

Independent variables	Triticale variety	'NS Paun' and 'Odisej' both in unmalted and malted forms
	Triticale ratio in the grist (%)	10, 30, 50, and 70
	Enzyme quantity (μL)	50, 10, and 5
	Mashing regime	Congress mashing [25]: Initial temperature of 50 °C was maintained for 40 min. The temperature was then increased to 63 °C and held for 45 min. After the addition of a further 100 mL of distilled water, temperature was raised to 70 °C and held for 30 min. Final temperature of 76 °C was maintained for an additional 10 min.
		Modified mashing [26]: Initial temperature of 45 °C was maintained for 30 min before it was increased up to 70 °C at 1 °C/min. After the addition of a further 100 mL of distilled water, the temperature was maintained at 70 °C for a further 60 min.
Dependent variables	Quality parameters	Wort extract content, wort viscosity, and FAN content

The variation in independent variables directly influenced key wort quality indicators, crucial in determining high-quality wort: extract content, viscosity, and FAN content of wort.

The aim of mashing is to yield as much extract content from the grain as possible, which is further related to the soluble fermentable sugars in wort and is correlated with high concentrations of ethanol produced during fermentation. Wort extract content, which refers to the compounds from finely milled malt that are brought into solution during a mashing process, was determined by MEBAK method 3.1.4.2.2 [25] using Pyknometers on the basis of the official Plato tables.

In this study, high wort viscosity was observed, which is a common occurrence when adjuncts are utilized. Therefore, the commercial enzyme Shearzyme® 500 L was added to the mashes. At the beginning of experiments, 50 μL of enzyme per beaker was used, according to the producer's recommendations. However, initial enzyme concentration significantly reduced wort viscosity (discussed below). Therefore, in further experiments, enzyme concentration was optimized, e.g., reduced to achieve an adequate wort viscosity as well as to bring down total production costs. The final enzyme concentration was 5 μL per beaker. The viscosity of the wort was measured using a Höppler falling ball viscometer (Brookfield Engineering Laboratories, Inc., Stoughton, MA, USA). Following the MEBAK method 3.1.4.4.1 [25], the amount of time required for a special ball to fall while sinking through a glass tube filled with a wort, between two marks, was measured.

FAN content has been regarded as a superior parameter that represents nitrogen compounds which may be assimilated or metabolized by yeast during fermentation. The problem arises when the most commonly used adjuncts usually decrease the FAN content of wort [27,28]. However, triticale stood out as a cereal that contributes a considerable amount of proteolytic enzyme activity, even in its native form [29]. FAN content in the wort, as reported in MEBAK method 3.1.4.5.5.1 [25], was measured by spectrophotometry. Sample was diluted 1:100. A total of 2 mL of sample was put in a test tube with 1 mL of color reagent and placed in a boiling water bath for exactly 16 min and then cooled to 20 °C for 20 min. Then, 5 mL of dilution solution was added to the test tube, mixed, and absorbance was measured at 570 nm against a reagent blank.

2.2.2. Mashing Process Modelling

In this study, an ANN is proposed as a modeling tool to capture the complex relationships between the input (independent) variables and output (dependent) variables in the mashing process for triticale wort production (Figure 1). For that purpose, Deep Learning Toolbox from Matlab was used.

ANNs are computational models inspired by the structure and functioning of the human brain. They consist of interconnected nodes, or neurons, organized in layers. ANNs learn from data by adjusting the weights of connections between neurons through a training process, enabling them to recognize patterns. The network utilizes mathematical functions, called activation functions, to compute the output of each neuron based on the weighted sum of its inputs.

In this research, a three-layer feed-forward ANN with sigmoid activation function in hidden neurons and linear activation function in output neurons was developed for predicting the characteristics of wort. Network inputs were carefully designed as four indicators with the greatest impact on prediction performance—triticale ratio, enzyme ratio, mashing regime index, and triticale variety index.

The grist included different triticale ratios: 10%, 30%, 50%, and 70%. The enzyme ratio, which denotes the percent value of the concentration of the commercial enzyme Shearzyme® 500 L, was scaled to 50 μL (100%). Based on that, 10 μL quantity was scaled as 20% and 5 μL quantity as 10%. Maintaining uniformity in the range of input data is crucial for better convergence and improved performance. The mashing regime index reflected two different mashing regimes (Congress–1 and Modified–2). The triticale variety index represented two triticale varieties in different forms utilized in the process (unmalted NS Paun–1, unmalted Odisej–2, malted NS Paun–3, and malted Odisej–4). Network outputs were carefully designed as three indicators for wort quality estimation—wort extract content, wort viscosity, and FAN content.

The number of neurons in the first hidden layer is denoted as n and the number of neurons in the second hidden layer is denoted as m. The optimal number of neurons in each layer m and n was determined through systematic experimentation. The selected configuration, obtained through iterative optimization on a validation set, strikes a balance between model complexity and generalization, ensuring optimal performance. The architecture of the model is illustrated in Figure 2.

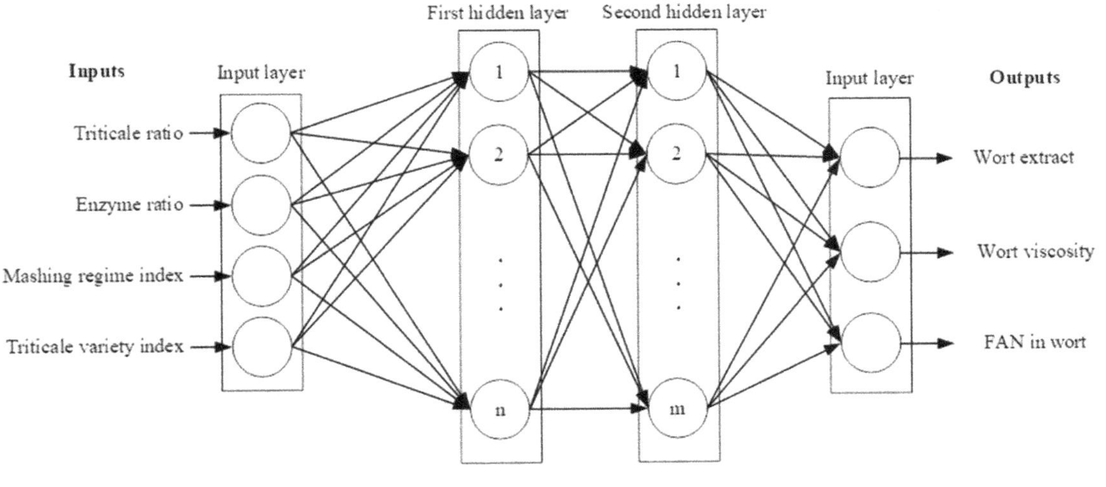

Figure 2. The architecture of the ANN model.

In this study, three different training functions, namely Levenberg–Marquardt, Bayesian regularization, and scaled conjugate gradient algorithm (implemented in Deep Learning Toolbox from Matlab), were employed to optimize the ANN model using the comprehensive and well-structured dataset.

The Levenberg–Marquardt training function is a widely used method in ANN training. It combines the advantages of both the gradient descent and Gauss–Newton methods, offering robustness and fast convergence. By adjusting the damping factor during the training process, the algorithm effectively balances the stability and speed of convergence, ensuring accurate predictions for the mashing process.

Another training function utilized in this study was Bayesian regularization. This method incorporates Bayesian principles to improve the generalization ability of ANN models. It employs prior knowledge to guide the learning process and prevent overfitting, where the model becomes overly specialized to the training data and fails to generalize to new instances.

The third training function employed in this study was the scaled conjugate gradient algorithm. It is a variant of the conjugate gradient method that adapts the learning rate dynamically during the training process. As a result, it optimizes ANN models by efficiently traversing the error surface, leading to faster convergence and improved prediction accuracy.

By employing three different training functions, the study aimed to thoroughly explore the optimization landscape and identify the most effective approach for training the ANN model. By comparing and evaluating the performance of the ANN models trained with these training functions, valuable insights can be gained regarding the most suitable approach for accurately modeling and predicting the mashing process in triticale wort production.

2.2.3. Mashing Process Model Optimization

In this study, the GA optimization method was proposed as a powerful approach to optimize the input variables in the mashing process for triticale wort production (Figure 1). For that purpose, Global Optimization Toolbox from Matlab was used.

A genetic algorithm is a search heuristic inspired by the process of natural selection. It starts with a population of potential solutions represented as individuals in a population. These individuals undergo evolution through processes such as selection, crossover (recombination), and mutation. The fittest individuals, as determined by a fitness function, are more likely to be selected and pass their genetic information to the next generation. Over successive generations, the algorithm converges towards optimal or near-optimal solutions, mimicking the principles of natural evolution to find solutions to optimization and search problems.

The GA efficiently explored the solution space and converged towards a set of solutions that optimized the desired wort quality objectives. The integration of the GA with the mashing process model, which incorporated ANNs, enabled the optimization of input variables based on their impact on wort quality, offering improved control and efficiency in triticale wort production.

To optimize the mashing process represented by the ANN model, a GA was employed to find the combination of input parameters that yielded the most suitable outputs satisfying the multi-objective function. Through the iterative process of GA optimization, the ANN model was able to converge on the most favorable combination of inputs, resulting in improved predictions of wort production characteristics.

In the optimization process, the input variables triticale ratio, enzyme ratio, mashing regime index, and triticale variety index were constrained only to their natural bounds (between minimum and maximum possible values).

To accurately reflect the importance of the model outputs in the optimization, we introduced penalties that scaled the output of the objective function. By incorporating these penalties, we ensured a balanced consideration of both the primary objective and the influence of the model outputs. Through this approach, our optimization procedure effectively captured the intricate relationships between the inputs, outputs, and penalties, allowing for an informed decision-making process that optimized the primary objective while appropriately accounting for the impact of the model outputs.

The penalties modified the fitness value of each chromosome in the GA optimization process. A higher penalty value indicated poorer fitness, discouraging the selection of chromosomes that violated constraints or deviated from preferred input variable combinations.

As there is no specifically defined range of wort extract content in Brewer's analytics, in this study, we aimed for wort extract to be in the range of the control sample—100% barley malt (8.57% *w/w*) and the highest achieved value in triticale-based wort (8.89% *w/w*—70% ratio of variety NS Paun malt without enzyme addition).

The desired value for wort extract was w_e* is 8.65% *w/w* (randomly chosen from obtained wort extract content results which complied with above mentioned criterion), so the penalty function P_e for wort extract is as follows:

$$P_e = \begin{cases} w_e > 8.89 & 10 \\ 8.57 \leq w_e \leq 8.89 & 1 \\ w_e < 8.57, & 10 \end{cases} \tag{1}$$

where w_e is the current real value of wort extract.

The recommended value for wort viscosity w_v ranges from 1.5 m·Pas to 1.6 m·Pas [25] and the desired value w_v* is 1.580 m·Pas (reference value obtained in 100% barley malt wort), so the penalty function P_v is as follows:

$$P_v = \begin{cases} w_v > 1.5 & 10 \\ 1.5 \leq w_v \leq 1.6 & 1 \\ w_v < 1.6, & 10 \end{cases} \tag{2}$$

where w_v is the current real value of wort viscosity.

The recommended value for FAN content in wort w_{FAN} ranges from 110 mg/L to 180 mg/L [25] and the desired value w_{FAN}* is 147.99 mg/L (reference value obtained in 100% barley malt wort), so the penalty function P_{FAN} is as follows:

$$P_{FAN} = \begin{cases} w_{FAN} > 110 & 10 \\ 110 \leq w_{FAN} \leq 180 & 1 \\ w_{FAN} < 180, & 10 \end{cases} \tag{3}$$

where w_{FAN} is the current real value of FAN content.

We formulated a multi-objective optimization problem incorporating three primary objectives. The initial objective was to align the wort extract content with its desired values as accurately as possible. This objective takes into account a penalty P_e and can be expressed as follows:

$$F_e = P_e|w_e - w_e^*| \tag{4}$$

The second objective was to closely align the wort viscosity with its desired values. This objective takes into account a penalty P_v and can be expressed as follows:

$$F_v = P_v|w_v - w_v^*| \tag{5}$$

The third objective was to achieve a close match of FAN content with its desired values. This objective takes into account a penalty P_{FAN} and can be expressed as follows:

$$F_{FAN} = P_{FAN}|w_{FAN} - w_{FAN}^*| \tag{6}$$

By minimizing these metrics during the optimization process, we strived to align the model outputs as closely as possible with their desired values.

By formulating an objective function with penalties, the GA efficiently searches for solutions that optimize the desired wort quality objectives while satisfying the imposed constraints and preferences. This approach provides a comprehensive optimization framework that balances multiple objectives and ensures the feasibility and desirability of the optimized input variable combinations for triticale wort production.

3. Results and Discussion

3.1. Influence of Triticale Characteristics on Wort Quality

Analysis of the wort was conducted in triplicate, encompassing a total of 384 samples. For the purpose discussing wort quality, the results of all produced triticale wort were compared to the control sample—100% barley malt wort (wort extract content—8.57% *w/w*; wort viscosity—1.580 mPa·s; FAN content—147.99 mg/L).

The wort results obtained from triticale mashing in the range of 10–70% in grist, applying both mashing regimes, are shown in Tables 2 and 3.

Table 2. Results of triticale-based wort in 10–70% ratios in grist obtained from Congress mashing.

	Variety	Enzyme Addition	Outputs Wort Extract Content (% *w/w*)	Wort Viscosity (mPa·s)	Wort FAN Content (mg/L)
Unmalted	NS Paun	Without enzyme	8.28–8.49	1.626–1.765	64.1–129.42
		50 µL (100%)	8.13–8.38	1.384–1.498	77.23–148.16
		10 µL (20%)	8.22–8.39	1.403–1.587	70.81–141.89
		5 µL (10%)	8.22–8.46	1.576–1.613	67.23–136.37
	Odisej	Without enzyme	8.27–8.47	1.639–2.220	56.0–120.78
		50 µL (100%)	8.22–8.33	1.386–1.495	74.62–139.99
		10 µL (20%)	8.22–8.43	1.401–1.588	67.33–130.03
		5 µL (10%)	8.22–8.45	1.538–1.629	64.24–124.55
Malted	NS Paun	Without enzyme	8.75–8.89	1.500–1.755	130.95–156.12
		50 µL (100%)	8.45–8.80	1.432–1.582	157.43–191.37
		10 µL (20%)	8.49–8.81	1.461–1.583	142.19–189.04
		5 µL (10%)	8.54–8.83	1.468–1.590	146.98–187.97
	Odisej	Without enzyme	8.69–8.78	1.564–1.833	125.80–142.91
		50 µL (100%)	8.39–8.71	1.440–1.570	143.37–186.35
		10 µL (20%)	8.42–8.74	1.498–1.685	140.50–184.46
		5 µL (10%)	8.46–8.77	1.500–1.783	139.76–183.59

Table 3. Results of triticale-based wort in 10–70% ratios in grist obtained from Modified mashing.

	Variety	Enzyme Addition	Outputs Wort Extract Content (% *w/w*)	Wort Viscosity (mPa·s)	Wort FAN Content (mg/L)
Unmalted	NS Paun	Without enzyme	8.21–8.41	1.596–1.650	71.28–134.67
		50 µL (100%)	8.12–8.36	1.320–1.432	79.49–151.82
		10 µL (20%)	8.12–8.37	1.408–1.491	72.29–147.16
		5 µL (10%)	8.28–8.38	1.419–1.527	69.70–140.07
	Odisej	Without enzyme	8.23–8.44	1.607–2.060	63.76–132.69
		50 µL (100%)	8.20–8.32	1.339–1.446	76.4–144.33
		10 µL (20%)	8.21–8.40	1.415–1.522	68.94–138.31
		5 µL (10%)	8.20–8.42	1.529–1.618	67.04–135.13
Malted	NS Paun	Without enzyme	8.69–8.79	1.518–1.713	141.27–159.19
		50 µL (100%)	8.42–8.74	1.414–1.502	162.38–198.62
		10 µL (20%)	8.48–8.75	1.433–1.529	154.18–192.72
		5 µL (10%)	8.52–8.77	1.430–1.562	150.30–188.19
	Odisej	Without enzyme	8.65–8.74	1.541–1.790	138.30–149.20
		50 µL (100%)	8.38–8.67	1.422–1.563	155.13–192.13
		10 µL (20%)	8.40–8.70	1.460–1.680	150.09–188.90
		5 µL (10%)	8.41–8.71	1.480–1.720	148.50–184.42

At a 10% ratio (as outlined in Table 2), the unmalted triticale sample of the NS Paun variety demonstrated the highest extract value of 8.49% *w/w*. Additionally, samples derived from a mixture of triticale malt and barley malt exhibited comparable or even higher extract content when compared to the control sample of barley malt (up to 8.89% *w/w* for 70%

triticale malt of the NS Paun variety without enzyme addition—Table 2). These findings are in accordance with the results obtained by Cioch-Skoneczny et al. (2019) [12] where a 50% ratio of triticale malt yielded an extract content higher than that obtained in 100% barley malt wort (8.79 and 8.57% *w*/*w*, respectively). The Congress mashing regime, compared to the modified one, showed higher extract content values in all examined samples. Certainly, using triticale in both forms notably influenced the concentration of the wort extract.

At the beginning of the experiments, when 50 μL of enzyme was used, viscosity in all produced wort was significantly reduced, and was lower than the prescribed values (1.5–1.6 m·Pas) [25]. Therefore, in further experiments, enzyme concentration was refined. The final amount was 5 μL, where desirable viscosity was obtained when unmalted and malted triticale were used in 10–50% grist. Overall, the Odisej variety obtained higher viscosity values compared to those produced from the NS Paun variety. It is important to emphasize that compared to the 100% barley malt, all examined samples, in both mashing regimes, had higher viscosity values if the commercial enzyme was not applied. Without the addition of Shearzyme® 500 L, the unmalted Odisej variety showed up to 2.22 mPa·s viscosity, while the unmalted NS Paun variety had much lower viscosity values, even when the highest triticale ratio was applied (1.765 mPa·s—Table 2). When employing unmalted triticale for wort production, the modified mashing process yielded lower viscosity values compared to the Congress mashing method. However, according to MEBAK [25], only the NS Paun variety at 10% ratio in the grist had acceptable values, while none of the wort produced from the Odisej variety fulfilled the above-mentioned range (1.5–1.6 mPa·s). As filtration of the wort is one of the most difficult steps in beer production, controlling the wort viscosity is a crucial moment for a successful brewing process. In this respect, the use of a commercial enzyme was necessary and, as expected, using Shearzyme® 500 L helped decrease viscosity.

Almost all examined malt triticale wort, prepared with the Congress mashing regime, contained FAN contents similar to those of 100% barley malt wort or even higher (Table 2). As expected, FAN content in malt wort, in both mashing regimes, was higher than in wort produced with native triticale, since, during malting, grain components degrade and are more susceptible to enzymes. In this study, the 50% ratio of NS Paun malt showed a similar concentration as the wort produced from 100% barley malt (147.97 and 147.99 mg/L, respectively), which is in correlation with the results obtained by Cioch-Skoneczny (2019) [12].

3.2. Dataset Analysis

ANNs operate directly on input–output data, so there must be a sufficient number of data to train the ANN and produce valid results [20]. In this study, a comprehensive dataset consisting of 384 samples was utilized to develop and evaluate the proposed models and optimization methods for the mashing process in triticale wort production. The dataset consisted of carefully collected information on various input (triticale ratio, enzyme ratio, mashing regime index, and triticale variety index) and output (wort quality indicators—extract content, viscosity, and FAN content) variables. Statistical parameters were calculated for all indicators and are given in Table 4.

Table 4. Statistical parameters of research data.

	Name of Parameter	Min.	Max.	Mean	Std. Dev.
Inputs	Triticale ratio (%)	10	70	40	22.39
	Enzyme ratio (%)	0	100	32.5	39.66
	Mashing regime index	1	2	1.5	0.5
	Triticale variety index	1	4	2.5	1.12
Outputs	Wort extract content (% *w*/*w*)	8.12	8.89	8.47	0.2
	Wort viscosity (mPa·s)	1.320	2.220	1.56	0.13
	FAN content (mg/L)	56.0	198.62	130.48	39.32

By utilizing this comprehensive and well-structured dataset, the proposed models and optimization methods could be effectively developed, validated, and evaluated.

The dataset was divided into appropriate subsets for training, validation, and testing. The training subset was used to train the models, allowing them to learn the relationships between the input variables and the corresponding output variables. The validation subset was employed to fine-tune the models and assess their generalization ability. Finally, the testing subset served as an independent evaluation set to objectively measure the performance of the trained models and optimization methods.

The dataset served as a valuable resource for capturing the complex relationships between the input variables and the output variables, facilitating the modeling and optimization of the mashing process for triticale wort production.

The careful curation and division of the dataset into subsets for training (340 samples), validation (36 samples), and testing (8 samples) ensured the reliability and accuracy of the proposed models and optimization methods. By leveraging this comprehensive dataset, the study aimed to advance our understanding of the mashing process and enhance the quality and efficiency of wort production in the brewing industry.

3.3. ANN Training Results

The training of ANN models is a critical step in developing reliable and accurate predictions for the mashing process in triticale wort production. The training process involved adjusting the weights and biases of the neurons iteratively using optimization algorithms to minimize the error between the predicted outputs and the actual values in the training dataset.

To evaluate the performance and generalization ability of the trained ANN model, a separate validation dataset was used. The validation dataset consisted of samples that were not included in the training dataset. The model's predictions for the validation dataset were compared to the actual values, and various metrics, such as mean squared error (MSE) or correlation coefficients, were computed to assess the model's accuracy and reliability.

A total of 231 different values of neurons in both hidden layers (n and m) were tested. In experiments, the number of neurons ranged in steps of five from ten to sixty neurons in the first hidden layer and from zero to thirty in the second layer (zero represents omitting second layer). ANN had four inputs, so it is not recommended to use a number of neurons in the hidden layer smaller than the number of inputs. Also, a very large number of neurons often leads to network overfitting and can be time-consuming during the training process. The training was repeated ten times for each combination of parameters (three training functions and 231 different numbers of neurons in hidden layers). This yielded 6.930 ($3 \times 231 \times 10$) treatments for the ANN. The best training function, the most appropriate number of neurons in the hidden layers, and the best combination of both were determined in these experiments.

The Levenberg–Marquardt training algorithm yielded optimal results for the ANN model, achieving peak performance with 45 neurons in the first hidden layer and 25 neurons in the second hidden layer. The training concluded after 180 epochs, with a final mean squared error (MSE) of 0.001 on the entire training dataset in the last epoch, as illustrated in Figure 3a.

Similarly, the scaled conjugate gradient training algorithm demonstrated its effectiveness with the ANN model featuring forty neurons in the first hidden layer and five neurons in the second hidden layer. The training process spanned 1000 epochs, and in the last epoch, the MSE on the complete training dataset was 0.04. The convergence pattern is depicted in Figure 3b.

For the Bayesian regularization training algorithm, optimal results were achieved with an ANN model comprising 30 neurons in the first hidden layer and 10 neurons in the second hidden layer. Training concluded after 500 epochs, and in the last epoch, the MSE on the entire training dataset was an impressively low 0.000084. The convergent graph for this algorithm is displayed in Figure 3c.

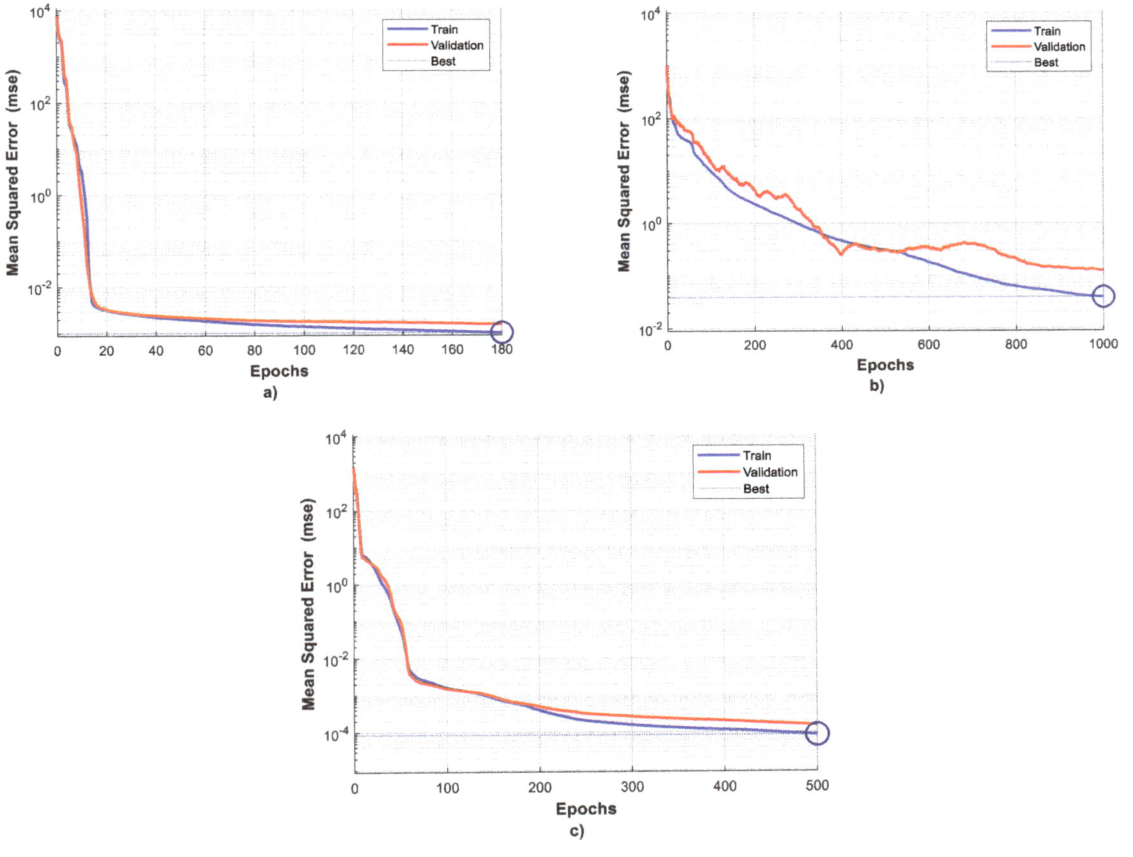

Figure 3. Training performances for the best ANN models obtained for three training algorithms separately; (**a**)—Levenberg–Marquardt training algorithm; (**b**)—scaled conjugate gradient training algorithm and (**c**)—Bayesian regularization training algorithm.

It is important to highlight that the Bayesian regularization algorithm exhibited superior performance when compared to the other two algorithms, ultimately yielding the most favorable outcomes.

3.4. ANN Test Results

To assess the performance and reliability of the trained ANN model, the testing phase was conducted using a set of eight samples. These samples represented real-world scenarios and were carefully selected to cover a diverse range of input variable combinations for the mashing process.

During the testing phase, the eight samples were fed into the trained ANN model, and the predicted output variables were compared with their respective actual values. The performance of the model was evaluated by calculating various metrics, such as its mean and minimum and maximum values. The obtained results can be seen in Figure 4.

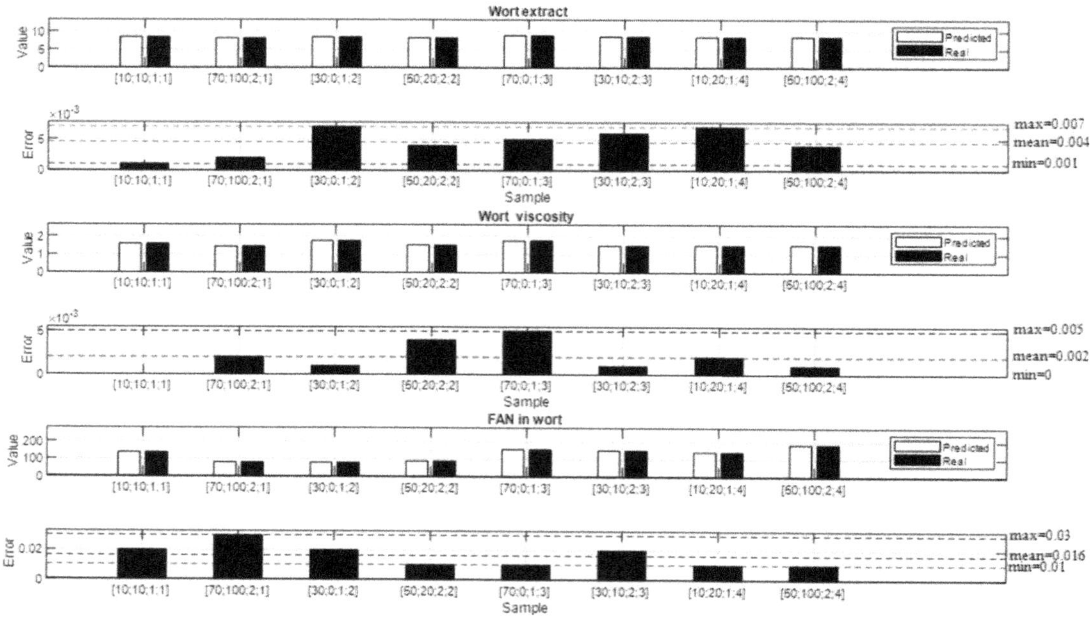

Figure 4. Test results for the proposed neural network model.

In Table 5, key metrics for the test results are presented, including minimum and maximum absolute errors, mean absolute error, mean square error, root mean square error, and mean absolute percentage error. These metrics offer a concise assessment of our model's accuracy, precision, and overall predictive performance on the test dataset.

Table 5. Test results metrics.

	Min. abs. Error	Max. abs. Error	Mean abs. Error	Mean Square Error	Root Mean Square Error	Mean abs. Percent. Error
Wort extract	0.001	0.007	0.004	0.00002	0.005	0.05
Wort viscosity	0	0.005	0.002	0.00001	0.002	0.13
FAN in wort	0.01	0.03	0.016	0.0003	0.018	0.015

The results obtained from the testing phase revealed a high degree of accuracy and precision in the predictions of the trained ANN model for all eight samples. The calculated error metrics indicated very small deviations between the predicted and actual values, emphasizing the model's ability to effectively capture the complex relationships within the mashing process and make accurate estimations of the desired wort quality indicators.

These findings demonstrated the robustness and generalizability of the trained ANN model. The model's consistent and accurate predictions across the eight samples suggest its reliability in providing precise estimations of the output variables for various input combinations. This indicates the potential of the trained ANN model to be applied in practical brewing scenarios to optimize the mashing process and enhance triticale wort production.

3.5. Optimization Results Applying GA

The genetic algorithm was employed to optimize the mashing process for triticale wort production, considering three conflicting objectives simultaneously. The objective functions aimed to address different aspects of the optimization problem, leading to a multi-objective optimization task. The chosen approach for assessing the best set of solutions involved evaluating the Pareto front, which represents the trade-off solutions between the conflicting objectives.

Specific parameters for the GA were configured to address the multi-objective nature of the optimization problem. The population size ranged from 40 to 80 individuals in steps of five and with a crossover rate of 0.8 and a mutation rate of 0.1. Tournament selection was utilized as the selection mechanism, selecting parents based on their fitness with respect to the multiple conflicting objectives.

The GA optimization process involved exploring the Pareto front to identify a set of solutions that provide a balanced trade-off among the conflicting objectives. The termination criteria were based on predefined conditions associated with the convergence of the Pareto front.

The optimization experiment was repeated five times for each combination of parameters, resulting in 40 (8×5) optimization attempts. Among the solutions on the Pareto front, one notable set of solutions was found with objective functions values of 0.0949, 0.0131, and 1.6812 for the three conflicting objectives, respectively. These values represent a trade-off that optimally balances the different aspects of the optimization problem.

The corresponding input values for this solution on the Pareto front are shown in Table 6. This solution, achieved with a population size of 50 after 70 generations, is recognized as the best solution among the optimization experiments, demonstrating the most favorable trade-off among the conflicting objectives.

Table 6. The optimized input variables obtained from the GA.

Inputs	Result
Triticale ratio (%)	23
Enzyme ratio (%)	9
Mashing regime index	Congress–1
Triticale variety index	Malted NS Paun–3

The analysis of the Pareto front provided insights into the solutions that achieved a balance among the conflicting objectives. The selected set of solutions demonstrated a substantial improvement in the overall objective values compared to the initial population, indicating the successful convergence of the GA towards solutions on the Pareto front.

As observed in Figure 5, the upper segment depicts the Pareto front, illustrating the trade-offs and optimal compromises between conflicting objectives. Meanwhile, the lower portion represents a score histogram in Figure 5, offering insights into the distribution of fitness values throughout the optimization process. Together, these visualizations provide a comprehensive understanding of the diverse solutions and the quantitative performance across specific fitness ranges.

The successful convergence of the GA towards optimal solutions is evident from both the Pareto front analysis and the score histogram. The results indicate a substantial improvement in the objective function compared to the initial population, underscoring the efficacy of the multi-objective optimization approach.

The obtained fitness values and the corresponding input–output values highlight the success of the GA in finding an optimized solution for the mashing process in triticale wort production. These results demonstrate the ability of the GA to identify input variable combinations that lead to improved wort quality, as reflected in the obtained output values.

The resulting outputs corresponding to these optimized input variables were measured as 8.65% *w*/*w* for wort extract content, 1.52 mPa·s for wort viscosity, and 148.32 mg/L for FAN content in wort. These values indicated the predicted levels of these wort quality indicators based on the optimized input variables obtained through the GA optimization process.

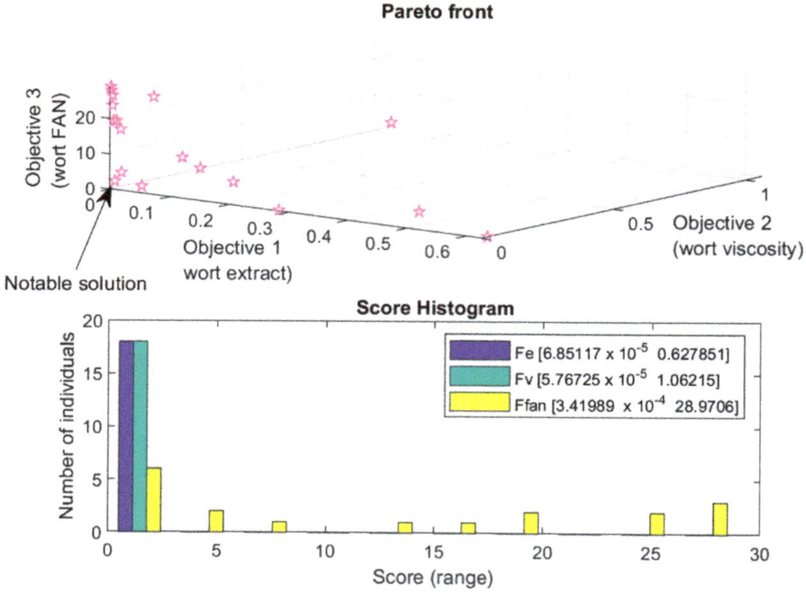

Figure 5. Pareto front for best optimization attempt. Asterisks denote solutions on the Pareto front.

3.6. Experimental Verification

After obtaining the optimization results for the input variables (Table 6), an experimental verification was performed using actual laboratory mashing to validate the effectiveness of the optimized input variable combinations. Wort analyses were performed again in triplicate and the corresponding results are presented in Table 7.

Table 7. Optimization results applying GA and results obtained from real mashing process.

	Wort Extract Content (% *w/w*)	Wort Viscosity (mPa·s)	FAN Content (mg/L)
GA optimization results	8.65	1.52	148.32
Real mashing process	8.63 ± 0.01	1.51 ± 0.02	148.88 ± 0.02

The experimental verification of the real mashing process served as a crucial step in validating the effectiveness of the optimization results and the ANN model. Results from Table 7 demonstrate a very close agreement between the GA-optimized values and the values obtained from the real mashing process and validate the effectiveness in predicting and achieving the desired wort quality.

4. Conclusions

The data presented in this work strongly indicate that triticale shows high potential and it is suitable to serve as a brewing adjunct, as it fulfills the most important requirements for wort production. From the presented parameters (Tables 2 and 3), it is noticeable that triticale influenced the concentration of the wort extract, especially in the form of malt, where extract content was similar to or even surpassed the control sample of barley malt used in research. Regarding wort viscosity, compared to the 100% barley malt, all examined samples had higher viscosity values if the commercial enzyme was not applied. Among the two varieties examined, NS Paun was identified as more suitable in that aspect. The use of a commercial enzyme was necessary and, as expected, using Shearzyme® 500 L helped decrease viscosity which aids wort filtration. FAN content in malt wort, in both mashing regimes, was higher than in wort produced with native triticale.

The utilization of ANNs in triticale wort production has demonstrated its effectiveness in modeling the mashing process. By leveraging the capabilities of ANNs, it becomes possible to accurately predict wort quality based on various input variables, such as triticale variety, triticale ratios, mashing regimes, and enzyme concentrations. The integration of ANNs with GAs further enhanced the optimization process, allowing for the identification of optimal input combinations that minimize specific objective functions related to wort quality. The application of ANNs in the brewing industry represents a significant advancement, enabling brewers to achieve improved control and efficiency in the production of high-quality triticale wort.

The results obtained from the optimization process using a GA and the subsequent experimental verification on the real mashing process have demonstrated the practical applicability of the proposed approach which confirms the potential to enhance the quality and efficiency of triticale wort production.

Author Contributions: M.P.: Conceptualization, writing—original draft preparation, methodology, data curation, investigation. I.K.: methodology, investigation, data curation, writing—original draft preparation, writing—review and editing. S.D.: investigation, writing—original draft preparation. M.M.: investigation, visualization. J.P.: validation, writing—review and editing. All authors have read and agreed to the published version of the manuscript.

Funding: This research was funded by the Ministry of Education, Science and Technological Development of the Republic of Serbia, University of Novi Sad—Faculty of Technology Novi Sad, and University of Belgrade—Faculty of Agriculture (Agreements No. 451-03-47/2023-01/200134, 451-03-47/2023-01/200116, and 451-03-1202/2021-09).

Institutional Review Board Statement: Not applicable.

Informed Consent Statement: Not applicable.

Data Availability Statement: Data is contained within the article.

Acknowledgments: The authors would like to thank Novozymes (Bagsværd, Denmark) for the donation of enzyme.

Conflicts of Interest: The authors declare no conflicts of interest.

References

1. Dabija, A.; Ciocan, M.E.; Chetrariu, A.; Codină, G.G. Maize and sorghum as raw materials for brewing, a review. *Appl. Sci.* **2021**, *11*, 3139. [CrossRef]
2. Materna, K.; Bernhäuserová, V.; Hasman, J.; Hána, D. How microbreweries flooded Europe: Mapping a new phenomenon in the beer industry. *J. Maps* **2022**, *18*, 18–25. [CrossRef]
3. Anderson, H.E.; Santos, I.C.; Hildenbrand, Z.L.; Schug, K.A. A review of the analytical methods used for beer ingredient and finished product analysis and quality control. *Anal. Chim. Acta* **2019**, *1085*, 1–20. [CrossRef] [PubMed]
4. Yorke, J.; Cook, D.; Ford, R. Brewing with unmalted cereal adjuncts: Sensory and analytical impacts on beer quality. *Beverages* **2021**, *7*, 4. [CrossRef]
5. Ozatay, S. Recent Applications of Enzymes in Food Industry. *J. Curr. Res. Eng. Sci. Technol.* **2020**, *6*, 17–30. [CrossRef]
6. Steiner, E.; Auer, A.; Becker, T.; Gastl, M. Comparison of beer quality attributes between beers brewed with 100% barley malt and 100% barley raw material. *J. Sci. Food Agric.* **2012**, *92*, 803–813. [CrossRef] [PubMed]
7. Gomaa, A.M. Application of Enzymes in Brewing. *J. Nutr. Food Sci. Forecast.* **2018**, *1*, 17–30.
8. Cadenas, R.; Caballero, I.; Nimubona, D.; Blanco, C.A. Brewing with starchy adjuncts: Its influence on the sensory and nutritional properties of beer. *Foods* **2021**, *10*, 1726. [CrossRef]
9. Rosa, R.S.; Lannes, S.C.d.S. Impact of the use of unmalted adjuncts on the rheological properties of beer wort. *Food Sci. Technol.* **2022**, *42*, 1–8. [CrossRef]
10. Kok, Y.J.; Ye, L.; Muller, J.; Ow, D.S.W.; Bi, X. Brewing with malted barley or raw barley: What makes the difference in the processes? *Appl. Microbiol. Biotechnol.* **2019**, *103*, 1059–1067. [CrossRef]
11. Ambriz-Vidal, T.N.; Mariezcurrena-Berasain, M.D.; Heredia-Olea, E.; Pinzon Martinez, D.L.; Gutierrez-Ibañez, A.T. Potential of Triticale (X Triticosecale Wittmack) Malts for Beer Wort Production. *J. Am. Soc. Brew. Chem.* **2019**, *77*, 282–286. [CrossRef]

12. Cioch-Skoneczny, M.; Zdaniewicz, M.; Pater, A.; Skoneczny, S. Impact of triticale malt application on physiochemical composition and profile of volatile compounds in beer. *Eur. Food Res. Technol.* **2019**, *245*, 1431–1437. [CrossRef]
13. Hu, S.; Dong, J.; Fan, W.; Yu, J.; Yin, H.; Huang, S.; Liu, J.; Huang, S.; Zhang, X. The influence of proteolytic and cytolytic enzymes on starch degradation during mashing. *J. Inst. Brew.* **2014**, *120*, 379–384. [CrossRef]
14. Lalor, E.; Goode, D. Brewing with Enzymes. In *Enzymes in Food Technology*, 2nd ed.; Whitehurst, R.J., van Oor, M., Eds.; Blackwell Publishing Ltd.: Clive, IA, USA, 2009; pp. 163–193.
15. Demeester, A.; Laureys, D.; Baillière, J.; Huys, J.; Vermeir, P.; De Leyn, I.; Vanderputten, D.; De Clippeleer, J. Comparison of Congress Mash with Final 65 °C Mash for Wort Production with Unmalted Barley, Tritordeum, and Quinoa, with or without Pregelatinization and/or Enzyme Addition. *J. Am. Soc. Brew. Chem.* **2023**, *81*, 66–75. [CrossRef]
16. Min, W.; Jiang, S.; Liu, L.; Rui, Y.; Jain, R. A Survey on Food Computing. *ACM Comput. Surv.* **2019**, *52*, 92. [CrossRef]
17. Gonzalez Viejo, C.; Fuentes, S.; Torrico, D.D.; Godbole, A.; Dunshea, F.R. Chemical characterization of aromas in beer and their effect on consumers liking. *Food Chem.* **2019**, *293*, 479–485. [CrossRef]
18. Takahashi, M.B.; Coelho de Oliveira, H.; Fernández Núñez, E.G.; Rocha, J.C. Brewing process optimization by artificial neural network and evolutionary algorithm approach. *J. Food Process Eng.* **2019**, *42*, e13103. [CrossRef]
19. Mane, S.U.; Rao, M.R.N. Many-Objective Optimization: Problems and Evolutionary Algorithms-A Short Review. *Int. J. Appl. Eng. Res.* **2017**, *12*, 9774–9793.
20. Imandi, S.B.; Karanam, S.K.; Nagumantri, R.; Srivastava, R.K.; Sarangi, P.K. Neural networks and genetic algorithm as robust optimization tools for modeling the microbial production of poly-β-hydroxybutyrate (PHB) from Brewers' spent grain. *Biotechnol. Appl. Biochem.* **2023**, *70*, 962–978. [CrossRef]
21. Becker, T.; Enders, T.; Delgado, A. Dynamic neural networks as a tool for the online optimization of industrial fermentation. *Bioprocess Biosyst. Eng.* **2002**, *24*, 347–354.
22. Dębska, B.; Guzowska-Świder, B. Application of artificial neural network in food classification. *Anal. Chim. Acta* **2011**, *705*, 283–291. [CrossRef] [PubMed]
23. Zhang, Y.; Jia, S.; Zhang, W. Predicting acetic acid content in the final beer using neural networks and support vector machine. *J. Inst. Brew.* **2012**, *118*, 361–367. [CrossRef]
24. Hassen, E.B.; Asmare, A.M. Predictive performance modeling of Habesha brewery wastewater treatment plant using artificial neural networks. *Chem. Int.* **2019**, *5*, 87–96.
25. Methodensammlung der Mitteleuropäischen Analysenkommission. *Raw Materials: Barley, Adjuncts, Malt, Hops and Hop Products*; MEBAK: Freising, Germany, 2011.
26. Glatthar, J.; Heinisch, J.; Senn, T. A study on the suitability of unmalted triticale as a brewing adjunct. *J. Am. Soc. Brew. Chem.* **2002**, *60*, 181–187. [CrossRef]
27. Black, K.; Tziboula-Clarke, A.; White, P.J.; Iannetta, P.P.M.; Walker, G. Optimised processing of faba bean (*Vicia faba* L.) kernels as a brewing adjunct. *J. Inst. Brew.* **2021**, *127*, 13–20. [CrossRef]
28. Puligundla, P.; Smogrovicova, D.; Mok, C.; Obulam, V.S.R. Recent developments in high gravity beer-brewing. *Innov. Food Sci. Emerg. Technol.* **2020**, *64*, 102399. [CrossRef]
29. Goode, D.L.; Arendt, E.K. Developments in the supply of adjunct materials for brewing. In *Brewing*; Bamforth, C.W., Ed.; Blackwell Publishing Ltd.: Cambridge, UK, 2006; pp. 30–67.

foods

MDPI

Article

Boosting *Purnica granatum* L. Seed Oil Yield: An Adaptive Neuro-Fuzzy Interference System Fuels SC-CO_2 Extraction Breakthrough

Padej Pao-la-or [1], Boonruang Marungsri [1], Pornariya Chirinang [2], Kakanang Posridee [3], Ratchadaporn Oonsivilai [3,4,*] and Anant Oonsivilai [1,*]

1 School of Electrical Engineering, Institute of Engineering, Suranaree University of Technology, Nakhon Ratchasima 30000, Thailand; padej@sut.ac.th (P.P.-l.-o.); bmshvee@sut.ac.th (B.M.)
2 Applied Food and Nutrition Division, Faculty of Science and Technology, Phetchaburi Rajabhat University, Phetchaburi 76000, Thailand; pornariya.chi@mail.pbru.ac.th
3 School of Food Technology, Institute of Agricultural Technology, Suranaree University of Technology, Nakhon Ratchasima 30000, Thailand; posridee.ka@gmail.com
4 Health and Wellness Research Group, Suranaree University of Technology, Nakhon Ratchasima 30000, Thailand
* Correspondence: roonsivi@g.sut.ac.th (R.O.); anant@sut.ac.th (A.O.); Tel.: +66-44224233 (R.O.); +66-914193541 (A.O.)

Abstract: This study used supercritical fluid extraction to successfully enhance the conditions for extracting oil from pomegranate seeds. To determine the optimal extraction conditions for maximizing pomegranate oil yield, the researchers employed a Box–Behnken design experimental strategy, involving three parameters with three levels each: extraction pressure, extraction temperature, and extraction time. To determine the optimal optimization conditions, the Response Surface Method (RSM) and the Artificial Neural Fuzzy Intelligent System (ANFIS) were also used. The results revealed a strong correlation with the experimental data, demonstrating that both strategies were helpful in optimizing the extraction process. The ideal extraction parameters, according to this study, were an extraction pressure of 40 MPa, an extraction temperature of 55 °C, and an extraction time of 120 min with a CO_2 flow rate of 21.3 L/h.

Keywords: Box–Behnken design; percentage yield; oil; response surface methodology; adaptive neuro-fuzzy interference system; pomegranate seed

1. Introduction

Pomegranate seed oil obtained through cold extraction is classified as an aromatic essential oil. It is often used as an additive in cosmetics due to its rich oil content comprising antioxidant compounds and flavonoids. It is also popular in supplements because it is rich in fatty acids. Additionally, the important phytoestrogens found in pomegranate seed oil may also relieve the symptoms of menopause. Moreover, pomegranate oil may help to alleviate night sweats, prevent depression, and reduce vaginal dryness (although scientific data confirming this is lacking) [1]. According to Promprom [2], pomegranate seed and peel extracts are rich in phytoestrogens and exhibit estrogenic properties in ovariectomized rats. Pomegranate seed and peel extracts increase uterine wet weight and promote cornification of the vagina and proliferation of the uterine endometrium in ovariectomized rats. Pomegranate seed and peel extracts (at a dosage of 1000 mg/kg B.W.) tend to improve bone mineral densities, as determined by dual energy X-ray absorptiometry. Furthermore, several studies have found that pomegranate seed oil has a higher antioxidant capacity than red wine and is comparable to green tea extract [3].

Pomegranate products, including pomegranate seed oil (PSO) and fermented juice (PFJ), contain potent antioxidants and eicosanoid enzyme inhibitors in the form of flavonoids.

Studies have shown that PSO flavonoids inhibit both COX-1 and LOX-1 activity, whereas PFJ flavonoids only inhibit LOX-1 activity. Furthermore, PSO has comparable antioxidant properties to butylated hydroxyanisole (BHA) and green tea. These findings suggest that pomegranate products may offer potential health benefits by safeguarding cells from oxidative damage and decreasing inflammation [4].

The process extracting oil from seeds involves complex stages that are time-consuming and typically result in a low yield. It is estimated that commercially marketed pomegranate seed oil, sold in one-pound increments, is derived from seeds weighing more than 200 pounds (91 kg). The oil can be stored in a cool and dry place for only 14 to 16 months. Therefore, the production of pomegranate oil is expensive. Currently, the sales price for commercial pomegranate seed oil is 15 baht per 220 cc. The countries that produce pomegranate seed oil include China, Egypt, India, and Turkey [1]. There are several methods to extract oil, such as solvent extraction, compression, etc. [5]. Supercritical CO_2 extraction shines as a green champion in the extraction industry, offering safer and cleaner alternatives to traditional methods. Furthermore, this method delivers high-purity, consumer-friendly extracts. Its rise signals a promising shift towards sustainable and responsible extraction practices for the future [6].

Despite its growing popularity, the health benefits of pomegranate seed oil are still emerging. Although promising research suggests that its potent antioxidants, particularly punicic acid, may offer cardioprotective, anti-inflammatory, and blood-sugar-regulating effects, further studies are needed to solidify these claims and establish safe dosages. Although it shows promise as a complementary therapy for certain conditions, maintaining a balanced diet and a healthy lifestyle remain the pillars of good health. Pomegranate seed oil could serve as a promising addition to explore under the guidance of your healthcare provider [7].

Superheated hexane extraction (SHHE) is a promising technique for obtaining high-quality pomegranate seed oil with a high yield. The oil yield increases with higher temperatures and longer extraction times. The highest yield of 21.3% was attained under conditions of 70 °C for 30 min. The resulting oil possesses desirable characteristics, such as a light-yellow color, low acid value, high refractive index, and a rich content of bioactive compounds. These qualities make it suitable for various applications in the food, cosmetic, and pharmaceutical industries [8].

Pomegranate seed oil (PSO) boasts a wealth of bioactive compounds and potent antioxidant activity, making it a sought-after ingredient in food, cosmetics, and pharmaceuticals. Although PSO is traditionally extracted through cold pressing, this method faces challenges due to its low yield and high cost. This study investigated the SFE extraction of pomegranate seed oil (PSO) through a combination of experimental and modeling approaches [9–11]. Supercritical CO_2 extraction (SC-CO_2) has emerged as a superior alternative method, providing a high oil yield (17.1%), abundant bioactive compounds (with a punicic acid content of 60.6% and γ-tocopherol at 325.2 mg/100 g), and potent antioxidant activity comparable to reference antioxidants [12]. The optimal SC-CO_2 parameters for PSO include a pressure of 35 MPa, a temperature of 60 °C, and an extraction time of 90 min. However, the longer extraction time may prompt the consideration of alternative methods, such as ultrasonic-assisted extraction (UAE) [13]. UAE delivers exceptional oil yields and quality, with a high bioactive content (punicic acid content of 65.3% and γ-tocopherol at 296.8 mg/100 g) and strong antioxidant activity, all achieved within a shorter 36-min timeframe. However, it necessitates higher energy consumption compared to other methods. Supercritical fluid extraction (SFE) also offers promise, with complex modeling approaches such as Sovová and Chrastil models enabling optimization and a high punicic acid content compared to traditional Soxhlet extraction [14]. Ultimately, the selection of the optimal extraction method depends on specific priorities. For achieving high yield, substantial bioactive content, and potent antioxidant activity, SC-CO_2 and UAE reign supreme. For shorter processing times, SFE or other methods might be preferable. Considering the trade-offs between yield, quality, efficiency, and sustainability is crucial for selecting the most suitable PSO extraction technique.

Supercritical fluid extraction (SFE) offers immense potential for extracting bioactive compounds from seed oils; however, optimizing its complex parameters can be challenging. The Adaptive Neuro-Fuzzy Inference System (ANFIS) is a powerful tool that shines in this domain. Studies by Zhang et al. [15] and Li et al. [16] demonstrate its effectiveness, with ANFIS models achieving R^2 values exceeding 0.99 for predicting both yield and antioxidant activity/purity. These models identified optimal SFE conditions for specific goals, such as maximizing the yield (35 MPa pressure, 45 °C temperature) or purity (30 MPa pressure, 40 °C temperature). ANFIS's strength lies in its ability to learn from data and adapt to the intricate nonlinear relationships between SFE parameters and the desired outcomes. This translates to improved extraction efficiency, resulting in higher quality oil with maximized bioactive content while minimizing impurities. The benefits extend beyond SFE, with ANFIS's optimization capabilities finding applications in diverse fields, such as control systems, finance, and manufacturing. Overall, the ANFIS model emerges as a game-changer for optimizing SFE, paving the way for the more efficient, sustainable, and cost-effective extraction of valuable bioactive compounds from seed oils [17].

Seed oil extraction processes are being revolutionized by sophisticated optimization techniques, such as Response Surface Methodology (RSM) and the Adaptive Neuro-Fuzzy Inference System (ANFIS). Although both effectively model the intricate relationships between the extraction parameters and desired outcomes, ANFIS often shines with its superior prediction accuracy and optimization capabilities. Studies across diverse seed oils, including crude rubber, sea buckthorn, black cumin, and Nigella sativa, demonstrate this advantage. ANFIS models consistently achieve R^2 values that exceed 0.98 for yield prediction in SFE and successfully identify optimal conditions for maximizing yield and bioactive compound content [18–22]. Despite variations in specific pressure, temperature, and time values depending on the oil type, a pressure range of 30–35 MPa, a temperature range of 40–45 °C, and a 60-min extraction time have emerged as the optimal conditions for achieving high yield and quality across different studies. This trend underscores the potential of ANFIS to streamline and optimize seed oil extraction processes, leading to increased efficiency, improved product quality, and ultimately, a more sustainable and cost-effective approach for extracting valuable bioactive compounds from these diverse sources. Mathematical modeling methods, notably RSM, are crucial for evaluating and optimizing the extraction process. RSM combines mathematical and statistical principles to develop a mathematical model that describes the process, analyzes the impact of independent variables, and optimizes the processing parameters. This research investigates the elaborate relationship between the extraction parameters, namely pressure, temperature, and time, and their impact on pomegranate seed oil yield. By employing sophisticated optimization techniques (RSM and ANFIS), we aimed to uncover the ideal extraction conditions that maximize pomegranate seed oil yield.

2. Materials and Methods

2.1. Sample Preparation

Pomegranate seeds, sourced from Nakhon Ratchasima, Thailand, were subjected to a drying process within a tray dryer (Kluaynumtaitowop, Bangkok, Thailand) maintained at a temperature of 60 °C. Subsequently, the dried seeds were pulverized into a fine powder and stored under vacuum conditions at a temperature of 4 °C until further use.

2.2. Proximate Analysis

The proximate composition of the dried pomegranate seed powder was determined using standard analytical methods, including analyzing moisture (AOAC Method 925.10), ash (AOAC Method 900.02 A), protein (AOAC Method 928.08), fat (AOAC Method 945.16), crude fiber (AOAC Method 978.10), and carbohydrates (AOAC Method 995.13) [23].

2.3. Supercritical Carbondioxide Extraction (SC-CO_2)

All SC-CO_2 extraction experiments were conducted using a Spe-ed® SFC supercritical fluid machine obtained from Applied Separations Inc., (Allentown, PA, USA). A precise amount of dried *Punica granatum* L. seed powder was carefully weighed and placed into the extraction chamber. Subsequently, the extracted PSO was weighed to determine the percentage yield of extraction using the formula below [5,6]:

$$\text{Yield percentage} = (\text{oil (g)}/\text{seed powder (g)}) \times 100 \quad (1)$$

2.4. Experimental Design

The SC-CO_2 conditions were optimized using Response Surface Methodology (RSM), a widely used optimization technique [24,25]. RSM was employed to determine the optimal levels of three extraction variables (pressure, temperature, and time), with respect to the extraction yield response (Table 1). The independent and dependent variables were as follows: extraction pressure (X_1; 20, 30, and 40 MPa), extraction temperature (X_2; 50, 55, and 60 °C), and extraction time (X_3; 60, 90, and 120 min). The experiments were designed according to the Box–Behnken design, and the order of the experiments was completely randomized. The data were analyzed using regression analyses [5,6].

Table 1. Encoded and coded levels of independent variables used in the experimental design.

Symbols	Independent Variables	Coded Levels		
		−1	0	1
X_1	Extraction pressure (MPa)	20	30	40
X_2	Extraction temperature (°C)	50	55	60
X_3	Extraction time (min)	60	90	120

2.5. Experimental Validation of the Optimal Conditions

The extraction yield response was optimized by maximizing its value according to the developed model. To validate the model's accuracy, three additional extraction experiments were conducted under the predicted optimal operating conditions, which differed from those used to develop the model.

2.6. ANFIS Modelling

ANFIS (Adaptive Neuro-Fuzzy Inference System) is a modeling technique that blends the strengths of fuzzy logic and neural networks to create a system capable of learning and making decisions based on input data. The ANFIS model comprises three crucial phases: fuzzification, inference engine, and defuzzification.

During the fuzzification phase, input values are converted into fuzzy sets using membership functions (MFs). Membership functions represent the degree to which a given input value belongs to each fuzzy set. This nonlinear mapping of inputs aids in capturing the linguistic information inherent in the data. Although various types of membership functions can be employed, the Gaussian shape was considered in this instance.

The inference engine phase involves constructing fuzzy rules to determine the output based on the fuzzy inputs. These rules typically follow an IF–THEN format, where the antecedent (IF) part outlines the conditions based on the fuzzy inputs, and the consequent (THEN) part dictates the output fuzzy set. The ANFIS model utilizes two methods to generate these rules: grid partitioning (GP) and subtractive clustering (SC). In this work, the SC method was adopted as it generates rules by clustering the data, resulting in a minimal number of rules.

Finally, the defuzzification phase transforms the fuzzy output into a crisp value that can be utilized for decision-making or further processing. Defuzzification methods determine the single output value based on the fuzzy sets and their respective degrees of membership. In this case, the weighted average method was employed.

Overall, the ANFIS model effectively combines fuzzy-logic-based fuzzy rules with the adaptive learning capabilities of neural networks. This powerful tool is well-suited for modeling complex systems and making decisions based on uncertain or imprecise data [18].

An example of the ANFIS rules is outlined below:

IF $\times$ is A_1 and y is B_1, then f 1 = g1(x, y)
IF $\times$ is A_2 and y is B_2, then f 2 = g2 (x, y)

Where, the As and Bs are the membership functions of the two inputs x and y. The output f is determined using the outputs of the two rules, f1 and f2 [18].

2.7. Fatty Acid Composition by GC Chromatography

Fatty acid methyl esters (FAMEs) were synthesized through a transmethylation reaction using boron trifluoride in methanol. The FAMEs were subsequently analyzed using an Agilent 7890 (Agilent, Santa Clara, CA, USA) a instrument equipped with a 100 m × 0.25 mm × 0.2 µm fused silica capillary column (SP2560, Supelco Inc., Bellefonte, PA, USA). A flame ionization detector (FID) was used, with both the injector and detector temperatures maintained at 260 °C. The column temperature was initially set at 70 °C and was then gradually increased at a rate of 13 °C/min until a temperature of 175 °C was reached. Following this, the temperature was further elevated at a rate of 4 °C/min until a temperature of 240 °C was reached. Helium was utilized as the carrier gas, flowing at a rate of 20 mL/min. A 1 µL aliquot of FAME was injected with a split ratio of 1:30. The identification and quantification of fatty acids were achieved by comparing the relative retention times of FAME peaks in the sample with those of standard compounds (37 component FAME Mix, catalog No. 47885-U, Supelco, Bellefonte, PA, USA) [6,26].

3. Results

3.1. Proximate Analysis

The pomegranate seed powder had a moisture content of 3.53%, an ash content of 1.56%, a fat content of 2.63%, a crude fiber content of 45.26%, and a protein content of 9.34%.

In comparing the proximate analysis results of dried pomegranate seed powder and dried winged bean seed powder, significant compositional differences were revealed (Table 2. Dried pomegranate seed powder had a lower moisture content (3.53%) than dried winged bean seed powder (9.22%), indicating more thorough drying of pomegranate seeds. Moreover, dried pomegranate seed powder had lower ash (1.56%) and fat (2.63%) levels compared to dried winged bean seed powder, which contained higher amounts of ash (4.91%) and fat (17.51%). However, dried winged bean seed powder had a higher protein content (33.83%) than dried pomegranate seed powder (9.34%). Additionally, dried winged bean seed powder contained significant amounts of crude fiber (12.23%) and carbohydrates (22.30%), whereas dried pomegranate seed powder had a considerably higher crude fiber content (45.26%) without specifying the carbohydrate content. These differences highlight the contrasting nutritional profiles of the two powders, suggesting their potential suitability for different applications in the food and nutraceutical industries [27].

Table 2. Proximate analysis.

Components	Content (%)
Moisture	3.53 ± 0.15
Ash	1.56 ± 0.09
Protein	9.34 ± 0.38
Fat	2.63 ± 0.24
Crude fiber	45.26 ± 0.76
Carbohydrate	37.68 ± 0.70

3.2. Supercritical Carbondioxide Extraction (SC-CO_2)

The optimization of pomegranate seed oil extraction was carried out using the same RSM methodology employed by Liu et al. [7]. However, the composition of the pomegranate seed oil in this study differed significantly from theirs, featuring a higher proportion of unsaturated fatty acids, particularly oleic acid, linoleic acid, and α-linolenic acid. It is confirm Table 3 summarizes the percentage yields (Y1) of pomegranate oil obtained from all the experiments, including the response surface method and ANFIS modeling.

Table 3. Experimental design and response of independent variables to the extract parameters.

Exp. No.[a]	Independent Variables			*Punica granatum* L.		
	Pressure (MPa) X_1	Temp. (°C) X_2	Time (min) X_3	% Yield (Y_1) Experimental	% Yield (Y_1) RSM-Predicted	% Yield (Y_1) ANFIS-Predicted
1	20	50	90	3.57 ± 0.38 [d]	3.57	3.57
2	20	55	60	3.73 ± 0.28 [bcd]	3.73	3.73
3	20	55	120	3.80 ± 0.18 [bcd]	3.8	3.80
4	20	60	90	3.70 ± 0.12 [cd]	3.70	3.70
5	30	50	60	4.00 ± 0.21 [abcd]	4.00	4.00
6	30	50	120	4.24 ± 0.31 [abc]	4.24	4.24
7	30	55	90	4.33 ± 0.10 [ab]	4.35	4.35
8	30	55	90	4.44 ± 0.06 [a]	4.35	4.35
9	30	55	90	4.28 ± 0.22 [abc]	3.93	3.93
10	30	60	60	3.93 ± 0.29 [abcd]	4.14	4.14
11	30	60	120	4.14 ± 0.04 [abc]	4.25	4.25
12	40	50	90	4.25 ± 0.64 [abc]	4.10	4.10
13	40	55	60	4.10 ± 0.62 [abcd]	4.50	4.50
14	40	55	120	4.50 ± 0.13 [a]	4.21	4.21
15	40	60	90	4.21 ± 0.36 [abc]	3.57	3.57

[a] Experiments were conducted in a random order. [b,c,d] Same letter show no significant difference at $p < 0.05$, different letter show significant difference ($p < 0.05$) in the same column.

The statistical analysis of the results was conducted using Analysis of Variance (ANOVA) to assess the significance of mean differences. Duncan's procedure was utilized at a significance level of $p \leq 0.05$ for the mean analysis. Additionally, Response Surface Methodology (RSM) was applied to determine the regression coefficients and to evaluate the statistical significance of the model terms. RSM enables the fitting of regression models to experimental data, facilitating the identification of an optimal region encompassing all studied response variables. Design-Expert® software (Version 8.0.7.1, Stat-Ease, Inc., Minneapolis, MN, USA) was employed to meticulously craft the experimental design, rigorously analyze the collected data, and optimize the process parameters. Moreover, the data presented in Table 3 was subjected to One-Way ANOVA to establish correlations between the independent variables and to obtain the optimal extraction conditions for achieving the highest percentage yield, as outlined in Table 3.

According to Table 3, the optimal extraction conditions for pomegranate oil production were determined to be a pressure of 40 MPa, a temperature of 55 °C, and an extraction time of 120 min, resulting in a maximum yield of 4.50%. After placing the optimal condition value into the regression equation produced, the expected yield value was 4.55%. The results of the experiment revealed that extraction pressure affected the percentage yield of oil ($p < 0.05$) more than extraction temperature and extraction duration.

Furthermore, the data were tested to determine correlations between the independent variables and the yield %. The statistical analysis revealed a significant association ($p < 0.01$). Liu et al. [9] studied the extraction of pomegranate seed oil using SC-CO_2 extraction. They found that the optimal extraction conditions were a pressure of 37.9 MPa, a temperature of 47.0 °C, and a CO_2 flow rate of 21.3 L/h. Under these conditions, they were able to achieve an oil yield of 156.3 g/kg (dry basis) [9].

Table 4 provides an overview of the regression coefficients and the regression model. According to Raymond et al. [28], an R^2 value of at least 0.80 indicates a good fit for a model. The high value of the coefficient of multiple determination (R^2 = 0.9788) suggests that the model adequately represents the experimental results. The two methods of pomegranate seed oil extraction, SC-CO_2 extraction and screw pressing, both have their own advantages and disadvantages. SC-CO_2 extraction can achieve higher oil yields, but it is more expensive and requires specialized equipment. Screw pressing is simpler and less expensive, but it may not be as efficient at extracting oil. The choice of extraction method depends on the specific needs of the producer. If a high oil yield is required, then SC-CO_2 extraction may be the best choice. However, if cost is a major concern, then screw pressing may be a more viable option. Response surfaces can be visualized through three-dimensional plots, illustrating the functional response of two factors while keeping the other factor constant. The relationship between the extraction pressure, extraction temperature, and extraction time and the highest yield of pomegranate seed oil is depicted in Figure 1.

Table 4. Analysis of variance (ANOVA) of independent variables for the extraction of oil from pomegranate seeds. R-squared = 0.9788. df, degree of freedom.

Source	*Purnica granatum* L.					
	Sum of Squares	**df**	**Mean Square**	**F Value**	***p*-Value Prob > F**	**Significant**
Model	1.09	9	0.1216	25.66	0.0012	significant
A-pressure	0.6384	1	0.6384	134.69	<0.0001	
B-temp	0.0008	1	0.0008	0.1688	0.6982	
C-time	0.1058	1	0.1058	22.32	0.0052	
AB	0.0072	1	0.0072	1.52	0.2718	
AC	0.0272	1	0.0272	5.74	0.0619	
BC	0.0002	1	0.0002	0.0475	0.8361	
A^2	0.1975	1	0.1975	41.66	0.0013	
B^2	0.1281	1	0.1281	27.02	0.0035	
C^2	0.0275	1	0.0275	5.79	0.0611	
Residual	0.0237	5	0.0047			
Lack of Fit	0.0103	3	0.0034	0.5124	0.7135	not significant
Pure Error	0.0134	2	0.0067			
Cor Total	1.12	14				
	Std.Dev = 0.0688 R-Squared = 0.9788 Mean = 4.08 R-Squared = 0.9407 C.V.% = 1.69 Adeq Precision = 15.9213					

3.3. Fitting Model

The regression coefficients were derived from the percentage yields of pomegranate oil obtained in each experiment. The resulting model was statistically significant, indicating that at least one of the extraction variables was responsible for the variation in the response variable. The regression coefficients and the regression model are presented in the equation for *Purnita granatum* L. seed oil.

$$Y1 = -23.03375 + 0.189X_1 + 0.8475X_2 + 0.015583X_3 - 8.5 * 10^{-4}\, X_1\, X_2 + 2.75 * 10^{-4}\, X_1\, X_3 - 5 * 10^{-5}\, X_2\, X_3 - 2.313 * 10^{-3}\, X_{12} - 7.45 * 10^{-3}\, X_{22} - 9.6 * 10^{-5}\, X_{32} \quad (2)$$

The model showed a strong correlation with the experimental data, as evidenced by the high R-squared value of 0.9788. Additionally, the *p*-value obtained from the lack-of-fit test confirmed the model's ability to accurately represent the experimental findings. These results collectively demonstrate the model's effectiveness in capturing and representing the observed experimental data.

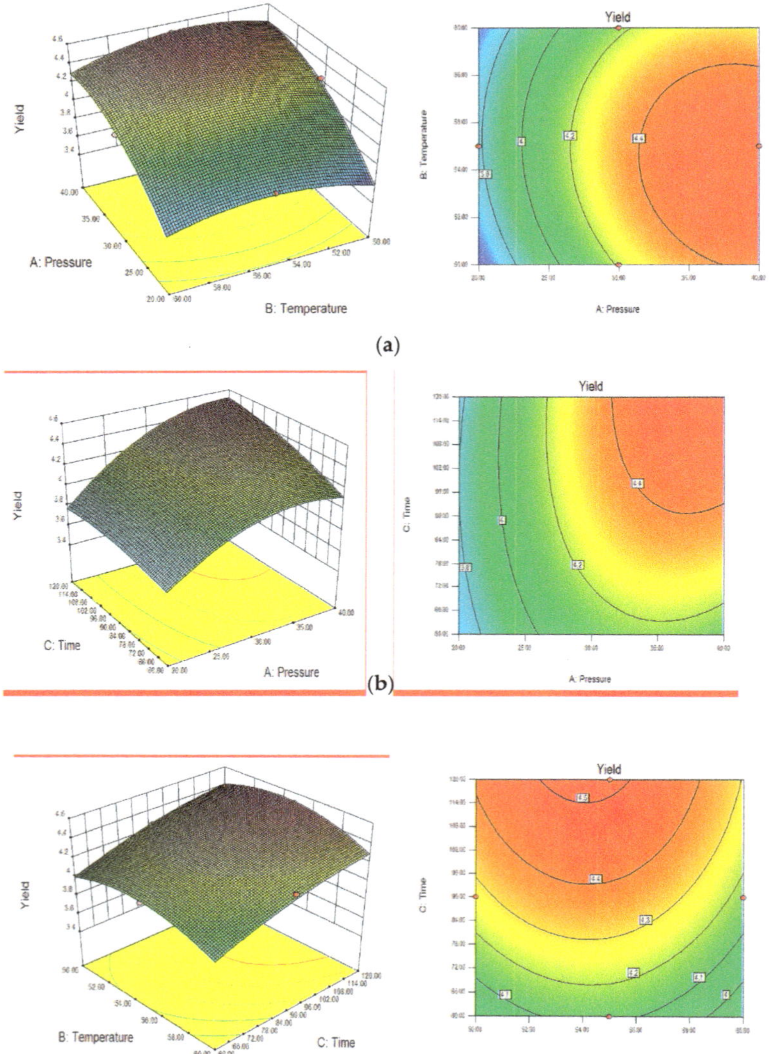

Figure 1. Response surface plots depicting the interactive effects of extraction pressure and time on the percentage yield of pomegranate oil (*Purnica granatum* L. oil) at a constant time of 120 min (**a**), a constant temperature of 55 °C (**b**), and a constant pressure of 40 MPa (**c**).

3.4. Analysis of Regression Coefficients

Increasing the extraction pressure considerably increased the oil extraction yield ($p < 0.0001$). Increasing the extraction time also tended to increase the yield ($p < 0.05$). However, after a certain point, further increasing the temperature reduced the yield, indicating an ideal temperature range (Figure 1) Reaction surfaces can be shown using three-dimensional plots that depict the reaction as a function of two factors while holding the third constant. The winged bean seed oil yield was strongly impacted by extraction pressure, temperature, and time (Figure 1). The response surface plot (Figure 1) reveals the interactive influence of extraction time and temperature on pomegranate seed oil yield when maintaining a constant extraction pressure of 40 MPa. The plot effectively demonstrates the synergistic relationship between these parameters in optimizing oil yield.

3.5. Validation of the Model

The most effective extraction method for pomegranate seed oil was determined by combining the extraction factors that resulted in the highest yield (22.29% and 4.50%). These conditions included an extraction pressure of 40 MPa, extraction temperatures of 50 °C and 55 °C, and extraction times of 90 and 120 min. Three extractions were performed under these conditions to verify the model's predictive capacity (Table 4). The predicted and experimental values of the percentage yield were in agreement, confirming the accuracy of the model. Furthermore, the data were analyzed to determine the correlation between independent variables and the percentage yield. The statistical analysis was significant, with a p-value less than 0.01.

3.6. Data Acquisition and Preprocessing for ANFIS

In the laboratory setting, determining the optimal extraction conditions can be a lengthy and complicated process, especially when dealing with three variables, each with three levels. To simplify this process and reduce the number of experiments, the Box–Behnken design is often employed. However, for achieving even better optimization, the Adaptive Neuro-Fuzzy Inference System (ANFIS) has proven to be a more effective approach compared to Response Surface Methodology (RSM). The Adaptive Neuro-Fuzzy Inference System (ANFIS) stands out as a valuable tool for optimization problems due to its ability to effectively combine the strengths of neural networks and fuzzy logic. The ANFIS model is particularly adept at handling intricate and nonlinear relationships between variables, making it a powerful instrument for optimizing various systems and processes. The performance of the ANFIS model is evaluated by comparing actual and predicted values and visually representing the linear regression analysis. Furthermore, as demonstrated in Figures 2 and 3, the ANFIS model offers valuable insights into its predictive accuracy compared to RSM.

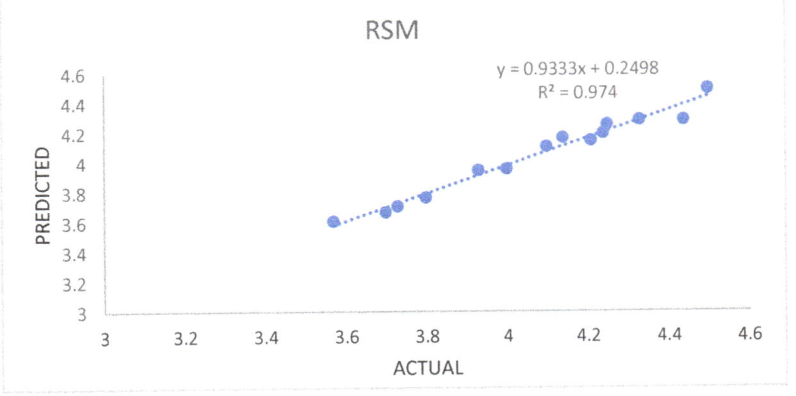

Figure 2. A plot of predicted and experimental values for the % yield of oil extracted from pomegranate seeds using RSM.

3.7. Fatty Acid Composition using GC Chromatography

The analysis of pomegranate seed oil using gas chromatography (GC) revealed the presence of various fatty acids, including saturated and unsaturated varieties. The saturated fatty acids identified in pomegranate seed oil include palmitic acid (C16:0), stearic acid (C18:0), and lignoceric acid (C24:0). Additionally, pomegranate seed oil contains unsaturated fatty acids, such as oleic acid (C18:1n9c), linoleic acid (C18:2n6c), and α-linoleic acid (C18:3n3). A detailed breakdown of the fatty acid profile of pomegranate seed oil is presented in Table 5 and Figure 4.

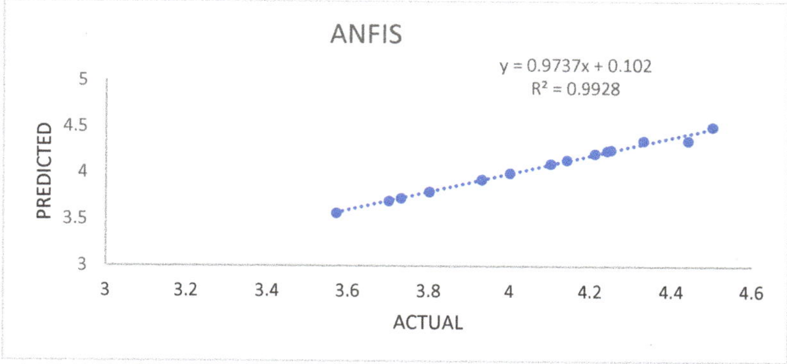

Figure 3. A plot of predicted and experimental values for the % yield of oil extracted from pomegranate seeds using the ANFIS model.

Table 5. Fatty acid profile of *Purnica granatum* L. oil prepared by SC-CO_2 extraction.

Ret Time (min)	Type	Area (Pa·s)	Amt/Area	Norm (%)	Grp	Name
27.676	BB	427.21164	9.55544×10^{-4}	9.990683	Palmitic acid sat	C16:0
34.103	BB	323.84680	9.45592×10^{-4}	7.494542	Stearic acid sat	C18:0
36.838	BV	837.20972	9.41893×10^{-4}	19.299116	Oleic acid w-9 FA	C18:1n9c
41.400	BB	796.17188	9.88680×10^{-4}	19.264782	Linoleic acid	C18:2n6c
45.734	BB	96.01950	1.23414×10^{-3}	2.900177	α-Linolenic acid ALA	C18:3n3
57.766	VB	1796.60498	9.33612×10^{-4}	41.050699	Lignoceric acid	C24:0
Total				100.000000		

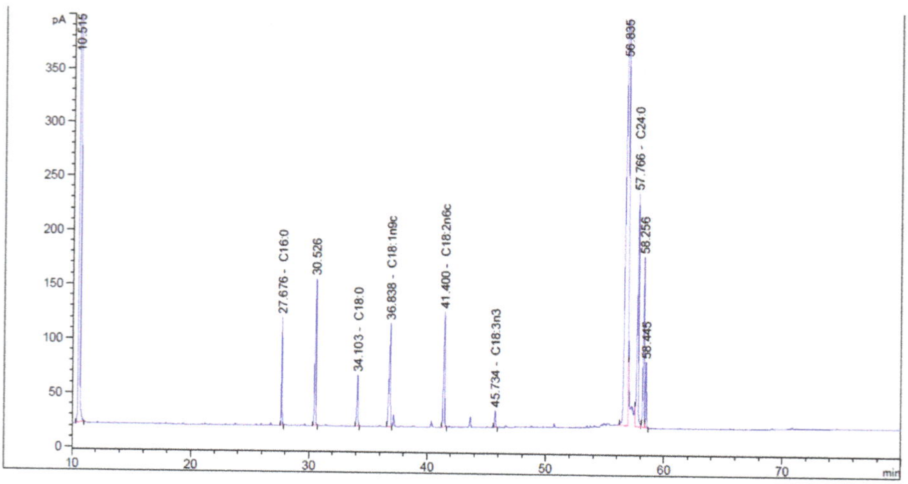

Figure 4. Gas chromatography of fatty acids from pomegranate seed oil using the SC-CO_2 extraction method.

4. Discussion

When comparing the models, both Kumar et al. [29] and Jisieike et al. [18] found the RSM and ANFIS models effective in predicting biodiesel yield with high R^2 values (>0.97). Jisieike et al. [18] further compared optimization algorithms, with ANFIS-PSO achieving the lowest final FFA content. These results suggest that both models are valuable tools, with ANFIS potentially offering superior optimization capability.

Abbasi et al. [30] compared several extraction methods and found that SC-CO_2 using hexane as the modifier provided the highest yield. However, they highlighted significant differences in the fatty acid profiles of the oil depending on the method used. Their results are not directly comparable to the seed oil profile presented in this study due to the use of an unknown extraction method.

The fatty acid profile of the seed oil provided in this research differs significantly from Abbasi et al.'s SC-CO_2 extraction with hexane in several key fatty acids, particularly palmitic, stearic, oleic, α-linolenic, and lignoceric acids. This implies that the extraction method used significantly impacts the fatty acid composition and potentially other oil properties.

Optimizing biodiesel production and maximizing the quality of pomegranate seed oil both demand strategic choices, including the selection of appropriate models for precise prediction and extraction methods that yield the desired composition. Although RSM excels in robust modeling, ANFIS, particularly when paired with PSO, holds promise for superior optimization in biodiesel production. Further research comparing their performance across diverse feedstocks and catalysts is crucial. For pomegranate seed oil, the stark differences in fatty acid profiles highlight the significant influence of the extraction methods on the properties of the oil. Future studies should investigate how method choice not only affects quality, but also the antioxidant content and potential health benefits. Ultimately, both biodiesel and pomegranate seed oil applications must consider cost-effectiveness and scalability for successful industrial implementation. This necessitates the careful consideration of both optimization strategies and extraction methods, ensuring a balance between the desired outcomes and economic feasibility.

5. Conclusions

The optimal conditions for extracting pomegranate seed oil with the highest percentage yield were determined using a randomized Box–Behnken design. The data were statistically analyzed using multiple regression. The conditions that yielded the highest percentage yield were an extraction pressure of 40 MPa, an extraction temperature of 55 °C, and an extraction time of 120 min, which resulted in a yield of 4.50%. Future research should focus on determining the fatty acid profile and bioactive compounds of pomegranate seed oil.

Author Contributions: Conceptualization, A.O. and R.O.; data curation, P.C.; methodology, P.C. and R.O.; formal analysis, P.C. and R.O.; investigation, A.O. and R.O.; resources, A.O.; writing—original draft preparation, R.O. and K.P.; writing—review and editing, R.O., A.O., P.P.-l.-o., and B.M. All authors have read and agreed to the published version of the manuscript.

Funding: This research was funded by the SUT Research and Development Fund.

Data Availability Statement: The data are contained within the article.

Acknowledgments: The authors would like to thank the Suranaree University of Technology Research and Development Fund.

Conflicts of Interest: The authors declare no conflicts of interest.

References

1. What Is Pomegranate Seed Oil? Available online: http://www.wisegeek.com/what-is-pomegranate-seed-oil.htm (accessed on 30 July 2010).
2. Promprom, W. Estrogenic Activity of Pomegranate (*Punica granatum*) Extract in Ovariectomized Rats. Ph.D. Thesis, School of Biology, Institute of Science, Suranaree University of Technology, Nakhon Ratchasima, Thailand, 2009.

3. Jurenka, J. Therapeutic Applications of Pomegranate (*Punica granatum* L.): A Review. *Altern. Med. Rev.* **2008**, *13*, 128–144.
4. Schubert, S.Y.; Lansky, E.P.; Neeman, I. Antioxidant and eicosanoid enzyme inhibition properties of pomegranate seed oil and fermented juice flavonoids. *J. Ethnopharmacol.* **1999**, *66*, 11–17. [CrossRef] [PubMed]
5. Chirinang, P.; Oonsivilai, R. Physicochemical properties, in-vitro binding capacities for lard, cholesterol, bile acids and assessment of prebiotic potential of dietary fiber from cassava pulp. *Int. Food Res. J.* **2018**, *25*, S63–S74.
6. Pao-la-or, P.; Marungsri, B.; Posridee, K.; Oonsivilai, R.; Oonsivilai, A. Supercritical CO_2 extraction of seed oil from *Psophocarpus tetragonolobus* (L.) DC: Optimization of operating conditions through response surface methodology and probabilistic neural network. *Processes* **2023**, *11*, 1949. [CrossRef]
7. Liu, G.; Xu, X.; Hao, Q.; Gao, Y. Supercritical CO_2 extraction optimization of pomegranate (*Punica granatum* L.) seed oil using response surface methodology. *LWT-Food Sci. Technol.* **2009**, *42*, 1491–1495. [CrossRef]
8. Eikani, M.H.; Golmohammad, F.; Homami, S.S. Extraction of pomegranate (*Punica granatum* L.) seed oil using superheated hexane. Food and Bioproducts Processing. *Food Bioprod. Process.* **2012**, *90*, 32–36. [CrossRef]
9. Sangwato, S.; Oonsivilai, A. Optimal power flow with interline power flow controller using hybrid genetic algorithm. *Int. Rev. Electr. Eng.* **2015**, *10*, 727–733. [CrossRef]
10. Oonsivilai, A.; El-Hawary, M.E. Self-organizing fuzzy power system stabilizer. In Proceedings of the Canadian Conference on Electrical and Computer Engineering, Waterloo, ON, Canada, 25–28 May 1998; pp. 197–200. [CrossRef]
11. Rodporn, S.; Kulworawanichpong, T.; Oonsivilai, A.; Uthitsunthorn, D.; Oonsivilai, R. Optimal coordination of over-current relays using differential evolution. In Proceedings of the 9th International Conference on Electrical Engineering/Electronics, Computer, Telecommunications and Information Technology, Phetchaburi, Thailand, 16–18 May 2012; pp. 1–4. [CrossRef]
12. Liu, G.; Xu, X.; Gong, Y.; He, L.; Gao, Y. Effects of supercritical CO_2 extraction parameters on chemical composition and free radical-scavenging activity of pomegranate (*Punica granatum* L.) seed oil. *Food Bioprod. Process.* **2012**, *90*, 573–578. [CrossRef]
13. Tian, Y.; Xu, Z.; Zheng, B.; Lo, Y.M. Optimization of ultrasonic-assisted extraction of pomegranate (*Punica granatum* L.) seed oil. *Ultrason. Sonochem.* **2013**, *20*, 202–208. [CrossRef]
14. Natolino, A.; Porto, C.D. Supercritical carbon dioxide extraction of pomegranate (*Punica granatum* L.) seed oil: Kinetic modelling and solubility evaluation. *J. Supercrit. Fluids* **2019**, *151*, 30–39. [CrossRef]
15. Zhang, Y.; Li, J.; Wang, J. Modeling and Optimization of Supercritical Fluid Extraction of Bioactive Compounds from Seed Oil Using Carbon Dioxide and ANFIS. *J. Chem. Technol. Biotechnol.* **2021**, *96*, 3610–3619. [CrossRef]
16. Li, H.; Li, T.; Zhao, C.; Liu, X. Application of Adaptive Neuro-Fuzzy Inference System (ANFIS) for modeling and optimization of supercritical fluid extraction of bioactive compounds from seed oil using carbon dioxide. *J. Food Eng.* **2022**, *322*, 108300. [CrossRef]
17. Salameh, T.; Rezk, H.; Issa, U.; Kamarudin, S.K.; Abdelkareem, M.A.; Olabi, A.G.; Alkasrawi, M. Boosting Biodiesel Production from Dairy-Washed Scum Oil Using Beetle Antennae Search Algorithm and Fuzzy Modelling. *Resources* **2023**, *12*, 131. [CrossRef]
18. Jisieike, C.F.; Ishola, N.B.; Latinwo, L.M.; Betiku, E. Crude rubber seed oil esterification using a solid catalyst: Optimization by hybrid adaptive neuro-fuzzy inference system and response surface methodology. *Energy J.* **2023**, *263*, 125734. [CrossRef]
19. Wang, L.; Li, H.; Zhao, C. Application of ANFIS for prediction and optimization of supercritical fluid extraction of bioactive compounds from Nigella Sativa seed oil using carbon dioxide. *J. Chem. Technol. Biotechnol.* **2023**, *98*, 432–441. [CrossRef]
20. Li, X.; Li, H.; Zhao, C.; Liu, X. Modeling and Optimization of Supercritical Fluid Extraction of Bioactive Compounds from Black Cumin Seed Oil Using ANFIS and Carbon Dioxide. *J. Food Eng.* **2023**, *335*, 111802. [CrossRef]
21. Zhang, Y.; Li, J.; Wang, J. Modeling and optimization of supercritical fluid extraction of bioactive compounds from sea buckthorn seed oil using ANFIS. *Food Bioproc. Tech.* **2023**, *16*, 53–62. [CrossRef]
22. Wang, L.; Li, H.; Zhao, C. Application of ANFIS for prediction and optimization of supercritical fluid extraction of bioactive compounds from seed oil using carbon dioxide. *J. Chem. Technol. Biotechnol.* **2023**, *98*, 213–222. [CrossRef]
23. Association of Official Analytical Chemists (AOAC). *Official Methods of Analysis Association of Official Analytical Chemists*; AOAC: Gaithersburg, MD, USA, 2019.
24. Posridee, K.; Oonsivilai, A.; Oonsivilai, R. Optimization of sweet cassava (*Manihot esculents crantz.*) crude extract with high maltodextrin level using Response Surface Methodology. *Int. Food Res. J.* **2018**, *25*, S51–S56.
25. Kachenpukdee, N.; Santerre, C.R.; Ferruzzi, M.G.; Oonsivilai, R. Enzymatic digestion optimization of dietary fiber from cassava pulp and their effect on mercury bioaccessibility and intestinal uptake from fish using an in vitro digestion/Caco-2 model. *Int. Food Res. J.* **2016**, *23*, 660–666.
26. Intarapichet, K.-O.; Suksombat, W.; Maikhunthod, B. Chemical compositions, fatty acid, collagen and cholesterol contents of Thai hybrid native and broiler chicken meats. *J. Poult. Sci.* **2008**, *45*, 7–14. [CrossRef]
27. Ananthasiri, T.; Pomloy, N.; Oonsivilai, R. Winged bean seed oil preparation and dietary supplement development from winged bean seed, pomegranate seed and moringa oleifera seed oil mixture. In Proceedings of the GMSARN International Conference on Smart Energy, Environment, and Development for Sustainable GMS, Luanprabang, Lao People's Democratic Republic, 27–29 November 2020.
28. Raymond, H.M.; Douglas, C.M.; Christine, M.A.-C. *Response Surface Methodology: Process and Product Optimization Using Designed Experiments*, 4th ed.; John Wiley & Sons: Hoboken, NJ, USA, 2016; 856p, ISBN 978-1-118-91601-8.

29. Kumar, S.; Jain, S.; Kumar, H. Application of adaptive neuro-fuzzy inference system and response surface methodology in biodiesel synthesis from jatropha–algae oil and its performance and emission analysis on diesel engine coupled with generator. *Energy J.* **2021**, *226*, 120428. [CrossRef]
30. Abbasi, H.; Rezaei, K.; Emamdjomeh, Z.; Mousavi, S.M.E. Effect of various extraction conditions on the phenolic contents of pomegranate seed oil. *Eur. J. Lipid Sci. Technol.* **2008**, *110*, 435–440. [CrossRef]

foods

Article

Development of a Low-Cost Artificial Vision System as an Alternative for the Automatic Classification of Persian Lemon: Prototype Test Simulation

Bridget V. Granados-Vega [1], Carlos Maldonado-Flores [2], Camila S. Gómez-Navarro [1], Walter M. Warren-Vega [1], Armando Campos-Rodríguez [2,*,†] and Luis A. Romero-Cano [1,*,†]

[1] Grupo de Investigación en Materiales y Fenómenos de Superficie, Departamento de Biotecnológicas y Ambientales, Universidad Autónoma de Guadalajara, Av. Patria 1201, Zapopan 45129, Mexico; bridget.granados@edu.uag.mx (B.V.G.-V.); camila.gomez@edu.uag.mx (C.S.G.-N.); wwarrenv12@gmail.com (W.M.W.-V.)

[2] Laboratorio de Innovación y Desarrollo de Procesos Industriales Sostenibles, Departamento de Biotecnológicas y Ambientales, Universidad Autónoma de Guadalajara, Av. Patria 1201, Zapopan 45129, Mexico; carlos.maldonado@edu.uag.mx

* Correspondence: armando.crodriguez@edu.uag.mx (A.C.-R.); luis.cano@edu.uag.mx (L.A.R.-C.)

† These authors contributed equally to this work.

Abstract: In the present research work, an algorithm of artificial neural network (ANN) has been developed based on the processing of digital images of Persian lemons with the aim of optimizing the quality control of the product. For this purpose, the physical properties (weight, thickness of the peel, diameter, length, and color) of 90 lemons selected from the company Esperanza de San José Ornelas SPR de RL (Jalisco, Mexico) were studied, which were divided into three groups (Category "extra", Category I, and Category II) according to their characteristics. The parameters of weight (26.50 ± 3.00 g), diameter/length (0.92 ± 0.08) and thickness of the peel (1.50 ± 0.29 mm) did not present significant differences between groups. On the other hand, the color (determined by the RGB and HSV models) presents statistically significant changes between groups. Due to the above, the proposed ANN correctly classifies 96.60% of the data obtained for each of the groups studied. Once the ANN was trained, its application was tested in an automatic classification process. For this purpose, a prototype based on the operation of a stepper motor was simulated using Simulink from Matlab, which is connected to three ideal switches powered by three variable pulse generators that receive the information from an ANN and provide the corresponding signal for the motor to turn to a specific position. Manual classification is a process that requires expert personnel and is prone to human error. The scientific development presented shows an alternative for the automation of the process using low-cost computational tools as a potential alternative.

Keywords: MATLAB Simulink model prototype; quality control; colorimetric chart; artificial neural networks; electronic eye

1. Introduction

In Mexico, two varieties of lemon are mainly cultivated, the Persian and the Mexican, corresponding to 30% and 70% of the total production. The Persian lemon is dedicated almost exclusively to export, so its quality controls are strict to satisfy the demands of the market [1]. According to information from national producers, lemons for export can be classified according to their characteristics in: (i) Category "extra", which must have a superior quality and must not present defects, except for very slight superficial ones as long as the general appearance of the product is not affected; (ii) Category I, which has slight defects in shape, color, and surface that do not exceed 1 cm^2; and (iii) Category II, those lemons that do not fit the previously mentioned categories and that present superficial defects that exceed 1 cm^2. Once the lemon is harvested, most farmers carry out this classification manually, which generates errors in the quality of the process, making it essential to use an artificial vision system that allows the process to be automated and avoid human errors [2].

In this sense, recent research has focused on proposing solutions based on the use of artificial vision and machine learning to solve this type of problem [3]. Through the use of artificial intelligence, such as the implementation of artificial neural networks, it has been possible to evaluate complex systems for the prediction and control of different areas in the food industry, such as extrusion [4], process conditions (formulations) [5], drying [6], or evaluating its functional content [7]. In the last few years, artificial neural networks have proven to be advantageous because they allow the evaluation of complex and multivariable systems, thus providing an indicator that demonstrates the quality of a product. For example, the analysis to estimate total chlorophyll content in apples has been studied by color analysis and spectra data [8], or measuring six color values from surface images to predict texture characteristics such as hardness and gumminess [9]. Another use of artificial neural networks is to classify products by their physicochemical properties and be discriminant with respect to location, composition, or quality [10–12]. In this sense, various studies have been reported for the evaluation of the quality of fruits using image analysis in combination with convolutional models of neural networks (CNN), among which studies in apples [13], lemons [14], grapevine [15], papaya [16], tomato [17], strawberry [18], blueberry [19], banana [20], and melon [21] stand out. In all cases, the studies are based on the acquisition of data using digital cameras, which are subsequently processed mathematically to obtain the RGB (RGB color model is an additive color model in which the red, green, and blue primary colors of light are added together in various ways to reproduce a broad array of colors) and/or HSV (the HSV model is an alternative representation of the RGB color model in which the hue, saturation, and value are represented) values in order to provide this information to the designed CNN algorithm. The data set obtained is entered into the network for training to obtain a result. After the analysis, the error between the given labels and the output predictions is calculated using a loss function to propagate the error throughout the network by means of a backpropagation algorithm. Finally, the weights are updated to minimize the error, and the procedure is repeated until it converges or reaches a limit of iterations [22].

However, although there is now a lot of information reported on the use of artificial intelligence to evaluate the quality of different fruits, it must be integrated into automation systems to achieve product control at an industrial level. In the food industry, the automation of processes becomes essential since its selection is more efficient than human work due to the minimization of errors [23–26]. Based on this, the objective of this research corresponds to the development of a prototype and its potential application in the automatic classification of the three categories that define the quality of a Persian lemon. To achieve these, an artificial neural network was developed and trained using the physicochemical properties of the fruit for subsequent programming in the prototype model and testing its operation in a computer-assisted simulation.

2. Materials and Methods

2.1. Samples

A batch of 90 Persian lemons (*Citrus* × *latifolia*) was selected from the company Esperanza de San José Ornelas SPR de RL (Jalisco, Mexico), which is dedicated to the production and distribution of various fruits (blueberry, corn, and Persian lemon). The samples were harvested and immediately classified by the expert workers of the company, who have more than 20 years of experience; then, within 1 day, the lemons were transported to our laboratory. Of these, 30 correspond to the Category "extra", 30 to Category I, and 30 to Category II (Figure 1).

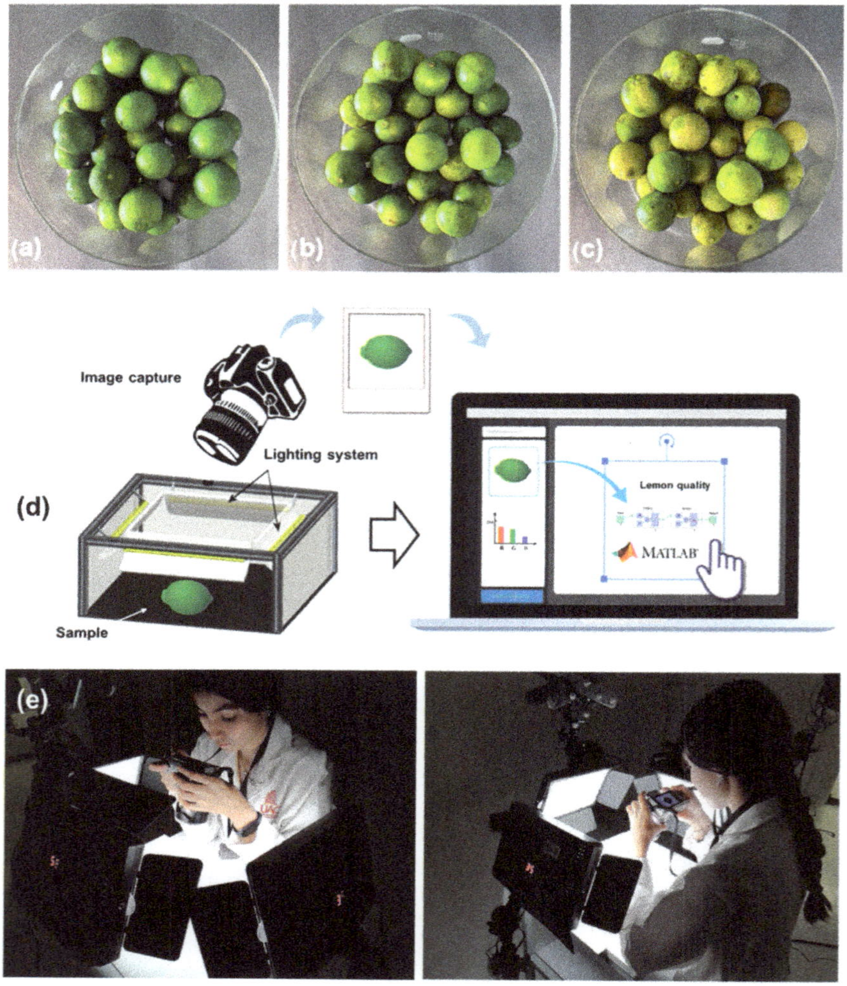

Figure 1. Photographs of the visual classification of lemons: (**a**) Category "extra", (**b**) Category I, and (**c**) Category II. (**d**) Visual artificial system to classify lemons; (**e**) experimental Set-up.

STATISTICA 10.0 software (StatSoft, Palo Alto, CA, USA) was used to analyze the experimental data. The descriptive statistics analysis was performed by determining the mean and standard deviation. Additionally, a one-way analysis of variance (ANOVA) was employed to determine the existence of statistically significant differences between the means of the groups (Category extra, Category I, and Category II) using a significance level of 95%.

2.2. Color Study Using Digital Image Processing

The digital image processing was carried out in a similar way as previously reported by Warren-Vega et al. [27]. It is important to highlight that one of the limitations of the study is related to the fact that the classification of fruits based on image processing is sensitive to the lighting conditions and the background color at the time of taking the photograph, so it is important to standardize the method. Due to the above, digital images were performed under controlled light conditions and were subjected to continuous and direct illumination from a white LED light source (45 W) at a power of 35%. All photos were taken at a distance of 30 cm between the object and the camera. Each sample was analyzed in duplicate, performing 3 repetitions to ensure the repeatability and reproducibility of the measures. The operating conditions of the camera were taken with the manual option, at a lens objective of 18–25 mm and a lens aperture in a range of 4 to 5 at an ISO 500. (Note: camera specifications do not limit the application of the method. However, the size of the photography area selected for RGB decomposition needs to be observed to contain a minimum number of pixels to provide the same analytical sensitivity). Using its CCD (Charge-Couple Device, Sony Corp., Bangkok, Thailand) detector, it is possible to capture digital images and convert them into a voltage sequence that can be translated into an analytical signal. The digital images obtained for each sample were decomposed using the RGB model (red, green, and blue), and Color Intensity (C.I.) was obtained from the sum of the three values (C.I. = red + green + Blue).

Image Analysis. The digital images obtained were saved as individual files in jpeg format. The average size per image was 3 MB (14.2 megapixel resolution, 4592 × 3056 pixels). Image processing consisted of selecting and clipping a region of interest (defined as at least 60% of the total area of the fruit resulting in a new image with a dimension of 843 × 880 pixels. The obtained images were decomposed according to the RGB color model into distribution histograms for each channel (red, green, and blue) using the MATLAB R2023a software. The results obtained were corroborated using the free app Color Name® (version 3.2, 2022). After that, the values of RGB were transformed to HSV (Hue, Saturation, and Value), which resembles color in human perception [28].

2.3. *Physical Properties Analysis in Lemons*

Lemon measurements were performed to obtain the following characteristics: thickness, weight, length, and diameter. For measurements in lemons, a NIMBUS analytical balance and a digital Vernier were employed.

2.4. *Artificial Neural Network Approach*

The MATLAB R2023a program was used to establish an artificial neural network algorithm that allows the classification of lemons based on quality standards (Category "extra", Category I, and Category II). In the case of the inputs, physical properties and colorimetric measurements (intensity of color, hue, saturation, and value) were used as classifiers to obtain the group to which they belonged (outputs). The optimal conditions for the artificial neural network architecture consisted of a total of 5 neurons with a sigmoid hidden layer and a softmax output layer to permit a correct classification. The network will be trained using the scaled conjugated gradient method. To evaluate the performance of the ANN, cross-entropy performance was used to establish the best architecture. The training of the neural network consisted as follows: 60 lemons (n = 20 lemons from each group) were used to obtain the algorithm, in which data was randomly divided into training (70%), validation (15%), and testing (15%). Subsequently, the remaining 30 lemons were used as a second set with the objective of establishing their effectiveness. The architecture of the neural network is shown in Figure 2.

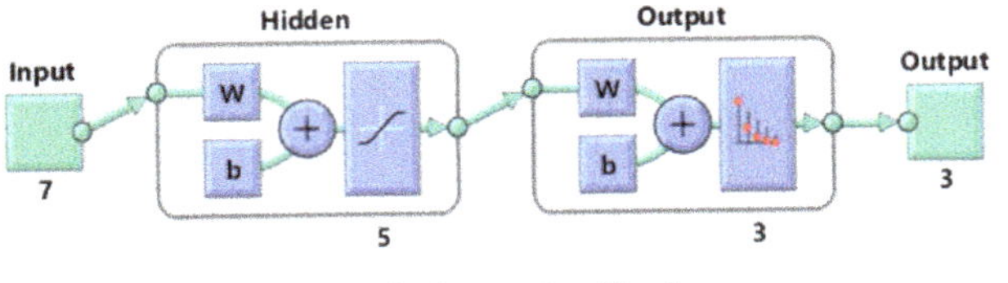

Figure 2. ANN architecture for lemon classification.

2.5. *MATLAB Simulink Model Prototype*

A model prototype was developed to evaluate the feasibility of the automatic classification of lemons using the proposed ANN. For this purpose, input and output interfaces were developed using control algorithms in Simulink. The proposed model consisted of connecting the ANN to three variable pulse generators, which, depending on the value returned by the ANN, activate an ideal switch connected to a stepper motor (motor type: variable reluctance, number of phases: 3, maximum winding inductance: 0.01, minimum winding inductance: 0.002, winding resistance 1.2 ohm, step angle: 120°, initial position: 0°, total inertia: 2×10^{-5} kg m^2, and total friction: 0.001 N m), which rotates at 0°, 120° (clockwise) or −120° (anticlockwise) depending on the category of lemon detected by the ANN. It is proposed that turning the motor will send the lemon to the corresponding basket according to the corresponding category.

3. Results and Discussion

3.1. Physicochemical Parameters of Persian Lemons: Analysis of Categories

Within the three different groups of lemons analyzed based on their quality, the RGB analysis was carried out to appreciate differences in the colors presented by the fruits. Figure 3 and Supplementary Figures S1–S3 present representative samples of the three categories analyzed. At least 60% of the total fruit area was studied in all cases. Once the digital images were obtained, they were decomposed into the primary colors red, green, and blue. The results are presented as an RGB color histogram, representing its distribution in the image. From this information, the maximum values of each of the analyzed colors were defined. For the category extra, the green color predominates (R = 90, G = 107, B = 8) and is characteristic of the fruit freshly harvested.

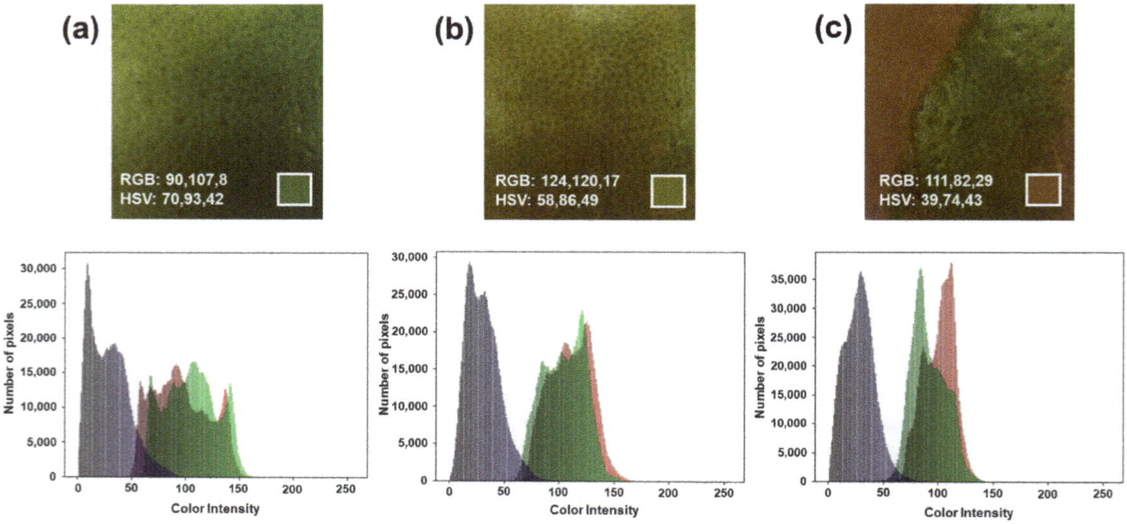

Figure 3. Representative sample of digital images and RGB color distribution for each of the categories of lemons analyzed: (**a**) Category extra, (**b**) Category I, and (**c**) Category II.

On the other hand, for categories I and II, there is a displacement of the red color, almost wholly overlapping with the green color in category I (R = 124, G = 120, B = 17), resulting in a yellow coloration characteristic of fruit ripening. Finally, category II stands out for the superposition of red and green color with a majority tendency toward the intensity of the red tone (R = 111, G = 82, B = 29), associated with the localized brown tone of the fruit caused by mechanical damage at harvest time. From this information, it is concluded that the area analyzed is representative of each of the fruits studied, evidencing that the proposed method satisfactorily identifies minor alterations.

Using the RGB analysis obtained from the decomposition of the images, the color intensity value was calculated as the first variable for the artificial neural network.

To demonstrate the color differences presented in the groups, Figure 4 was made to demonstrate the variability that the camera sensor detected and how it was perceived. In the color chart, the extra category of lemons is mainly green, because that is the way consumers prefer it in North America. At this stage, the fruit has not yet reached maturity, so the main pigment in the rind corresponds to chlorophyll. As the fruit begins to ripen, its color tones change from light green tones (category I) to yellow tones (category II), in which the pigment of the rind is mainly carotenoids [29]. For this reason, the changes in the color of the fruit are related to its maturity, in which the degradation of chlorophylls occurs in four stages: (i) synthesis of Mg-protoporphyrin IX from glutamic acid through a series of reactions and then conversion to chlorophyll a; (ii) interconversion of chlorophyll a and chlorophyll b, also known as the chlorine leaf plant cycle; (iii) chlorophyll binding; and (iv) degradation of chlorophyll [30].

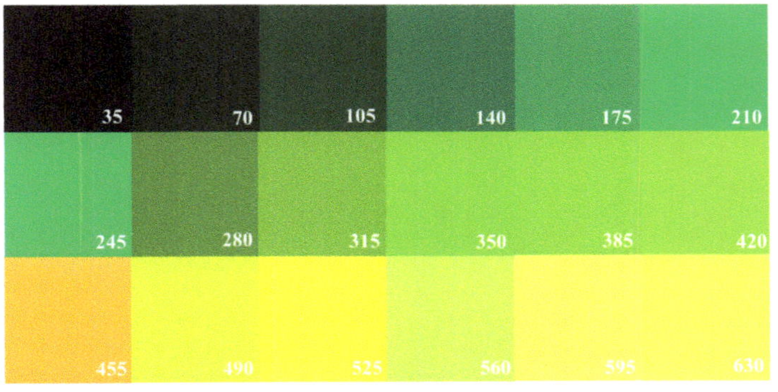

Figure 4. Color chart of the lemons analyzed based on their quality criteria. The number corresponds to Color Intensity (CI). From top to bottom: Category "extra" CI (35–210), Category I CI (245–420), and Category II CI (455–630).

Finally, the RGB data was converted to another color space, HSV, which has demonstrated better results in evaluating fruit properties by machine learning methods [29]. In Figure 5 and Supplementary Figure S4, the statistical analysis of hue, saturation, value, color intensity, red, green, and blue can be appreciated. Saturation was the only property that demonstrated a not-significant difference between the excellent and medium groups. With this information, it is concluded that there are differences in the colors, thus establishing significant variables for the design of the ANN architecture, as has been reported by other authors [31,32].

To complete the characterization of the lemons studied, the properties of weight, sphericity (Diameter/Length, D/L) and thickness of the peel were evaluated. The results are presented in Table 1 and Supplementary Figure S5. It is observed that there are no significant differences between the properties studied in the categories of fruit quality (weight p-value = 0.00017 < 0.05; D/L p-value = 0.181327 > 0.05; thickness p-value = 0.22179 > 0.05). For all categories, the weight was 25 g with a peel thickness of 1.5 mm on average, resulting in spherical fruits (D/L = 0.9). The foregoing shows that the consumer's perception of the quality of the Persian lemon depends exclusively on the maturity of the fruit, which can be measured from image analysis.

From this information, the physicochemical parameters of lemons are going to be used to feed the ANN for Persian lemon classification.

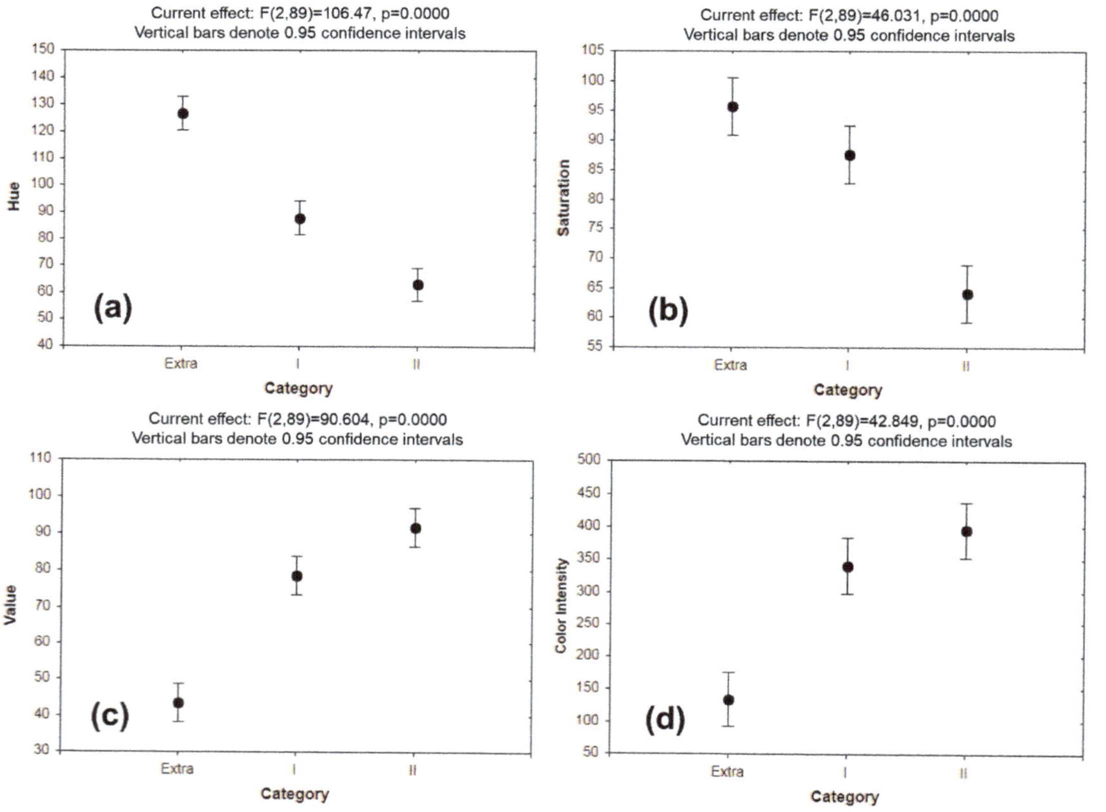

Figure 5. Statistical analysis of the values of (**a**) hue, (**b**) saturation, (**c**) value, and (**d**) color intensity in Persian lemons. Where: Category "extra" = high quality, Category I = medium quality, and Category II = low quality.

Table 1. Input variables in an artificial neural network for the classification of lemons.

Properties	Category "Extra"	Category I	Category II
Hue	126.83 ± 7.23	87.83 ± 25.60	60.95 ± 9.08
Saturation	95.60 ± 7.21	87.85 ± 11.07	62.44 ± 17.36
Value	42.39 ± 20.85	78.65 ± 12.05	91.84 ± 4.88
Color intensity	127.43 ± 56.62	341.2 ± 68.54	538.55 ± 57.49
Weight (g)	26.50 ± 3.00	25.69 ± 3.02	23.05 ± 3.08
Diameter/Length	0.92 ± 0.08	0.91 ± 0.06	0.97 ± 0.07
Thickness (mm)	1.50 ± 0.29	1.48 ± 0.32	1.29 ± 0.33

3.2. Construction and Training of the Proposed ANN

Having this, the neural network was used to classify between the different groups, having within the training a 100% classification within the confusion matrix using 60 lemons, as shown in Figure 6.

To evaluate the feasibility of the ANN, it was proposed to perform a second set, in which the remaining 30 lemons were employed to verify that the algorithm would allow classifying future cases. The results are shown in Figure 7. It can be appreciated that a 93.3% correct classification could be achieved with two misclassified lemons as Category II being Category I. In the case of the ROC curve, it can be appreciated that the true positive rate is 1 for Category "extra", then 0.8 for Category I, and finally 0.7 for Category II.

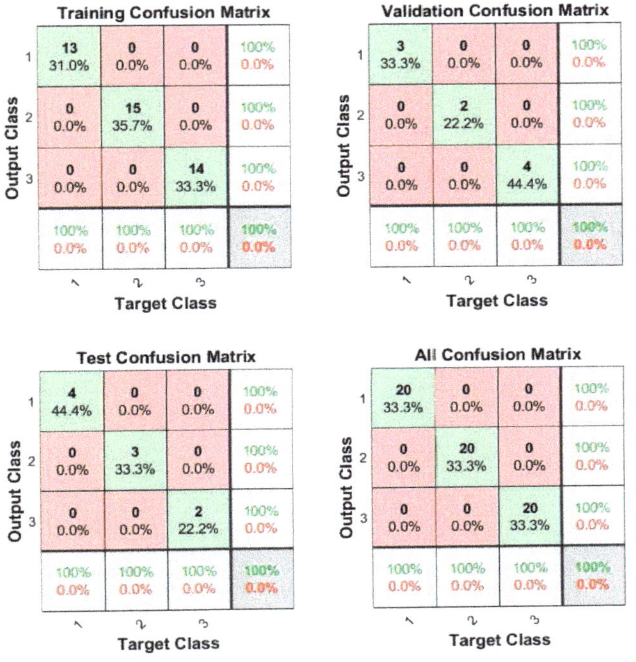

Figure 6. Confusion matrix for lemon classification (outputs: 1: Category "extra", 2: Category I, 3: Category II).

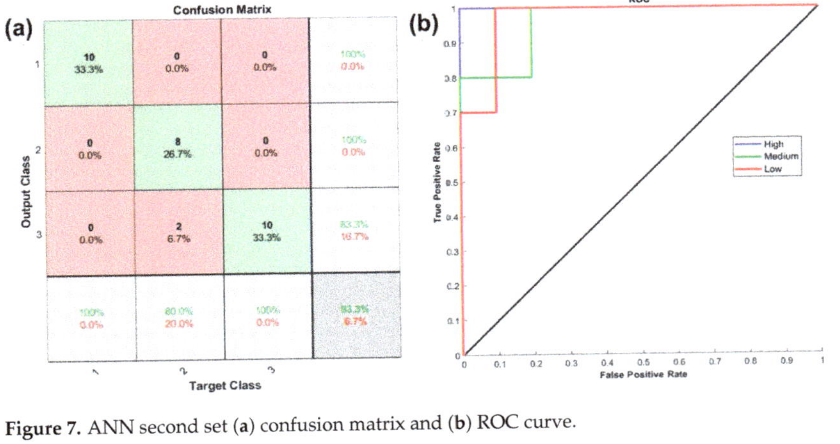

Figure 7. ANN second set (**a**) confusion matrix and (**b**) ROC curve.

With the result of both validations, a value of 96.60% of the correct classification is obtained using the physical and colorimetric properties of the analyzed fruits.

The results obtained show that the use of artificial neural networks for the classification of lemon quality is feasible. Similar results have been previously reported showing that the use of machine learning is helpful for fruit classification (Table 2); however, in all cases, the studies focus on the development of basic principles, which, according to the Technology Readiness Level (TRL), classify the technology at level 1 (TRL1). In order to show an advance with respect to the previously reported studies, proofs of concept were carried out by connecting the proposed artificial neural networks to a stepper motor (Section 3.3) to demonstrate the usefulness of the technology in developing an automatic classification system at an industrial level. Based on the TRL, these results classify the proposed technology at a TRL3 level.

Table 2. Bibliographic comparison against outstanding works previously reported in the literature.

Fruit	Study	Goal	Ref.
Apple	Laser-induced light backscattering imaging and convolutional neural network	Automatic identification and detection of apple defects achieving a recognition rate of 92.5%	[13]
Lemon	Stochastic Pooling Mechanism in Deep Convolutional Neural Networks	Detect defects and grade them achieving a recognition rate of 100%	[14]
Papaya	CNNs * and Convolutional Block Attention Modules	disease detector with a mean average precision of 86.2%	[33]
Tomato	Deep CNN *	nutritional deficiencies detection in the cropping process with a mean average precision of 91%	[17]
Strawberry	Deep CNN *	quality classification achieving a recognition rate of 96.49%	[18]
Blueberry	CNN *, ResNet *, ResNeXt *	Detection of internal mechanical damage achieving a recognition rate of 89.5%	[19]
Banana	CNN *	Fine-grained classification of banana's ripening stages achieving a recognition rate of 94.4%	[20]
Melon	CNN *	Skin lesion achieving an accuracy rate of 98.5%	[21]

* CNN—Convultional Neural Network; ResNet—Residual Network; ResNeXt—Improved version of ResNet.

3.3. MATLAB Simulink Model Prototype: Test Simulation

In order to show the technological maturity of the proposed prototype, the simulation of the device was carried out using the Simulink model presented in Figure 8. The simulation results are shown in Figure 9.

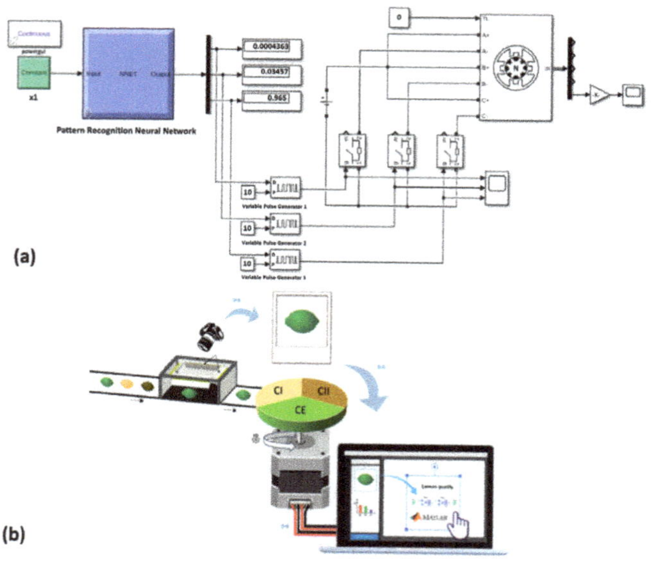

Figure 8. (**a**) MATLAB Simulink model prototype; (**b**) proposed diagram for the automatic classification of lemons.

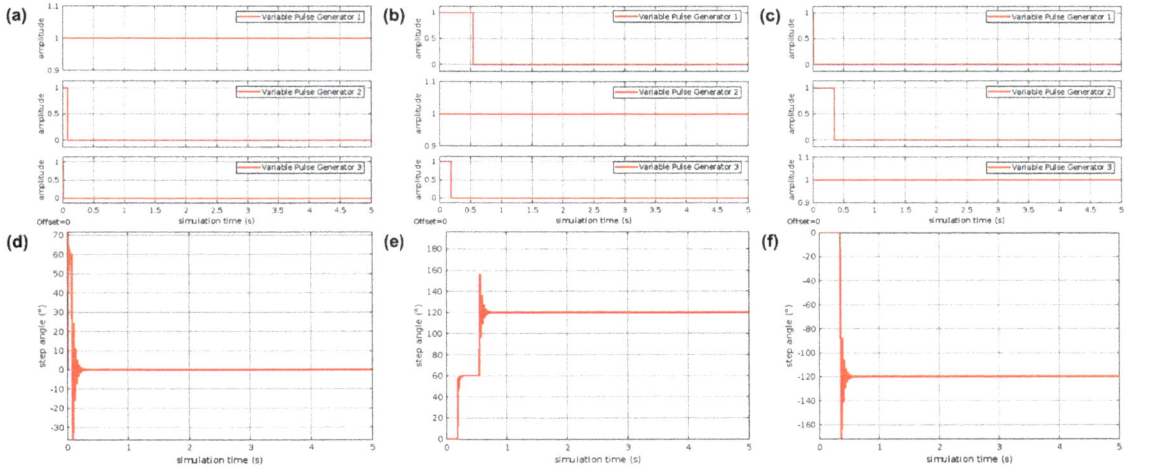

Figure 9. Pulse outputs (**a**–**c**) and step angles (**d**–**f**) with different delay times based on the classification of lemons using the trained ANN.

Figure 9a–c shows typical forms of amplitude modulated pulse output at 0 and 1 corresponding to the input duty cycle by changing the on/off delay time of the switches. It is observed that depending on the signal sent by the ANN, only one of the three pulse generators gives an amplitude equal to 1, which is translated into the rotation of the stepper motor (Figure 9d–f) depending on the category of lemon detected. It should be noted that the turn response time is dependent on the probability value thrown by the ANN; that is, for values close to 1.0, the turn response is fast and decreases as the value of the probability of success of the ANN decreases. However, in none of the cases tested, the response time was greater than 1 s, so the proposed prototype is feasible for use.

Based on the above, the system shown in Figure 7b is proposed as a prototype of an automatic lemon quality classifier, which consists of taking a photograph that is decomposed according to the RGB and HSV models. This information is sent to the MATLAB software in which the previously trained ANN is used, and the resulting information is sent to the controller (stepper motor), which transforms the digital signal into a mechanical rotation of the axis capable of diverting the lemon to the corresponding basket according to your category. From the results obtained, it is concluded that the technological maturity of the proposed device is at a Technology Readiness Level (TRL) of 3 since a proof-of-concept has been satisfactorily evaluated.

4. Conclusions

The present study shows the basis for the development of a simple, low-cost, and fast technology for the automatic classification of Persian lemons according to their quality parameters to preserve the standards requested by the market. It is possible to use a low-cost device to analyze digital images and, coupled with an ANN, make decisions to differentiate between categories. The algorithm developed to analyze digital images shows 96.6% accuracy for product classification, showing itself as a potential application to automate manual fruit classification processes, helping to reduce errors made by personnel and increasing the speed of the process with the aim of having low costs. The proposed prototype model has been evaluated through simulations, evidencing its feasibility of use. As a perspective, it is contemplated to evaluate the proposed prototype on a pilot scale under controlled conditions in order to show the proposal with a level of technological maturity of TRL4 (technology validated in the lab).

Supplementary Materials: The following supporting information can be downloaded at: https://www.mdpi.com/article/10.3390/foods12203829/s1. Supplementary Figure S1. Representative sample of 10 lemons analyzed for category "extra". Note: The differences in the dimensions of the images correspond to the condition of analyzing at least 60% of the total area of the fruit. Supplementary Figure S2. Representative sample of 10 lemons analyzed for category I. Note: The differences in the dimensions of the images correspond to the condition of analyzing at least 60% of the total area of the fruit. Supplementary Figure S3. Representative sample of 10 lemons analyzed for category II. Note: The differences in the dimensions of the images correspond to the condition of analyzing at least 60% of the total area of the fruit. Supplementary Figure S4. Statistical analysis of the values of (a) red, (b) green, and (c) blue in Persian lemons according to the category. Supplementary Figure S5. Statistical analysis of the values of (a) weight, (b) diameter/length, and (c) thickness in Persian lemons according to the category.

Author Contributions: B.V.G.-V. and C.S.G.-N., Investigation, Data curation, Methodology, Formal analysis. C.M.-F., Validation, Data curation. W.M.W.-V., Conceptualization, Data Curation, Formal analysis, Methodology, Software, Original writing—draft, Writing: proofreading and editing. A.C.-R. and L.A.R.-C. Conceptualization, Supervision, Methodology, Project administration, Resources, Original writing—draft, Writing—proofreading and editing. All authors have read and agreed to the published version of the manuscript.

Funding: The APC was funded by the Dirección de Investigación—Universidad Autónoma de Guadalajara.

Data Availability Statement: The data that support the findings of this study are available from the corresponding author upon reasonable request.

Acknowledgments: B.V.G.-V. and C.M.-F. thanks the Universidad Autónoma de Guadalajara for the scholarship received for the completion of undergraduate studies. C.S.G.-N. and L.A.R.-C. thanks the "Dirección de Investigación—UAG" for the support received through the "Convocatoria de Proyectos de Investigación para incidir en el entorno UAG 2022". The authors thank Arturo Padilla-Padilla and the "Centro de Producción Audiovisual—UAG" for the support received during the analysis of digital images. The authors appreciate the support provided by the company Esperanza de San José Ornelas SPR de RL for carrying out this research work.

Conflicts of Interest: The authors declare no conflict of interest.

References

1. Raddatz-Mota, D.; Franco-Mora, O.; Mendoza-Espinoza, J.A.; Rodríguez-Verástegui, L.L.; Díaz de León-Sánchez, F.; Rivera-Cabrera, F. Effect of Different Rootstocks on Persian Lime (*Citrus latifolia* T.) Postharvest Quality. *Sci. Hortic.* **2019**, *257*, 108716. [CrossRef]
2. Mukhtar, T.; Jamil, S.; Arif, U.; Razzaq, W.; Wasif, M. Lemon Grading and Sorting Using Computer Vision. *Eng. Proc.* **2021**, *12*, 55. [CrossRef]
3. Nayak, J.; Vakula, K.; Dinesh, P.; Naik, B.; Pelusi, D. Intelligent Food Processing: Journey from Artificial Neural Network to Deep Learning. *Comput. Sci. Rev.* **2020**, *38*, 100297. [CrossRef]
4. Cubeddu, A.; Rauh, C.; Delgado, A. Hybrid Artificial Neural Network for Prediction and Control of Process Variables in Food Extrusion. *Innov. Food Sci. Emerg. Technol.* **2014**, *21*, 142–150. [CrossRef]
5. Batista, L.F.; Marques, C.S.; Pires, A.C.d.S.; Minim, L.A.; Soares, N.d.F.F.; Vidigal, M.C.T.R. Artificial Neural Networks Modeling of Non-Fat Yogurt Texture Properties: Effect of Process Conditions and Food Composition. *Food Bioprod. Process.* **2021**, *126*, 164–174. [CrossRef]
6. Zalpouri, R.; Singh, M.; Kaur, P.; Kaur, A.; Gaikwad, K.K.; Singh, A. Drying Kinetics, Physicochemical and Thermal Analysis of Onion Puree Dried Using a Refractance Window Dryer. *Processes* **2023**, *11*, 700. [CrossRef]
7. da Silva Sauthier, M.C.; da Silva, E.G.P.; da Silva Santos, B.R.; Silva, E.F.R.; da Cruz Caldas, J.; Cavalcante Minho, L.A.; dos Santos, A.M.P.; dos Santos, W.N.L. Screening of *Mangifera indica* L. Functional Content Using PCA and Neural Networks (ANN). *Food Chem.* **2019**, *273*, 115–123. [CrossRef]
8. Pourdarbani, R.; Sabzi, S.; Hernández-Hernández, M.; Hernández-Hernández, J.L.; Gallardo-Bernal, I.; Herrera-Miranda, I. Non-Destructive Estimation of Total Chlorophyll Content of Apple Fruit Based on Color Feature, Spectral Data and the Most Effective Wavelengths Using Hybrid Artificial Neural Network—Imperialist Competitive Algorithm. *Plants* **2020**, *9*, 1547. [CrossRef]
9. Fan, F.H.; Ma, Q.; Ge, J.; Peng, Q.Y.; Riley, W.W.; Tang, S.Z. Prediction of Texture Characteristics from Extrusion Food Surface Images Using a Computer Vision System and Artificial Neural Networks. *J. Food Eng.* **2013**, *118*, 426–433. [CrossRef]
10. Debska, B.; Guzowska-Swider, B. Application of Artificial Neural Network in Food Classification. *Anal. Chim. Acta* **2011**, *705*, 283–291. [CrossRef]

11. Cimpoiu, C.; Cristea, V.M.; Hosu, A.; Sandru, M.; Seserman, L. Antioxidant Activity Prediction and Classification of Some Teas Using Artificial Neural Networks. *Food Chem.* **2011**, *127*, 1323–1328. [CrossRef]
12. da Silva, C.E.T.; Filardi, V.L.; Pepe, I.M.; Chaves, M.A.; Santos, C.M.S. Classification of Food Vegetable Oils by Fluorimetry and Artificial Neural Networks. *Food Control* **2015**, *47*, 86–91. [CrossRef]
13. Wu, A.; Zhu, J.; Ren, T. Detection of Apple Defect Using Laser-Induced Light Backscattering Imaging and Convolutional Neural Network. *Comput. Electr. Eng.* **2020**, *81*, 106454. [CrossRef]
14. Jahanbakhshi, A.; Momeny, M.; Mahmoudi, M.; Zhang, Y.D. Classification of Sour Lemons Based on Apparent Defects Using Stochastic Pooling Mechanism in Deep Convolutional Neural Networks. *Sci. Hortic.* **2020**, *263*, 109133. [CrossRef]
15. Barré, P.; Herzog, K.; Höfle, R.; Hullin, M.B.; Töpfer, R.; Steinhage, V. Automated Phenotyping of Epicuticular Waxes of Grapevine Berries Using Light Separation and Convolutional Neural Networks. *Comput. Electron. Agric.* **2019**, *156*, 263–274. [CrossRef]
16. Munasingha, L.V.; Gunasinghe, H.N.; Dhanapala, W.W.G.D.S. Identification of Papaya Fruit Diseases Using Deep Learning Approach. In Proceedings of the 4th International Conference on Advances in Computing and Technology (ICACT–2019), Colombo, Sri Lanka, 29 July 2019.
17. Tran, T.T.; Choi, J.W.; Le, T.T.H.; Kim, J.W. A Comparative Study of Deep CNN in Forecasting and Classifying the Macronutrient Deficiencies on Development of Tomato Plant. *Appl. Sci.* **2019**, *9*, 1601. [CrossRef]
18. Sustika, R.; Subekti, A.; Pardede, H.F.; Suryawati, E.; Mahendra, O.; Yuwana, S. Evaluation of Deep Convolutional Neural Network Architectures for Strawberry Quality Inspection. *Int. J. Eng. Technol.* **2018**, *7*, 75–80.
19. Wang, Z.; Hu, M.; Zhai, G. Application of Deep Learning Architectures for Accurate and Rapid Detection of Internal Mechanical Damage of Blueberry Using Hyperspectral Transmittance Data. *Sensors* **2018**, *18*, 1126. [CrossRef]
20. Zhang, Y.; Lian, J.; Fan, M.; Zheng, Y. Deep Indicator for Fine-Grained Classification of Banana's Ripening Stages. *EURASIP J. Image Video Process* **2018**, *2018*, 46. [CrossRef]
21. Tan, W.; Zhao, C.; Wu, H. Intelligent Alerting for Fruit-Melon Lesion Image Based on Momentum Deep Learning. *Multimed. Tools Appl.* **2016**, *75*, 16741–16761. [CrossRef]
22. Naranjo-Torres, J.; Mora, M.; Hernández-García, R.; Barrientos, R.J.; Fredes, C.; Valenzuela, A. A Review of Convolutional Neural Network Applied to Fruit Image Processing. *Appl. Sci.* **2020**, *10*, 3443. [CrossRef]
23. Pandey, V.K.; Srivastava, S.; Dash, K.K.; Singh, R.; Mukarram, S.A.; Kovács, B.; Harsányi, E. Machine Learning Algorithms and Fundamentals as Emerging Safety Tools in Preservation of Fruits and Vegetables: A Review. *Processes* **2023**, *11*, 1720. [CrossRef]
24. Zhu, L.; Spachos, P.; Pensini, E.; Plataniotis, K.N. Deep Learning and Machine Vision for Food Processing: A Survey. *Curr. Res. Food Sci.* **2021**, *4*, 233–249. [CrossRef]
25. Zhang, Y.; Deng, L.; Zhu, H.; Wang, W.; Ren, Z.; Zhou, Q.; Lu, S.; Sun, S.; Zhu, Z.; Gorriz, J.M.; et al. Deep Learning in Food Category Recognition. *Inf. Fusion* **2023**, *98*, 101859. [CrossRef]
26. Kakani, V.; Nguyen, V.H.; Kumar, B.P.; Kim, H.; Pasupuleti, V.R. A Critical Review on Computer Vision and Artificial Intelligence in Food Industry. *J. Agric. Food Res.* **2020**, *2*, 100033. [CrossRef]
27. Warren-Vega, W.M.; Contreras-Atrisco, Z.A.; Ramírez-Quezada, M.F.; Romero-Cano, L.A. A Novel Approach of Artificial Intelligence for the Study of the Relation of Physicochemical Profile and Color Acquired by Tequila 100% Agave in Its Maturation Process. *J. Food Compos. Anal.* **2023**, *123*, 105533. [CrossRef]
28. Heil, J.; Marschner, B.; Stumpe, B. Digital Photography as a Tool for Microscale Mapping of Soil Organic Carbon and Iron Oxides. *Catena* **2020**, *193*, 104610. [CrossRef]
29. Rodrigo, M.J.; Alquézar, B.; Alós, E.; Lado, J.; Zacarías, L. Biochemical Bases and Molecular Regulation of Pigmentation in the Peel of Citrus Fruit. *Sci. Hortic.* **2013**, *163*, 46–62. [CrossRef]
30. Tanaka, A.; Tanaka, R. Chlorophyll Metabolism. *Curr. Opin. Plant Biol.* **2006**, *9*, 248–255. [CrossRef]
31. Cho, B.H.; Koseki, S. Determination of Banana Quality Indices during the Ripening Process at Different Temperatures Using Smartphone Images and an Artificial Neural Network. *Sci. Hortic.* **2021**, *288*, 110382. [CrossRef]
32. Mukherjee, A.; Sarkar, T.; Chatterjee, K.; Lahiri, D.; Nag, M.; Rebezov, M.; Shariati, M.A.; Miftakhutdinov, A.; Lorenzo, J.M. Development of Artificial Vision System for Quality Assessment of Oyster Mushrooms. *Food Anal. Methods* **2022**, *15*, 1663–1676. [CrossRef]
33. de Moraes, J.L.; de Oliveira Neto, J.; Badue, C.; Oliveira-Santos, T.; de Souza, A.F. Yolo-Papaya: A Papaya Fruit Disease Detector and Classifier Using CNNs and Convolutional Block Attention Modules. *Electronics* **2023**, *12*, 2202. [CrossRef]

foods

MDPI

Article

Deep Learning-Based Near-Infrared Hyperspectral Imaging for Food Nutrition Estimation

Tianhao Li [1,2,†], Wensong Wei [3,4,†], Shujuan Xing [3,4], Weiqing Min [1,2,*], Chunjiang Zhang [3,4,*] and Shuqiang Jiang [1,2]

1 The Key Laboratory of Intelligent Information Processing, Institute of Computing Technology, Chinese Academy of Sciences, Beijing 100190, China
2 University of Chinese Academy of Sciences, Beijing 100049, China
3 Institute of Food Science and Technology, Chinese Academy of Agricultural Sciences, Beijing 100193, China
4 Key Laboratory of Agro-Products Processing, Ministry of Agriculture and Rural Affairs, Beijing 100193, China
* Correspondence: minweiqing@ict.ac.cn (W.M.); chjiang1976@126.com (C.Z.)
† These authors contributed equally to this work.

Abstract: The limited nutritional information provided by external food representations has constrained the further development of food nutrition estimation. Near-infrared hyperspectral imaging (NIR-HSI) technology can capture food chemical characteristics directly related to nutrition and is widely used in food science. However, conventional data analysis methods may lack the capability of modeling complex nonlinear relations between spectral information and nutrition content. Therefore, we initiated this study to explore the feasibility of integrating deep learning with NIR-HSI for food nutrition estimation. Inspired by reinforcement learning, we proposed OptmWave, an approach that can perform modeling and wavelength selection simultaneously. It achieved the highest accuracy on our constructed scrambled eggs with tomatoes dataset, with a determination coefficient of 0.9913 and a root mean square error (RMSE) of 0.3548. The interpretability of our selection results was confirmed through spectral analysis, validating the feasibility of deep learning-based NIR-HSI in food nutrition estimation.

Keywords: deep learning; near-infrared hyperspectral imaging; food nutrition estimation; wavelength selection

1. Introduction

Diet plays a crucial role in maintaining health. Poor diet stands as a prominent instigator of diseases on a global scale, contributing to over a quarter of preventable fatalities across the world [1]. Food nutrition estimation is a way to ensure a healthy diet by allowing individuals to understand the specific nutritional value of foods [2]. Traditionally, it has relied on laboratory analysis, a time-consuming, costly, and labor-intensive process requiring specialized skills and equipment. While professional dietitians also offer a means of estimating food nutrition, their estimations may have slightly lower accuracy and may not fully align with the public's increasing demand for convenient and reliable methods of daily food nutrition estimation.

In recent years, as a subdomain of food computing [3], food nutrition estimation techniques have emerged as promising alternatives. Most existing methods rely on food's external representation to estimate its nutrition [4], including from the chewing sounds [5], food images [6–8], or ingredient text [9]. However, these methods are highly dependent on datasets and lack a comprehensive analysis of the chemical properties of the food, which is crucial for accurate nutrition estimation. A more promising approach is near-infrared hyperspectral imaging (NIR-HSI) which offers a direct and precise prediction of food nutrition by analyzing functional groups and molecular structures. NIR-HSI has been shown to be an effective tool for assessing various food quality parameters, including carbohydrate, fat, and protein composition [10,11]. Hu et al. leveraged off-the-shelf near-infrared emitting diodes and a photodiode to develop a portable calorie estimation system, demonstrating the great potential of near-infrared technology for food nutrition estimation in everyday life [12]. However, existing works in this field have mainly focused on simple foods such as staple foods, drinks, and fruits, which exhibit higher inter-class similarities in terms of their nutritional composition. Yet more complex food items, such as Chinese dishes, may exhibit larger variations in nutritional content due to differences in ingredient proportions. The accurate estimation of nutrition in such diverse food items remains a significant challenge. Therefore, to address this research gap and explore the potential of combining deep learning and NIR-HSI technology for food nutrition estimation, we present a case study on predicting the protein content in scrambled eggs with tomatoes. Protein plays a vital role in food nutrition as it provides essential amino acids and serves as a fundamental building block for human cells, tissues, and organs [13]. Consequently, accurately estimating the protein content in food is of great importance for nutritional assessment. In the context of Chinese dishes, scrambled eggs with tomatoes, a popular dish, has been selected as the focal point of this study. This choice is motivated by the dish's simplicity, consisting of only two main ingredients, and the significant impact that ingredient proportions can have on its protein content. Therefore, analyzing the protein content in scrambled eggs with tomatoes serves as an illustrative case to explore the potential of combining deep learning and NIR-HSI technology for accurate food nutrition estimation.

The main contributions of this study are as follows: (1) Construct a near-infrared spectral dataset of scrambled eggs with tomatoes using meticulous measurements and processing. (2) Propose a novel deep learning method called OptmWave, which integrates two neural networks to simultaneously predict protein content and select wavelengths. (3) The comparison with conventional approaches demonstrates the effectiveness of our proposed deep learning methods. (4) The interpretability of our selection results is confirmed through near-infrared spectral analysis.

2. Material and Methods

2.1. Sample Preparation

In this study, tomatoes were purchased from Beijing Xingfurongyao supermarket, China, and eggs were bought from Beijing Hongyuan Technology Company, Beijing, China. The tomatoes were blanched and peeled using boiling water and then cut into 8 g chunks. Meanwhile, the eggs were evenly stirred using an eggbeater. The total weight of each sample was 560 g, consisting of raw materials (500 g) and seasonings (60 g). Based on the Chinese dietary guidelines, the seasonings used in the dish included oil (50 g), salt (5 g), and sugar (5 g). The dish incorporated varying proportions of eggs and tomatoes as raw materials: 50 g egg with 450 g tomatoes, 100 g egg with 400 g tomatoes, 150 g egg with 350 g tomatoes, 200 g egg with 300 g tomatoes, and 250 g egg with 250 g tomatoes. To diversify the sample types, the study also included samples containing only eggs or tomatoes.

The scrambled egg with tomatoes was prepared according to a traditional Chinese recipe: eggs and salt were evenly mixed, added to 160 °C oil, and fried for 1–1.5 min before the addition of tomatoes and sugar. The cooking time was 3.5 min, using a cooking power of 2100 W. Once cooled to an ambient temperature of 24 °C, the samples were homogenized using a KENWOOD-AT320B cooking-chef machine at third gear for 30 s. The resulting mixture was poured into Petri dishes with a diameter of 3 cm and a height of 0.5 cm and filled to a weight of 8 g before being smoothed out.

2.2. Hyperspectral Image Acquisition and Spectra Extraction

2.2.1. Hyperspectral Imaging System

The reflectance spectra of the scrambled eggs with tomatoes samples were recorded using a near-infrared (NIR) system provided by American Ocean Optics Company. The NIR system comprised a FLAME-S-VIS-NIR-ES spectrometer with a wavelength range of 900 to 1700 nm, an HL-2000-FHSA light source, a diffuse reflection standard plate, a reflection probe bracket, and a laboratory-level reflection probe with a diameter of 400 μm. For spectrum acquisition, OceanView 1.6.7 software was employed, which allowed for precise control and measurement of the reflectance spectra. The spectral resolution achieved with this setup was 3.1 nm, ensuring detailed characterization of the samples. The FLAME-S-VIS-NIR-ES spectrometer served as the primary instrument for capturing the reflectance spectra. It utilized the HL-2000-FHSA light source to illuminate the samples, and the reflected light was collected using the reflection probe. The probe, securely held in place by the reflection probe bracket, maintained consistent positioning during the measurements. The diffuse reflection standard plate was employed as a reference to calibrate the system and ensure accurate measurements. The process of hyperspectral image acquisition is illustrated in the Figure 1.

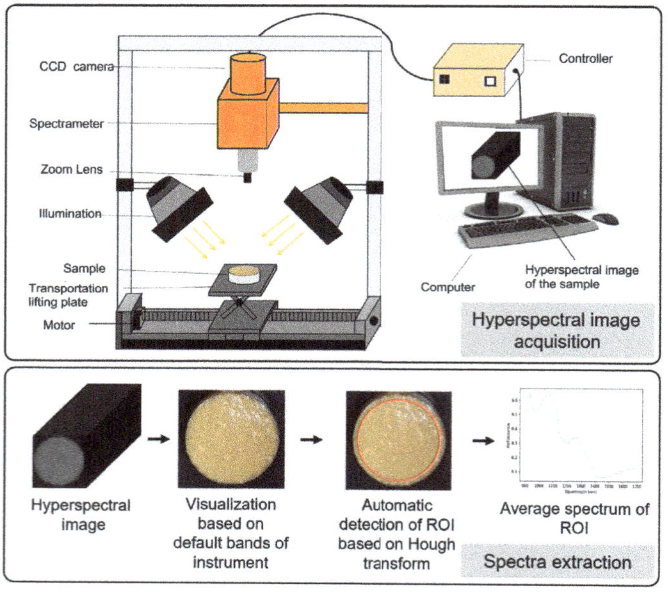

Figure 1. The process of hyperspectral image acquisition spectra extraction.

2.2.2. Hyperspectral Image Correction

To minimize the impact of solar altitude angle variations on spectral measurements, sample spectra were obtained between 10:00 and 14:00 Beijing time. Throughout this time frame, the solar altitude angle remained above 45 degrees [14]. Prior to data collection, both the light source and spectrometer were preheated for 30 min, and the sample was maintained at room temperature (24 degrees Celsius). After cooling, the following parameters were set for collecting spectra of scrambled eggs with tomatoes: an integration time of 196 ms, 10 scanning repetitions, and 5 smoothing points. The spectral data were captured within the range of 900 to 1700 nm, with a spectral interval of approximately 3.1 nm. A diffuse reflectance standard served as the white reference, while the dark current was recorded by deactivating the light source. Furthermore, to achieve black–white calibration, the following formula was employed:

$$R_{\text{calibrated}} = \frac{R_{\text{measured}} - R_{\text{dark}}}{R_{\text{white}} - R_{\text{dark}}} \tag{1}$$

where $R_{\text{calibrated}}$ represents the calibrated hyperspectral image, R_{measured} represents the measured hyperspectral image, R_{dark} corresponds to the dark current hyperspectral image, and R_{white} corresponds to the white reference intensity.

2.2.3. Spectra Extraction

After correction, the spectra of each sample were extracted by utilizing the default band parameters of the instrument (1453.51 nm, 1246.54 nm, 1123.34 nm) as the RGB channels for hyperspectral image visualization, followed by the application of the Hough transform for sample position detection. This computer vision technique, commonly used for object detection in images, works by transforming the image into a parameter space to identify peaks corresponding to the presence of a particular shape. The detected position information was then utilized to automatically delineate the region of interest (ROI) for each sample. Finally, the sample's spectral information was extracted by averaging the data within the ROI. The process of spectra extraction is illustrated in Figure 1.

2.3. Reference Measurement of Protein Content

The Kjeldahl method was used, as specified in GB 5009.5-2016, for measuring the protein content of the scrambled eggs with tomatoes. A thoroughly mixed sample weighing 0.1 g was added to 0.4 g of copper sulphate, 6 g of potassium sulphate, and 20 mL of sulphuric acid in a digestion furnace for digestion. After digestion, the green and transparent liquid in the digestive tube was cooled and automatically dosed, distilled, titrated, and recorded using an automatic Kjeldahl nitrogen determinator (FOSS-Kjeltec 2300). The protein content of the sample was calculated using the following equation:

$$Z = \frac{(V_1 - V_2) \times c \times 0.014}{m \times V_3/100} \tag{2}$$

where Z represents the protein content of the sample (g/100 g); V_1 is the volume of the reagent blank that consumes the hydrochloric acid standard titration solution (mL); V_2 is the volume of the standard titration solution of sulphuric acid or hydrochloric acid consumed by the reagent blank (mL); c is the concentration of the hydrochloric acid standard titration solution (mol/L); 0.014 is the mass of nitrogen equivalent to hydrochloric acid [c(HCl) = 1.000 mol/L] standard titration solution (g); m is the mass of the sample (g); V_3 is the volume of absorbed digestive juice (mL); and F is the reduction coefficient (6.25), where 100 is the conversion factor.

2.4. A Deep Learning Framework for Wavelength Selection and Regression Simultaneously

2.4.1. Overall Network Architecture

To gain deeper insights into the correlation between the wavelength and protein content, we present the OptmWave framework, a novel approach for effective wavelength selection in spectral data. Inspired by INVASE [15], OptmWave integrates two neural networks: the Selection Probability Generation Network (SPGN) and the Prediction Network (PN). This framework addresses the challenge of accurately identifying the most informative wavelengths and predicting protein content. By leveraging the SPGN and PN, OptmWave streamlines the wavelength selection process and enhances protein content predictions in a spectral data analysis. The architecture of our framework is illustrated in Figure 2a.

Our framework also inherits the general design of INVASE. The SPGN could be any neural network with the same input and output dimensions, while the PN could be any regression network. However, considering the specific characteristics of the dataset in this case study, it is still necessary to perform a detailed design for the SPGN and PN. A Multilayer Perceptron (MLP) is chosen as the foundational structure. An MLP is a type of artificial neural network that consists of an input layer, one or more hidden layers, and an output layer. The input layer receives the input data, and the hidden layers perform the computation to extract features and representations of the input data. The output layer produces the final prediction result. The principle of an MLP is based on the concept of artificial neurons, which are modeled after the biological neurons in the human brain. Each neuron receives inputs from the previous layer and produces an output signal by applying a nonlinear activation function. The weights and biases of the neurons are learned through the training process to minimize the prediction error.

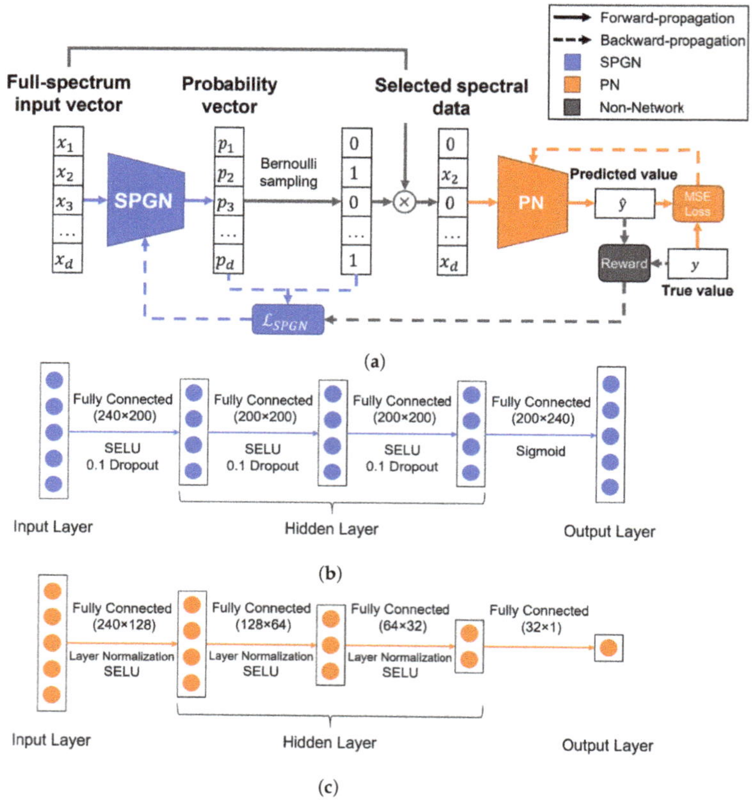

Figure 2. (**a**) The architecture of our proposed OptmWave framework. (**b**) Detailed design of SPGN. (**c**) Detailed design of PN.

The detailed design of the SPGN adopted in this work is illustrated in Figure 2b. It consists of an MLP with three hidden layers, incorporating dropout layers in both the input and hidden layers to enhance its robustness. The output layer is connected to a sigmoid function to ensure that the output values are bounded between 0 and 1. The activation function is the Scaled Exponential Linear Unit (SELU), which is a self-normalizing activation function that ensures that the activations of each layer remain close to the mean and variance of the inputs, regardless of the network's depth, stabilizing the training process and leading to faster convergence [16].

The detailed design of the PN adopted in this work is illustrated in Figure 2c. The PN has three hidden layers of size 128, 64, and 32, respectively, with layer normalization (LN) [17] layers connected after the input and hidden layers. LN is a technique that normalizes the activations of each neuron across the features dimension, rather than the batch dimension, like in batch normalization. This technique has been shown to be effective in stabilizing the training process and improving the generalization performance of deep neural networks. The activation function used is the SELU. It is worth noting that the PN can be trained as an MLP regression model independently to predict protein content. Therefore, it can be combined with conventional wavelength selection methods. Consequently, in the experimental section, we will also evaluate the design of this part separately.

2.4.2. Joint Training Strategy

OptmWave includes our proposed joint training strategy similar to the critic–actor architecture in reinforcement learning that optimizes both neural networks. The SPGN operates by receiving a full-spectrum input vector **x** of dimension d. This input vector represents the spectral data, with each dimension corresponding to a different wavelength. Then, it generates an output vector **p** of the same dimension, with each value of **p** representing a probability between 0 and 1, indicating the likelihood of selecting each corresponding wavelength in the input vector. The probability vector **p** is used for Bernoulli sampling to obtain a selection result vector **s** of dimension d, consisting of only 0 s and 1 s. A value of 0 signifies that the corresponding wavelength is not selected, while a value of 1 denotes its selection. The selection result vector **s** is element-wise multiplied with the input vector **x**, yielding the selected spectral data $\mathbf{x}^*$.

The PN receives the selected spectral data $\mathbf{x}^*$ and performs one iteration of training. To estimate the performance of the current selection result **s**, the trained PN is then evaluated on $\mathbf{x}^*$. The prediction result is used with the true values to calculate the R^2 score as an evaluation metric. The R^2 score measures the goodness of fit between the predicted and true values and is calculated as 1 min the ratio of the sum of squared errors of the predicted values to the sum of squared errors of the mean of the true values. A reward is defined to quantify the performance of the SPGN:

$$\text{Reward} = \alpha(\mathrm{R}^2 - \varphi) \quad (3)$$

where R^2 represents the R^2 score on the validation set, φ denotes a baseline R^2 score that serves as a reference point for comparison, and α is a scaling factor employed to adjust the R^2 score to a suitable range. The reward can be either positive or negative, depending on the performance of the PN after training with the selected spectral data $\mathbf{x}^*$. A positive reward indicates that **s** is a good selection result, as it leads to favorable prediction performance on the training set. In contrast, a negative reward implies that the selection result may not be suitable and should be avoided for future iterations. Then, the SPGN can perform one iteration of training according to the reward.

After each iteration of joint training, where the SPGN and PN are trained according to the proposed strategy, we evaluate the OptmWave's performance on the validation set. The training process continues until the SPGN converges, but there is no guarantee that the PN will also converge. Therefore, we fix the SPGN's weight and train the PN separately. At the start, we train the PN with the selected spectral data obtained from the SPGN in a single iteration. After this training, we evaluate the PN's performance on the validation set. If the PN reaches convergence, we stop training and select the optimal model based on its performance on the validation set. If the PN does not reach convergence, we continue training with the same spectral data from the SPGN until we achieve satisfactory performance. When both the SPGN and PN converge, they can be taken out and used separately. The SPGN prints the same mask, representing the final selection result. The PN can predict the protein content using spectral data after the wavelength selection.

2.4.3. Loss Functions

To successfully implement the proposed joint training strategy, well-designed loss functions are essential. In the case of OptmWave, where the PN functions as a regression model, we have opted for the mean squared error (MSE) as the chosen loss function. Widely used for regression problems, the MSE calculates the average squared difference between predicted and actual values, providing a suitable measure for training the PN.

Regarding the SPGN loss function, denoted as $\mathcal{L}_{\mathrm{SPGN}}$, it comprises two main components: $\mathcal{L}_{\mathrm{Selection}}$ and $\mathcal{L}_{\mathrm{Sparsity}}$. These components are effectively balanced through the hyperparameter λ. During the training process, the combined loss function is utilized to update the SPGN, taking into consideration the associated reward. By incorporating both selection and sparsity considerations, this approach ensures the effectiveness and efficiency of the SPGN's training process. The loss function of the SPGN is defined as follows:

$$\mathcal{L}_{\text{SPGN}} = \left(\mathcal{L}_{\text{Selection}} + \lambda \mathcal{L}_{\text{Sparsity}}\right) \times \text{Reward} \quad (4)$$

$$\mathcal{L}_{\text{Selection}} = -\sum_{i=1}^{d} s_i \log(p_i) + (1 - s_i)\log(1 - p_i) \quad (5)$$

$$\mathcal{L}_{\text{Sparsity}} = \sum_{i=1}^{d} p_i \quad (6)$$

$\mathcal{L}_{\text{Selection}}$ quantifies the discrepancy between the selection result vector $\mathbf{s}$ and the probability vector $\mathbf{p}$. It is calculated via the cross-entropy loss function and is thus suitable for optimization problems involving probabilities. A smaller value signifies a better match between the selection policy and the actual wavelength selection, as the probability vector $\mathbf{p}$ is closer to the selection result vector $\mathbf{s}$. $\mathcal{L}_{\text{Sparsity}}$ encourages wavelength selection sparsity by penalizing the sum of probabilities in the probability vector $\mathbf{p}$. A smaller value corresponds to a sparser selection, reducing the number of selected wavelengths for more efficient computations and better generalization performance. The hyperparameter λ balances the contributions of these two loss components, controlling the trade-off between matching the selection policy to the actual wavelength selection and promoting sparsity in the selection process. The optimization process for $\mathcal{L}_{\text{SPGN}}$ is influenced by the reward, which depends on the performance of the PN on the training set after training with the selected spectral data $\mathbf{s}$. A positive reward signifies that the current wavelength selection results in good prediction performance. The optimizer minimizes $\mathcal{L}_{\text{SPGN}}$ to encourage the SPGN to maintain or improve the wavelength selection policy. This update enhances the match between the probability vector $\mathbf{p}$ and the selection result vector $\mathbf{s}$ while promoting wavelength sparsity. A negative reward suggests that the current wavelength selection leads to poor performance. The optimizer maximizes $\mathcal{L}_{\text{SPGN}}$, guiding the SPGN to update its weights to avoid the detrimental wavelength selection policy and explore alternative strategies by reducing sparsity in the selected wavelengths. The optimizer adjusts the SPGN's weights based on the reward sign, aiming to minimize $\mathcal{L}_{\text{SPGN}}$ when the reward is positive and maximize it when the reward is negative. This adaptive learning process refines the SPGN's wavelength selection policy to achieve better prediction performance and sparsity. The overall loss function, $\mathcal{L}_{\text{SPGN}}$, is obtained by adding the weighted selection and sparsity loss components and then multiplying the sum by the reward. This ensures that the SPGN is updated based on its performance, allowing it to learn an effective wavelength selection policy while considering both prediction performance and the sparsity of the selected wavelengths.

2.5. Conventional Approaches

2.5.1. Common Regression Models

In the field of near-infrared (NIR) spectroscopy analysis, there are several conventional data analysis approaches that are commonly used for quantitative prediction of the physical and chemical properties of samples based on their NIR spectra. Two of the most common regression models in NIR analysis are partial least squares (PLS) [18] and support vector regression (SVR) [19]. PLS is a linear regression technique that seeks to model the relationship between the NIR spectra and the property of interest by constructing a set of orthogonal latent variables, which are also known as latent factors. The latent variables are constructed in such a way so as to maximize the covariance between the NIR spectra and the property of interest. The regression model is then built based on the latent variables instead of the original NIR spectra. SVR is a type of machine learning algorithm that can be used for regression problems. It aims to find the optimal boundary in the feature space that separates the samples into two classes and then maps the samples to a high-dimensional feature space where a linear regression model can be applied. The regression model is built based on a subset of the training samples, which are known as support vectors, and the optimization of the model is performed by minimizing the distance between the support vectors and the boundary.

2.5.2. Effective Wavelength Selection

Efficient wavelength selection constitutes a crucial phase in NIR data analysis, given the potential hindrance posed by high-dimensional, uninformative, and redundant variables to the precise interpretation of wavelength-related information. There are several effective wavelength selection strategies, such as the successive projections algorithm (SPA) and competitive adaptive reweighted sampling (CARS). The SPA operates by projecting one variable onto the others to identify candidate wavelengths. In combination with the weighted regression coefficient analysis, the SPA assists in determining the optimal wavelengths for use in multivariate regression models [20]. One PLS model is created via Monte-Carlo sampling in the CARS. The CARS method involves creating a PLS model via Monte-Carlo sampling. This process produces regression coefficient magnitudes [21], which are associated with each wavelength. A decaying exponential function is then used to exclude wavelengths with the smallest magnitudes. By employing adaptive reweighted sampling, candidate subsets can be derived. After that, the subset that aligns with the PLS models showing the lowest RMSE during cross-validation is considered as the optimal wavelength set.

2.6. Model Evaluation

All the procedures described were implemented using Python. The Spectral Python library was employed for processing hyperspectral imaging data, while OpenCV and NumPy were utilized for spectra extraction. For conventional data analysis approaches, sklearn was used, whereas PyTorch was employed for deep learning approaches.

The performances of the models were assessed using three evaluation metrics: coefficient of determination (R^2), root mean square error (RMSE), and predictive residual deviation (RPD). These metrics are defined as follows: R^2 measures the proportion of the variance in the dependent variable that can be explained by the independent variable(s). It provides an indication of how well the model fits the observed data, with values ranging from 0 to 1. A higher R^2 value indicates a better fit. The formula for R^2 is given by

$$R^2 = 1 - \frac{\sum_{i=1}^{n}(y_i - \hat{y}_i)^2}{\sum_{i=1}^{n}(y_i - \bar{y})^2} \tag{7}$$

where y_i represents the observed values of the dependent variable, $\hat{y}_i$ represents the predicted values of the dependent variable, $\bar{y}$ represents the mean of the observed values of the dependent variable, and n represents the number of data points.

The RMSE measures the average deviation between the predicted values and the observed values. The RMSE provides an estimate of the model's accuracy, with lower values indicating better performance. The formula for RMSE is given by

$$\text{RMSE} = \sqrt{\frac{\sum_{i=1}^{n}(y_i - \hat{y}_i)^2}{n}} \tag{8}$$

where y_i represents the observed values of the dependent variable, $\hat{y}_i$ represents the predicted values of the dependent variable, and n represents the number of data points.

The RPD is a measure of the ratio between the standard deviation of the reference values and the standard error of prediction. The RPD is commonly used in spectroscopy to evaluate the precision of models. Higher RPD values indicate better predictive performance. The formula for RPD is given by

$$\mathrm{RPD} = \frac{\text{Standard Deviation of Reference Values}}{\text{Standard Error of Prediction}} \tag{9}$$

The optimal model is determined based on the model that achieves the highest R^2 and RPD values while also minimizing the RMSE value.

3. Results

3.1. Spectral Profiles

The preprocessed reflectance spectra of the samples are illustrated in Figure 3a. To eliminate obvious noise at the head and end of the spectra, only the range of 900–1700 nm was investigated. As shown in Figure 3b, the spectral data for all samples demonstrated a consistent pattern of peaks and valleys. The major absorptions were observed at valleys around 970, 1206, and 1440 nm. The local absorption maxima at 970 nm and 1440 nm (O-H stretching second and first overtones, respectively) were attributed to the presence of water in the sample [22]. Also, the valley around 1206 nm, which corresponds to the second overtone of C-H stretching, can be attributed to the fat content of the sample [23,24].

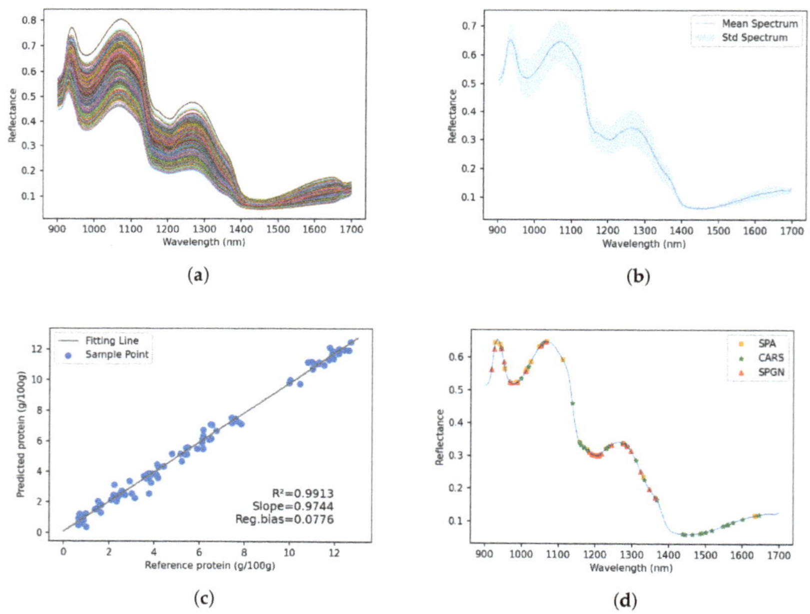

Figure 3. Spectral characterization for the dataset and the visualization of experiment results. (**a**) Reflectance spectra of all samples. (**b**) Mean spectrum and standard deviation of the spectral data. The blue line shows the mean spectrum, while the shaded area represents the standard deviation of the reflectance values at each wavelength. (**c**) Reference versus predicted values for protein content from OptmWave. (**d**) Visualization of effective wavelength selection result by SPA, CARS, and our proposed SPGN.

3.2. Sample Set Split

A total of 487 distinct samples were collected and measured. Before partitioning the dataset, spectral anomalies were detected using the Hotelling T^2 test and subsequently removed. The remaining samples were randomly shuffled and divided into training, validation, and test sets with a ratio of 6:2:2. The final training, validation, and test sets consisted of 292, 98, and 97 samples, respectively. Table 1 presents the minimum value, maximum value, mean, and standard deviation of the protein content for each data split.

Table 1. The statistical summaries of the protein content for each data split.

Data Split	Minimum (g/100 g)	Maximum (g/100 g)	Mean (g/100 g)	Standard Deviation (g/100 g)
Training set	0.6258	12.9541	5.4654	3.7056
Validation set	0.6016	13.1858	5.5092	3.6868
Test set	0.6505	12.6843	5.3266	3.8062

3.3. Experimental Setup

3.3.1. Deep Learning Approaches

For the case of using the PN only, the learning rate was set to 10^{-4}. And for the OptmWave, the hyperparameters were set as follows: the learning rate of the SPGN was 5×10^{-5}, the learning rate of the PN was 2×10^{-3}, α was 100, φ was 0.95, and λ was 0.1. The hyperparameters were manually tuned based on the performance on the validation set, and φ was set based on the prediction performance of the PLS. Compared to the case of using the PN only, the joint training strategy requires the PN to adapt to continuously changing training data, thus necessitating a larger learning rate.

3.3.2. Conventional Approaches

In this study, we utilized PLS and SVR as conventional prediction models and employed the SPA and CARS methods to select effective wavelengths. To optimize the performance of each model, we employed different strategies. For the PLS model, we utilized a grid search approach. To elaborate, a grid search strategy incorporating a 5-fold cross-validation was utilized. The grid search is a brute-force approach that exhaustively searches over a predefined parameter space to find the optimal set of hyperparameters. The search parameters involved were the maximum number of iterations with values of 500, 1000, and 2000; the tolerance with values of 10^{-4}, 10^{-3}, and 10^{-2}; and the number of components ranging from two to the specified maximum number of components. For the SVR model, we used an automated approach called Autosklearn [25] to search for the optimal hyperparameters. Autosklearn is an automated machine learning tool that applies Bayesian optimization to find the optimal hyperparameters for a given dataset. The search time was set to two hours to find the best SVR hyperparameters.

3.4. Result Analysis

The experimental results are summarized in Table 2. For the conventional approaches, the PLS and SVR models using full-spectra and selected wavelengths all obtained good results. The R^2 score of all the models is over 0.9, and the RPD is over 2.5. The comparison of the SVR and PLS models using the same full-spectra and wavelength selection dataset revealed that the SVR outperformed the PLS in the training, validation, and testing sets, possibly due to SVR's ability to model nonlinear relationships in the data.

Among the PLS models, the CARS method yielded the best prediction accuracy of 0.9558. However, the results of applying CARS to the SVR model were not as satisfactory, possibly due to the algorithm's coupling with PLS. In contrast, the SPA showed slightly better results than using the full-spectra data, likely because of its ability to reduce multicollinearity in the data.

And for the case of using the PN only, the experimental results showed that the accuracy on both the full-spectrum data and the wavelength selection data was higher than that of the conventional models. This is because the deep learning model can capture more complex nonlinear relationships in the data, as demonstrated by its higher accuracy on the training set. However, this also leads to a decrease in accuracy on the wavelength selection data, as the reduction in wavelength selection reduces the amount of information available for prediction. The CARS results were higher than those of the PLS, possibly because CARS selected more wavelengths.

Table 2. Prediction results of protein content.

Models	Data Type	N.V. *	Training Set		Validation Set			Test Set		
			R^2	RMSE	R^2	RMSE	RPD	R^2	RMSE	RPD
PLS	Full	240	0.9624	0.7187	0.9689	0.6497	5.6742	0.9487	0.8620	4.4154
	CARS	37	0.9673	0.6697	0.9745	0.5887	6.2628	0.9558	0.8004	4.7553
	SPA	19	0.9535	0.7990	0.9638	0.7015	5.2556	0.9507	0.8449	4.5051
	SPGN	25	0.9569	0.7691	0.9645	0.6943	5.3104	0.9504	0.8478	4.4897
SVR	Full	240	0.9872	0.4189	0.9833	0.4767	7.7335	0.9772	0.5748	6.6219
	CARS	37	0.9799	0.5247	0.9781	0.5453	6.7605	0.9649	0.7131	5.3372
	SPA	19	0.9835	0.4762	0.9807	0.5116	7.2059	0.9773	0.5740	6.6314
	SPGN	25	0.9722	0.6182	0.9749	0.5845	6.3075	0.9653	0.7090	5.3684
PN	Full	240	0.9922	0.3267	0.9834	0.4748	7.7644	0.9876	0.4238	8.9802
	CARS	37	0.9926	0.3191	0.9814	0.5027	7.3334	0.9822	0.5073	7.5021
	SPA	19	0.9814	0.5061	0.9720	0.6167	5.9779	0.9754	0.5971	6.3740
	SPGN	25	0.9945	0.2761	0.9840	0.4666	7.9021	0.9852	0.4633	8.2159
OptmWave		25	0.9938	0.2920	0.9796	0.5263	7.0045	0.9913	0.3548	10.7278

* N.V. is the number of variables.

OptmWave reached the highest prediction performance. Figure 3c shows the visualization of the reference versus predicted values for the protein content from the OptmWave. It can be seen that the protein content can be well-predicted for all samples in the test set. The experiments were also designed to assess the efficacy of the wavelength selection outcomes between OptmWave, CARS, and the SPA. Figure 3d shows the visualization of the effective wavelength selection result by the SPA, CARS, and our method SPGN. In detail, 917.4, 924.03, 943.93, 950.57, 967.16, 970.48, 987.08, 1010.32, 1050.19, 1063.48, 1183.25, 1189.91, 1193.24, 1199.9, 1203.23, 1206.56, 1209.89, 1216.55, 1243.21, 1276.55, 1286.55, 1296.56, 1323.25, and 1346.61 nm were selected. From the experimental results in Table 2, compared with the wavelengths selected by the SPA and CARS, our method achieved competitive results: on the PLS model, the prediction performance was close to that of the SPA and higher than that of the full-spectrum data; the prediction performance of the SVR model was better than CARS and the number of selected wavelengths was less than CARS. Good results were achieved even if the prediction was made directly using the selected wavelengths from the SPGN without using the joint training strategy.

We also try to explain the selection result by the vibration of chemical bonds. The selected wavelengths at 970.48, 987.08, and 1010.32 nm correspond to the N-H stretch second overtone locations associated with proteins [26]. In addition, wavelengths at 1276.55, 1286.55, 1296.56, 1323.25, and 1346.61 nm are also associated with proteins. They can be attributed to a combination of the first overtone of a specific N-H vibration (in Fermi resonance with N-H in-plane bend) with the fundamental vibrations of the N-H in-plane bend and C-N stretch with the N-H in-plane bend [27]. Other wavelengths selected may be indirectly related to the protein content. For example, the wavelengths around 1200 nm (1183.25, 1189.91, 1193.24, 1199.9, 1203.23, 1206.56, 1209.89, and 1216.55 nm) correspond to the C-H stretching second overtone [22,27,28]; 1363.3 nm is related to the C-H combination second overtone [29]; and 917.4 nm and 924.03 nm correspond to the C-H stretching third overtone from methyl and methylene, respectively [26]. And these wavelengths are associated with aliphatic compounds. From 950 nm to 1050 nm, the wavelengths at 943.43, 950.57, and 967.16 nm are related to the third overtone of C-H and the third overtone region of O-H from oil nutrient [30–32]. In this study, it has been observed that the variation among the samples is primarily attributed to the proportion of tomato and egg, which display significant differences in their aliphatic compounds and oil content. It is possible that deep learning models could use the spectra associated with these compounds to uncover their hidden relationship with protein content. Furthermore, our method does not include wavelengths beyond 1400 nm, as shown in Figure 3b,d, where reflectance is low and variance is small within the range of 1400 nm to 1700 nm. Protein information around 1500 nm, specifically the N-H stretching first overtone, may be masked by water information [22]. This limited ability to discriminate may have resulted in the exclusion of this spectral region in our method selection.

3.5. Ablation Study on Deep Learning Approaches

Ablation studies are conducted on deep learning approaches to evaluate the effectiveness of specific design choices. The experimental results for this part are summarized in Table 3. For the case of using the PN only, the performance without using the SELU, LN, or both was tested. In the absence of the SELU, a replacement was carried out using the ReLU activation function, while the omission of all the layer normalization layers was implemented in the absence of LN. This led to observed declines in varying degrees in the prediction results, thus affirming the efficacy of these designs.

Table 3. The results of ablation study on proposed deep learning approaches.

Method	Training Set		Validation Set			Test Set		
	R^2	RMSE	R^2	RMSE	RPD	R^2	RMSE	RPD
PN	0.9922	0.3267	0.9834	0.4748	7.7644	0.9876	0.4238	8.9802
w/o SELU	0.9902	0.3667	0.9811	0.5075	7.2643	0.9773	0.5731	6.6409
w/o LN	0.9566	0.7717	0.9564	0.7694	4.7917	0.9574	0.7859	4.8428
w/o LN and SELU	0.9475	0.8488	0.9599	0.7379	4.9965	0.9460	0.8845	4.3032
OptmWave	0.9938	0.2920	0.9796	0.5263	7.0045	0.9913	0.3548	10.7278
w/o joint training strategy	0.9945	0.2761	0.9840	0.4666	7.9021	0.9852	0.4633	8.2159
w/o $\mathcal{L}_{SPGN}$	0.9363	0.9349	0.8011	1.6444	2.2420	0.8248	1.5929	2.3894

Additionally, the performance of OptmWave without the proposed joint training strategy was evaluated. This involved training the PN directly using the selected wavelength data from the SPGN in Table 2. The reduction in prediction accuracy highlighted the importance of the joint training strategy. Our approach enables the MLP to gradually adapt to changes from the full-spectrum training data to the selected wavelengths, which has an effect similar to fine-tuning on a pre-trained model. The performance of OptmWave was also evaluated without our designed $\mathcal{L}_{SPGN}$. Instead, the loss function design from the implementation of INVASE was adopted [15], where $\mathcal{L}_{Sparsity}$ is not affected by the value of the reward. This resulted in poor results because the optimizer continuously tried to reduce the number of selected wavelengths without considering the prediction performance. That is mainly because INVASE is a method used for classification tasks to obtain the importance of input variables at the instance level. In contrast, our method is used for global variable selection on regression tasks.

4. Discussion

In the present study, the feasibility of deep learning-based NIR-HSI methods in predicting the protein content of specific foods is confirmed. This confirmation was established through a comparison of the accuracy and interpretability of various conventional methods and near-infrared spectroscopy analysis. The results demonstrate the significant potential of deep learning-based NIR-HSI in predicting food nutrition.

Establishing and analyzing models in the field of NIR-HSI holds significant importance [33]. According to Table 2, conventional methods have displayed underfitting in the training set: the models established could not fully fit the training set data. This is typically due to insufficient model complexity. Given identical data distribution, the better the results on the training set, the better they will perform on the test set. This is presumably the primary reason why deep learning methods outperformed conventional methods on our constructed dataset. Additionally, the SPGN and PN are not limited to our proposed design. For instance, the SPGN could be any neural network with the same input and output dimensions, while the PN could be any regression network. This design offers the possibility of extending to other tasks in food science. Researchers can design an SPGN and a PN according to their needs, and use the OptmWave method for training, thus obtaining models suitable for their tasks. We believe that this will promote the cross-disciplinary development of food science and artificial intelligence to some extent.

Despite the satisfactory results, our method does have some limitations. Firstly, deep learning methods require a certain amount of data. With a smaller dataset, the results might be comparable to, or even worse than, conventional methods [24]. Secondly, data distribution should be consistent, especially when partitioning into training, validation, and test sets. If there is a significant difference in distribution, it could lead to severe overfitting. Finally, the sampling mechanism in OptmWave increases the randomness of the method. If the random seed is not fixed, the results may vary.

Based on our findings, future research could further explore new paradigms in food nutrition assessment. For example, combining deep learning models capable of recognizing foods [34] with NIR-HSI methods for specific foods to accurately predict food nutrition. Furthermore, future research could focus on creating a large-scale benchmark dataset for food spectra [35]. Deep learning methods could then be applied to various foods, thereby establishing a large model for food spectra, empowering various downstream tasks in food science.

5. Conclusions

In this study, the feasibility of using deep learning approaches in determining the protein content of a specified food using near-infrared hyperspectral imaging is explored. A framework is proposed to achieve better prediction performance and effective wavelength selection. It is compared to conventional prediction approaches, such as PLS and SVR, and effective wavelength selection approaches, such as the SPA and CARS. The results demonstrated the effectiveness of deep learning approaches and showed that our proposed framework and wavelength selection method outperformed the conventional approaches. Our study highlights the great potential of deep learning-based near-infrared hyperspectral imaging in predicting nutrient content in food. The findings support the application of deep learning methods in food science and offer new avenues for food quality control and nutrition monitoring. Considering the practical implications of our research, the application of deep learning techniques can lead to increased automation and reduced human involvement. This could contribute to the development of more portable and affordable devices for food assessment. Looking ahead, our future studies will focus on investigating the application of these deep learning approaches in assessing nutrient content in a wider array of foods. The goal is to further enhance the prediction performance and broaden the range of practical applications.

Author Contributions: Conceptualization, W.W., W.M. and S.J.; methodology, T.L., W.W. and W.M.; software, T.L.; validation, T.L.; formal analysis, T.L.; investigation, T.L. and S.X.; resources, W.W. and S.X.; data curation, T.L. and S.X.; writing—original draft preparation, T.L.; writing—review and editing, W.W., W.M., C.Z. and S.J.; visualization, T.L.; supervision, W.W., W.M., C.Z. and S.J.; project administration, W.M.; funding acquisition, C.Z. All authors have read and agreed to the published version of the manuscript.

Funding: This work was supported in part by the National Natural Science Foundation of China (Grant Nos. 62125207, 32201952) and the Open Project Program of the Key Laboratory of Intelligent Information Processing of the Chinese Academy of Sciences, Institute of Computing Technology, Chinese Academy of Sciences. This work was also supported by the CAAI-Huawei MindSpore Open Fund; the Key R&D Program of Shandong Province, China (Grant No. 2022TZXD0021); and the Agricultural Science and Technology Innovation Program of Institute of Food Science and Technology, Chinese Academy of Agricultural Sciences (CAAS-ASTIP-G2022-IFST-08).

Data Availability Statement: The data used to support the findings of this study can be made available by the corresponding author upon request.

Acknowledgments: We would like to thank Qiang Wang for his support and assistance on the near-infrared spectrometer.

Conflicts of Interest: The authors declare no conflicts of interest.

Abbreviations

The following abbreviations are used in this manuscript:

NIR-HSI	Near-infrared hyperspectral imaging
RMSE	Root mean square error
NIR	Near infrared
ROI	Region of interest
SPGN	Selection probability generation network
PN	Prediction network
MLP	Multilayer perceptron
SELU	Scaled exponential linear unit
LN	Layer normalization
MSE	Mean squared error
PLS	Partial least squares
SVR	Support vector regression
SPA	Successive projections algorithm
CARS	Competitive adaptive reweighted sampling
RPD	Predictive residual deviation

References

1. Miller, V.; Webb, P.; Cudhea, F.; Shi, P.; Zhang, J.; Reedy, J.; Erndt-Marino, J.; Coates, J.; Mozaffarian, D. Global dietary quality in 185 countries from 1990 to 2018 show wide differences by nation, age, education, and urbanicity. *Nat. Food* **2022**, *3*, 694–702. [CrossRef] [PubMed]
2. O'Hearn, M.; Erndt-Marino, J.; Gerber, S.; Lauren, B.N.; Economos, C.; Wong, J.B.; Blumberg, J.B.; Mozaffarian, D. Validation of Food Compass with a healthy diet, cardiometabolic health, and mortality among US adults, 1999–2018. *Nat. Commun.* **2022**, *13*, 7066.
3. Min, W.; Jiang, S.; Liu, L.; Rui, Y.; Jain, R. A survey on food computing. *ACM Comput. Surv. CSUR* **2019**, *52*, 1–36. [CrossRef]
4. Wang, W.; Min, W.; Li, T.; Dong, X.; Li, H.; Jiang, S. A review on vision-based analysis for automatic dietary assessment. *Trends Food Sci. Technol.* **2022**, *122*, 223–237. [CrossRef]
5. Mirtchouk, M.; Merck, C.; Kleinberg, S. Automated estimation of food type and amount consumed from body-worn audio and motion sensors. In Proceedings of the 2016 ACM International Joint Conference on Pervasive and Ubiquitous Computing, Heidelberg Germany, 12–16 September 2016; pp. 451–462.
6. Ma, P.; Lau, C.P.; Yu, N.; Li, A.; Liu, P.; Wang, Q.; Sheng, J. Image-based nutrient estimation for Chinese dishes using deep learning. *Food Res. Int.* **2021**, *147*, 110437. [CrossRef]
7. Shao, W.; Hou, S.; Jia, W.; Zheng, Y. Rapid Non-Destructive Analysis of Food Nutrient Content Using Swin-Nutrition. *Foods* **2022**, *11*, 3429. [CrossRef]
8. Shao, W.; Min, W.; Hou, S.; Luo, M.; Li, T.; Zheng, Y.; Jiang, S. Vision-based food nutrition estimation via RGB-D fusion network. *Food Chem.* **2023**, *424*, 136309. [CrossRef]
9. Ma, P.; Zhang, Z.; Li, Y.; Yu, N.; Sheng, J.; McGinty, H.K.; Wang, Q.; Ahuja, J.K. Deep learning accurately predicts food categories and nutrients based on ingredient statements. *Food Chem.* **2022**, *391*, 133243. [CrossRef]
10. Lau, E.; Goh, H.J.; Quek, R.; Lim, S.W.; Henry, J. Rapid estimation of the energy content of composite foods: The application of the Calorie Answer. *Asia Pac. J. Clin. Nutr.* **2016**, *25*, 18–25.
11. Ahn, D.; Choi, J.Y.; Kim, H.C.; Cho, J.S.; Moon, K.D.; Park, T. Estimating the composition of food nutrients from hyperspectral signals based on deep neural networks. *Sensors* **2019**, *19*, 1560. [CrossRef]
12. Hu, H.; Zhang, Q.; Chen, Y. NIRSCAM: A Mobile Near-Infrared Sensing System for Food Calorie Estimation. *IEEE Int. Things J.* **2022**, *9*, 18934–18945. [CrossRef]
13. Wu, G. Amino acids: Metabolism, functions, and nutrition. *Amino Acids* **2009**, *37*, 1–17. [CrossRef]
14. Li, H.; Sun, H.; Li, M. Detection of vitamin C content in head cabbage based on visible/near-infrared spectroscopy. *Trans. Chin. Soc. Agric. Eng.* **2018**, *34*, 269–275.
15. Yoon, J.; Jordon, J.; van der Schaar, M. INVASE: Instance-wise variable selection using neural networks. In Proceedings of the International Conference on Learning Representations, Vancouver, BC, Canada, 30 April–3 May 2018.

16. Klambauer, G.; Unterthiner, T.; Mayr, A.; Hochreiter, S. Self-normalizing neural networks. *Adv. Neural Inf. Process. Syst.* **2017**, *30*.
17. Ba, J.L.; Kiros, J.R.; Hinton, G.E. Layer normalization. *arXiv* **2016**, arXiv:1607.06450.
18. Geladi, P.; Kowalski, B.R. Partial least-squares regression: A tutorial. *Anal. Chim. Acta* **1986**, *185*, 1–17. [CrossRef]
19. Platt, J. Probabilistic outputs for support vector machines and comparisons to regularized likelihood methods. *Adv. Large Margin Classif.* **1999**, *10*, 61–74.
20. Araújo, M.C.U.; Saldanha, T.C.B.; Galvao, R.K.H.; Yoneyama, T.; Chame, H.C.; Visani, V. The successive projections algorithm for variable selection in spectroscopic multicomponent analysis. *Chemom. Intell. Lab. Syst.* **2001**, *57*, 65–73. [CrossRef]
21. Li, H.; Liang, Y.; Xu, Q.; Cao, D. Key wavelengths screening using competitive adaptive reweighted sampling method for multivariate calibration. *Anal. Chim. Acta* **2009**, *648*, 77–84. [CrossRef]
22. ElMasry, G.; Sun, D.W.; Allen, P. Non-destructive determination of water-holding capacity in fresh beef by using NIR hyperspectral imaging. *Food Res. Int.* **2011**, *44*, 2624–2633. [CrossRef]
23. Wu, D.; Sun, D.W. Potential of time series-hyperspectral imaging (TS-HSI) for non-invasive determination of microbial spoilage of salmon flesh. *Talanta* **2013**, *111*, 39–46. [CrossRef]
24. Zhang, C.; Wu, W.; Zhou, L.; Cheng, H.; Ye, X.; He, Y. Developing deep learning based regression approaches for determination of chemical compositions in dry black goji berries (*Lycium ruthenicum* Murr.) using near-infrared hyperspectral imaging. *Food Chem.* **2020**, *319*, 126536. [CrossRef]
25. Feurer, M.; Klein, A.; Eggensperger, K.; Springenberg, J.; Blum, M.; Hutter, F. Efficient and Robust Automated Machine Learning. In Proceedings of the Advances in Neural Information Processing Systems 28 (2015), Montreal, QC, Canada, 7–12 December 2015; pp. 2962–2970.
26. Workman, J., Jr.; Weyer, L. *Practical Guide and Spectral Atlas for Interpretive Near-Infrared Spectroscopy*; CRC Press: Boca Raton, FL, USA, 2012.
27. Daszykowski, M.; Wrobel, M.; Czarnik-Matusewicz, H.; Walczak, B. Near-infrared reflectance spectroscopy and multivariate calibration techniques applied to modelling the crude protein, fibre and fat content in rapeseed meal. *Analyst* **2008**, *133*, 1523–1531. [CrossRef]
28. Hayati, R.; Zulfahrizal, Z.; Munawar, A.A. Robust prediction performance of inner quality attributes in intact cocoa beans using near infrared spectroscopy and multivariate analysis. *Heliyon* **2021**, *7*, e06286. [CrossRef] [PubMed]
29. Workman, J., Jr.; Springsteen, A. *Applied Spectroscopy: A Compact Reference for Practitioners*; Academic Press: Cambridge, MA, USA, 1998.
30. Weinstock, B.A.; Janni, J.; Hagen, L.; Wright, S. Prediction of oil and oleic acid concentrations in individual corn (*Zea mays* L.) kernels using near-infrared reflectance hyperspectral imaging and multivariate analysis. *Appl. Spectrosc.* **2006**, *60*, 9–16. [CrossRef] [PubMed]
31. Balbino, S.; Vincek, D.; Trtanj, I.; Egređija, D.; Gajdoš-Kljusurić, J.; Kraljić, K.; Obranović, M.; Škevin, D. Assessment of pumpkin seed oil adulteration supported by multivariate analysis: Comparison of GC-MS, colourimetry and NIR spectroscopy data. *Foods* **2022**, *11*, 835. [CrossRef] [PubMed]
32. Wang, Y.; Xiong, F.; Zhang, Y.; Wang, S.; Yuan, Y.; Lu, C.; Nie, J.; Nan, T.; Yang, B.; Huang, L.; et al. Application of hyperspectral imaging assisted with integrated deep learning approaches in identifying geographical origins and predicting nutrient contents of *Coix seeds*. *Food Chem.* **2023**, *404*, 134503. [CrossRef]
33. Manley, M. Near-infrared spectroscopy and hyperspectral imaging: Non-destructive analysis of biological materials. *Chem. Soc. Rev.* **2014**, *43*, 8200–8214. [CrossRef] [PubMed]
34. Min, W.; Wang, Z.; Liu, Y.; Luo, M.; Kang, L.; Wei, X.; Wei, X.; Jiang, S. Large scale visual food recognition. *IEEE Trans. Pattern Anal. Mach. Intell.* **2023**, *45*, 9932–9949. [CrossRef] [PubMed]
35. Xia, X.; Liu, W.; Wang, L.; Sun, J. HSIFoodIngr-64: A Dataset for Hyperspectral Food-Related Studies and a Benchmark Method on Food Ingredient Retrieval. *IEEE Access* **2023**, *11*, 13152–13162. [CrossRef]

foods MDPI

Article

A Novel Foodborne Illness Detection and Web Application Tool Based on Social Media

Dandan Tao [1], Ruofan Hu [2], Dongyu Zhang [2], Jasmine Laber [2], Anne Lapsley [2], Timothy Kwan [2,3], Liam Rathke [3], Elke Rundensteiner [2,3,*] and Hao Feng [4,*]

1 Vanke School of Public Health, Tsinghua University, Beijing 100084, China; dandantao@mail.tsinghua.edu.cn
2 Data Science Program, Worcester Polytechnic Institute, Worcester, MA 01609, USA; rhu@wpi.edu (R.H.)
3 Department of Computer Science, Worcester Polytechnic Institute, Worcester, MA 01609, USA
4 College of Agricultural & Environmental Sciences, North Carolina A & T State University, Greensboro, NC 27411, USA
* Correspondence: rundenst@wpi.edu (E.R.); hfeng@ncat.edu (H.F.)

Abstract: Foodborne diseases and outbreaks are significant threats to public health, resulting in millions of illnesses and deaths worldwide each year. Traditional foodborne disease surveillance systems rely on data from healthcare facilities, laboratories, and government agencies to monitor and control outbreaks. Recently, there is a growing recognition of the potential value of incorporating social media data into surveillance systems. This paper explores the use of social media data as an alternative surveillance tool for foodborne diseases by collecting large-scale Twitter data, building food safety data storage models, and developing a novel frontend foodborne illness surveillance system. Descriptive and predictive analyses of the collected data were conducted in comparison with ground truth data reported by the U.S. Centers for Disease Control and Prevention (CDC). The results indicate that the most implicated food categories and the distributions from both Twitter and the CDC were similar. The system developed with Twitter data could complement traditional foodborne disease surveillance systems by providing near-real-time information on foodborne illnesses, implicated foods, symptoms, locations, and other information critical for detecting a potential foodborne outbreak.

Keywords: social media data; text mining; machine learning; bid data; foodborne diseases; Twitter; PostgreSQL; foodborne outbreaks

1. Introduction

Foodborne diseases, arising from the consumption of contaminated food, pose a significant public health concern and have a severe impact on human well-being. Annually, these diseases contribute to a staggering number of illnesses worldwide, leading to 600 million cases and 420,000 deaths [1,2]. Foodborne disease surveillance plays a crucial role in safeguarding human health by monitoring and controlling foodborne diseases and identifying potential foodborne threats. Traditionally, foodborne disease surveillance has relied on data from various sources, such as healthcare facilities, laboratories, and government agencies. The CDC's National Outbreak Reporting System (NORS), Foodborne Disease Active Surveillance Network (FoodNet), and PulseNet are several foodborne disease surveillance tools used in the United States [3–5]. For instance, NORS plays a crucial role in detecting and responding to public health threats by monitoring and reporting foodborne outbreaks in a timely manner. Local health departments report individual cases or clusters of illnesses to NORS, providing demographic information, symptoms, onset dates, exposure history, laboratory results, and other relevant details. Then, NORS integrates data from various sources, performs data analysis, and generates outbreak reports summarizing the findings, including the implicated pathogens, affected population, geographic distribution, potential sources of contamination, and recommended control measures. NORS has been utilized as an important data source in the United States for evaluating the impact of foodborne outbreaks and the associated risk factors with both statistical analysis and data mining methods [6–8].

Recently, with the advent of social media and the widespread use of online platforms, there is a growing recognition of the potential value of incorporating social media data into foodborne disease surveillance systems [9–14]. Social media platforms, such as Twitter, Facebook, and Instagram, have become popular channels for individuals to share their thoughts, experiences, and daily activities. These platforms have also become spaces where users express their health concerns, seek health-related information, and discuss experiences with diseases and outbreaks [10]. As a result, social media data offer a unique opportunity to tap into the collective wisdom and sentiments of the public, providing valuable insights into public health trends, behaviors, and perceptions. For example, Twitter and Yelp were utilized as tools for detecting unreported cases of foodborne illnesses in various local public health departments across the United States. These methods were implemented and evaluated in cities including Chicago, New York, and Las Vegas [11–14].

However, analyzing social media data poses significant challenges due to their inherent noise, ambiguity, and unstructured nature. Social media platforms provide users with the freedom to express themselves in an unrestricted manner. Consequently, the data generated on these platforms often contain informal language, abbreviations, slang, misspellings, grammatical errors, and sentiment-laden expressions [15]. This "noisy" nature of social media data makes it challenging to extract meaningful insights and gain an accurate understanding from the vast volumes of information available. The analysis of unstructured social media data necessitates the application of advanced natural language processing (NLP) technologies. NLP encompasses a set of computational techniques that enable machines to understand, interpret, and generate human language. It aims to bridge the gap between human communication and machine understanding, facilitating the extraction of valuable information from unstructured text. NLP technologies have made significant advancements in addressing the challenges associated with noisy social media data. These technologies encompass a wide range of techniques, including text preprocessing, tokenization, part-of-speech tagging, named entity recognition, sentiment analysis, topic modeling, and language generation [16]. These advancements have enabled researchers and practitioners to extract meaningful patterns, sentiments, and insights from social media data, supporting various applications such as opinion mining, trend analysis, social network analysis, and public sentiment tracking. In the field of public health, typical NLP techniques have been widely employed to identify potential cases of public threats such as COVID-19 diseases [17]. More recently, the language model BERTweet, a variant of BERT (Bidirectional Encoder Representations from Transformers) was trained to classify relevant Foodborne illness cases from Twitter data [9]. Also, state-of-the-art single- and multi-task deep learning models such as RoBERTa and BiLSTM have been trained to extract critical entities related to foodborne illnesses on Twitter data [18].

This paper aims to explore the potential use of social media data in developing an alternative foodborne disease surveillance system. We discuss the methodologies, tools, and techniques employed in leveraging Twitter data and NORS data for detecting and evaluating foodborne illnesses in the United States, including natural language processing, machine learning, database system, and website development. Furthermore, we address the challenges associated with the use of social media data. By examining the current surveillance system and the potential of social media, this paper highlights the opportunities for developing a complementary and enhanced foodborne disease surveillance system that can leverage the real-time and wide-reaching nature of social media data. One practical advantage of such a system would be the substantial reduction in labor required from experts in the field as they would no longer need to manually examine and extract essential information from social media. Instead, the system could automatically transform the unstructured social media data into a more organized format, focusing on important entities crucial for identifying potential outbreaks of foodborne illnesses.

Overall, the major contributions of this work include the following:

1. We collected a large volume of Twitter data related to foodborne illness and transformed them into a more structured dataset with critical 3W information (PostgreSQL) using the pretrained machine learning models.
2. We compared the descriptive statistics of foodborne illness cases from Twitter data and official NORS data in multiple aspects (numbers, place, time, and food) and built predictive models for predicting cases based on Twitter and/or NORS data.
3. We developed the frontend applications based on the two sources of data for assisting the early detection of foodborne outbreaks.

2. Related Work

Foodborne diseases pose significant risks to public health, necessitating effective surveillance systems for early detection, rapid response, and prevention. Traditional foodborne disease surveillance systems primarily rely on data from healthcare facilities, laboratories, and government agencies. However, these systems face challenges such as underreporting, time delays, and limited coverage. The Internet serves as a valuable data source for disease surveillance, enabling the early detection of food safety and food fraud hazards and more digitalized supply chain management in the food industry [19,20]. Across the globe, various information systems have been developed to leverage internet data retrieval and text mining techniques, aiming to enhance early warning capabilities. For example, a Japanese group constructed a database of food safety documents by conducting keyword searches on Google web pages [21]. In Singapore, the National Environment Agency collaborated with IBM Research to establish the Food Safety Information System (FoodSIS), which proactively monitors emerging food safety issues by extracting relevant content from the Internet [22]. In China, a database system of food safety information was created in 2016, utilizing food safety news from media and government websites to facilitate the efficient assessment of food safety concerns [23]. Additionally, a food fraud reporting system MeDISys-FF was developed based on an infrastructure MeDISys that gathers worldwide reports published in the media [24]. In addition to utilizing online Internet information, applications that allow the positive reporting of unpleasant dining experiences can provide another means to record food safety issues. For example, iwaspoisoned.com is an online platform where individuals can voluntarily report incidents of foodborne illnesses they have experienced. Users can provide details about the location, date, symptoms, and the suspected food establishment. The platform aggregates and analyzes these data, providing insights into detecting potential foodborne outbreaks [25]. Database data and text-based data reported in more structured formats are major data sources of these related foodborne illness surveillance systems.

In recent years, the widespread use of social media platforms has provided an opportunity to explore alternative approaches to foodborne disease surveillance. Social media data are notorious for their unstructured characteristics and for being difficult to analyze, and they have been widely recognized as a potential data source for the early detection of public health threats [10]. In the food safety field, classification models were typically used to identify relevant foodborne illness incidents from Twitter posts [11–14]. These studies present basic NLP methods and evidence showing that Twitter can provide additional insights into detecting foodborne illness cases in a sentence classification perspective. In addition to sentence-level classification to detect if a tweet indicates a foodborne illness, our previous work constructed token-level models to extract valuable information from Twitter with high accuracy [18]. The availability of what, where, and when (3W) information about people's everyday lives on social media websites has proven to be valuable for predicting the flu well before outbreaks formally have been reported by the CDC and for preventing public health crises [26]. In the food safety scenario, what refers to the content of the tweet describing a potential food safety incident, e.g., the food product and the complaints about it, while where and when encode the geolocation and the timeframe of the incidents, respectively. Therefore, critical entities such as food and symptoms related to a food safety incident and the location of the incidents are valuable information for detecting a potential foodborne outbreak and, thus, should not be ignored in the full use of Twitter data. The potential value of social media data food safety surveillance has been explored in many previous studies [10–15]. Detecting foodborne illness cases via sentence-level text classification models and further examinations of whether positive-predicted incidences can indicate foodborne outbreaks with the assistance of epidemiologists are the major focuses of these work. In contrast, our work, for the first time, attempts to develop a system that can automatically transform the unstructured social media data to a more structured format with 3W information such as food, symptoms, and location, which is essential for the examination of its relevance to foodborne outbreaks.

3. Materials and Methods

3.1. Data Collection and the Pipeline

Twitter allows researchers who are interested in studying historical tweets to visit its full archive data using Twitter academic accounts. A pipeline (Figure 1) was created to perform the following tasks in sequence. First, to extract the tweets most likely related to foodborne illness, the Tweetkit Python package was used to create a specific query to run on the Twitter archive. The Tweetkit package was used to fit it into our pipeline and was customized to collect all the data associated with a tweet [27]. The pipeline efficiently retrieves tweets from the Twitter archive that may indicate a case of foodborne illness with keywords query: '#foodpoisoning', '#stomachache', 'food poison', 'food poisoning', 'stomachache', 'vomit', 'puke', 'diarrhea', 'the runs', 'nausea', 'stomach cramp', and 'nauseous'. An automatic system was set up to collect the tweets cyclically in 24 h intervals. Appropriate data cleaning was conducted before data were inserted into the PostgreSQL database. Specifically, only tweets in English were retained for further processing. Duplicates, retweets, and tweets with less than four tokens were removed. Then, the tweets were normalized by converting user mentions and URL links to @USER and HTTPURL, respectively. Since emojis may carry important information, we keep emojis and translate the icons into text strings. These tweets are passed through a refined BERTweet machine learning model, which will provide a prediction on whether the tweet indicates a foodborne illness case.

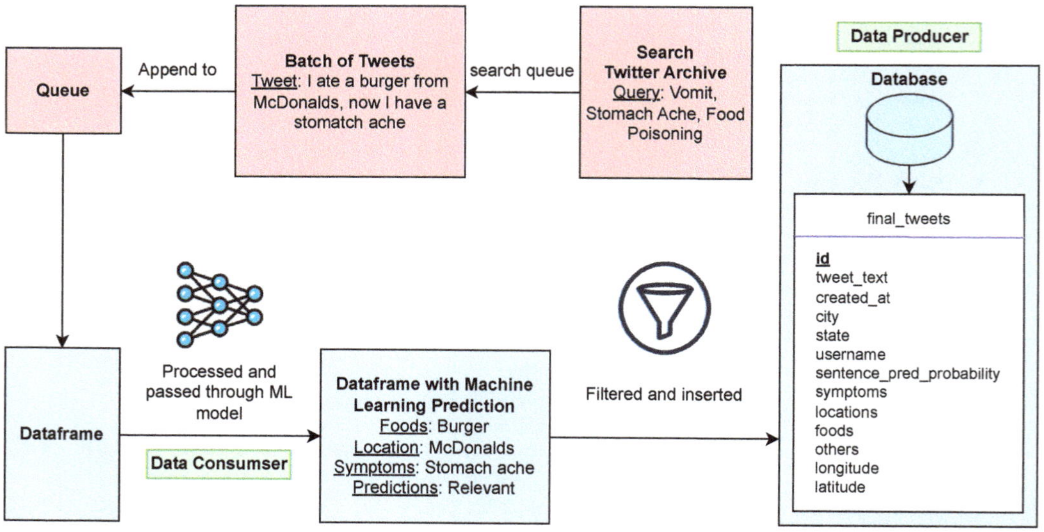

Figure 1. Diagram of the pipeline for Twitter data collection and storage.

Finally, the tweets and their predictions are then fed into the PostgreSQL database with a simple schema to store raw tweets and the attributes, including symptom and food entities extracted by the machine learning model. Tweets related to foodborne illnesses from 2017 to 2021 and with geolocations in the U.S. were collected and stored in the database. The ground truth data of real-world foodborne outbreaks collected from the National Outbreak Reporting System (NORS, https://www.cdc.gov/nors/index.html, accessed on 24 June 2023) were employed to see if the trends observed in the Twitter data correlate to real-world data. The data that are available for each outbreak include year, month, state, primary mode, etiology, serotype or genotype, etiology status, setting, illnesses, hospitalizations, and deaths.

3.2. Information Extraction and Language Models

Twitter provides a unique opportunity to monitor food-safety-related incidents and the temporal and spatial patterns in a near-real-time fashion. Besides time, location, and information embedded in the structured meta data, other valuable information may be expressed in the unstructured text such as the name of the restaurant, grocery store, or dish. In recent years, deep learning models, especially pretrained BERT [28] models, are promising techniques for sequence tagging and named entity recognition (NER) tasks due to their ability to learn from the context surrounding the words in a sequence. In our previous study, single- and multi-task deep learning models including BERTweet [29], RoBERTa [30], BiLSTM [31], and MGADE [32] were trained to extract the critical entities related to foodborne illnesses on Twitter data [18]. As shown in Table 1, important 3W information, such as food, location, and symptoms, that was useful for the analysis of a foodborne illness was extracted by the deep learning models and inserted in the database. The 3W properties together form the core entities essential for food safety outbreak monitoring and prevention. However, with social media data being big data, meaning a high volume of data arriving with high velocity, it is not realistic to conduct reliable analysis on the fly. Instead, we propose to merge the 3W information extracted from external sources into a unified data manager.

Table 1. Examples of tweets processed by the language model.

Tweets	Food	Location	Symptom
Biggest regret of #2013: eating sushi from safeway's refrigerated section #foodpoisoning #projectilevomitingatmidnight	Sushi	Safeway	#projectilevomitingatmidnight
@USER I ate like 3 tacos from jackinbox lastnight in when I woke up my stomach been hurting every since I think I got food poison	Tacos	Jackinbox	Stomach been hurting
@USER @USER Bought family meal in Rustenburg on my way from mahikeng last weekend. All four of us had severe cramps and running stomach the next day. Went to the doctor and was told it's food poisoning. Never again will I buy nandos HTTPURL	Nando's meal	Rustenburg	Severe cramps, running stomach
Panda Express got me@and three others sick. Orange chicken.. we all ate different day at different locations!! What the hell is going on? I get my days off till next week. Thanks Panda I am hoping I don't puke on my staff [EMOJI_disappointed_face] [EMOJI_disappointed_face][EMOJI_disappointed_face] #pandaexpress #foodsafety #foodpoisoning	Orange chicken	Panda Express	Sick, puke

3.3. Data Storage

A PostgreSQL database was designed to store a large amount of data that could be easily and quickly accessed and queried for our frontend visualizations. The PostgreSQL database allows to store complex data types such as the array data type, which is necessary to store the 'symptoms', 'locations', 'foods', and 'others' values as these all contain multiple entity words per tweet. All the collected data were stored in one table named "final_tweets". This table contains 13 fields, each of which we consider useful for querying on our frontend. For each of these fields, the appropriate data type was selected so that it would be most efficient for frontend use, as well as database storage. The 'id' is the primary key of our final_tweets table as it uniquely identifies each individual tweet. Querying the count of ids can be used to obtain tweet counts based on different factors. The 'id' column was stored as the bigint type as the integer type is ideal for storing ids in a database. The 'tweet_text' column is stored as the text type as we found issues trying to limit storage to a certain number of characters. The 'created_at' column stores tweet timestamps as PostgreSQL timestamps so that we can easily query data over certain time intervals. The 'city', 'state', and 'username' columns are all of the varchar type to ensure that these values will be stored without taking up too much space in our database. The 'sentence_pred_prob' is stored as a numeric type as it has many digits past the decimal point that we want to preserve for precision. As mentioned above, the 'foods', 'symptoms', 'locations', and 'others' entities were stored as text arrays. Lastly, 'longitude' and 'latitude' were both stored as numeric types in order to account for precision. These 13 fields were retained in our database in order to optimize queries for our frontend. The columns here contain data that are needed to display frontend visualizations or conduct simple statistics. Before inserting the collected data into the database for frontend use, the data were filtered through the pretrained machine learning model so that the tweets having a sentence prediction equal to 1/indicating foodborne illness were included in the database used for frontend interface design. In order to perform a later analysis of all the tweets that ran through the collection process, we implemented raw data storage that contains all the fields initially collected from Twitter.

3.4. Frontend Website Design

Next.js (Next.js, 2023), a frontend web development framework based on the React (React, 2023) library, was chosen as our frontend web framework. The implementation of Next.js in the system ensures a responsive user interface and boosts system performance [33]. An additional benefit of Next.js is that it leverages the React frontend library with well-supported React resources online. D3.js (D3.js, 2023) was chosen as our visualization library for the frequency graph, which is widely used and considered by many to be an industry standard. Mapbox (Mapbox, 2023), a hosted GIS mapping service, was used to connect the tweets in our database to an interactive Mapbox map, allowing users to interact with the tweets on a US map. Finally, Tailwind CSS (Tailwind CSS, 2023) was chosen as our UI library as it is small and efficient while also providing significant flexibility to developers.

4. Results and Discussion

4.1. Statistics of the Collected Data

The pipeline collected around 430,000 geolocated tweets from the beginning of 2017 to the end of 2022, with 110,000 predicted as positive for indicating a foodborne illness and stored in the database after passing through a machine learning model. Figure 2 shows the number of tweets by state and the number of tweets per capita. This visualization allows one to visually compare the difference between the number of identified tweets related to foodborne illnesses from each state to the number of tweets by state per capita.

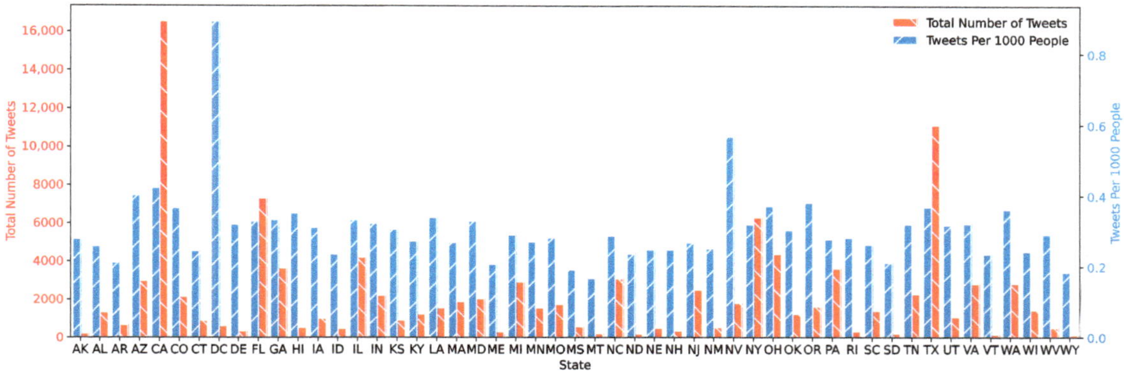

Figure 2. Graph of collected tweets compared to per capita tweets from 2017–2021.

One potential issue is the repeated author IDs, which could possibly mean a Twitter account dedicated to foodborne illnesses and tweet foodborne illness information rather than instances of foodborne illness, such as news from the CDC, FDA, and associated agencies. This would not be an issue if the tweets were relevant to the research goal, for example, the official Twitter account iWasPoisioned posts about instances of foodborne illness and, thus, should not be excluded from the raw data. Luckily, in searching samples of collected tweets for repeated author IDs, the only one that did appear to come up often was the iWasPoisioned Twitter account, so we deemed that not an issue since the iWasPoisoned Twitter account most likely offers cases of foodborne illness. After filtering the raw data with the pretrained machine learning model, less than 0.5% (less than 300 tweets) of over 56,000 tweets collected from iWasPoisoned were predicted not related to a case of foodborne illness. Meanwhile, the other repeated Twitter accounts had few tweets that were predicted to be related to foodborne illness incidences, posing little impact on the whole dataset.

4.2. Descriptive Analysis

As mentioned above, the pretrained machine learning model extracted 'food' entities from a tweet. These entities are identified words that most likely indicate a key food or ingredient. To evaluate the validity of the collected tweets in indicating real foodborne illness cases, the most frequent food entities extracted by the model were compared with the real food vehicles involved in historical foodborne outbreaks reported in the NORS data (Figure 3). As shown in Figure 3a, the top 15 most frequent food entities that appeared in the Twitter database from 2017–2021 were "chicken", "sandwich", "salad", "cheese", "pizza", "fries", "burger", "burrito", "shrimp", "beef", "steak", "rice", "bacon", "meat", and "cream". "Chicken" was identified as the most frequent food entity mentioned in tweets related to foodborne illnesses, with a much greater number compared to the rest of the other food entities. Figure 3b shows the top 15 most frequent food entities mentioned in the NORS reports of foodborne outbreaks from 2017–2021. It was noticed that some of the food entities retrieved from NORS were not necessarily food related. As shown in the figure, "chicken" also appears as the most frequent food entity in the NORS reports. The other frequent food entities are "oysters", "salad", "fish", "beef", "tuna", "rice", "pork", "sandwich", "turkey", "milk", "cheese", "lettuce", "beans", and "pizza". Aquatic products appeared more frequently in the NORS reports than in the Twitter dataset. It should be noticed that less frequent food entities such as "lettuce", "sprouts", "tomato", and "lettuce" belong to the category of "vegetable". Therefore, a comparison of food categories might help to identify patterns in related food entities.

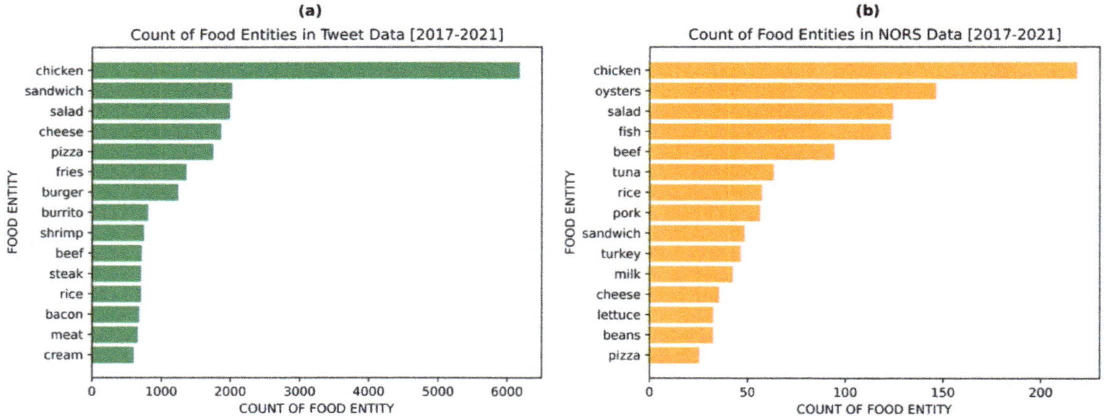

Figure 3. Graph displaying the 15 most frequent food entities found in collected tweets (**a**) and NORS data (**b**) from 2017–2021.

The Interagency Food Safety Analytics Collaboration (IFSAC) Food Categorization Scheme, used to categorize food sources of contamination in an outbreak and perform attribution analysis, was created by three federal agencies: the CDC, the U.S. Food and Drug Administration (FDA), and the U.S. Department of Agriculture Food Safety and Inspection Services (USDA-FSIS) [34]. Based on the IFSAC category, identified foods are labeled into one of seventeen categories. Identified foods causing historical foodborne outbreaks were labeled with an IAFSC category in the NORS database. Because of the sheer number of unique food entities, grouping the foods into general categories for analysis could provide valuable insight. In order to automatically convert foods extracted from the Twitter data into corresponding categories, we hand-labeled the top 250 common food entities into their respective categories based on the IFSAC Food Categorization Scheme. These labels were put into a JSON dictionary, which could be easily read to convert entities to categories.

Figure 4 shows the total percentage breakdown of each category for Twitter data and NORS data. The volume of food categories between the NORS dataset and the Twitter dataset were relatively similar, which is promising for an accurate comparison. The percentage of meat, poultry, fruit/nut, and fish categories in both datasets are close to each other, while NORS reports include a significantly higher percentage of vegetable category than Twitter data. Dairy, oil/sugar, and grain/bean are mentioned more frequently than in the NORS data after the percentage breakdown. Researchers have found similar trends in the percentage composition of food categories in comparison to Yelp data and NORS data collected in 2006–2011 [35]. With a small number of expert evaluation (labeling) data, the pretrained machine learning model combined with an automatic category conversion mechanism found that foods implicated in foodborne illness posts on Twitter correlated with foods implicated in reports from the CDC, indicating that Twitter posts could complement traditional surveillance systems by providing near-real-time information on foodborne illnesses and the implicated foods.

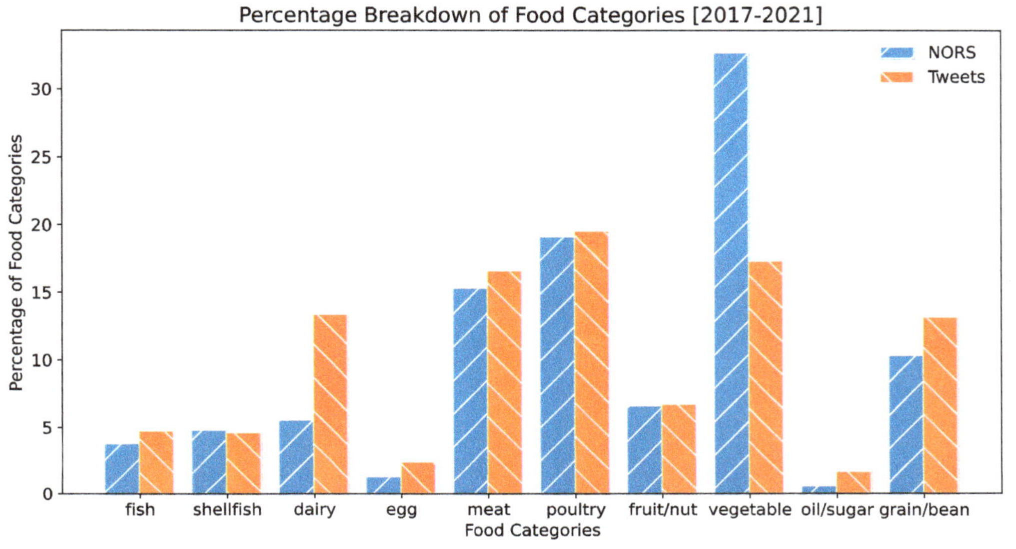

Figure 4. Graph displaying the percentage of food categories in NORS reports and Tweets data.

4.3. Outbreak Forecasting Model

Traditional epidemiological models for outbreak forecasting include a variety of regression models, such as timeseries regression, multivariant regression for the prediction of case numbers, and multinomial regression, binary regression, and logistic regression for the prediction of a class, whether it is high/low risk classification or multiple outbreak/sporadic outbreak classification. In the foodborne illness outbreak forecasting scenario, ARIMA and Gaussian distribution models were used to conduct a timeseries analysis of foodborne outbreaks and predict potential outbreaks in India [36]. In addition to timeseries regression models, a multivariate regression model was used for the prediction of cases of Salmonella enterica serovar Enteritidis infections [37]. Regression models can also be regarded as classification models when the goal of prediction turns out to be a class. For example, a multinomial regression model was used to predict caustic pathogens of food poisoning cases for assisting outbreak analysis and forecasting possible pathogens of contamination in future outbreaks [38]. A binary regression model was used to classify the restaurants as high risk or low risk from Twitter data [13]. Logistic regression, a classical statistical regression model in which the response variable is categorical, was employed to classify Yelp reviews indicating "sick" and "multiple outbreaks" [14]. Similarly, logistic regression was used as the classification method in this study with the goal to classify the response variable as "outbreak"–1 or "not outbreak"–0 by establishing a threshold. The threshold is the number of cases in a specific time period, a hyperparameter that could be modified during modeling. This is useful to show days that have more than a certain number of cases, indicating when the higher risk times are. The logistic regression equation resembles the linear regression equation; however, it generates a value ranging from 0 to 1. The predicted label is determined by its proximity to either end of the range.

The accuracy score of a logistic model indicates the number of correct predictions made by the model. When Twitter data was used in combination with the NORS data, a best accuracy score of 0.82 was obtained on the validation set (Figure 5a). Since NORS data are not as timely as Twitter Data, we further explored the effectiveness of the models built upon tweets only. For tweets-only modeling, the best accuracy score reached 0.8 on the validation set (Figure 5b), which shows a great chance of employing tweets to detect significant foodborne illness outbreak. In previous studies, classifiers with performance scores of 0.74, 0.84, and 0.64 were obtained with fairly good performance when predicting foodborne illness cases [13,14]. The logistic regression model appears to be quite accurate in predicting whether the NORS and/or Twitter cases will exceed 200. A foodborne outbreak was defined as an event when two or more people get sick after eating the same food [39]. While most of the foodborne outbreaks are sporadic and only affect a small number of people, some outbreaks especially multistate outbreaks often cause a significant number of sicknesses. The prediction model helps to identify time periods with high risk (prediction with the class "outbreak" or "1") when the number of cases exceeds the threshold.

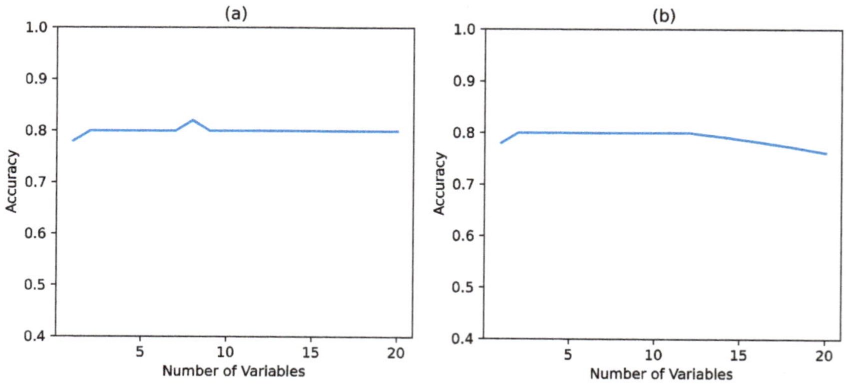

Figure 5. Accuracy of logistic regression with tweets and NORS data (**a**) and with tweets (**b**).

4.4. Frontend Visualizations

A website was created, and made publicly available at https://usda-foodpoisoning.wpi.edu/, to communicate the results of our work to both policymakers and ordinary people. Three core data sections are included, each presenting a different aspect of our data collection. Firstly, on the homepage, the user is presented with several meaningful statistics about the collected tweets. The number of tweets collected, the number of tweets that we are more than 90% confident were correctly identified by the algorithm, the number of tweets with symptom entities, the number of tweets with city information, and the top five states ranked by tweet volume (California, Texas, Florida, New York, and Ohio) are displayed. The purpose of this section is to provide quick insights into the extensive volume of analyzed tweets. Secondly, a graph is displayed illustrating the top symptoms over time for key foodborne illness-related keywords. This data is generated in real time from the backend since users will have the ability to filter results within a specific time range. An increase in frequency for a particular keyword could indicate the occurrence of a foodborne illness event. Finally, an interactive map (as shown in Figure 6) was created to display the geocoordinates of every tweet containing location data in a heatmap format. Dark shades of red indicate areas with the highest number of tweets about foodborne illness, while teal represents areas with lower foodborne illness tweet density. The heatmap is essential for accurately presenting the data due to a limitation in the geodata received. In most cases, the geolocation provided is a city or neighborhood rather than exact coordinates matching the original user's location when tweeting. Consequently, multiple tweets often have the same geocoordinates, which are plotted on the map. For instance, there are over 700 tweets with the same coordinates in different parts of New York City. Fortunately, with the properties of the heatmap, an increase in tweets with the same coordinates results in a darker shade of red, enabling users to identify significant tweet volume associated with a specific point.

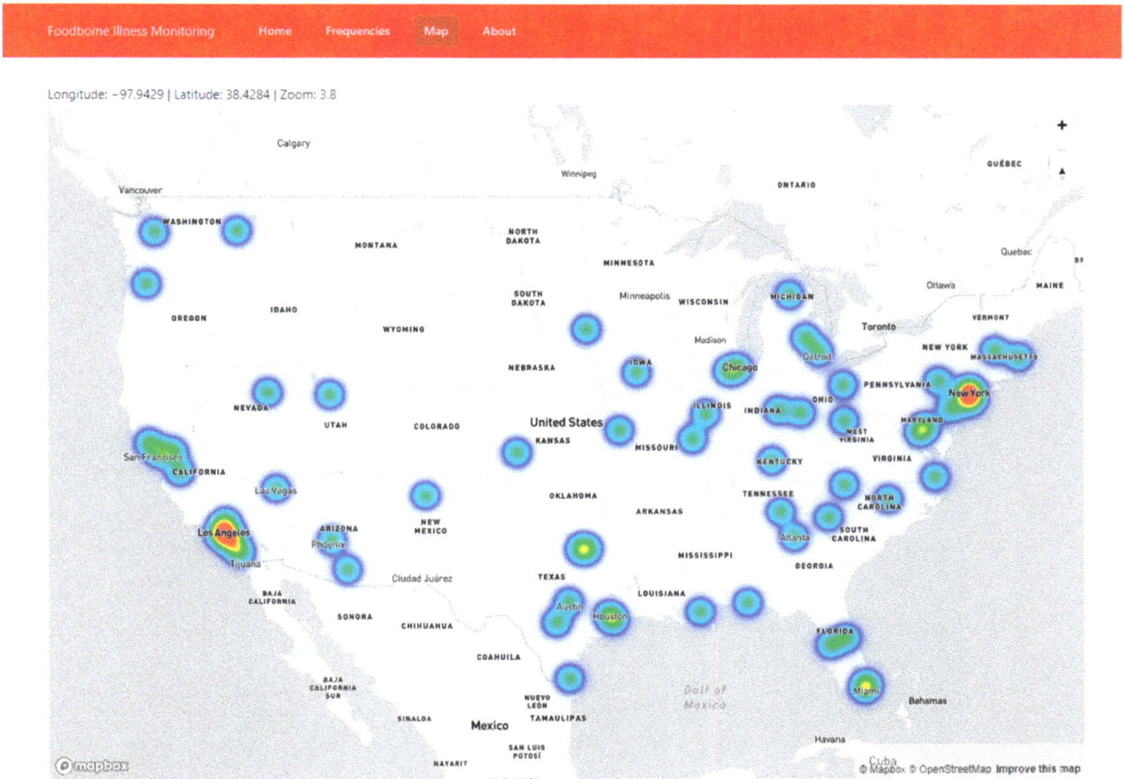

Figure 6. The heat map of tweets within the U.S.

In addition to zooming in and out, the map supports click actions. When a user taps on a location with tweet density, a popup appears displaying the text of the tweet. This feature is crucial in validating the results of our model as users can independently assess the raw text of a tweet that we have predicted to be related to foodborne illness and form their own conclusion about the accuracy of our prediction. In cases in which multiple tweets are stacked on top of each other, users are provided with "Previous" and "Next" buttons to cycle through potentially hundreds of tweets associated with a specific location.

The Python module Dash [40] was used to create an interactive dashboard (Figure 7) for the frequency module. An interactive dashboard was developed using the React framework. The focus of this prototype was on interactivity, allowing users to modify date ranges and explore specific food groups. The interactivity of the dashboard can be advantageous for both government officials and the general public, enabling them to gain a broader perspective on potential foodborne illness outbreaks.

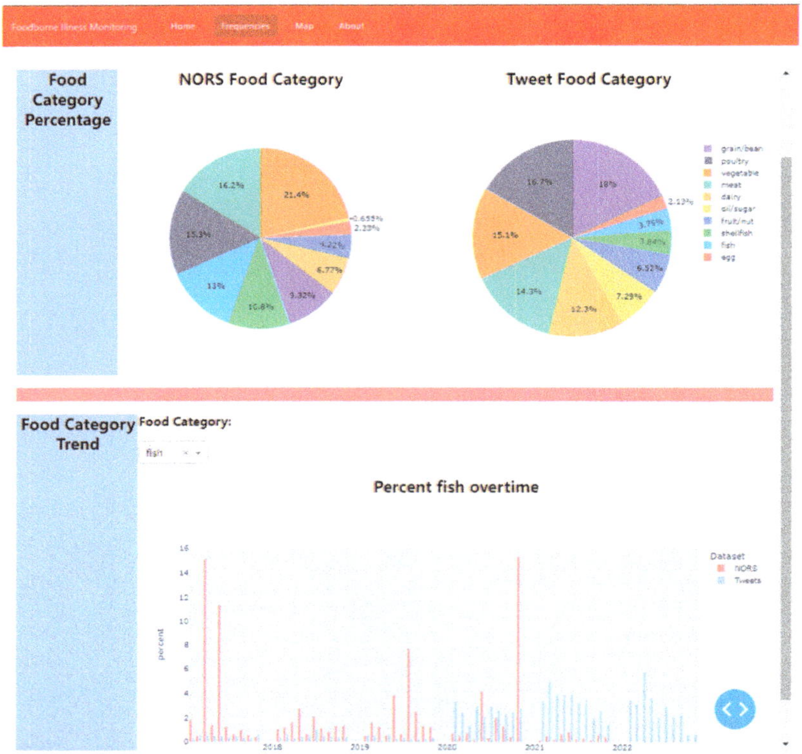

Figure 7. The frequencies of most common food categories related to foodborne illnesses.

4.5. Limitations and Implications of This Study

In this study, we successfully developed a pipeline that efficiently collects Twitter data, identifies foodborne illness cases, and extracts important information from those cases. The frontend visualizations also allow users to observe the geographical distribution of potential foodborne illnesses on a map, compare common food categories associated with historical outbreaks (based on CDC data) with the categories detected in positive-predicted cases from Twitter data, and track changes over time for different food categories in both datasets. Overall, our system offers an alternative approach to assist foodborne outbreak surveillance. However, there are certain limitations to consider, as described below.

Firstly, our data collection mechanism relies on sampling Twitter data, which means that a substantial number of foodborne illness cases may be missing from our dataset. Additionally, we only include tweets with geolocation information, which may result in the exclusion of informative tweets lacking such location data. Secondly, when classifying newly captured tweets, we utilize a pre-trained RoBERTa model that demonstrates high accuracy but may have limited generalization capability for unseen tweets. Finally, the existence of multiple platforms can lead to redundancies. For instance, websites like iwaspoisoned.com offer similar functionalities to our system but utilize structured data positively reported by users. In contrast, we aimed to leverage the largely untapped potential of unstructured data from social media, although a portion of our collected data does come from that platform. Despite appearing redundant, these two types of systems can complement each other in various ways, such as evaluating model performance using different datasets and leveraging diverse data sources to assist in identifying foodborne outbreaks.

5. Conclusions

A unique information system for monitoring foodborne illnesses was developed by utilizing social media data. By converting unorganized Twitter data into a structured database format that includes essential 3W information, the system enables more efficient detection of potential outbreaks related to foodborne illnesses. One valuable outcome of implementing this system is the reduction in human efforts needed to extract important entities from social media, while also preparing for the analysis of large-scale datasets in close to real time. Through descriptive and predictive analyses of the collected data, we compared the results with the ground truth data reported by the CDC. Our study revealed that the most implicated food categories and their distributions in both the Twitter dataset and NORS dataset were similar. This finding suggests that our system, developed using Twitter data, can complement traditional surveillance systems by providing near-real-time information on foodborne illnesses and implicated foods. The outbreak forecasting model also helps to identify time periods with a high risk of causing a certain number of cases. By incorporating social media data into surveillance systems, we can benefit from timely and comprehensive insights, allowing for quicker responses and more effective control measures. In conclusion, this research highlights the potential of social media as a valuable resource for enhancing foodborne disease surveillance and ultimately improving public health outcomes. The continued exploration and refinement of such alternative surveillance approaches will contribute to the development of more robust and proactive systems for monitoring and preventing foodborne diseases.

Author Contributions: Conceptualization, D.T., H.F. and E.R.; methodology, J.L., A.L., T.K., L.R., R.H. and D.Z.; data collection and curation, J.L., A.L., T.K., L.R., R.H. and D.Z.; writing—original draft preparation, D.T., R.H. and D.Z.; writing—review and editing, H.F. and E.R.; visualization, R.H. and D.Z.; supervision, E.R.; project administration, H.F. and E.R.; funding acquisition, H.F. and E.R. All authors have read and agreed to the published version of the manuscript.

Funding: This work was supported by the Agriculture and Food Research Initiative (AFRI) award No. 2020-67021-39133 from the U.S. Department of Agriculture (USDA) National Institute of Food and Agriculture (NIFA).

Data Availability Statement: The datasets collected and analyzed during the current study are available from the corresponding author on reasonable request. In addition, we would like to clarify that the terms of service of Twitter were followed in order to collect the data used in our study. The CDC dataset of historical foodborne outbreaks would be available from a CDC National Outbreak Reporting System (NORS) data request application by emailing the NORS Dashboard mailbox.

Acknowledgments: We express our gratitude to the Worcester Polytechnic Institute MQP team, including Katy Hartmann, Isabel Alvarado Blanco Uribe, John Carroll, David Leandres, Cole Noreika, and Nick Vachon, for their valuable assistance in processing the tweet data, conducting model validation, and designing the frontend interface. We specially thank the NORS team for sharing the NORS dataset and Cassie Hartley's kind assistance with data interpretation and answering any questions.

Conflicts of Interest: The authors declare no conflict of interest.

References

1. Scallan, E.; Hoekstra, R.M.; Angulo, F.J.; Tauxe, R.V.; Widdowson, M.A.; Roy, S.L.; Jones, J.L.; Griffin, P.M. Foodborne illness acquired in the United States—Major pathogens. *Emerg. Infect. Dis.* **2011**, *17*, 7–15. [CrossRef]
2. Havelaar, A.H.; Kirk, M.D.; Torgerson, P.R.; Gibb, H.J.; Hald, T.; Lake, R.J.; Praet, N.; Bellinger, D.C.; De Silva, N.R.; Gargouri, N.; et al. World Health Organization global estimates and regional comparisons of the burden of foodborne disease in 2010. *PLoS Med.* **2015**, *12*, e1001923. [CrossRef] [PubMed]
3. Swaminathan, B.; Barrett, T.J.; Hunter, S.B.; Tauxe, R.V.; Force, C.P.T. PulseNet: The molecular subtyping network for foodborne bacterial disease surveillance, United States. *Emerg. Infect. Dis.* **2001**, *7*, 382–389. [CrossRef]
4. Jones, T.F.; Scallan, E.; Angulo, F.J. FoodNet: Overview of a decade of achievement. *Foodborne Pathog. Dis.* **2007**, *4*, 60–66. [CrossRef]

5. White, A.E.; Tillman, A.R.; Hedberg, C.; Bruce, B.B.; Batz, M.; Seys, S.A.; Dewey-Mattia, D.; Bazaco, M.C.; Walter, E.S. Foodborne illness outbreaks reported to national surveillance, United States, 2009–2018. *Emerg. Infect. Dis.* **2022**, *28*, 1117–1127. [CrossRef]
6. Thakur, M.; Olafsson, S.; Lee, J.S.; Hurburgh, C.R. Data mining for recognizing patterns in foodborne disease outbreaks. *J. Food Eng.* **2010**, *97*, 213–227. [CrossRef]
7. Sanchez, E.; Simpson, R.B.; Zhang, Y.; Sallade, L.E.; Naumova, E.N. Exploring risk factors of recall-associated foodborne disease outbreaks in the United States, 2009–2019. *Int. J. Environ. Res. Public Health* **2022**, *19*, 4947. [CrossRef]
8. Simpson, R.; Sallade, L.; Sanchez, E.; Zhang, Y.; Naumova, E. Analysing foodborne illness outbreak severity in the USA, 2009–19. *Lancet Glob. Health* **2022**, *10*, S5. [CrossRef]
9. Tao, D.; Zhang, D.; Hu, R.; Rundensteiner, E.; Feng, H. Crowdsourcing and machine learning approaches for extracting entities indicating potential foodborne outbreaks from social media. *Sci. Rep.* **2021**, *11*, 21678. [CrossRef]
10. Paul, M.J.; Dredze, M. Social monitoring for public health. In *Synthesis Lectures on Information Concepts, Retrieval, and Services*; Morgan & Claypool: San Raferl, CA, USA, 2017; Volume 9, pp. 1–183.
11. Harris, J.K.; Mansour, R.; Choucair, B.; Olson, J.; Nissen, C.; Bhatt, J. Health department use of social media to identify foodborne illness—Chicago, Illinois, 2013–2014. *Morb. Mortal. Wkly. Rep.* **2014**, *63*, 681–685.
12. Harrison, C.; Jorder, M.; Stern, H.; Stavinsky, F.; Reddy, V.; Hanson, H.; Waechter, H.; Lowe, L.; Gravano, L.; Balter, S. Using online reviews by restaurant patrons to identify unreported cases of foodborne illness—New York City, 2012–2013. *Morb. Mortal. Wkly. Rep.* **2014**, *63*, 441–445.
13. Sadilek, A.; Kautz, H.; DiPrete, L.; Labus, B.; Portman, E.; Teitel, J.; Silenzio, V. Deploying nEmesis: Preventing foodborne illness by data mining social media. *Ai Mag.* **2017**, *38*, 37–48. [CrossRef]
14. Effland, T.; Lawson, A.; Balter, S.; Devinney, K.; Reddy, V.; Waechter, H.; Gravano, L.; Hsu, D. Discovering foodborne illness in online restaurant reviews. *J. Am. Med. Inform. Assoc.* **2018**, *25*, 1586–1592. [CrossRef] [PubMed]
15. Tao, D.; Yang, P.; Feng, H. Utilization of text mining as a big data analysis tool for food science and nutrition. *Compr. Rev. Food Sci. Food Saf.* **2020**, *19*, 875–894. [CrossRef]
16. Zhai, C.; Massung, S. *Text Data Management and Analysis: A Practical Introduction to Information Retrieval and Text Mining*; Morgan & Claypool: San Rafael, CA, USA, 2016.
17. Al-Garadi, M.A.; Yang, Y.C.; Sarker, A. The Role of Natural Language Processing during the COVID-19 Pandemic: Health Applications, Opportunities, and Challenges. *Healthcare* **2022**, *10*, 2270. [CrossRef]
18. Hu, R.; Zhang, D.; Tao, D.; Hartvigsen, T.; Feng, H.; Rundensteiner, E. TWEET-FID: An Annotated Dataset for Multiple Foodborne Illness Detection Tasks. *arXiv* **2022**, arXiv:2205.10726.
19. Waldner, C. *Big Data for Infectious Diseases Surveillance and the Potential Contribution to the Investigation of Foodborne Disease in Canada*; National Collaborating Centre for Infectious Diseases: Winnipeg, MB, Canada, 2017.
20. Subramaniyam, M.; Halim-Lim, S.A.; Mohamad, S.F.B.; Priyono, A. Digital supply chain in the food industry: Critical success factors and barriers. In Proceedings of the 2021 IEEE International Conference on Industrial Engineering and Engineering Management (IEEM), Singapore, 13–16 December 2021; pp. 404–410.
21. Maeda, Y.; Kurita, N.; Ikeda, S. An early warning support system for food safety risks. In *New Frontiers in Artificial Intelligence: Joint JSAI 2005 Workshop Post-Proceedings*; Springer: Berlin/Heidelberg, Germany, 2006; pp. 446–457.
22. Kate, K.; Chaudhari, S.; Prapanca, A.; Kalagnanam, J. FoodSIS: A text mining system to improve the state of food safety in singapore. In Proceedings of the 20th ACM SIGKDD International Conference on Knowledge Discovery and Data Mining, New York, NY, USA, 24–27 August 2014; pp. 1709–1718.
23. Chen, S.; Huang, D.; Nong, W.; Kwan, H.S. Development of a food safety information database for Greater China. *Food Control* **2016**, *65*, 54–62. [CrossRef]
24. Bouzembrak, Y.; Steen, B.; Neslo, R.; Linge, J.; Mojtahed, V.; Marvin, H.J.P. Development of food fraud media monitoring system based on text mining. *Food Control* **2018**, *93*, 283–296. [CrossRef]
25. Quade, P.; Nsoesie, E.O. A platform for crowdsourced foodborne illness surveillance: Description of users and reports. *JMIR Public Health Surveill.* **2017**, *3*, e7076. [CrossRef]
26. Aramaki, E.; Maskawa, S.; Morita, M. Twitter catches the flu: Detecting influenza epidemics using Twitter. In Proceedings of the 2011 Conference on Empirical Methods in Natural Language Processing, Edinburgh, UK, 27–31 July 2011; pp. 1568–1576.
27. Tweetkit. Tweetkit 0.2.8. Available online: https://pypi.org/project/tweetkit/ (accessed on 24 February 2023).
28. Devlin, J.; Chang, M.W.; Lee, K.; Toutanova, K. Bert: Pre-training of deep bidirectional transformers for language understanding. *arXiv* **2018**, arXiv:1810.04805.
29. Nguyen, D.Q.; Vu, T.; Nguyen, A.T. BERTweet: A pre-trained language model for English Tweets. *arXiv* **2020**, arXiv:2005.10200.
30. Liu, Y.; Ott, M.; Goyal, N.; Du, J.; Joshi, M.; Chen, D.; Levy, O.; Lewis, M.; Zettlemoyer, L.; Stoyanov, V. Roberta: A robustly optimized bert pretraining approach. *arXiv* **2019**, arXiv:1907.11692.
31. Graves, A.; Mohamed, A.R.; Hinton, G. Speech recognition with deep recurrent neural networks. In Proceedings of the 2013 IEEE International Conference on Acoustics, Speech and Signal Processing, Vancouver, BC, Canada, 26–31 May 2013; pp. 6645–6649.
32. Wunnava, S.; Qin, X.; Kakar, T.; Kong, X.; Rundensteiner, E. A dual-attention network for joint named entity recognition and sentence classification of adverse drug events. In Proceedings of the Findings of the Association for Computational Linguistics: EMNLP 2020, Online, 16–20 November 2020; pp. 3414–3423.

33. Lazuardy, M.F.S.; Anggraini, D. Modern front end Web Architectures with React. Js and Next. Js. *Res. J. Adv. Eng. Sci.* **2022**, *7*, 132–141.
34. Richardson, L.C.; Bazaco, M.C.; Parker, C.C.; Dewey-Mattia, D.; Golden, N.; Jones, K.; Klontz, K.; Travis, C.; Kufel, J.Z.; Cole, D. An updated scheme for categorizing foods implicated in foodborne disease outbreaks: A tri-agency collaboration. *Foodborne Pathog. Dis.* **2017**, *14*, 701–710. [CrossRef] [PubMed]
35. Nsoesie, E.O.; Kluberg, S.A.; Brownstein, J.S. Online reports of foodborne illness capture foods implicated in official foodborne outbreak reports. *Prev. Med.* **2014**, *67*, 264–269. [CrossRef]
36. Bisht, A.; Kamble, M.P.; Choudhary, P.; Chaturvedi, K.; Kohli, G.; Juneja, V.K.; Sehgal, S.; Taneja, N.K. A surveillance of food borne disease outbreaks in India: 2009–2018. *Food Control* **2021**, *121*, 107630. [CrossRef]
37. Rojas, F.; Ibacache-Quiroga, C. A forecast model for prevention of foodborne outbreaks of non-typhoidal salmonellosis. *PeerJ* **2020**, *8*, e10009. [CrossRef] [PubMed]
38. Inoue, H.; Suzuki, T.; Hyodo, M.; Miyake, M. Evaluation of multinomial logistic regression models for predicting causative pathogens of food poisoning cases. *J. Vet. Med. Sci.* **2018**, *80*, 1223–1227. [CrossRef]
39. Greig, J.D.; Ravel, A. Analysis of foodborne outbreak data reported internationally for source attribution. *Int. J. Food Microbiol.* **2009**, *130*, 77–87. [CrossRef]
40. Dash 2.11.1. 2014, pp. 1709–1718. Available online: https://pypi.org/project/dash/ (accessed on 30 June 2023).

foods

MDPI

Article

The Estimation of Chemical Properties of Pepper Treated with Natural Fertilizers Based on Image Texture Parameters

Ewa Ropelewska * and Justyna Szwejda-Grzybowska

Fruit and Vegetable Storage and Processing Department, The National Institute of Horticultural Research, Konstytucji 3 Maja 1/3, 96-100 Skierniewice, Poland; justyna.grzybowska@inhort.pl
* Correspondence: ewa.ropelewska@inhort.pl

Abstract: The cultivar and fertilization can affect the physicochemical properties of pepper fruit. This study aimed at estimating the content of α-carotene, β-carotene, total carotenoids, and the total sugars of unfertilized pepper and samples treated with natural fertilizers based on texture parameters determined using image analysis. Pearson's correlation coefficients, scatter plots, regression equations, and coefficients of determination were determined. For red pepper Sprinter F_1, the correlation coefficient (R) reached 0.9999 for a texture from color channel B and −0.9999 for a texture from channel Y for the content of α-carotene, −0.9998 (channel a) for β-carotene, 0.9999 (channel a) and −0.9999 (channel L) for total carotenoids, as well as 0.9998 (channel R) and −0.9998 (channel a) for total sugars. The image textures of yellow pepper Devito F_1 were correlated with the content of total carotenoids and total sugars with the correlation coefficient reaching −0.9993 (channel b) and 0.9999 (channel Y), respectively. The coefficient of determination (R^2) of up to 0.9999 for α-carotene content and the texture from color channel Y for pepper Sprinter F_1 and 0.9998 for total sugars and the texture from color channel Y for pepper Devito F_1 were found. Furthermore, very high coefficients of correlation and determination, as well as successful regression equations regardless of the cultivar were determined.

Keywords: red pepper; yellow pepper; α-carotene; β-carotene; total carotenoids; total sugars; image processing; regression

1. Introduction

Nowadays, the consequences associated with conventional agricultural production are observed, such as the disruption of natural diversity, soil and water pollution by fertilizers and plant protection products, reduced biomass production, and an increase in diseases. Therefore, there is an increasing trend in agricultural production based on the agroecological concept. With the ever-increasing nutritional awareness of society, the group of consumers looking for organically produced food is growing, and thus the demand for products with high health-promoting value and free of pesticides is increasing. The use of chemical plant protection products is an easy way of combating organisms that are harmful to plants, but it carries a risk to the health of people, animals, and the environment. More and more modern preparations of natural origin for use in plant cultivation are appearing on the market, which have a beneficial effect on plant-life processes and, at the same time, can influence the composition and content of bioactive and nutritional compounds and protect plants from diseases and pests [1,2].

Sweet pepper (*Capsicum annuum* L.), otherwise known as the annual pepper plant, is classified as a plant of the Solanaceae family. About 39 pepper species have been identified, including five domesticated species, *C. annuum*, *C. baccatum*, *C. chinense*, *C. frutescens*, and *C. pubescens* [3]. It is a vegetable with low energy value and low index and glycemic load. It is one of the vegetables characterized by high health-promoting compounds, carotenoids (especially beta-carotene, α-carotene, and lutein); phenolic compounds; and vitamins, such as ascorbic acid, vitamins (A, E, and K). It is also a good source of folic acid, minerals (potassium, manganese, iron, and magnesium), and dietary fiber [4–6]. In previous studies, a 100 g serving of peppers was found to cover 100% of the recommended daily intake of vitamin C, 5–10% of tocopherol, and 5–10% of provitamin A [6].

Nutritional and health-promoting compounds contained in pepper fruits play an important role in the prevention of cardiovascular diseases and cancer [7–9]. Among these compounds, the carotenoids—pigments present in the seed coat of pepper fruits—are of great importance. Scientists are increasingly paying attention to the health-promoting value of carotenoids and their role in disease prevention. Among other things, these compounds are part of the body's defense system against free radicals, an excess of which accelerates cellular aging processes. These processes can become the cause of the development of chronic diseases, such as cancer and cardiovascular diseases. As the selected are precursors of vitamin A, they also counteract diseases associated with vitamin A deficiency. Vitamin A is responsible for vision and is involved in processes in the retina. It also plays a very important role in the proliferation of bone tissue and epithelial cells, especially of the oral mucosa, gastrointestinal tract, urinary tract, respiratory tract, as well as the organ of vision. In plants, carotenoids are pigments that play an important role in protecting plants from photo-oxidative effects. They are effective antioxidants and participate in scavenging singlet oxygen and superoxide radicals [10].

Among the compounds that largely determine the nutritional value of pepper fruits are the sugars that give them their sweet taste. Their content is genetically determined and can be modified by agrotechnical and climatic conditions [11–14]. The sugar content also depends on the ripeness of the pepper fruit [13,15,16]. These compounds mainly accumulate in the pericarp of peppers and their content may vary from variety to variety [17]. In addition, sugars, together with specific volatile compounds, can affect the overall perception of taste and are vitamin C stabilizers [18]. Vegetables and fruits that have a high vitamin C content in their composition have been proven to have high levels of sugars [13,19].

The pepper fruit cultivars differ in color, size, shape, and flavor. The color and shape of pepper fruit at the stage of consumer maturity depend on the presence of a gene system. Pepper fruits synthesize and accumulate a variety of carotenoids in tissues with a wide range of content. Even within the same cultivar, the composition and number of carotenoids can vary considerably depending on maturity stage, growing region, and growing conditions [20–23]. The first level of regulation of carotenoid biosynthesis in pepper fruit is the transcriptional control of gene biosynthesis. It is the main determinant of carotenoid production in pepper during fruit ripening in response to developmental cues [24,25]. Various studies confirm that capsanthin, β-carotene, and β-cryptoxanthin are the carotenoids responsible for the red color of peppers [26,27]. In contrast, yellow pepper varieties contain lutein, violaxanthin, antheraxanthin, and zeaxanthin. The orange color of pepper fruits is determined by the presence of α- and β-carotene, β-cryptoxanthin, zeaxanthin, and lutein. Brown varieties mainly contain red carotenoid pigments, such as red pepper fruits [26]. The shape of the pepper fruit is also an important varietal characteristic, ranging from very long and narrow, through spherical to fruit in the form of a short or elongated prism. The wall thickness of the pepper fruit can be 5–12 mm depending on the variety and growth conditions. The average weight of pepper fruit in large-flowered sweet varieties reaches 250–350 g, and often more [13]. Flesh and skin texture, juiciness and postharvest shelf life are determined by both genetic traits and growing conditions [28].

As mentioned above, the physicochemical properties depend on the pepper cultivar and can be affected by fertilization. Research involving image analysis allows for an objective assessment of the quality of food products with the reduction in the time, costs, and labor consumption of analyzes [29]. Due to image processing, the textural, color, and morphological parameters in the images can be computed and then interpreted [30]. The image texture parameters can be considered a function of the spatial variation of the brightness pixel intensity and carry information about the structure of the samples, whereas analyses of image textures provide insights into product quality. Image textures can be analyzed using specialized software packages, for example, MaZda, which enables quantitative analysis, including texture parameters computation, as well as texture selection and extraction, data classification and visualization, and image segmentation tools. MaZda provides options for the transformation of color images to grayscale and conversions to individual color channels (components) [31–33]. Several channels are most commonly used. There are color channels *R* (red), *G* (green), and *B* (blue) from the RGB color space; *L* (lightness component from black to white), *a* (red for positive and green for negative values), and *b* (yellow for positive and blue for negative values) from the Lab color space; and *X* (component with color information), *Y* (lightness), and *Z* (component with color information) from the XYZ color space [29,34].

The objective of this study was to estimate the chemical properties of peppers belonging to two cultivars treated with natural fertilizers in a fast and inexpensive manner based on image texture parameters. A great novelty of the present study is related to the estimation of α-carotene, β-carotene, total carotenoids, and total sugars of untreated red pepper Sprinter F_1 and yellow pepper Devito F_1 and samples subjected to different organic fertilization, such as NaturalCrop® SL, Bio-algeen S90, and nettle fertilizer. The innovative regression equations including features selected from a set of 1629 image textures from nine color channels were determined. The developed procedure could enable the selection of fruit with the desired properties based on the image features without the need to perform more labor-intensive and expensive measurements.

2. Materials and Methods

2.1. Experimental Design

The experimental material consisted of pepper growing in greenhouses at the Laboratory of Cultivation of Vegetable and Edible Mushrooms of the National Institute of Horticultural Research in Skierniewice in Poland. Peppers belonged to two cultivars of Sprinter F_1 characterized by red fruit and Devito F_1 with yellow fruit. The experiment was carried out between March and October 2021. During the experiment, daily climatological data were collected inside the greenhouse, using a thermo-hygrometer. Organic pepper seeds were used and potted seedlings were planted in the soil in an ecological greenhouse under the principles of organic farming. The experiment was set up in the soil with a pH of 6.0. Before planting, the soil was fertilized with organic plant compost at a dose of 300 kg per 100 m^2, obtaining the soil fertility at the level (mg dm^{-3}) of N-230, P-310, K-400, and Mg-350. The seedlings were planted on April 20, in a row-row system, at a spacing of 35 cm × 60 cm (3.6 plants per 1 m^2). Plants were grown in a four-shoot system. Pepper fertigation was carried out with the use of a drip system and fertilizers approved for organic cultivation depending on the current humidity and nutritional requirements of the pepper in particular growth stages [35]. The fruit was harvested at the stage of colorful ripeness.

2.2. Preparations

In research natural preparation and fertilizers approved for organic farming was performed such as: nettle fertilizer (NE/427/2018) (as a substance with biocidal and nutritional effect) based on N-NO_3 about 0.5–0.7 mg·L^{-1}, N-NH4 >100 mg·L^{-1}, P about 20 mg·L^{-1}, K-about 650 mg·L^{-1}, F—0.2 mg·L^{-1}, pH 6.3–6.7 (20% solution was used); Bio-algeen S90 (organic fertilizer) containing 90 groups of chemicals compounds, including amino acids; vitamins; alginic acid and other unexplored active components of marine algae; N—0.02%; P—0.006%; K—0.096%; Ca—0.31%; MG—0.021% and B—16 mg·kg^{-1}; Fe—6.3 mg·kg^{-1}; Cu—0.2 mg·kg^{-1}; Mn—0.6 mg·kg^{-1}; Zn −1.0 mg·kg^{-1} (0.3% solution was used); NaturalCrop® SL (growth and development stimulator)—enzymatic concentrate of peptides; and 16 L-amino acids most important for plants (GLY, GLU, HIS, LYS) of plant origin >50% (605 g·L^{-1}), including free amino acids 0.2% (242 g·L^{-1}), amino nitrogen 9%

(109 g·L^{-1}), organic carbon 24.5% (296.5 g·L^{-1}) (0.3% solution was used), and the control (unfertilized plots).

2.3. Image Acquisition and Processing

The image acquisition was performed using an Epson Perfection V600 flatbed scanner. The scanner was placed in a black box to obtain a black background of images. The images of pepper slices (cross-sections) with a thickness of 10 mm obtained using a slicing machine were considered. In the case of each cultivar and treatment, images of 50 slices were acquired, as follows:

- Fifty slices of a control sample of Sprinter F_1 pepper,
- Fifty slices of Sprinter F_1 pepper treated with NaturalCrop® SL,
- Fifty slices of Sprinter F_1 pepper treated with Bio-algeen S90,
- Fifty slices of Sprinter F_1 pepper treated with nettle fertilizer,
- Fifty slices of a control sample of Devito F_1 pepper,
- Fifty slices of Devito F_1 pepper treated with NaturalCrop® SL,
- Fifty slices of Devito F_1 pepper treated with Bio-algeen S90,
- Fifty slices of Devito F_1 pepper treated with nettle fertilizer.

The acquired images were saved in a TIFF format at 800 dpi resolution. Before processing in the Mazda software (Łódź University of Technology, Institute of Electronics, Łódź, Poland) [31–33], the pepper slice images were converted to a BMP format. Then, texture parameter extraction was carried out for each of the color channels *L*, *a*, *b*, *R*, *G*, *B*, *X*, *Y*, and *Z*. The sample images are presented in Figure 1. The image segmentation into pepper slices and the background was performed manually based on the pixel brightness intensity. Each pepper slice (the area of flesh without skin) was considered as a region of interest (ROI). For each ROI, in total 1629 image textures, including 181 textures for each color channel were computed.

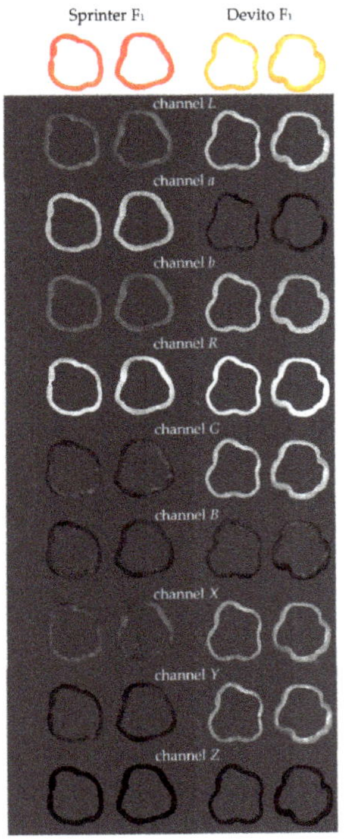

Figure 1. Original color images of pepper slices and images in different color channels.

2.4. Chemical Properties

The samples for chemical analysis were prepared from 10 pepper fruits from each repetition. Pepper fruits were cut into quarters, frozen in liquid nitrogen, crushed in a Blixer homogenizer and stored at −20 °C.

2.4.1. HPLC Analysis of Sugars

An HPLC analysis of total sugars (sucrose, glucose, and fructose) was determined by high-performance liquid chromatography (Agilent 1200 HPLC system, equipped with a differential refractometric detector), using Aminex HPX-87C (300 mm × 7.5 mm) with a precolumn according to European Standard EN 12630. The isocratic flow was 0.6 $mL \cdot min^{-1}$, column temperature was 80 °C, and mobile phase-edetate calcium disodium (Ca-EDTA). The samples were dissolved in redistilled water, homogenized, and purified on a Waters SepPak PLUS C18 filter. The sugars were quantified by calibration curve for sucrose, glucose and fructose, and the results were expressed as mg 100 g^{-1} fresh mass (f.m.).

2.4.2. Carotenoid Extraction

Carotenoid content was determined by the method according to Bohoyo-Gil et al. [36]. The sample was homogenized in the extraction solution (hexane:acetone 6:4) and filtered through a Büchner funnel under reduced pressure. Next, the extract was transferred to a separating funnel and shaken with the addition of water. After phase separation, the water–acetone phase was discarded. The acetone rinsing operation was repeated until the lower phase was free of acetone and the upper hexane phase containing carotenoids was filtered through a filter paper containing anhydrous sodium sulphate into an evaporation flask. Hexane was evaporated to dryness in a vacuum evaporator at 40 °C, the dry residue was quantitatively transferred to a 25 mL flask with a solution of acetonitrile:methanol:ethyl acetate 55:25:20 + 0.1% BHT + 1 mL TEA and 4 mL of hexane. The flask extract was filtered with a 45 µm PTFE filter into an amber bottle and analyzed by HPLC.

2.4.3. HPLC Analysis of Carotenoids

An HPLC analysis of total carotenoids in pepper samples was determined by high-performance liquid chromatography (HPLC), using a Kinetex C-18 column (250 mm × 4.6 mm; 5 µm) on an Agilent 1200 HPLC system equipped with a DAD detector. The elution conditions were as follows, 0.7 mL min^{-1}, temperature 28 °C, wavelength 450 nm, and 472 nm; in the mobile phase, acetonitrile (A), methanol + 1 mL TEA + 1 g BHT (B), ethyl acetate (C); and in gradient flow, 0–10 min, 95% A, 2% B, 3% C; 10–25 min, 55% A, 20% B, 25% C; 25–35 min, 95% A, 2% B, and 3% C. The calculations were quantified by calibration with the standards of β-carotene and α-carotene (Sigma-Aldrich, Darmstadt, Germany). The carotenoid content was expressed in mg 100 g^{-1} f.m.

2.5. Statistical Analysis

The linear relationships between the chemical properties and image textures of pepper samples were determined using STATISTICA 13.1 (Dell Inc., Tulsa, OK, USA, StatSoft Polska, Kraków, Poland). The analysis was performed separately for samples of Sprinter F_1 and Devito F_1 and then for a dataset combining both cultivars Sprinter F_1 and Devito F_1. In the case of Sprinter F_1 pepper, among the chemical properties, α-carotene, β-carotene, total carotenoids, and total sugars were considered. For Devito F_1, α-carotene was not detected and the differences in the content of β-carotene were very small. Therefore, the analyses for Devito F_1 and a dataset combining Sprinter F_1 and Devito F_1 were performed only for total carotenoids and total sugars. In the first step, Pearson's correlation coefficients (R) were determined at a significance level of $p < 0.05$, and scatter plots were created. In the case of image textures from color channels L, a, b, R, G, B, X, Y, and Z for which the correlations occurred, the highest value of R for positive correlation and the highest value of R for negative correlation were selected to be presented in this paper. Then, regression equations for the estimation of chemical properties of pepper untreated and treated with natural fertilizers using image texture parameters were determined and coefficients of determination (R^2) were computed.

3. Results and Discussion

3.1. Relationship between Chemical Properties and Image Texture Parameters of Red Pepper Sprinter F_1

In the case of red pepper Sprinter F_1, correlations between α-carotene, β-carotene, total carotenoids, and total sugars with selected image textures were observed (Table 1). Very strong positive and negative correlations were determined for each of the considered organic chemical compounds reaching 0.9999 and −0.9999 for α-carotene and total carotenoids. In the case of α-carotene, statistically significant correlation coefficients were found for textures of images in color channels *a*, *R*, *G*, *B*, *Y*, and *Z*. The correlation coefficient (R) of 0.9999 was found for one image texture from color channel *B* and −0.9999 for a texture from channel *Y*. None of the image textures from channels *L*, *b*, and *X* correlated with α-carotene. Therefore, these channels turned out to be useless for the estimation of α-carotene content. For total carotenoids, both positive and negative relationships with image textures from all color channels *L*, *a*, *b*, *R*, *G*, *B*, *X*, *Y*, and *Z* were observed, and the values of R equal to 0.9999 and −0.9999 were determined for color channels *a* and *L*, respectively. In the case of β-carotene, the positive correlations with image textures from color channels *a*, *R*, *G*, *B*, *Y*, and *Z* were observed, reaching 0.9994 for a texture from channel *B*. Whereas negative correlations with textures from color channels *a*, *R*, *G*, *B*, *X*, *Y*, and *Z* were observed, and the highest value was −0.9998 for a texture from channel *a*. It was found that the content of total sugars was positively correlated only with textures of images in color channels *R*, *G*, and *X* with the coefficient of up to 0.9998 in the case of color channel *R*. Negative correlations with textures from color channels *L*, *a*, *R*, *G*, *X*, *Y*, and *Z* reached the value of −0.9998 for channel *a*.

Table 1. Correlation coefficients (R) between α-carotene (mg 100 g^{-1} f.m.), β-carotene (mg 100 g^{-1} f.m.), total carotenoids (mg 100 g^{-1} f.m.), and total sugars (%) with selected image texture parameters for control and treated with NaturalCrop®, Bio-algeen S90, and nettle fertilizer red pepper Sprinter F_1 samples; $p < 0.05$.

α-Carotene (mg 100 g^{-1} f.m.)		β-Carotene (mg 100 g^{-1} f.m.)		Total Carotenoids (mg 100 g^{-1} f.m.)		Total Sugars (%)	
Texture Parameter	Correlation Coefficient	Texture Parameter	Correlation Coefficient	Texture Parameter	Correlation Coefficient	Texture Parameter	Correlation Coefficient
BS5SN3Entropy	0.9999	BS5SN3Entropy	0.9994	aS5SH3DifEntrp	0.9999	RS5SH5SumOfSqs	0.9998
ZS4RZFraction	0.9996	aS5SV5Entropy	0.9991	GS5SV5Contrast	0.9998	XS5SH3SumOfSqs	0.9708
aS5SV5Entropy	0.9989	ZS4RZFraction	0.9985	XS5SZ3SumOfSqs	0.9933	GS4RVFraction	0.9685
GS5SN1Entropy	0.9964	GS5SV5Entropy	0.9984	ZS5SH3DifEntrp	0.9984		
YS5SN5DifEntrp	0.9847	YS5SN5DifEntrp	0.9763	RS5SN3Entropy	0.9984	aS5SN1SumVarnc	−0.9998
RS5SN5DifEntrp	0.9766	RS5SN5DifEntrp	0.9666	BS5SH3Entropy	0.9920	ZHDomn01	−0.9997
				YS5SH5DifEntrp	0.9825	XS5SV5SumEntrp	−0.9993
YS4RVLngREmph	−0.9999	aS5SV1AngScMom	−0.9998	bS5SN5Entropy	0.9507	YHDomn10	−0.9993
aS5SV3AngScMom	−0.9996	GS5SV1AngScMom	−0.9993	LS5SN5DifEntrp	0.9505	RHMaxm01	−0.9981
RS5SZ5SumAverg	−0.9975	YS4RVLngREmph	−0.9987			LHDomn10	−0.9979
GS5SV1AngScMom	−0.9972	RS5SZ5SumAverg	−0.9971	LS5SZ5SumVarnc	−0.9999	GS5SZ5SumVarnc	−0.9790
BS5SV5AngScMom	−0.9968	BS5SV5AngScMom	−0.9939	GS5SV3InvDfMom	−0.9999		
ZS5SV5InvDfMom	−0.9920	ZS5SV5InvDfMom	−0.9859	RS5SZ3InvDfMom	−0.9999		
		XS5SN5InvDfMom	−0.9631	bHPerc10	−0.9998		
				aS5SN1InvDfMom	−0.9996		
				BS4RZLngREmph	−0.9986		
				YS5SH3InvDfMom	−0.9909		
				ZS5SH5Correlat	−0.9847		
				XS5SZ5InvDfMom	−0.9724		

The first letter in the texture parameter name means the color channel *L*, *a*, *b*, *R*, *G*, *B*, *X*, *Y*, or *Z*; Entropy—entropy; Fraction—fraction of image in runs; DifEntrp—difference entropy; LngREmph —long run emphasis; AngScMom—angular second moment; SumAverg—sum average; InvDfMom—inverse difference moment; Contrast—contrast; SumOfSqs—sum of squares; SumVarnc—sum variance; Perc—percentile; Domn—dominant; SumEntrp—sum entropy; and Maxm—maximum of moments.

The scatter plots for α-carotene, β-carotene, total carotenoids, total sugars with selected image texture parameters, for which the highest values coefficient R for positive and negative correlations were determined, confirmed the strong relationship between chemical properties and image textures of control red pepper Sprinter F_1 and samples treated with natural fertilizers (Figure 2). The positive relationships between α-carotene and the texture from color channel *B*, β-carotene and the texture from color channel *B*, total carotenoids, and the texture from color channel *a*, as well as total sugars and the texture from color channel *R*, are presented. The strongest negative correlations between α-carotene with the texture from color channel *Y*, β-carotene with the texture from color channel *a*, total carotenoids with the texture from color channel *L*, and total sugars with the texture from color channel *a* are shown.

3.2. Relationship between Chemical Properties and Image Texture Parameters of Yellow Pepper Devito F_1

The texture parameters of flesh images of yellow pepper Devito F_1 were correlated with the content of total carotenoids and total sugars (Table 2). Positive correlations were found between total carotenoids and image textures from color channels *L*, *a*, *b*, *B*, *X*, and *Y* with the value of correlation coefficient reaching 0.9899 for the texture from channel *X*, as well as total sugars and image textures from color channels *L*, *a*, *b*, *G*, *B*, *X*, and *Y* reaching 0.9999 for the texture from channel *Y*; while the content of total carotenoids was negatively correlated with image textures from color channels *L*, *b*, *R*, *B*, and *X* with a correlation coefficient of up to −0.9993 for the texture from channel *b*. The negative correlations between the content of total sugars and image textures from channels *L*, *a*, *G*, *B*, *X*, and *Y* reaching −0.9971 for the texture from channel *G* were determined.

Table 2. Correlation coefficients (R) between total carotenoids (mg 100 g^{-1} f.m.) and total sugars (%) with selected image texture parameters for control and treated with NaturalCrop®, Bio-algeen S90, and nettle fertilizer yellow pepper Devito F_1 samples; $p < 0.05$.

Total Carotenoids (mg 100 g^{-1} f.m.)		Total Sugars (%)	
Texture Parameter	**Correlation Coefficient**	**Texture Parameter**	**Correlation Coefficient**
XS5SV1DifEntrp	0.9899	YS5SH1Entropy	0.9999
aSGNonZeros	0.9840	bS5SH3SumEntrp	0.9997
bATeta4	0.9783	GSGMean	0.9997
YATeta2	0.9716	LS5SZ5Contrast	0.9995
BS5SV1DifVarnc	0.9708	XS5SN3Contrast	0.9990
LS5SV3DifEntrp	0.9595	BS5SH1DifVarnc	0.9915
		aS5SZ5Correlat	0.9752
bHPerc50	−0.9993		
RSGKurtosis	−0.9948	GS5SH1InvDfMom	−0.9971
XS5SV1InvDfMom	−0.9909	XS5SZ5InvDfMom	−0.9894
GS4RHLngREmph	−0.9765	BS5SZ5SumVarnc	−0.9875
BS5SH5SumVarnc	−0.9696	aS5SN5DifVarnc	−0.9870
LATeta2	−0.9681	LS5SH1InvDfMom	−0.9808
		YS5SV1AngScMom	−0.9553

The first letter in the texture parameter name means the color channel *L*, *a*, *b*, *R*, *G*, *B*, *X*, *Y*, or *Z*; DifEntrp—difference entropy; GNonZeros—percentage of pixels with nonzero gradient; Teta4—parametr θ4; Teta2—parametr θ2; DifVarnc—difference variance; Perc—percentile; Kurtosis—histogram's kurtosis; InvDfMom—inverse difference moment; LngREmph—long run emphasis; SumVarnc—sum variance; Entropy—entropy; SumEntrp—sum entropy; GMean—absolute gradient mean; Contrast—contrast; and Correlat—correlation.

The relationships between total carotenoids and total sugars with selected Image texture parameters of control and fertilized yellow pepper Devito F_1 are presented in Figure 3 in the form of scatter plots. The graphs for textures characterized by the strongest positive and negative correlations, such as textures from color channels *X* and *b* for total carotenoids, and textures from color channels *Y* and *G* for total sugars are shown.

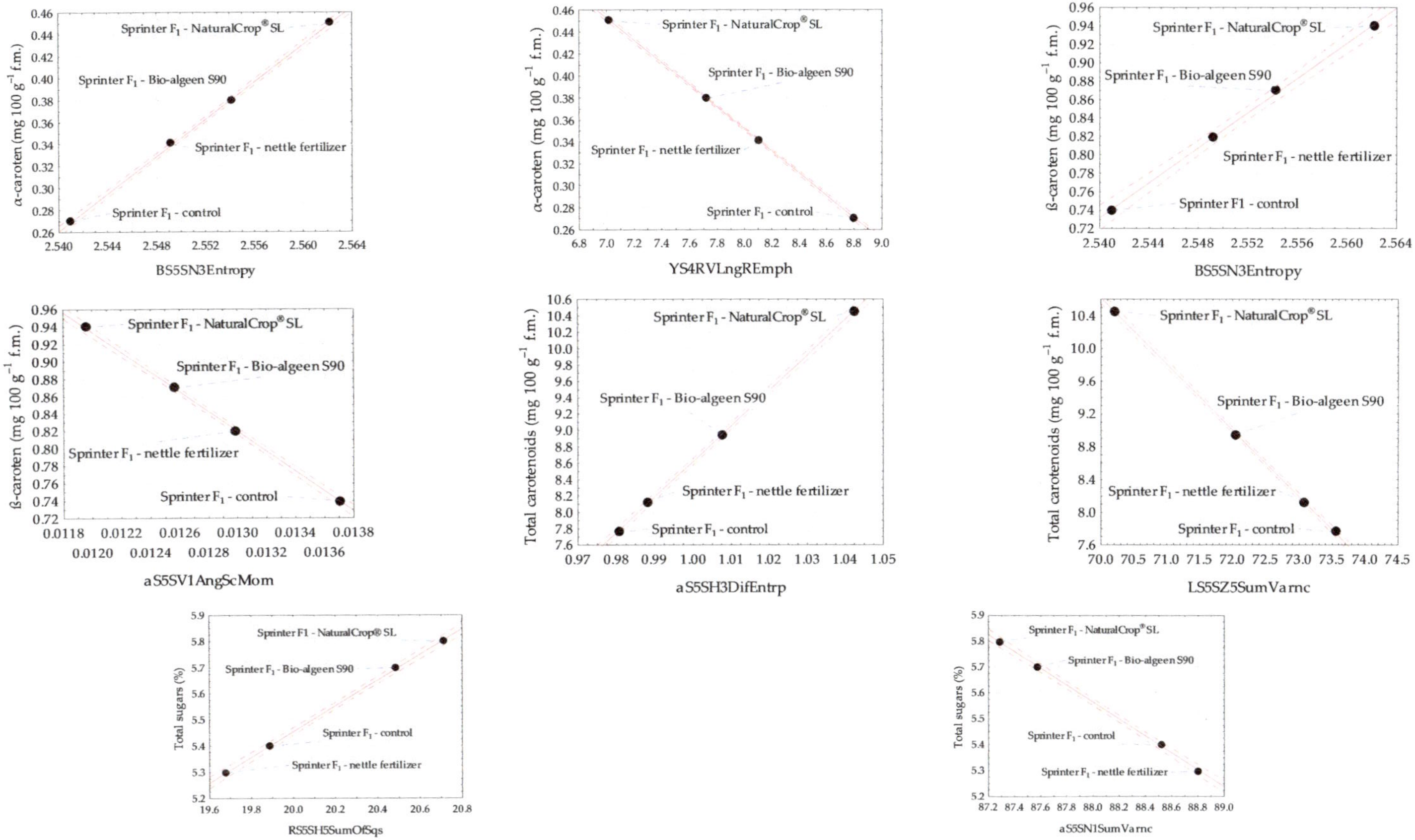

Figure 2. Scatter plots for α-carotene, β-carotene, total carotenoids, and total sugars with selected image texture parameters of control and treated with NaturalCrop®, Bio-algeen S90, and nettle fertilizer red pepper Sprinter F_1 samples. Black dot—mean value; blue dashed line—line connecting the sample name to the mean value; red dashed line—confidence interval (95%); and solid red line—regression line.

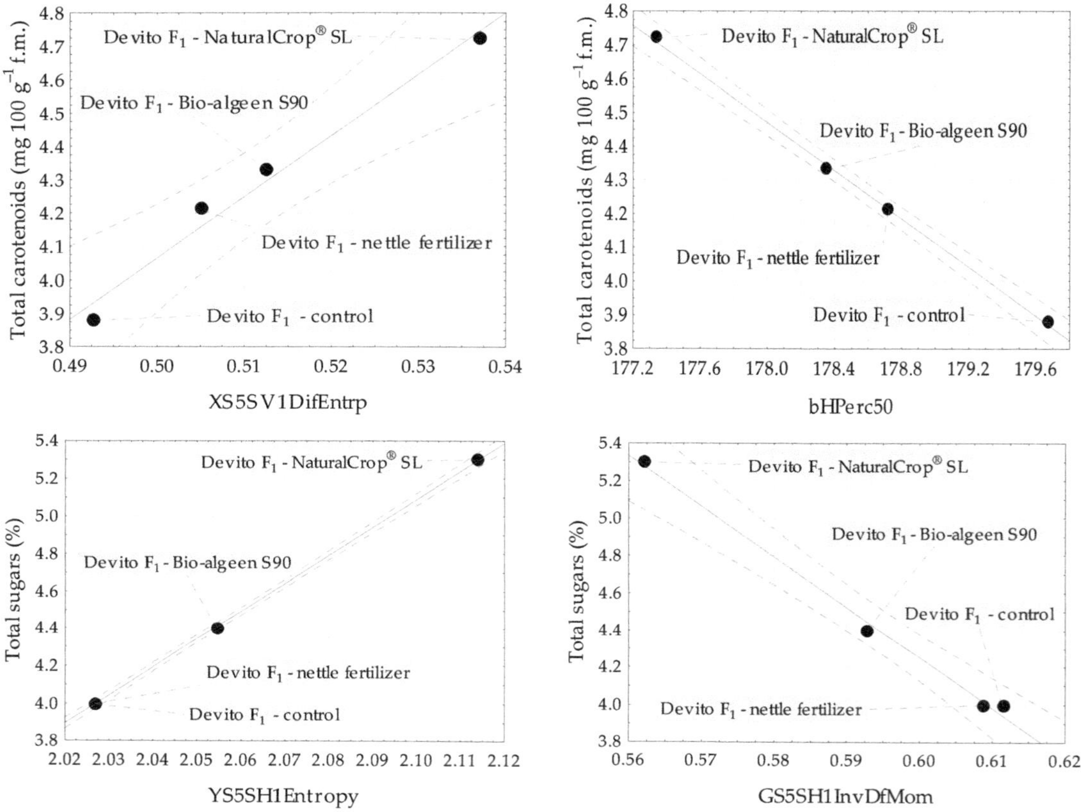

Figure 3. Scatter plots for total carotenoids and total sugars with selected image texture parameters of control and treated with NaturalCrop®, Bio-algeen S90, and nettle fertilizer yellow pepper Devito F_1. Black dot—mean value; blue dashed line—line connecting the sample name to the mean value; red dashed line—confidence interval (95%); and solid red line—regression line.

3.3. Relationship between Combined Chemical Properties and Image Texture Parameters of Red Pepper Sprinter F_1 and Yellow Pepper Devito F_1

In the next step of the analysis, correlations between a set of combined image texture parameters and chemical properties of red pepper Sprinter F_1 and yellow pepper Devito F_1 belonging to the unfertilized group and samples treated with NaturalCrop®, Bio-algeen S90, and nettle fertilizer were determined (Table 3). The content of total carotenoids was positively and negatively correlated with image texture parameters from color channels *L*, *a*, *b*, *B*, *G*, *X*, *Y*, and *Z*. The highest positive and negative correlation coefficients were 0.9956 and −0.9935 for textures from color channel *Y*. In the case of total sugars, the positive correlations with image textures from channels *L*, *a*, *b*, *G*, *X*, and *Y* were observed and the correlation coefficient reached 0.9768 for the texture from channel *Y*; whereas the negative correlations with image textures from channels *L*, *a*, *b*, *G*, *B*, *X*, and *Y* with the highest value of −0.9583 for the texture from color channel *G* were determined.

The scatter plots for total carotenoids and total sugars confirmed the strong relationships with selected image texture parameters of red pepper Sprinter F_1 and yellow pepper Devito F_1 belonging to control and samples treated with NaturalCrop®, Bio-algeen S90, and nettle fertilizer (Figure 4). The strongest positive and negative correlations between total carotenoids with textures from color channel Y and total sugars with textures from color channels Y and G are presented.

Table 3. Correlation coefficients (R) between total carotenoids (mg 100 g^{-1} f.m.) and total sugars (%) with selected image texture parameters for red pepper Sprinter F_1 and yellow pepper Devito F_1 belonging to control and samples treated with NaturalCrop®, Bio-algeen S90, and nettle fertilizer; $p < 0.05$.

Total Carotenoids (mg 100 g^{-1} f.m.)		Total Sugars (%)	
Texture Parameter	**Correlation Coefficient**	**Texture Parameter**	**Correlation Coefficient**
YS5SH5Contrast	0.9956	YS5SH5Entropy	0.9768
GS5SV5Contrast	0.9941	XS5SZ5Entropy	0.9562
LS5SN5Contrast	0.9883	LS5SH3Entropy	0.9461
aS5SH5Contrast	0.9832	GS5SN5Entropy	0.9443
XS5SZ5DifEntrp	0.9773	bS5SZ5SumEntrp	0.9375
bS5SV5SumEntrp	0.9415	aSGNonZeros	0.9309
ZSGVariance	0.8966		
BS5SZ3Contrast	0.8953	GHPerc99	−0.9583
		YS5SH1AngScMom	−0.9537
YS5SH5Correlat	−0.9935	LS5SN5InvDfMom	−0.9536
GS5SH5Correlat	−0.9921	XS5SH3AngScMom	−0.9494
LS5SH5Correlat	−0.9901	aHMaxm01	−0.9210
aS5SH1AngScMom	−0.9882	bHPerc99	−0.8803
XS4RNLngREmph	−0.9756	BS5SZ3Correlat	−0.8565
bHDomn01	−0.9663		
BS5SH3Correlat	−0.9202		
ZS5SH3Correlat	−0.9109		

The first letter in the texture parameter name means the color channel L, a, b, R, G, B, X, Y, or Z; Contrast—contrast; DifEntrp—difference entropy; SumEntrp—sum entropy; Variance—histogram's variance; Correlat—correlation; AngScMom—angular second moment; LngREmph—long run emphasis; Domn—dominant; GNonZeros—percentage of pixels with nonzero gradient; Perc—percentile; InvDfMom—inverse difference moment; and Maxm—maximum of moments.

Additionally, the regression equations for estimating chemical properties of unfertilized and fertilized pepper samples, as well as coefficients of determination (R^2) separately for red pepper Sprinter F_1, yellow pepper Devito F_1, and a set combining parameters for Sprinter F_1 and Devito F_1 were determined (Table 4). The higher R^2, the better the data fit the regression model. Image textures whose values were the most different between groups of peppers were the most suitable for building regression models. In the case of each chemical parameter, one regression equation for a texture for which the strongest positive correlation and one equation for the strongest negative correlation were set up. In the case of pepper Sprinter F_1, all values of R^2 were higher than 0.9980, reaching 0.9999 for α-carotene content and the texture YS4RVLngREmph from color channel Y. It indicated that the texture YS4RVLngREmph most differentiated all groups of pepper Sprinter F_1, such as control and samples treated with NaturalCrop® SL, Bio-algeen S90, and nettle fertilizer. Pepper Devito F_1 was also characterized by very high R^2 of up to 0.9998 for the content of total sugars and the texture YS5SH1Entropy from color channel Y. It meant that the control sample of pepper Devito F_1, and samples treated with three different natural preparations were the most different in terms of the texture YS5SH1Entropy that was most useful for developing the regression model; whereas the values of R^2 reached 0.9912 for total carotenoids and the texture YS5SH5Contrast from color channel Y for a set combining parameters for both cultivars. This confirmed that the textures of color channel Y of images are the most successful for estimating chemical properties of untreated peppers and samples treated with natural fertilizers.

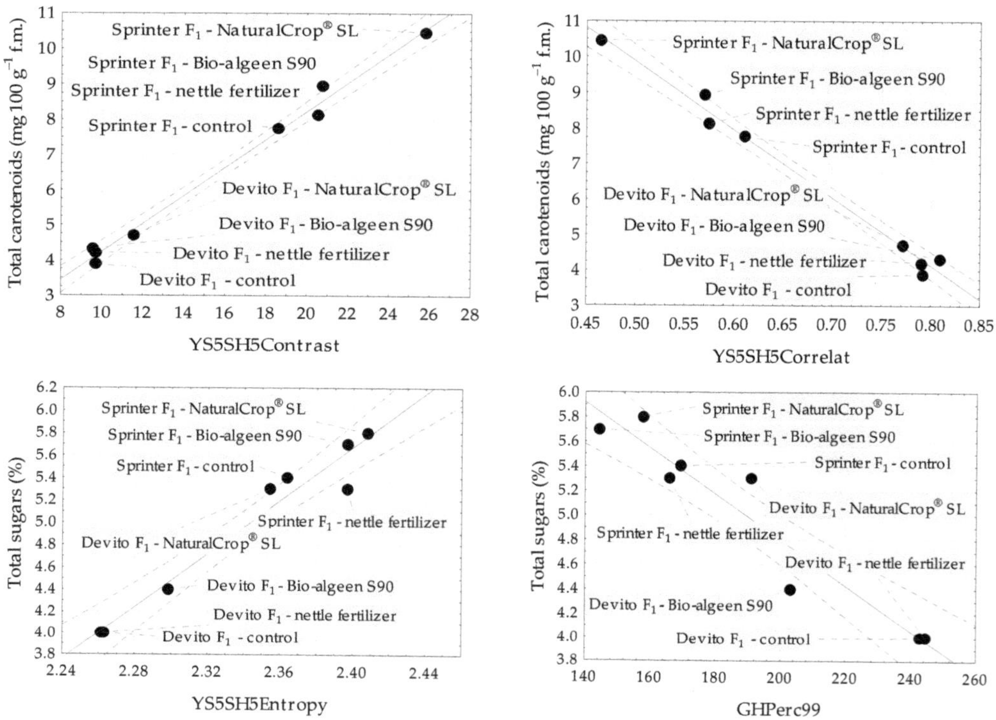

Figure 4. Scatter plots for total carotenoids and total sugars with selected image texture parameters of red pepper Sprinter F_1 and yellow pepper Devito F_1 belonging to control and samples treated with NaturalCrop®, Bio-algeen S90, and nettle fertilizer. Black dot—mean value; blue dashed line—line connecting the sample name to the mean value; red dashed line—confidence interval (95%); and solid red line—regression line.

Table 4. The regression equations for chemical properties based on image parameters and coefficients of determination for red pepper Sprinter F_1 and yellow pepper Devito F_1 belonging to control samples and samples treated with NaturalCrop®, Bio-algeen S90, and nettle fertilizer; $p < 0.05$.

Regression Equation	Coefficient of Determination (R^2)
Sprinter F_1	
α-carotene (mg 100 g^{-1} f.m.) = −21.28 + 8.4805 × BS5SN3Entropy	0.9998
α-carotene (mg 100 g^{-1} f.m.) = 1.1611 − 0.1012 × YS4RVLngREmph	0.9999
ß-carotene (mg 100 g^{-1} f.m.) = −23.35 + 9.4824 × BS5SN3Entropy	0.9988
ß-carotene (mg 100 g^{-1} f.m.) = 2.3051 − 114.3 × aS5SV1AngScMom	0.9996
Total carotenoids (mg 100 g^{-1} f.m.) = −34.78 + 43.407 × aS5SH3DifEntrp	0.9998
Total carotenoids (mg 100 g^{-1} f.m.) = 67.258 − 0.8091 × LS5SZ5SumVarnc	0.9998
Total sugars (%) = −4.350 + 0.49032 × RS5SH5SumOfSqs	0.9996
Total sugars (%) = 34.265 − 0.3261 × aS5SN1SumVarnc	0.9996
Devito F_1	
Total carotenoids (mg 100 g^{-1} f.m.) = −5.122 + 18.372 × XS5SV1DifEntrp	0.9798
Total carotenoids (mg 100 g^{-1} f.m.) = 68.248 − 0.3583 × bHPerc50	0.9986
Total sugars (%) = −26.24 + 14.916 × YS5SH1Entropy	0.9998
Total sugars (%) = 20.481 − 27.04 × GS5SH1InvDfMom	0.9942

Table 4. *Cont.*

Regression Equation	Coefficient of Determination (R^2)
Sprinter F_1 and Devito F_1	
Total carotenoids (mg 100 g^{-1} f.m.) = 0.24275 + 0.39985 × YS5SH5Contrast	0.9912
Total carotenoids (mg 100 g^{-1} f.m.) = 19.407 − 19.10 × YS5SH5Correlat	0.9870
Total sugars (%) = −22.91 + 11.908 × YS5SH5Entropy	0.9536
Total sugars (%) = 8.5299 − 0.0187 × GHPerc99	0.9166

The first letter in the texture parameter name means the color channel *L*, *a*, *b*, *R*, *G*, *B*, *X*, *Y*, or *Z*; Entropy—entropy; LngREmph—long run emphasis; AngScMom—angular second moment; DifEntrp—difference entropy; SumVarnc—sum variance; SumOfSqs—sum of squares; Perc—percentile; InvDfMom—inverse difference moment; Contrast—contrast; and Correlat—correlation.

The performed studies revealed strong correlations between the chemical properties and selected image texture parameters of red and yellow pepper samples. Very correct regression equations turn out to be useful for the estimation of the content of carotenoids and sugars based on pepper image parameters. Additionally, available literature reported the relationships between image or spectral features and chemical characteristics. In our previous research, the correlations between lycopene content and image textures of tomato fruit were determined with the correlation coefficient (R) reaching −0.99 for selected texture parameters of images in channels *G*, *b*, and *Y* and coefficients of determination (R^2) of 0.99 for image texture from channel *G* [37]. Pace et al. [38] observed relationships of the visual appearance and color parameters of images of fresh-cut nectarines with R^2 of up to 0.7622. The sugar content of potatoes was predicted based on hyperspectral imaging data by Rady et al. [39] with R reaching 0.97 for glucose. The sugar content was also estimated for apples with a high R parameter (0.8861) using multispectral imaging [40] and wine grapes with $R^2 > 0.8$ based on visible and near infrared and the short-wave infrared point spectroscopic data [41]. Furthermore, the prediction of soluble solid content of apples with $R > 0.8$ was possible based on features extracted from laser light backscattering images [42]. Our results presented in this paper were a significantly novel compared to the available literature, and included the determination of correlations and models for predicting the content of chemical properties based on selected parameters from a very large set of over 1600 textures from nine color channels. It allowed for the selection of textures providing correlation coefficients of up to 0.9999 and −0.9999, as well as very high coefficients of determination reaching 0.9999. The successful results may prompt the use of the developed procedure for other species and varieties of fruit and vegetables and the testing of the approach for other chemical parameters.

4. Conclusions

The correlations between chemical properties and image textures from different color channels of red and yellow peppers untreated and treated with natural fertilizers were determined. In the case of red pepper Sprinter F_1, statistically significant correlation coefficients were found between α-carotene, β-carotene, total carotenoids, and total sugars with selected image texture parameters. Yellow pepper Devito F_1, as well as a combined dataset for both cultivars were characterized by the correlations between total carotenoids and total sugars with selected textures. Furthermore, the high coefficients of determination confirmed the possibility of estimation of chemical properties based on image texture parameters from different color channels for individual cultivars, as well as regardless of the cultivar. The highest values of R^2 were for regression equations set up based on selected textures from images in color channel *Y*.

Due to the achievement of promising results, research aimed at estimating chemical properties based on image features can be continued and further research directions can be set. Further research may include more pepper cultivars, as well as other species of fruit and vegetables. Furthermore, more natural fertilizers can be used to provide more groups in an experimental design. The developed procedure can be of great practical importance to estimate the chemical properties of samples treated with various fertilizers based on selected image texture parameters.

Author Contributions: Conceptualization, E.R.; methodology, E.R. and J.S.-G.; software, E.R.; validation, E.R.; formal analysis, E.R. and J.S.-G.; investigation, E.R. and J.S.-G.; resources, E.R. and J.S.-G.; data curation, E.R. and J.S.-G.; writing—original draft preparation, E.R. and J.S.-G.; writing—review and editing, E.R. and J.S.-G.; visualization, E.R.; supervision, E.R.; project administration, J.S.-G.; funding acquisition, J.S.-G. All authors have read and agreed to the published version of the manuscript.

Funding: This study was conducted as part of a comprehensive study financed by the Ministry of Science and Higher Education, as part of the research subsidy granted to the National Institute of Horticultural Research, Poland (Grant No. ZPiPOiW/1/2020 entitled: "The influence of natural preparations on the quality of pepper in organic greenhouse production").

Data Availability Statement: The data presented in this study are available on request from the corresponding author.

Conflicts of Interest: The authors declare no conflict of interest.

References

1. Baker, B.P.; Green, T.A.; Loker, A.J. Biological Control and Integrated Pest Management in Organic and Conventional Systems. *Biol. Control* **2020**, *140*, 104095. [CrossRef]
2. Hamid, B.; Zaman, M.; Farooq, S.; Fatima, S.; Sayyed, R.Z.; Baba, Z.A.; Sheikh, T.A.; Reddy, M.S.; Enshasy, H.E.; Gafur, A.; et al. Bacterial Plant Biostimulants: A Sustainable Way towards Improving Growth, Productivity, and Health of Crops. *Sustainability* **2021**, *13*, 2856. [CrossRef]
3. Bosland, P.W.; Votava, E.J. *Peppers: Vegetable and Spice Capsicums*, 2nd ed.; CABI: London, UK, 2010.
4. Sreeramulu, D.; Raghunath, M. Antioxidant activity and phenolic content of roots, tubers and vegetables commonly consumed in India. *Food Res. Int.* **2010**, *43*, 1017–1020. [CrossRef]
5. Chen, L.; Kang, Y.H. Anti-inflammatory and antioxidant activities of red pepper (*Capsicum annum* L.) stalk extracts: Comparison of pericarp and placenta extract. *J. Funct. Food* **2013**, *5*, 1724–1731. [CrossRef]
6. Wahyuni, Y.; Ballester, A.R.; Sudarmonowati, E.; Bino, R.J.; Bovy, A.G. Secondary metabolites of Capsicum species and their importance in the human diet. *J. Nat. Prod.* **2013**, *76*, 783–793. [CrossRef]
7. Asnin, L.; Park, S.W. Isolation and Analysis of Bioactive Compounds in Capsicum Peppers. *Crit. Rev. Food Sci. Nutr.* **2015**, *55*, 254–289. [CrossRef] [PubMed]
8. Yang, L.; Wen, K.S.; Ruan, X.; Zhao, Y.X.; Wei, F.; Wang, Q. Response of plant secondary metabolites to environmental factors. *Molecules* **2018**, *23*, 762. [CrossRef]
9. Zou, L.; Tan, W.K.; Du, Y.; Lee, H.W.; Liang, X.; Lei, J.; Striegel, L.; Weber, N.; Rychlik, M.; Ong, C.N. Nutritional metabolites in *Brassica rapa* subsp. *chinensis* var. *parachinensis* (choy sum) at three different growth stages: Microgreen, seedling and adult plant. *Food Chem.* **2021**, *357*, 129535. [CrossRef]
10. Igielska-Kalwat, J.; Gościańska, J.; Nowak, I. Carotenoids as natural antioxidants. *Postepy Hig. I Med. Dosw. (Online)* **2015**, *69*, 418–428.
11. Szafirowska, A.; Elkner, K. The comparison of yielding and nutritive value of organic and conventional pepper fruits. *Veget. Crops Reser. Bull.* **2009**, *71*, 111–121. [CrossRef]
12. Ozgur, M.; Akpinar-Bayizit, A.; Ozcan, T.; Yilmaz-Ersan, I. Functional compounds and antioxidant properties of dried green and red peppers. *Afr. J. Agric. Res.* **2011**, *6*, 5638–5644. [CrossRef]
13. Russo, V.M. *Pepper, Botany, Production and Uses*; CABI: London, UK, 2012.
14. Jamiołkowska, A. *Preparaty Biotechniczne i Biologiczne w Ochronie Papryki Słodkiej (Capsicum annuum L.) Przed Grzybami Chorobotwórczymi i Indukowaniu Reakcji Obronnych Roślin*; Ser. Rozpr. Nauk., 376; Wydawnictwo Uniwersytetu Przyrodniczego w Lublinie: Lublin, Poland, 2013; p. 117.
15. Flores, P.; Navarro, J.M.; Garrido, C.; Rubio, J.S.; Martinez, V. Influence of Ca^{2+}, K^{+} and NO^{3-} fertilisation on nutritional quality of pepper. *J. Sci. Food Agric.* **2004**, *84*, 569–574. [CrossRef]
16. Michałojć, Z.; Dzida, K. Yielding and biological value of sweet pepper fruits depending on foliar feeding using calcium. *Acta Sci. Pol. Hortorum Cultus* **2012**, *11*, 255–264.
17. Zamljen, T.; Jakopic, J.; Hudina, M.; Veberic, R.; Slatnar, A. Influence of intra and inter species variation in chilies (*Capsicum* spp.) on metabolite composition of three fruit segments. *Sci. Rep.* **2021**, *11*, 4932. [CrossRef]
18. Eggink, P.M.; Maliepaard, C.; Tikunov, Y.; Haanstra, J.P.W.; Pohu-Flament, L.M.M.; De Wit-Maljaars, S.C.; Willeboordse-Vos, F.; Bos, S.; Waard, C.B.-D.; Leeuwen, P.J.D.G.-V. Prediction of sweet pepper (*Capsicum annuum*) flavor over different harvests. *Euphytica* **2012**, *187*, 117–131. [CrossRef]

19. Jamiołkowska, A.; Buczkowska, H.; Thanoon, A. Effect of biological preparations on the health state of pepper fruits and content of saccharides. *Acta Sci. Pol. Hortorum Cultus* **2016**, *15*, 95–107.
20. Ribes-Moya, A.M.; Adalid, A.M.; Raigón, M.D.; Hellín, P.; Fita, A.; Rodríguez-Burruezo, A. Variation in flavonoids in a collection of peppers (*Capsicum* spp.) under organic and conventional cultivation: Effect of the genotype, ripening stage, and growing system. *J. Sci. Food Agric.* **2019**, *100*, 2208–2223. [CrossRef]
21. Russo, V.M.; Howard, L.R. Carotenoids in pungent and non-pungent peppers at various developmental stages grown in the field and glasshouse. *J. Sci. Food Agric.* **2002**, *82*, 615–624. [CrossRef]
22. Tripodi, P.; Cardi, T.; Bianchi, G.; Migliori, C.A.; Schiavi, M.; Rotino, G.L.; Lo Scalzo, R. Genetic and environmental factors underlying variation in yield performance and bioactive compound content of hot pepper varieties (*Capsicum annuum* L.) cultivated in two contrasting Italian locations. *Eur. Food Res. Technol.* **2018**, *244*, 1555–1567. [CrossRef]
23. Bhandari, S.R.; Jung, B.D.; Baek, H.Y.; Lee, Y.S. Ripening-dependent changes in phytonutrients and antioxidant activity of red pepper (*Capsicum annuum* L.) fruits cultivated under open-field conditions. *Hort. Sci.* **2013**, *48*, 1275–1282. [CrossRef]
24. Guzman, I.; Hamby, S.; Romero, J.; Basland, P.; O'Connell, M. Variability of carotenoid biosynthesis in orange colored *Capsicum* spp. *Plant Sci.* **2010**, *179*, 49–59. [CrossRef] [PubMed]
25. Yuan, H.; Zhang, J.; Nageswaran, D.; Li, L. Carotenoid metabolism and regulation in horticultural crops. *Hortic. Res.* **2015**, *2*, 15036. [CrossRef]
26. Gomez-Garcia, M.; Ochoa-Alejo, N. Biochemistry and Molecular biology of Carotenoid biosynthesis In Chili pepper (*Capsicum* ssp.). *Int. J. Mol. Sci.* **2013**, *14*, 19025–19053. [CrossRef] [PubMed]
27. Chávez-Mendoza, C.; Sanchez, E.; Muñoz-Marquez, E.; Sida-Arreola, J.P.; Flores-Cordova, M.A. Bioactive compounds and antioxidant activity in different grafted varieties of bell pepper. *Antioxidants* **2015**, *4*, 427–446. [CrossRef] [PubMed]
28. Li, P.; Zhang, X.; Liu, Y.; Xie, Z.; Zhang, R.; Zhao, K.; Lv, J.; Wen, J.; Deng, M. Characterization of 75 Cultivars of Four Capsicum Species in Terms of Fruit Morphology, Capsaicinoids, Fatty Acids, and Pigments. *Appl. Sci.* **2022**, *12*, 6292. [CrossRef]
29. Ropelewska, E. The Application of Machine Learning for Cultivar Discrimination of Sweet Cherry Endocarp. *Agriculture* **2021**, *11*, 6. [CrossRef]
30. Bhargava, A.; Bansal, A. Fruits and vegetables quality evaluation using computer vision: A review. *J. King Saud Univ.-Comput. Inf. Sci.* **2021**, *33*, 243–257. [CrossRef]
31. Strzelecki, M.; Szczypinski, P.; Materka, A.; Klepaczko, A. A software tool for automatic classification and segmentation of 2D/3D medical images. *Nucl. Instrum. Methods Phys. Res. Sect. A Accel. Spectrometers Detect. Assoc. Equip.* **2013**, *702*, 137–140. [CrossRef]
32. Szczypiński, P.M.; Strzelecki, M.; Materka, A.; Klepaczko, A. MaZda-A software package for image texture analysis. *Comput. Methods Programs Biomed.* **2009**, *94*, 66–76. [CrossRef]
33. Szczypiński, P.M.; Strzelecki, M.; Materka, A. Mazda-A software for texture analysis. In Proceedings of the 2007 International Symposium on Information Technology Convergence (ISITC 2007), Jeonju, Republic of Korea, 23–24 November 2007; IEEE: Piscataway, NJ, USA, 2007; pp. 245–249.
34. Ibraheem, N.A.; Hasan, M.M.; Khan, R.Z.; Mishra, P.K. Understanding Color Models: A Review. *ARPN J. Sci. Technol.* **2012**, *2*, 265–275.
35. Stępowska, A.; Grudzień, K.; Adamicki, E.; Dobrzański, A.; Robak, J.; Szwejda, J. *Ekologiczne Metody Uprawy Papryki w Gruncie i Pod Osłonami*; Kaniszewski, S., Ed.; KCRE-RCDRRiOW: Radom, Poland, 2004; p. 44.
36. Bohoyo-Gil, D.; Dominguez-Valhondo, D.; García-Parra, J.J.; González-Gómez, D. UHPLC as a suitable methodology for the analysis of carotenoids in food matrix. *Eur. Food Res. Technol.* **2012**, *235*, 1055–1061. [CrossRef]
37. Ropelewska, E.; Szwejda-Grzybowska, J. Relationship of Textures from Tomato Fruit Images Acquired Using a Digital Camera and Lycopene Content Determined by High-Performance Liquid Chromatography. *Agriculture* **2022**, *12*, 1495. [CrossRef]
38. Pace, B.; Cefola, M.; Renna, F.; Attolico, G. Relationship between visual appearance and browning as evaluated by image analysis and chemical traits in fresh-cut nectarines. *Postharvest Biol. Technol.* **2011**, *61*, 178–183. [CrossRef]
39. Rady, A.; Guyer, D.; Lu, R. Evaluation of Sugar Content of Potatoes using Hyperspectral Imaging. *Food Bioprocess Technol.* **2015**, *8*, 995–1010. [CrossRef]
40. Tang, C.; He, H.; Li, E.; Li, H. Multispectral imaging for predicting sugar content of 'Fuji' apples. *Opt. Laser Technol.* **2018**, *106*, 280–285.
41. Kalopesa, E.; Karyotis, K.; Tziolas, N.; Tsakiridis, N.; Samarinas, N.; Zalidis, G. Estimation of Sugar Content in Wine Grapes via In Situ VNIR–SWIR Point Spectroscopy Using Explainable Artificial Intelligence Techniques. *Sensors* **2023**, *23*, 1065. [CrossRef]
42. Qing, Z.; Ji, B.; Zude, M. Predicting soluble solid content and firmness in apple fruit by means of laser light backscattering image analysis. *J. Food Eng.* **2007**, *82*, 58–67. [CrossRef]

foods

Article

Health-Aware Food Recommendation Based on Knowledge Graph and Multi-Task Learning

Yi Chen [1,*], Yandi Guo [1], Qiuxu Fan [1], Qinghui Zhang [1] and Yu Dong [2,*]

[1] Beijing Key Laboratory of Big Data Technology for Food Safety, Beijing Technology and Business University, Beijing 100048, China; ddyyg@st.btbu.edu.cn (Y.G.); fanqx_19@st.btbu.edu.cn (Q.F.); zhangqinghui@st.btbu.edu.cn (Q.Z.)
[2] School of Computer Science, University of Technology Sydney, Sydney, NSW 2008, Australia
* Correspondence: chenyi@th.btbu.edu.cn (Y.C.); yu.dong-3@student.uts.edu.au (Y.D.)

Abstract: Current food recommender systems tend to prioritize either the user's dietary preferences or the healthiness of the food, without considering the importance of personalized health requirements. To address this issue, we propose a novel approach to healthy food recommendations that takes into account the user's personalized health requirements, in addition to their dietary preferences. Our work comprises three perspectives. Firstly, we propose a collaborative recipe knowledge graph (*CRKG*) with millions of triplets, containing user–recipe interactions, recipe–ingredient associations, and other food-related information. Secondly, we define a score-based method for evaluating the healthiness match between recipes and user preferences. Based on these two prior perspectives, we develop a novel health-aware food recommendation model (*FKGM*) using knowledge graph embedding and multi-task learning. *FKGM* employs a knowledge-aware attention graph convolutional neural network to capture the semantic associations between users and recipes on the collaborative knowledge graph and learns the user's requirements in both preference and health by fusing the losses of these two learning tasks. We conducted experiments to demonstrate that *FKGM* outperformed four competing baseline models in integrating users' dietary preferences and personalized health requirements in food recommendations and performed best on the health task.

Keywords: health; food recommendation; knowledge graph; graph convolution network; multi-task learning

1. Introduction

A nutritious diet matters as the cornerstone of good health, and unhealthy dietary habits can lead to a host of preventable illnesses. According to a statistical report [1] from the World Health Organization in 2016, over 1.9 billion adults and 41 million children under the age of five were clinically overweight, with 650 million living with obesity. Poor dietary habits [2], particularly diets high in calories, sugar, and fat, and low in fiber, are primarily responsible for this epidemic. Improving the dietary structure is considered an effective approach to tackling this issue [3]. However, making healthy dietary choices can be challenging for individuals due to their hectic lifestyles and limited knowledge of food and nutrition.

In recent years, with the blossoming research in the food field [4–7], food recommendation systems [8] have garnered considerable attention from the public, promising tailor-made food choices to users [9–11]. However, current systems are constrained by two main factors. Firstly, they frequently overlook crucial food-related information such as recipe ingredients, food categories, and dietary restrictions when building recommendation models. This omission can significantly impede the system's ability to establish connections between users and foods, resulting in less accurate and varied recommendations. Secondly, most algorithms disregard users' health needs when defining their objectives, relying solely on dietary preferences. For instance, conventional food recommendation systems will offer recommendations based on a user's preference for high-fat and high-sugar diets, irrespective of potential health risks. Thus, food recommendation systems must account for both dietary preferences and health requirements to offer effective recommendations [12,13].

In this paper, we propose a novel health-aware food recommendation model, namely *FKGM*, which leverages a knowledge graph and multi-task learning approach. The *FKGM* takes into account both the user's dietary preferences and the health impact of four critical nutrients in recipes: sodium, fat, sugar, and saturated fat. By integrating these considerations into the recommendation process, *FKGM* offers users personalized food recommendations that are in line with their dietary preferences and health requirements.

The main contributions of this study are summarized into three oriented aspects as follows:

- Data-oriented: We created a collaborative recipe knowledge graph (*CRKG*) by merging the user–recipe bipartite graph and the recipe knowledge graph [14]. The *CRKG* includes user–recipe interactions and food-related details such as ingredients, food types, and cuisines. The *CRKG* provides the foundation of the recommendation model with attribute-based and association-based information on users and recipes, enabling the model to acquire a more comprehensive understanding of users and recipes that better reflects their semantic similarity.
- Score-oriented: We defined a score-based method for evaluating the healthiness match between recipes and user preferences. First, we defined a Nutrient Content Score in recipes and a user Nutrient Intake Score based on the nutrition criteria recommended by the UK Food Standards Agency to assess the healthiness of recipes and user preferences, respectively. We then defined a Nutrient Discrepancy Score between recipes and users to determine the healthiness match between recipes and user preferences, which is used to calculate the loss of the health learning task in the multi-task learning layer of the recommendation model.
- Model-oriented: We developed a new health-aware food recommendation model, *FKGM*, which embeds entities and relationships on the collaborative recipe knowledge graph using *TransD* [15] and updates entity representations through message passing using a knowledge-aware attention graph convolutional neural network. The model was trained using multi-task learning, which involves preference learning and health learning tasks. Our experimental results demonstrate that *FKGM* outperformed four existing baseline models in striking a balance between preference needs and health requirements.

The remainder of this paper is organized as follows. Section 2 describes related work, Section 3 introduces the construction of the collaborative knowledge graph, Section 4 introduces the methods for evaluating recipe and user preferences, Section 5 presents the proposed model, Section 6 details the experimental setup and results, and Section 7 provides a summary of this study.

2. Related Work

In this section, we first review knowledge graph-based recommendation methods, then review preference-based and health-aware food recommendation methods. In the field of food recommendation, preference-based and health-aware methods have their own advantages and limitations. Therefore, it is a worthwhile and challenging problem to effectively combine the two and balance users' preferences and health. Consequently, the work of our paper focuses on food recommendation that combines user preferences and health requirements.

2.1. Knowledge Graph-Based Recommendation

Knowledge graphs organize entities and relations in the real world in the form of graphs, which can effectively express the semantic relations between entities and have been applied to recommendation systems. The methods of knowledge graph-based recommendation systems mainly fall into two categories: path-based methods and embedding-based methods.

Path-based methods. These methods mainly use paths in knowledge graphs to make recommendations, which can intuitively represent users' preference propagation. Hu et al. [16] proposed a context-aware recommendation method based on meta-paths, which uses knowledge graphs to describe the relationships between users and items and uses meta-paths to capture these relationships. Fan et al. [17] proposed a meta-path-guided heterogeneous graph neural network for intent recommendation tasks. However, these methods usually rely heavily on the design of meta-paths.

Embedding-based methods. These methods usually use knowledge graph embedding [18] to learn a low-dimensional representation vector for each entity and relation in knowledge graphs, while preserving the structural information of knowledge graphs [19–21], and then use the vectors obtained by knowledge graph embedding to enhance the representation of users or items in recommendation systems. Zhang et al. [22] proposed *CKE*, which uses *TransR* for knowledge graph embedding and extracts semantic features from structured knowledge to enrich item representation. Many studies in embedding-based methods use graph neural networks to leverage their powerful modeling ability for graph data to improve the recommendation performance. Wang et al. [23] proposed *KGAT*, which explicitly models high-order connections in knowledge graphs in an end-to-end manner and attempts to use message-passing mechanisms to exploit structural knowledge for the first time. Ma et al. [24] addressed the problems of error propagation and weak interpretability in recommendation systems that combine graph neural networks and knowledge graphs and proposed *KR-GCN*, which designed a transition-based triplet scoring method and introduced a path-level self-attention mechanism to distinguish the contributions of different selection paths and predict interaction probabilities, achieving results better than baselines.

Our model belongs to embedding-based methods, which use knowledge graph embedding to learn entity and relation embeddings and use the message passing mechanism of a graph neural network to capture high-order relational information for entities and update entity representation.

2.2. Preference-Based Food Recommendation

In this section, we review the existing methods for preference-based food recommendation, which aim to recommend foods to users based on their taste and dietary preferences. We categorize these methods into three main types: collaborative filtering, content-based, and hybrid methods. However, each method has its strengths and weaknesses, and there are also challenges and open problems for future research.

Collaborative filtering methods. These methods learn user preferences from user–food interactions, such as ratings and tags, and assume that users who have similar interactions with foods will have similar preferences. For example, Ge et al. [25] combined user ratings and tags in food recommendations and used matrix factorization to generate food recommendations. Khan et al. [26] proposed a feature recognition technique based on *EnsTM* to effectively model user preferences. Collaborative filtering methods can capture the diversity and personalization of user preferences, but they also suffer from data sparsity and cold start problems, which limit their scalability and applicability. Therefore, it is important to explore other methods to overcome these limitations.

Content-based methods. These methods use food content to learn user preferences, such as ingredients and food images, and assume that users will prefer foods that have similar content to the foods they have liked before. For example, Freyne et al. [27] decomposed recipes into individual ingredients and constructed user profiles composed of ingredients that the user likes based on recipe ratings containing these ingredients, thus improving the recommendation performance. Yang et al. [28] incorporated food image information into food recommendations and learned user preferences based on food image content. Their work demonstrated the importance of food image information in learning food preferences. Content-based methods can overcome the data sparsity and cold start problems of collaborative filtering methods, but they also face the challenges of extracting and representing food content features, as well as dealing with the semantic gap between low-level features and high-level preferences. In addition, these methods may not capture the diversity and novelty of food recommendations as effectively as collaborative filtering methods.

Hybrid methods. These methods combine both collaborative filtering and content-based methods to leverage their advantages and mitigate their disadvantages. Gao et al. [29] considered food recommendation as a multimedia task and proposed a hierarchical attention-based method, *HAFR*, which simultaneously considers factors such as user–recipe interactions, food images, and food ingredients in modeling. Experimental results show that the average recommendation performance of the proposed method is 12% better than that of the baseline model. Gao et al. [30] considered the correlations between a user–recipe, recipe–ingredient, and ingredient–ingredient, providing richer auxiliary information for food recommendation models and helping to better model user preferences. They used graph convolutional neural networks to model the relationships between these foods. This method outperformed the state-of-the-art method in the recommendation performance by 5.4%. Consequently, hybrid methods have become a popular approach for food recommendation due to their ability to capture diverse and personalized user preferences.

Compared to the above methods, our method further considers the complex relationships between foods and uses knowledge embedding and a knowledge graph attention network to model the vector representations of entities in the food knowledge graph. In this way, the model can fully utilize the rich semantic information contained in *CRKG*, explore users' high-level preferences, and increase recommendation diversity. However, there are also challenges in designing and optimizing knowledge-based models, such as dealing with the sparsity and noise in knowledge graphs, as well as the complexity of the model training and inference process. Therefore, future research can focus on addressing these challenges and developing more effective and efficient knowledge-based food recommendation methods.

2.3. Health-Aware Food Recommendation

The healthy food recommendation based on the user's health status, nutritional needs, and dietary restrictions aims to recommend beneficial foods or recipes to improve users' health levels and quality of life. However, it may not consider users' personal preferences and diversity. There are two main methods for existing healthy food recommendations: rule-based and data-driven.

Rule-based methods. These methods rely on specific rules [31] or criteria, such as filtering, sorting, or classification, often limited by the usage scenario and based on domain knowledge, expert opinions [32], or logical reasoning. Shandilya et al. [33] proposed a food recommendation system named MATURE-Food that recommends suitable foods based on users' current mandatory requirements, using a real food item dataset [34] and medical records of chronic kidney disease (CKD) patients from University of Iowa Hospitals and Clinics (UIHC), mainly for medical and health fields rather than general food recommendation scenarios. Ribeiro et al. [35] presented a mobile catering recommendation system named *SousChef*, targeting elderly people and creating personalized meal plans based on users' information including body measurements, personal preferences, and activity levels. The nutritional suggestions and application are designed for elderly people, featuring a user-friendly interface and following the guidelines of nutritionists. User testing was conducted to determine the applicability of recipes and nutrition plans and the usability of the mobile application. The results showed that more than 70% of elderly participants were satisfied with the simplicity of the meal plan recommendations and the *SousChef* application. Ge et al. [36] considered calorie balance and proposed a calorie balance function based on the difference between user's needs and recipe calories, where users can define whether food recommendations should focus more on taste or health. However, rule-based methods typically use general or fixed rules without considering users' personalized needs.

Data-driven methods. With the development of the data era, data-driven methods are becoming mainstream. These methods can utilize a large amount of data to learn users' preferences and needs and generate more accurate and personalized recommendations based on them. They usually use machine learning or statistical techniques to analyze data and extract useful information from it. Wang et al. [37] proposed a personalized health food recommendation scheme. The scheme consists of three parts: the recipe retrieval, user health analysis, and health food recommendation. The authors describe users' health conditions by capturing text health-related information crawled from social networks, and propose a novel recommender based on category-aware hierarchical memory networks to learn health-aware user–recipe interactions, to better perform food recommendations. Chen et al. [38] proposed a new framework called *NutRec*, which aims to provide users with healthy recipe recommendations by simulating the interactions and proportions of ingredients in recipes. The framework consists of three main parts: ingredient prediction, ingredient quantity prediction, and healthy recipe recommendation. The authors conducted experiments on two recipe datasets, and the results supported the intuition of the framework and demonstrated its ability to retrieve healthier recipes. Li et al. [39] introduced a novel approach to recipe recommendation that incrementally shifts users towards healthier recipe options while respecting their past preferences. The authors proposed a model that jointly learns recipe representations via a graph over two graphs extracted from a large-scale Food KG, capturing different semantic relationships across the preferences and healthiness aspects. Experimental results on two large real-world recipe datasets showcase the model's ability to recommend tasty as well as healthy recipes to users. However, this approach only considers the healthiness of the food itself and does not take into account users' personalized health requirements.

Our work belongs to the data-driven method. Compared with the methods mentioned above, our method can represent the complex relationship between users and food more comprehensively by learning rich semantic information on *CRKG*. Furthermore, our method simultaneously learns user preferences and personalized health requirements. By providing corresponding constraints in the recommendation process, our work aims to offer food recommendations that not only satisfy users' tastes but also meet their personalized health requirements.

3. Collaborative Recipe Knowledge Graph

In this paper, the collaborative recipe knowledge graph (*CRKG*) refers to a collaborative knowledge graph that combines the user–recipe bipartite graph and recipe knowledge graph. It contains user–recipe interaction relationships and various food-related information, such as recipe–ingredient and recipe–cuisine relations. This provides food recommendation models with a better understanding of the semantic relationships between users and recipes and allows for the exploration of implicit associations between users and recipes. To clarify the construction process of *CRKG*, we first provide a formal description of the user–recipe bipartite graph, recipe knowledge graph, and *CRKG*, and then introduce the method for constructing *CRKG*.

3.1. Formal Definition

- **User–Recipe Bipartite Graph G_1:** We defined the set of m users as $U = \{u_1, u_2, \ldots, u_m\}$ and the set of n recipes as $RP = \{rp_1, rp_2, \ldots, rp_n\}$. The user–recipe interaction data are represented as a bipartite graph $G_1 = \{(u, interact, rp) \mid u \in U, rp \in RP\}$, where $interact_{u,rp} = 1$ if there exists an interaction history between user u and recipe rp, and $interact_{u,rp} = 0$ otherwise.
- **Recipe Knowledge Graph G_2:** The constructed recipe knowledge graph is represented as $G_2 = \{(h, r, t) | h \in E, r \in R, t \in E\}$, which consists of many entity-

relation-entity triplets (h, r, t), where h, r, t, respectively, represent the head entity, relation, and tail entity, describing the relationship between the head entity h and the tail entity t through the relation r. For instance, consider the triplet is $(pizza, food.dish.ingredients, mozzarella\ cheese)$. In this case, '$pizza$' is the head entity, '$food.dish.ingredients$' is the relation, and '$mozzarella\ cheese$' is the tail entity. This triplet indicates that mozzarella cheese is one of the ingredients of pizza. E and R represent the sets of entities and relations in the recipe knowledge graph G_2, respectively.

- **Collaborative Recipe Knowledge Graph $CRKG$:** We integrated the user–recipe bipartite graph G_1 and the recipe knowledge graph G_2 into a unified graph $CRKG$. Specifically, for any recipe $rp \in RP$ in G_1, a corresponding entity $e \in E$ can be found in G_2. For example, Roast Chicken with Rosemary is a recipe in both G_1 and G_2. G_1 and G_2 are aligned and fused into a unified graph $CRKG$ based on the matched recipes, denoted as $CRKG = \{(h, r, t) | h \in E', r \in R', t \in E'\}$, where $E' = E \cup U$ and $R = R' \cup \{interact\}$.

3.2. Construction Method

In this study, we first constructed a user–recipe bipartite graph G_1 using the widely used public dataset Allrecipes for food recommendation. The Allrecipes dataset contains user–recipe interaction data, ingredients, and nutritional information. We extracted user–recipe interaction data from Allrecipes to construct the user–recipe bipartite graph G_1, which includes 48,111 users, 38,115 recipes, and 1,267,176 interactions between them.

Then, the recipe knowledge graph G_2 was constructed using the large-scale knowledge base Freebase to provide recommendation models with recipe-related relational information. Freebase [40] is a large-scale knowledge base consisting of nodes and edges, where each node represents an entity (such as a person, place, organization, etc.), and each edge represents a relationship between two entities. It was proposed by Google in 2007 and covers knowledge in various fields such as arts, history, and food. It has been widely used in research on knowledge graphs, natural language processing, intelligent question answering, and other fields. This study used food-related knowledge from Freebase to construct the recipe knowledge graph G_2. Specifically, entity linking technology was used to construct G_2. This process first linked food entities in the Allrecipes dataset to their corresponding entities in Freebase and extracted the other entities that the corresponding entity is linked to in Freebase and the relationships between them to obtain triplets to form G_2. Then, the obtained triplets were refined by filtering out low-frequency relations and entities. Finally, a recipe knowledge graph G_2 was constructed, which contains 69,799 entities, 10 types of relations, and 2,500,801 triplets. G_2 includes 10 food-related relationships, encompassing various aspects such as the food dish-type (recipe-type), compatibility of ingredients with dietary restrictions (recipe–dietary restrictions), cuisine of food dishes (recipe–cuisine), ingredient compatibility with dietary restrictions (ingredient–dietary restrictions), narrower categorization of ingredients (ingredient categories), cuisine associated with an ingredient (ingredient–cuisine), dishes served with a specific dish (served-with), regions associated with a dish (region), and recipe–ingredient relationship.

After obtaining the user–recipe bipartite graph G_1 and recipe knowledge graph G_2, the two are fused into $CRKG$, as shown in Figure 1; the nodes of different colors in the figure represent different types of entities in $CRKG$. The green node represents a user, the orange node is a recipe, and the connection between the user and the recipe indicates that there is an interaction between them. In the figure, the recipe–ingredient association relation is shown as an example. The purple nodes are recipe ingredient nodes, and the recipe nodes and ingredient nodes are connected to indicate that the recipe is composed of these ingredients. The blue nodes are other types of entities in the collaborative knowledge graph. The nodes in $CRKG$ are called entities, and the edges between entities are represented as relations.

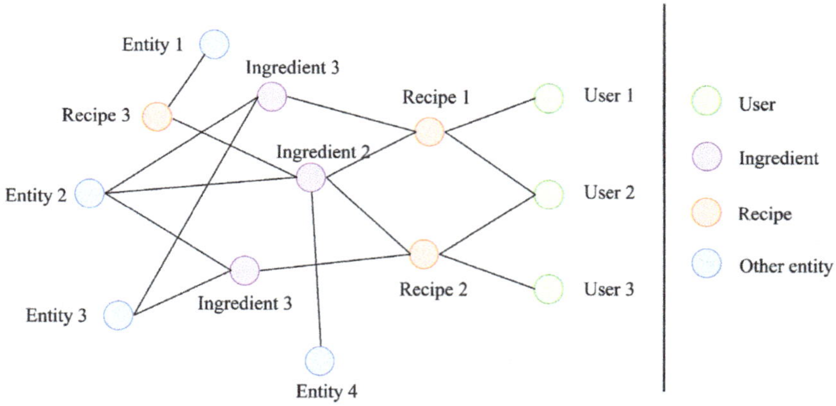

Figure 1. A toy example of collaborative recipe knowledge graph. The bipartite graph formed by connecting the user nodes and recipe nodes in Figure 1 is denoted as G_1, while the graph formed by the recipe nodes and the remaining nodes is denoted as G_2.

4. Assessment of Recipe and User Preference Healthiness

In this section, we first propose a calculation method for assessing the healthiness of recipes and the healthiness of user preferences and then define a method for calculating the healthiness match between recipes and user preferences.

4.1. Recipe Nutrient Content Score (NCS)

Before making healthy food recommendations, a standard needs to be developed to measure the healthiness of a recipe. In this paper, we refer to the nutrient rating system proposed by the UK Food Standards Agency (FSA) [41] to evaluate the healthiness of recipes. This rating assesses the healthiness of a recipe based on the content of four nutrients: sodium, fat, sugar, and saturated fat, as the long-term excessive intake of any of these nutrients can lead to health risks. We defined the set of these four nutrients as the $nutrient\ set = sodium, fat, sugar, saturatedfat$. The FSA rating evaluates the nutrient content of recipes and classifies them into three levels: healthy, medium, and unhealthy. Table 1 shows the healthy threshold values for the four nutrients (the salt content is converted to sodium content using a conversion factor of 0.388 g of sodium per 1 g of salt), and a recipe is considered unhealthy if the content of any nutrient exceeds its corresponding threshold value.

Table 1. Health threshold values for each nutrient in FSA rating.

Nutrient	Sodium	Fat	Sugar	Saturated Fat
Threshold value (g/portion)	0.7	21.0	27.0	6.0

In this study, the nutrient content of each recipe was obtained from Allrecipes and Table 2 shows examples of the nutrient content of several recipes in the dataset.

Table 2. Examples of nutrient content in recipes (partial).

Recipe	Sodium (g)	Fat (g)	Sugar (g)	Saturated Fat (g)
Homemade Bacon	2.01	23.58	0.09	7.73
Foolproof Rosemary Chicken Wings	0.76	23.62	0.23	5.68
Cranberry Pork Chops II	0.40	6.31	26.46	2.73
Chinese Pot Roast	2.81	19.97	12.16	7.11

In this study, the Nutrient Content Score (NCS) of a recipe was defined with reference to the FSA rating. The healthiness of a recipe was assessed based on the content of the recipe on four nutrients, which was calculated as shown in Equation (1). $NCS_k(rp)$ denotes the Nutrient Content Score of recipe rp on the k-th nutrient. k is the k-th nutrient in the nutrient set $k \in nutrient\ set$. $threshold_k$ is the limit value of nutrient k out of the health range in the FSA rating. $content(rp,k)$ is the content value of nutrient k in recipe rp.

$$NCS_k(rp) = \frac{content(rp,k) \times 10}{threshold_k} \quad (1)$$

The $NCSs$ of a recipe calculated by Equation (1) for the four nutrients are continuous values and unified to the same scale. Table 3 shows the $NCSs$ calculated by Equation (1) for the recipes in Table 2. According to the NCS scoring system, the scores can be categorized into three intervals: the healthy range is defined as [0,4), the moderate range is defined as [4,10), and the unhealthy range is defined as [10,+∞).

Table 3. Nutrient Content Scores calculated for the recipes in Table 2.

Recipe	NCS_{sodium}	NCS_{fat}	NCS_{sugar}	$NCS_{saturated\ fat}$
Homemade Bacon	28.81	11.23	0.03	12.90
Foolproof Rosemary Chicken Wings	10.90	11.25	0.09	9.47
Cranberry Pork Chops II	5.84	3.00	9.80	3.89
Chinese Pot Roast	40.14	9.51	4.50	11.86

4.2. User-Preferred Nutrient Intake Score (NIS)

The core of this study's recommendation task is to recommend healthier recipes based on users' dietary habits. Since different users have different dietary habits, their health requirements are also different. Users' dietary habits reflect not only their dietary preferences but also their health status. Therefore, we defined the Nutrient Intake Score to calculate users' high and low intake levels in four nutrient categories. In calculating the Nutrient Intake Score (NIS), a grouped weighted average method is used instead of directly averaging a certain nutrient in the user's interaction history to effectively avoid the influence of extreme values. The formula for calculating the Nutrient Intake Score for a certain nutrient is shown in Equation (2), where $k \in nutrient\ set$.

$$NIS_k(u) = \frac{c_1 \times med_1 + c_2 \times med_2 + c_3 \times med_3}{Count} \quad (2)$$

The specific calculation process is as follows:

Input: User's recipe history interaction data.

Procedure:

Step 1. First, all the recipes in the user's historical recipe interaction data are counted, and they are divided into three groups based on the NCS of each recipe in nutrient k, where [0,4) is the low level group, [4,10) is the medium level group and [10,+∞) is the high level group.

Step 2. Calculate the frequency of recipes in each group and calculate the median NCS in each group. The frequency of each group is denoted as c_1, c_2, and c_3; the median NCS of recipes in each group is denoted as med_1, med_2, and med_3; and $Count$ is the total number of user recipe interaction records.

Step 3. Apply Equation (2) to obtain $NIS_k(u)$.

Output: User's Nutrient Intake Score in nutrient k.

The NIS is calculated for the user in terms of sodium, fat, sugar, and saturated fat according to Equation (2). Typically, if a user's score for a certain nutrient intake is high, it indicates that the user has consumed too much of that nutrient in their diet, and therefore restrictions should be added to that nutrient in food recommendations.

4.3. Nutrient Discrepancy Score (NDS)

After defining the calculation method for recipe healthiness and user preference for healthiness, we defined the Nutrient Discrepancy Score (NDS) of a recipe for a user, which measures the degree of match between a recipe and a user in terms of healthiness. It is calculated based on the Nutrient Intake Score of the user and the Nutrient Content Score of the recipe. The higher the NDS, the less suitable the recipe is for the user's health.

The specific calculation process is as follows:

Input: Nutrient Content Score of recipe rp, Nutrient Intake Score of user u.

Procedure:

Step 1. Modifying the original $softplus$ function [42] as shown in Equation (3), the modified function will map values less than four to close to zero. Using this equation, the values of each item in NCS and NDS are mapped, and small values are mapped to values close to zero, while larger values are preserved. The purpose of this modification is to focus on the unhealthy nutrients in NCS or NIS, and ignore the healthy nutrients, when calculating the Nutrient Discrepancy Score between the recipe and the user in the next step. The Nutrient Content Score $NCS_k(rp)$ of recipe rp and the Nutrient Intake Score $NIS_k(u)$ of user u are processed using Equation (3).

$$softplus_mod(x) = log\left(1 + e^{x-4}\right) \tag{3}$$

Step 2. Based on the $softplus_mod(NIS_k(u))$ and $softplus\_mod(NCS_k(rp))$ obtained above, the Nutrient Discrepancy Score $NDS(u, rp)$ of recipe rp and user u is calculated as shown in Equation (4). When the NIS of a user on a certain nutrient is equal to 10 and the NCS of a recipe on a certain nutrient is equal to 10, the NDS of both is calculated by Equation (4) $\approx$ 36. Therefore, in general, when the $NDS(u, rp)$ exceeds 36, the recipe is considered unfavorable to the health of the user.

$$NDS(u, rp) = \sum_{k \in nutrient\ set} softplus_mod(NIS_k(u)) \times softplus_mod(NCS_k(rp)) \tag{4}$$

Output: Nutrient Discrepancy Score (NDS) of recipe rp for user u, denoted as $NDS(u, rp)$.

The following is an example to illustrate the calculation method of NDS. Assume that the Nutrient Intake Scores of two users (u_1 and u_2) and the Nutrient Content Scores of two recipes (rp_1 and rp_2) are known, as shown in Table 4.

Table 4. The NISs and the NCSs in case study.

k	Sodium	Fat	Sugar	Saturated Fat
$NIS_k(u_1)$	5.11	12.33	3.03	6.40
$NIS_k(u_2)$	10.39	4.50	13.14	6.24
$NCS_k(rp_1)$	10.90	11.25	0.09	9.47
$NCS_k(rp_2)$	4.63	5.80	11.38	5.51

Table 5 shows the results of users u_1 and u_2 with recipes rp_1 and rp_2, respectively, to calculate the Nutrient Discrepancy Scores.

Table 5. NDSs in case study.

$NDS(u_1, rp_1)$	$NDS(u_1, rp_2)$	$NDS(u_2, rp_1)$	$NDS(u_2, rp_2)$
74.55	16.40	44.51	68.99

From Table 5, it can be analyzed that for user u_1, recipe rp_1 is not conducive to their health, while recipe rp_2 is more suitable for user u_1. This is because user u_1 has an excessive intake in terms of fat and has unhealthy eating habits, while recipe rp_1 is too high in terms of fat content and is shown to be unhealthy, while recipe rp_2 is low in fat content. For user u_2, both recipes rp_1 and rp_2 are not conducive to their health. This is because user u_2 has unhealthy dietary habits in terms of sodium and sugar, and recipe rp_1 is too high in sodium and is shown to be unhealthy, while recipe rp_2 is too high in sugar and is also shown to be unhealthy. Therefore, both recipe rp_1 and recipe rp_2 have a high NDS for user u_2.

5. Recommendation Model FKGM

We define the task of personalized healthy food recommendation as follows: given the $CRKG$, the goal is to learn a prediction function $F(u, rp|\Theta, CRKG)$ with parameters Θ that can capture both the user's preference and health requirements, and output the matching degree between user u and recipe rp.

5.1. Preliminary

Before delving into the details of our proposed model, we introduce the data preprocessing steps and how the model works in this section. This will help readers understand the underlying processes.

When constructing the positive and negative sample sets for user preference, for each user, the recipes that have been interacted with are considered as the positive sample set P_u^+, while the recipes that have not been interacted with are considered as the negative sample set P_u^-.

When constructing the positive and negative sample sets for user health, the NDS calculation method proposed in Section 4 is first applied to calculate the NDS between all users and recipes in the recommendation dataset. Based on the NDS values between the user and the recipe, the positive and negative sample sets for healthy recipes, H_u^+ and H_u^-, are defined for each user. If the NDS between a user and a recipe is greater than or equal to 36, the recipe is added to the negative sample set H_u^- for that user. The positive and negative sample sets for user health defined here will be used for the health learning task of the model.

The multi-task learning stage of $FKGM$ learns the user's preference and health needs through the preference learning task and the health learning task. In both learning tasks, the Bayesian Personalized Ranking [43] loss is used as the loss function [44] to optimize the model parameters. The optimization goal is to make the model's recommendation results closer to the recipes in the positive sample set. In deep learning, the loss function is a function that measures the gap between the model's prediction results and the true results. During the training process, the optimization algorithm adjusts the model's parameters continuously, making the value of the loss function gradually decrease, thereby making the model's prediction results closer and closer to the target results.

The overall framework and operation process of the model is introduced below. The framework of the healthy food recommendation model $FKGM$ is illustrated in Figure 2, which consists of three components: (1) the embedding layer, which vectorizes each entity and relation in $CRKG$ into a vector through knowledge graph embedding; (2) the message passing layer, which utilizes a knowledge-aware attention graph convolutional neural network to perform iterative updates on the vector representation of each entity by receiving messages from its neighborhood; and (3) the multi-task learning layer, performing the preference learning task and health learning task, respectively.

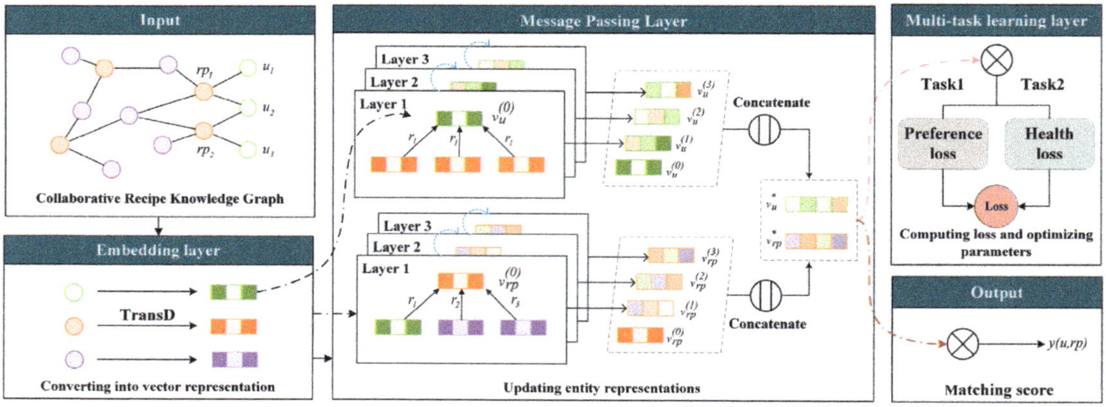

Figure 2. Illustration of the proposed *FKGM* model. The two curves outputted by the message passing layer in the graph, with the light pink curve representing the calculation of inner products between entities for computing loss during the model training phase, and the dark red curve representing the calculation of inner products between entities for outputting predicted matching scores during the model inference phase.

The input of the *FKGM* is *CRKG* constructed in Section 3. Firstly, in the embedding layer, all the entities and relations on the *CRKG* are embedded into vector [45] forms so that the computer can better understand and process this data. The vectors corresponding to entities or relations are also known as their representation. Then, in the message passing layer, each entity on the *CRKG* iteratively receives information from its neighboring entities that are connected to itself and assigns different weights to the information from different neighboring entities according to the relation attention coefficients. This process is also known as message passing. Through multiple layers of message passing, each entity can obtain extensive neighborhood information, thereby better capturing the complex relations in the graph structure. Figure 2 shows an example of three-layer message passing, where an entity is updated to a new representation after each layer of message passing. The final representation of an entity is obtained by concatenating its vector representation before message passing and its vector representations after each layer of updating.

Then, the model predicts the matching score between a user and a recipe by taking the inner product of vectors [46], a method commonly used in recommendation or classification tasks to calculate the similarity between two vectors. During the optimization phase of the model, which is the multi-task learning layer, the preference learning task calculates the loss of the model in terms of preference prediction and the health learning task calculates the loss of the model in predicting health requirements. The two losses are then weighted and summed to obtain the final loss used for optimizing the model parameters.

5.2. Embedding Layer

Embedding is the process of transforming entities and edges into vectors, which are usually called embedding vectors. Embedding vectors can map entities and edges from their original symbolic form (such as strings or IDs) to a continuous vector space, which facilitates computation and processing. We first used xavier_uniform to initialize the vectors of entities and relations. Then we used knowledge graph embedding to update the vector representations. Knowledge graph embedding aims to learn the latent representations of entities and relationships in the knowledge graph while preserving the structural information of the graph. *CRKG* is a heterogeneous graph with rich semantic information. We adopted *TransD* [15] as the method of knowledge graph embedding, which embeds entities and edges in *CRKG* into continuous vector space while retaining its structural information. *TransD* employs a method of dynamically constructing mapping matrices. Given a triplet (h, r, t), its vector includes v_h, v_{h_p}, v_r, v_{r_p}, v_t, and v_{t_p}, where the subscript p indicates the projected vector, $v_h, v_{h_p}, v_t, v_{t_p} \in \mathbb{R}^d$ and $v_r, v_{r_p} \in \mathbb{R}^z$. As shown in Figure 3, *TransD* maps the head entity h and tail entity t into a common space constructed by the entity and relation in the triplet using two mapping matrices, $\boldsymbol{M}_{rh}$ and $\boldsymbol{M}_{rt}$.

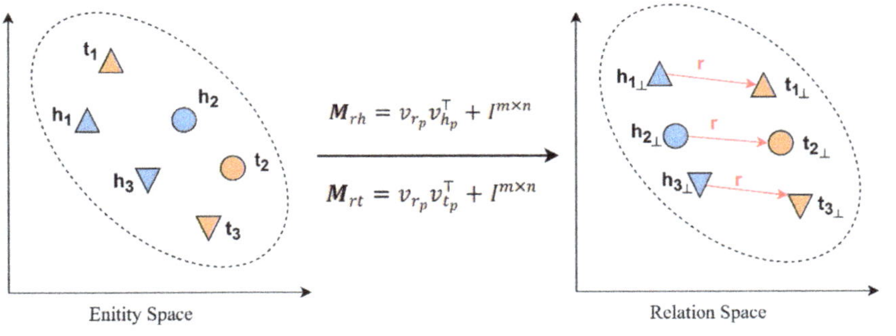

Figure 3. Simple illustration of *TransD*.

The head entity h and tail entity t are mapped to the relation space by constructing the projection matrix, as shown in Equation (5).

$$v_{h_\perp} = \boldsymbol{M}_{rh} v_h,\ v_{t_\perp} = \boldsymbol{M}_{rt} v_t \tag{5}$$

The symbols $v_{h_\perp}$ and $v_{t_\perp}$ in Equation (5) indicate the representations of the head entity h and the tail entity t in the relation space, respectively. The plausibility score $g(h, r, t)$ of a triplet is defined by Equation (6). *TransD* uses embedding scores to measure the plausibility of a triplet's embedding in a knowledge graph. The embedding score of a triplet (h, r, t) is defined by Equation (6), where a higher score indicates a more plausible embedding and a lower score indicates a less plausible one. The *TransD* scoring function first projects the head and tail entities into vector spaces that correspond to the relation and then calculate the distance between the projected head entity and the projected tail entity plus the relation vector. The smaller the distance, the higher the score, indicating the plausibility of the triplet. In other words, if the projected head entity is close to the projected tail entity plus the relation vector, then it is more likely that the triplet (h, r, t) is valid in the knowledge graph.

$$g(h, r, t) = -\|v_{h\perp} + v_r - v_{t\perp}\|_2^2 \tag{6}$$

TransD adopts the Bayesian Personalized Ranking loss, which aims to maximize the margin between positive and negative samples. The loss function is shown in Equation (7).

$$L_{kg} = \sum_{(h,r,t)\in\varepsilon^+} \sum_{(h,r,t')\in\varepsilon^-} -ln\sigma(g(h, r, t') - g(h, r, t)) \tag{7}$$

In Equation (7), ε^+ represents the set of true triplets in the *CRKG*, which has positive samples, while ε^- represents the negative samples. Negative samples are constructed by randomly selecting an entity t' to replace the true tail entity t in the triplet (h, r, t). σ denotes the sigmoid function.

5.3. Message Passing Layer

In this layer, the model constructs a knowledge-aware attention graph convolutional neural network, which updates entity representations through message passing [47]. In the message passing layer, each entity in *CRKG* broadcasts its representation to its first-order neighborhood entities that are directly connected to it and receive information from its neighborhood entities. The first-order neighborhood of an entity refers to the entities that are directly connected to it, while the high-order neighborhood refers to the entities that are indirectly connected to it. In one layer of message passing, all entities in *CRKG* integrate the received neighborhood information to generate their new representations. By stacking multiple layers of such operations, each entity can receive information from high-order neighborhoods, thus capturing the high-order similarity between entities. Inspired by *KGAT* [23], we introduced a knowledge-aware attention mechanism in the message-passing process, which can better distinguish the influence of different relations on entities in *CRKG*.

The first stage of message passing is to receive information from neighboring entities. The computation method for propagating neighborhood information for an entity e on *CRKG* is defined as Equation (8).

$$v_{N(e)} = \sum_{(r,t)\in N(e)} \pi_r(e,t)v_t \tag{8}$$

Here, $v_{N(e)}$ represents the neighborhood information of entity e, $N(e)$ represents the set of all neighboring entities of entity e, v_t denotes the representation of the neighboring entity t, and $\pi_r(e,t)$ is the normalized relation attention coefficient between entity e and entity t under relation r. When entity e and entity t are mapped to relation space r and their distance is closer to r, it indicates that entity t is more important to entity e under relation space r. In this step, the representations of all neighboring entities of entity e are multiplied by the normalized relation attention coefficients and summed to obtain the neighborhood information of entity e. The definition of relation attention coefficient $\hat{\pi}_r(e,t)$ is shown in Equation (9).

$$\hat{\pi}_r(e,t) = \left(v_{t_\perp}\right)^T tanh(v_{e_\perp} + v_r) \tag{9}$$

We used the *softmax* function to normalize the relation attention coefficients for all neighboring entities connected to entity e, as shown in Equation (10). The relation attention coefficient distinguishes the different importance levels of neighboring entities when entity e receives their information.

$$\pi_r(e,t) = \frac{\exp(\hat{\pi}_r(e,t))}{\sum_{(r',t')\in N(e)} \exp(\hat{\pi}_{r'}(e,t'))} \tag{10}$$

After obtaining all the information from the neighborhood of entity e, the next step is to aggregate the information. In this step, the neighborhood information of entity e and the representation of entity e itself are aggregated to update the representation of entity e. We adopted the bi-interaction approach to aggregate the neighborhood information, which can comprehensively capture and process the interaction information between entities. The specific aggregation formula is shown in Equation (11).

$$f_{Aggregator} = LeakyReLU\left(W_1(v_e + v_{N(e)})\right) + LeakyReLU\left(W_2(v_e \odot v_{N(e)})\right) \tag{11}$$

The matrices $W_1, W_2 \in \mathbb{R}^d$ are trainable parameter matrices, and the $\odot$ operator denotes element-wise multiplication between vectors. The *LeakyReLU* activation function was used. After one layer of message passing, the entity representation is updated, which is abstracted as Equation (12).

$$v_e^{(l)} = f\left(v_e^{(l-1)}, v_{N(e)}^{(l-1)}\right) \tag{12}$$

Here, $v_e^{(l)}$ represents the representation of entity e in the *l-th* layer of message passing. After one layer of message passing, the entity updates its representation based on the aggregation function and neighborhood information, enabling it to obtain information from the first-order neighborhood. By stacking more layers, entities on graph *CRKG* can capture information from higher-order neighborhoods, and thus mine users' latent preferences.

After L layers of message passing, entities on $CRKG$ are updated to a new representation at each layer. The final representation of entities on $CRKG$ is obtained by concatenating their representations across all layers.

$$v_e^* = v_e^{(0)} \Big|\Big| \ldots \Big|\Big| v_e^{(L)} \tag{13}$$

The model outputs the predicted matching score by calculating the inner product between the user entity representation v_u^* and the recipe entity representation v_{rp}^*. The larger the inner product between the user entity representation v_u^* and the recipe entity representation v_{rp}^*, the higher the degree of match between them predicted by the model.

$$y(u, rp) = v_u^{*T} \cdot v_{rp}^* \tag{14}$$

5.4. Multi-Task Learning Layer

The task of the $FKGM$ model is to learn the user's preferences and health requirements, so two learning tasks are set at the multi-task learning layer, namely the preference learning task and the health learning task.

5.4.1. Health Learning Task

In Section 4.3, we define the Nutrient Discrepancy Score of a recipe for a user, where a higher $NDS(u, rp)$ means that in terms of health, recipe rp is less suitable for user u. Based on the Nutrient Discrepancy Score, we defined a healthy positive sample set H^+ and a healthy negative sample set H^- for each user. In the context of the health learning task, it is anticipated that the value of $y(u, rp)$ will be higher when the recipe rp is better aligned with the health requirements of user u, and conversely, lower when the recipe is less suitable for user u's health needs. We used the idea of the Bayesian Personalized Ranking loss, and for the health learning task, the loss function L_{health} is shown in Equation (15).

$$L_{health} = \sum_{(u, rp_h^+) \in H^+, (u, rp_h^-) \in H^-} -ln\sigma\left(\left(NDS\left(u, rp_h^-\right) - NDS\left(u, rp_h^+\right)\right)\left(y\left(u, rp_h^+\right) - y\left(u, rp_h^-\right)\right)\right) \tag{15}$$

In Equation (15), $NDS\left(u, rp_h^+\right)$ and $NDS\left(u, rp_h^-\right)$ represent the NDS between user u and recipe rp_h^+ and between user u and recipe rp_h^-, respectively. σ is the sigmoid function. Recipe rp_h^+ is randomly selected from user u's healthy positive sample set H^+, while recipe rp_h^- is randomly selected from user u's healthy negative sample set H^-. For user u, the NDS of recipes in H^+ is lower than that of recipes in H^-. The difference in NDS between the healthy positive sample rp_h^+ and the healthy negative sample rp_h^- is used as the weight for the score difference of the predicted matching between the healthy positive and negative samples. When the difference in NDS between the two samples is larger, the match between these two recipes and user u's health requirements is further apart, and the loss in the health learning task is greater.

5.4.2. Preference Learning Task

For the preference learning task, the expectation is that $y\left(u, rp_p^+\right)$ should be greater than $y\left(u, rp_p^-\right)$ when there exists an interaction between user u and recipe rp_p^+ but not between user u and recipe rp_p^-. We used the Bayesian Personalized Ranking loss as the loss function $L_{preference}$ to optimize the model parameters, which is formulated as Equation (16).

$$L_{preference} = \sum_{(u, rp_p^+) \in P^+, (u, rp_p^-) \in P^-} -ln\sigma\left(y\left(u, rp_p^+\right) - y\left(u, rp_p^-\right)\right) \tag{16}$$

In Equation (16), P^+ is the set of positive samples of user interactions, i.e., the interactions between user u and recipe rp_p^+ that exist in the interaction history. P^- is the set of negative samples of user interactions, i.e., the interactions between user u and recipe rp_p^- that do not exist.

5.4.3. Multi-Task Loss Combination

The final objective function is a weighted combination of L_{health} and $L_{preference}$, as defined in Equation (17), where λ_h is the health loss weight of the health learning task and $\lambda||\Theta||_2^2$ is the regularization term [48] used to prevent overfitting.

$$Loss = \lambda_h L_{health} + L_{preference} + \lambda||\Theta||_2^2 \quad (17)$$

We used mini-batch Adam [49] to optimize the prediction loss and update the model's parameters. Adam is an algorithm used for gradient descent optimization, typically used for training deep learning models. Adam has the characteristic of an adaptive learning rate and performs well in handling large-scale datasets and high-dimensional parameter spaces.

6. Performance Evaluation

We evaluated the proposed healthy food recommendation model, aiming to answer the following questions.

- RQ1: How does the performance of the *FKGM* compare to several baseline models in balancing the performance on both preference and health requirements in the food recommendation task?
- RQ2: How should the health loss weight of the health learning task be set to balance the preference needs and health requirements?
- RQ3: How do different hyper-parameters affect the performance of the *FKGM* model?

6.1. Datasets and Baselines

We conducted experiments on the recipe dataset Allrecipes, which has been described in Section 3. The experimental dataset includes 48,111 users, 38,115 recipes, and 1,267,176 interactions between them. The interaction history of each user was divided into a training set and a test set according to a 9:1 ratio.

We compared our proposed *FKGM* with four baseline models on the task of healthy food recommendation.

CKE [22]: Collaborative knowledge base embedding is a recommendation algorithm that combines collaborative filtering with knowledge graph embedding. It uses TransR to extract information from the knowledge graph and extract semantic features from structured knowledge.

CFKG [50]: Collaborative filtering with knowledge graphs is a recommendation model that enhances the accuracy of recommendations by integrating collaborative filtering and knowledge graphs. It maps the relationships among users, relations, and items to a triplet prediction to achieve more precise recommendations.

KGAT [23]: A knowledge graph attention network is a recommendation model tailored to knowledge-aware personalized recommendation. Built upon the graph neural network framework, *KGAT* explicitly models the high-order relations in collaborative knowledge graphs to provide better recommendations with item side information.

BPRMF [43]: Bayesian Personalized Ranking matrix factorization is a recommendation algorithm based on matrix factorization, which uses the Bayesian Personalized Ranking loss function to consider both the similarity of users' interests and the relative level of interest in items.

The hyper-parameters setting of *FKGM* is presented in Table 6.

Table 6. Hyper-parameters setting.

Parameter Name	Values	Parameter Name	Values
Embedding Dimension	64	Parameter_update	Adam
Message Passing Layer	3	Learning_rate	0.0001
Batch_size	2048	Message_dropout	0.1
Epoch	300	λ_h	0.1

6.2. Evaluation Metrics

To evaluate the effectiveness of the model's recommendation results in meeting users' preferences and health requirements, separate evaluation metrics were set for preference and health requirements.

For preference evaluation, we used the commonly used recommender system evaluation metric, *Recall*. The equation is shown as Equation (18).

$$Recall = \frac{\sum_{u \in U} |R(u) \cap T(u)|}{\sum_{u \in U} T(u)} \tag{18}$$

$R(u)$ represents the set of recipes recommended to a user, and $T(u)$ represents the set of recipes that the user has interacted with in the test set. *Recall* [46] measures the proportion of recommended recipes in the final recommendation list that the user has interacted with in the past, reflecting the relevance of the recommended results to the user's historical interaction list. The higher the recall value, the higher the relevance of the recommended results. *Recall@K* represents the recall value of the top *K* recommended recipes.

The goal of this study for evaluating users' health requirements is to constrain the recommendations provided by the food recommendation model for users who have excessive intakes of nutrients such as sodium, fat, sugar, and saturated fat to achieve nutritional balance. To evaluate whether the model can meet users' health requirements, we ranked users based on their Nutrient Intake Scores in sodium, fat, sugar, and saturated fat in the dataset. In each ranking group, we selected the top 500 users, which were then divided into four groups: high-sodium, high-fat, high-sugar, and high-saturated fat. These four groups represent populations with excessive nutrient intake in the corresponding areas and should reduce their intake of these nutrients. This paper will use the average content of the corresponding nutrients in the top *K* recommended recipes generated by the model as the evaluation metric, in grams, represented as Sodium@*K*, Fat@*K*, Sugar@*K*, and Saturated Fat@*K*, respectively, to evaluate the recipes recommended to these four groups of users. The lower the value of this metric, the better the recommendation.

6.3. Model Comparison (RQ1)

To validate the effectiveness of our approach in combining preferences and health requirements, we compared the performance of *FKGM* with several baseline models in Top-N food recommendation, where ranking position *K* is set to 20. The experimental results are shown in Table 7.

Table 7. Performance of compared models.

Models	Test Set	High-Sodium	High-Fat	High-Sugar	High-Saturated Fat
	Recall@20	**Sodium@20**	**Fat@20**	**Sugar@20**	**Saturated Fat@20**
CKE	0.087	0.812	14.756	16.400	6.051
CFKG	0.086	0.772	13.799	16.414	5.858
KGAT	**0.090**	1.264	15.837	18.592	6.319
BPRMF	0.080	1.004	14.933	16.143	5.879
FKGM (ours)	0.074	**0.451**	**12.539**	**15.731**	**5.325**

Table 7 shows the performance of *FKGM* and four baseline models in terms of the preference evaluation metric *Recall@*20 and four health evaluation metrics. The analysis shows that although *FKGM* did not achieve an optimal performance in dietary preference recommendation, it outperformed other models that only focus on preferences in terms of meeting user health requirements. A further analysis of the comparison between *FKGM* and other models in terms of health recommendation with the ranking parameter *K* set to {20, 40, 60, 80, 100} is shown in Figure 4.

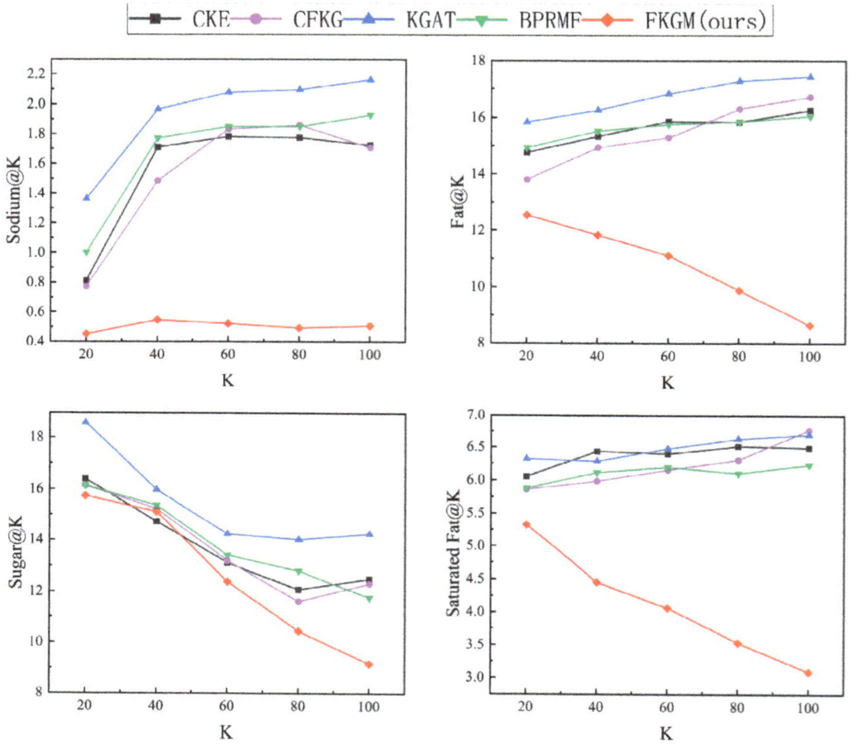

Figure 4. Comparison of performance of five models in health recommendation.

After analyzing the experimental results shown in Figure 4, it can be concluded that our proposed model consistently outperformed other baseline models in health requirement recommendation, demonstrating the effectiveness of our approach in combining preference needs and health requirements.

6.4. Health Loss Weight Study (RQ2)

In this section, we study the impact of health loss weight on balancing preference needs and health requirements in food recommendations. By setting different health loss weights for experiments and comparing the model performance in health requirement recommendations, the results are shown in Table 8.

Table 8. Performance comparison when *FKGM* sets different health loss weights.

Health Loss Weight	Recall@20	Sodium@20	Fat@20	Sugar@20	Saturated Fat@20
$\lambda_h = 0$	**0.089**	1.167	16.186	17.567	6.316
$\lambda_h = 0.05$	0.077	0.884	14.147	17.517	5.889
$\lambda_h = 0.1$	0.074	0.451	12.539	15.731	5.325
$\lambda_h = 0.15$	0.042	0.653	9.262	9.632	3.881
$\lambda_h = 0.2$	0.032	**0.175**	**4.329**	**7.734**	**1.757**

Experimental results show that with the increase of the health loss parameter λ_h, the performance of the model in health recommendation also increases. However, if λ_h is too large, the performance of the model in dietary preference recommendation will be compromised. To balance the preference and health requirements of users in food recommendation, we suggest setting the health loss weight λ_h to 0.1 for the model. Through experiments, it can be found that the model can control whether the recommendation results are more biased towards preferences or health requirements by adjusting the value of the health loss weight.

6.5. *Hyper-Parameter Studies (RQ3)*

To investigate the impact of different hyper-parameters on the performance of the *FKGM* model, we conducted multiple experiments and analyzed the performance of *FKGM* under different settings of embedding dimensions and message propagation layers.

6.5.1. The Impact of FKGM Embedding Dimension

To investigate the impact of embedding dimensions on the performance of the *FKGM* model, we conducted multiple experiments. We set the embedding dimensions of the model to 8, 16, 32, 64, and 128, kept other parameters constant, and compared the experimental results. Table 9 shows the performance of *FKGM* under different embedding dimensions.

Table 9. Performance comparison when model sets different embedding dimensions.

Dimension	Recall@20	Sodium@20	Fat@20	Sugar@20	Saturated Fat@20
8	0.055	1.026	15.793	17.011	6.181
16	0.063	0.735	15.855	**14.631**	6.293
32	0.071	0.812	14.756	18.382	5.850
64	**0.074**	**0.451**	**12.539**	15.731	**5.325**
128	0.067	0.571	13.331	16.235	5.855

The experimental results indicate that with the increase in embedding dimensions, the model's performance in preference recommendation first increases and then decreases. When the embedding dimension is set to 64, the model achieves the best performance in preference recommendation. In terms of healthy recommendation, the embedding dimension of 16 achieves the best performance in regulating sugar intake, while the other three nutrient constraints perform best when the embedding dimension is set to 64. Overall, the model performs best when the embedding dimension is set to 64.

6.5.2. The Impact of the Number of Message Passing Layers

To investigate the impact of message passing layer depth on the performance of the *FKGM* model, we conducted experiments by setting the message passing layer depth to 1–5 and keeping other parameters constant. Table 10 presents the performance of *FKGM* under different layer depths.

Table 10. Performance comparison when model sets different message passing layers.

Layers	Recall@20	Sodium@20	Fat@20	Sugar@20	Saturated Fat@20
FKGM-1	0.063	0.698	14.815	17.950	5.688
FKGM-2	0.071	0.816	14.755	16.400	**5.188**
FKGM-3	**0.074**	**0.451**	**12.539**	**15.731**	5.325
FKGM-4	0.065	0.871	14.426	16.358	5.566
FKGM-5	0.056	1.060	15.299	17.414	6.117

The experimental results show that the performance of the $FKGM$ model in preference recommendation initially increases with the depth of the message passing layer, but then starts to decrease. Stacking too many message passing layers does not always lead to an improvement in the recommendation performance of the model. This implies that overly deep message passing layers can lead to overfitting and loss of the model performance. Overall, the $FKGM$ model achieved the best performance when the message passing layer depth was three.

7. Discussion

In this section, we discuss the main findings and implications of our proposed model $FKGM$. We also acknowledge the limitations of our study and suggest some directions for future research.

7.1. Main Findings and Implications

Our study aimed to address the issue of personalized healthy food recommendations by considering both user dietary preferences and health requirements. We constructed a collaborative recipe knowledge graph ($CRKG$) that contains user–recipe interactions and various food-related information. We also proposed a method for calculating the healthiness match between recipes and user preferences based on the Nutrient Content Score (NCS) of recipes and the Nutrient Intake Score (NIS) of users. To achieve this, we developed a novel health-aware food recommendation model ($FKGM$) that uses knowledge graph embedding and a knowledge-aware attention graph convolutional neural network to capture the semantic associations between users and recipes on $CRKG$, and learns the user's requirements in both preference and health by fusing two losses of these two learning tasks.

We conducted experiments on the Allrecipes dataset and compared our model with four baseline models: CKE [22], $CFKG$ [50], $KGAT$ [23], and $BPRMF$ [43], aiming to show that the $FKGM$ model can learn users' dietary preferences while considering their health requirements. The experimental results show that the $FKGM$ model did not achieve the best performance in terms of preference; however, in the health recommendation task, our model achieved a significant advantage and outperformed the other baseline models. This suggests that our proposed $FKGM$ can make balanced food recommendations in terms of preference and health requirements, rather than simply focusing on users' preferences or the healthiness of the food itself. We believe that this advantage in terms of health has important practical application value in the future food recommendation field.

Furthermore, we conducted a study on the health loss weight in the multi-task learning layer, which directly affects the model's performance on the preference recommendation and health recommendation tasks. The health loss weight can be adjusted to decide whether the model is more inclined to consider health or preference.

Our study has several implications for both research and practice. For research, our study contributes to the literature on food recommendation by proposing a new solution that integrates user preferences and health requirements. Our study also demonstrates the usefulness of knowledge graphs and multi-task learning techniques for food recommendation. For practice, our study provides a practical tool for food-related applications and services that aim to promote healthy eating habits among users. Our model can help users discover new recipes that suit their tastes and health requirements.

7.2. Limitations and Future Work

Despite the significant contributions of our study, we recognize certain limitations that require attention in future research. Firstly, we only considered four nutrients (sodium, fat, sugar, and saturated fat) as indicators of healthiness, which may not encompass all aspects of healthiness. Future research could integrate additional nutrients or other factors, such as calories, allergies, and dietary restrictions, in health assessments. Secondly, we evaluated our model using only one dataset (Allrecipes), which may restrict the generalizability of our results. Future research could apply our model to alternative domains or datasets, such as restaurant reviews or online grocery shopping. Thirdly, we did not conduct user studies or surveys to validate user satisfaction and acceptance of our model. Future research could gather user feedback or ratings to assess the user experience and perceived usefulness of our model.

8. Conclusions

This work applies food-related knowledge to food recommendation and proposes a healthy food recommendation model $FKGM$ that considers both health requirements and user dietary preferences. $FKGM$ is a knowledge graph-based multi-task learning model that learns semantic information between users and recipes through knowledge graph embedding and message-passing mefchanisms. It also emphasizes users' unhealthy eating habits by learning their historical dietary behavior. This work constructed a large-scale collaborative recipe knowledge graph that contains user–recipe, recipe–ingredient, and other information for multi-task food recommendation, and extensive experiments were conducted on it. The results show that $FKGM$ outperformed current competing baseline methods in the task of healthy food recommendation. The proposed health-aware food recommendation model is expected to have significant practical application value in the future food recommendation field.

Author Contributions: Conceptualization, Y.C. and Y.G.; methodology, Y.C. and Y.G.; software, Y.G.; investigation, Y.C., Y.G. and Q.F.; resources, Y.C.; writing-original draft preparation, Y.C. and Y.G.; writing-review and editing, Y.C., Y.G., Q.F., Q.Z. and Y.D.; funding acquisition, Y.C. All authors have read and agreed to the published version of the manuscript.

Funding: This work was supported by the National Natural Science Foundation of China (Grant No. 61972010, 42101470), the National Key Research and Development Program of China (Grant No. 2022YFF1100905), and the 2023 Postgraduate Research Capability Improvement Program Project (Grant No. 19008023027).

Data Availability Statement: Data sharing is inapplicable for this article.

Acknowledgments: The authors would like to thank Haifeng Dou, Hanqiang Li, and Zhiying Luo for their advice and excellent technical support.

Conflicts of Interest: The authors declare no conflict of interest.

References

1. World Health Organization. Obesity and Overweight. Available online: https://www.who.int/news-room/fact-sheets/detail/obesity-and-overweight (accessed on 5 April 2022).
2. Ng, M.; Fleming, T.; Robinson, M.; Thomson, B.; Graetz, N.; Margono, C.; Mullany, E.C.; Biryukov, S.; Abbafati, C.; Abera, S.F.; et al. Global, Regional, and National Prevalence of Overweight and Obesity in Children and Adults during 1980–2013: A Systematic Analysis for the Global Burden of Disease Study 2013. *Lancet* **2014**, *384*, 766–781. [CrossRef] [PubMed]
3. Ornish, D.; Brown, S.E.; Billings, J.H.; Scherwitz, L.W.; Armstrong, W.T.; Ports, T.A.; McLanahan, S.M.; Kirkeeide, R.L.; Gould, K.L.; Brand, R.J. Can Lifestyle Changes Reverse Coronary Heart Disease?: The Lifestyle Heart Trial. *Lancet* **1990**, *336*, 129–133. [CrossRef] [PubMed]
4. Chen, Y.; Dong, Y.; Sun, Y.; Liang, J. A Multi-Comparable Visual Analytic Approach for Complex Hierarchical Data. *J. Vis. Lang. Comput.* **2018**, *47*, 19–30. [CrossRef]
5. Luo, Z.; Chen, Y.; Li, H.; Li, Y.; Guo, Y. TreeMerge: A Visual Comparative Analysis Method for Food Classification Tree in Pesticide Residue Maximum Limit Standards. *Agronomy* **2022**, *12*, 3148. [CrossRef]

6. Chen, Y.; Dou, H.; Chang, Q.; Fan, C. PRIAS: An Intelligent Analysis System for Pesticide Residue Detection Data and Its Application in Food Safety Supervision. *Foods* **2022**, *11*, 780. [CrossRef]
7. Rabbi, M.; Hane Aung, M.; Choudhury, T. Towards Health Recommendation Systems: An Approach for Providing Automated Personalized Health Feedback from Mobile Data. In *Mobile Health: Sensors, Analytic Methods, and Applications*; Springer: Berlin/Heidelberg, Germany, 2017; pp. 519–542. [CrossRef]
8. Anderson, C. A Survey of Food Recommenders. *arXiv* **2018**, arXiv:1809.02862.
9. Trattner, C.; Elsweiler, D. Food Recommender Systems: Important Contributions, Challenges and Future Research Directions. *arXiv* **2017**, arXiv:1711.02760.
10. Çelik Ertuğrul, D.; Elçi, A. A Survey on Semanticized and Personalized Health Recommender Systems. *Expert Syst.* **2020**, *37*, e12519. [CrossRef]
11. Thongsri, N.; Warintarawej, P.; Chotkaew, S.; Saetang, W. Implementation of a Personalized Food Recommendation System Based on Collaborative Filtering and Knapsack Method. *Int. J. Electr. Comput. Eng.* **2022**, *12*, 630–638. [CrossRef]
12. Padhiar, I.; Seneviratne, O.; Chari, S.; Gruen, D.; McGuinness, D.L. Semantic Modeling for Food Recommendation Explanations. In Proceedings of the 2021 IEEE 37th International Conference on Data Engineering Workshops (ICDEW), Chania, Greece, 19–21 April 2021; pp. 13–19.
13. Agapito, G.; Simeoni, M.; Calabrese, B.; Caré, I.; Lamprinoudi, T.; Guzzi, P.H.; Pujia, A.; Fuiano, G.; Cannataro, M. DIETOS: A Dietary Recommender System for Chronic Diseases Monitoring and Management. *Comput. Methods Programs Biomed.* **2018**, *153*, 93–104. [CrossRef]
14. Min, W.; Liu, C.; Xu, L.; Jiang, S. Applications of Knowledge Graphs for Food Science and Industry. *Patterns* **2022**, *3*, 100484. [CrossRef] [PubMed]
15. Ji, G.; He, S.; Xu, L.; Liu, K.; Zhao, J. Knowledge Graph Embedding via Dynamic Mapping Matrix. In Proceedings of the 53rd Annual Meeting of the Association for Computational Linguistics and the 7th International Joint Conference on Natural Language Processing, Beijing, China, 26–31 July 2015; Long Papers. Volume 1, pp. 687–696.
16. Hu, B.; Shi, C.; Zhao, W.X.; Yu, P.S. Leveraging Meta-Path Based Context for Top-N Recommendation with A Neural Co-Attention Model. In Proceedings of the 24th ACM SIGKDD International Conference on Knowledge Discovery & Data Mining, New York, NY, USA, 19–23 August 2018; ACM: London, UK, 2018; pp. 1531–1540.
17. Fan, S.; Zhu, J.; Han, X.; Shi, C.; Hu, L.; Ma, B.; Li, Y. Metapath-Guided Heterogeneous Graph Neural Network for Intent Recommendation. In Proceedings of the ACM SIGKDD International Conference on Knowledge Discovery & Data Mining, Anchorage, AK, USA, 25 July 2019; ACM: New York, NY, USA, 2019; pp. 2478–2486.
18. Wang, Q.; Mao, Z.; Wang, B.; Guo, L. Knowledge Graph Embedding: A Survey of Approaches and Applications. *IEEE Trans. Knowl. Data Eng.* **2017**, *29*, 2724–2743. [CrossRef]
19. Chen, Y.; Zhang, Q.; Guan, Z.; Zhao, Y.; Chen, W. GEMvis: A Visual Analysis Method for the Comparison and Refinement of Graph Embedding Models. *Vis. Comput.* **2022**, *38*, 3449–3462. [CrossRef]
20. Wu, C.; Chen, Y.; Dong, Y.; Zhou, F.; Zhao, Y.; Liang, C.J. VizOPTICS: Getting Insights into OPTICS via Interactive Visual Analysis. *Comput. Electr. Eng.* **2023**, *107*, 108624. [CrossRef]
21. Chen, Y.; Guan, Z.; Zhang, R.; Du, X.; Wang, Y. A Survey on Visualization Approaches for Exploring Association Relationships in Graph Data. *J. Vis.* **2019**, *22*, 625–639. [CrossRef]
22. Zhang, F.; Yuan, N.J.; Lian, D.; Xie, X.; Ma, W.-Y. Collaborative Knowledge Base Embedding for Recommender Systems. In Proceedings of the 22nd ACM SIGKDD International Conference on Knowledge Discovery and Data Mining, San Francisco, CA, USA, 13–17 August 2016; ACM: New York, NY, USA, 2016; pp. 353–362.
23. Wang, X.; He, X.; Cao, Y.; Liu, M.; Chua, T.-S. Kgat: Knowledge Graph Attention Network for Recommendation. In Proceedings of the 25th ACM SIGKDD International Conference on Knowledge Discovery & Data Mining, Anchorage, AK, USA, 4–8 August 2019; pp. 950–958.
24. Ma, T.; Huang, L.; Lu, Q.; Hu, S. Kr-Gcn: Knowledge-Aware Reasoning with Graph Convolution Network for Explainable Recommendation. *ACM Trans. Inf. Syst.* **2023**, *41*, 1–27. [CrossRef]
25. Ge, M.; Elahi, M.; Fernaández-Tobías, I.; Ricci, F.; Massimo, D. Using Tags and Latent Factors in a Food Recommender System. In Proceedings of the 5th International Conference on Digital Health, Florence, Italy, 18–20 May 2015; pp. 105–112.
26. Khan, M.A.; Smyth, B.; Coyle, D. Addressing the Complexity of Personalized, Context-Aware and Health-Aware Food Recommendations: An Ensemble Topic Modelling Based Approach. *J. Intell. Inf. Syst.* **2021**, *57*, 229–269. [CrossRef]
27. Freyne, J.; Berkovsky, S. Intelligent Food Planning: Personalized Recipe Recommendation. In Proceedings of the 15th International Conference on Intelligent User Interfaces, Hong Kong, China, 7–10 February 2010; ACM: New York, NY, USA, 2010; pp. 321–324.
28. Yang, L.; Cui, Y.; Zhang, F.; Pollak, J.P.; Belongie, S.; Estrin, D. Plateclick: Bootstrapping Food Preferences through an Adaptive Visual Interface. In Proceedings of the 24th ACM International on Conference on Information and Knowledge Management, Melbourne, Australia, 18–23 October 2015; pp. 183–192.
29. Gao, X.; Feng, F.; He, X.; Huang, H.; Guan, X.; Feng, C.; Ming, Z.; Chua, T.-S. Hierarchical Attention Network for Visually-Aware Food Recommendation. *IEEE Trans. Multimed.* **2019**, *22*, 1647–1659. [CrossRef]
30. Gao, X.; Feng, F.; Huang, H.; Mao, X.-L.; Lan, T.; Chi, Z. Food Recommendation with Graph Convolutional Network. *Inf. Sci.* **2022**, *584*, 170–183. [CrossRef]

31. Bianchini, D.; De Antonellis, V.; De Franceschi, N.; Melchiori, M. PREFer: A Prescription-Based Food Recommender System. *Comput. Stand. Interfaces* **2017**, *54*, 64–75. [CrossRef]
32. Espín, V.; Hurtado, M.V.; Noguera, M. Nutrition for Elder Care: A Nutritional Semantic Recommender System for the Elderly. *Expert Syst.* **2016**, *33*, 201–210. [CrossRef]
33. Shandilya, R.; Sharma, S.; Wong, J. MATURE-Food: Food Recommender System for MAndatory FeaTURE Choices a System for Enabling Digital Health. *Int. J. Inf. Manag. Data Insights* **2022**, *2*, 100090. [CrossRef]
34. Food Surveys Research Group. Food and Nutrient Dataset for Dietary Studies. 2022. Available online: https://www.ars.usda.gov/northeast-area/beltsville-md-bhnrc/beltsville-human-nutrition-research-center/food-surveys-research-group/docs/fndds-download-databases/ (accessed on 28 April 2023).
35. Ribeiro, D.; Machado, J.; Ribeiro, J.; Vasconcelos, M.J.M.; Vieira, E.F.; De Barros, A.C. SousChef: Mobile Meal Recommender System for Older Adults. In Proceedings of the ICT4AgeingWell, Porto, Portugal, 28–29 April 2017; pp. 36–45.
36. Ge, M.; Ricci, F.; Massimo, D. Health-Aware Food Recommender System. In Proceedings of the 9th ACM Conference on Recommender Systems, Vienna, Austria, 16–20 September 2015; pp. 333–334.
37. Wang, W.; Duan, L.-Y.; Jiang, H.; Jing, P.; Song, X.; Nie, L. Market2Dish: Health-Aware Food Recommendation. *ACM Trans. Multimed. Comput. Commun. Appl. (TOMM)* **2021**, *17*, 1–19. [CrossRef]
38. Chen, M.; Jia, X.; Gorbonos, E.; Hoang, C.T.; Yu, X.; Liu, Y. Eating Healthier: Exploring Nutrition Information for Healthier Recipe Recommendation. *Inf. Process. Manag.* **2020**, *57*, 102051. [CrossRef]
39. Li, D.; Zaki, M.J.; Chen, C. Health-Guided Recipe Recommendation over Knowledge Graphs. *J. Web Semant.* **2023**, *75*, 100743. [CrossRef]
40. Bollacker, K.; Evans, C.; Paritosh, P.; Sturge, T.; Taylor, J. Freebase: A Collaboratively Created Graph Database for Structuring Human Knowledge. In Proceedings of the 2008 ACM SIGMOD International Conference on Management of Data, Vancouver, Canada, 9–12 June 2008; pp. 1247–1250.
41. UK Government Front-of-Pack Nutrition Labelling Guidance. Available online: https://www.gov.uk/government/publications/front-of-pack-nutrition-labelling-guidance (accessed on 10 April 2022).
42. Dugas, C.; Bengio, Y.; Bélisle, F.; Nadeau, C.; Garcia, R. Incorporating Second-Order Functional Knowledge for Better Option Pricing. *Adv. Neural Inf. Process. Syst.* **2000**, *13*, 472–478.
43. Rendle, S.; Freudenthaler, C.; Gantner, Z.; Schmidt-Thieme, L. BPR: Bayesian Personalized Ranking from Implicit Feedback. *arXiv* **2012**, arXiv:1205.2618.
44. Tian, Y.; Su, D.; Lauria, S.; Liu, X. Recent Advances on Loss Functions in Deep Learning for Computer Vision. *Neurocomputing* **2022**, *497*, 129–158. [CrossRef]
45. Bengio, Y.; Courville, A.; Vincent, P. Representation Learning: A Review and New Perspectives. *IEEE Trans. Pattern Anal. Mach. Intell.* **2013**, *35*, 1798–1828. [CrossRef]
46. Isinkaye, F.O.; Folajimi, Y.O.; Ojokoh, B.A. Recommendation Systems: Principles, Methods and Evaluation. *Egypt. Inform. J.* **2015**, *16*, 261–273. [CrossRef]
47. Kipf, T.N.; Welling, M. Semi-Supervised Classification with Graph Convolutional Networks. *arXiv* **2016**, arXiv:1609.02907.
48. Tian, Y.; Zhang, Y. A Comprehensive Survey on Regularization Strategies in Machine Learning. *Inf. Fusion* **2022**, *80*, 146–166. [CrossRef]
49. Kingma, D.P.; Ba, J. Adam: A Method for Stochastic Optimization. *arXiv* **2014**, arXiv:1412.6980.
50. Zhang, Y.; Ai, Q.; Chen, X.; Wang, P. Learning over Knowledge-Base Embeddings for Recommendation. *arXiv* **2018**, arXiv:1803.06540.

foods

MDPI

Review

Combining AI Tools with Non-Destructive Technologies for Crop-Based Food Safety: A Comprehensive Review

Hind Raki [1,*,†], Yahya Aalaila [1,†], Ayoub Taktour [2] and Diego H. Peluffo-Ordóñez [1,†]

1 College of Computing, University Mohammed VI Polytechnic, Ben Guerir 43150, Morocco; yahya.aalaila@um6p.ma (Y.A.); peluffo.diego@um6p.ma (D.H.P.-O.)
2 Materials Sciences and Nanotechnoloy (MSN), University Mohammed VI Polytechnic, Ben Guerir 43150, Morocco; ayoub.taktour@um6p.ma
* Correspondence: hind.raki@um6p.ma
† Current address: SDAS Research Group, Ben Guerir 47963, Morocco.

Abstract: On a global scale, food safety and security aspects entail consideration throughout the farm-to-fork continuum, considering food's supply chain. Generally, the agrifood system is a multiplex network of interconnected features and processes, with a hard predictive rate, where maintaining the food's safety is an indispensable element and is part of the Sustainable Development Goals (SDGs). It has led the scientific community to develop advanced applied analytical methods, such as machine learning (ML) and deep learning (DL) techniques applied for assessing foodborne diseases. The main objective of this paper is to contribute to the development of the consensus version of ongoing research about the application of Artificial Intelligence (AI) tools in the domain of food-crop safety from an analytical point of view. Writing a comprehensive review for a more specific topic can also be challenging, especially when searching within the literature. To our knowledge, this review is the first to address this issue. This work consisted of conducting a unique and exhaustive study of the literature, using our *TriScope Keywords-based Synthesis* methodology. All available literature related to our topic was investigated according to our criteria of inclusion and exclusion. The final count of data papers was subject to deep reading and analysis to extract the necessary information to answer our research questions. Although many studies have been conducted, limited attention has been paid to outlining the applications of AI tools combined with analytical strategies for crop-based food safety specifically.

Keywords: chemometrics; food contaminants; food processes; machine learning; spectroscopy; sustainability

1. Introduction

One Health is a very distinct concept unifying soil, plant, and human health for a flourishing and sustainable ecosystem. This approach can also be applied on ensuring food safety and security. Ending hunger and ensuring access to safe, nutritious, and sufficient food all year round by all people are part of the Sustainable Development Goals (SDGs) to be achieved by 2030, according to the United Nations SDGs agenda, specifically SDG3 and SDG2, respectively (Good Health and Well-being, Zero Hunger). Nonetheless, multiple challenges and issues are increasingly making this mission impossible to accomplish [1]. As a matter of fact and according to statistical studies, humanity is expecting an enormous increase in the global population, reaching 9.7 billion individuals by 2025. Surely, ensuring that they all have access to safe and nutritious food becomes more challenging. The increasing demand for crop-based food in the market is due to multiple factors, mainly the propaganda of plant-based diets [2], the transition of human diets in society, and the factor of poor diversity in crop-food consumption while thousands of crops exist. These factors have impacted diversity concerning crops, and lately, scientists have been shedding light on orphan crops, which are crops not traded internationally, but would have been important for regional food security [3], hence our objective to target food of crop origins.

Concretely, it has always been a big challenge; the food system dates back several decades, with a tendency to evolve continuously, depending on its economic, social, cultural, and environmental factors, along with several external and internal variables [4]. These complex interdependent and interconnected factors impact the global agrifood system, encompassing the farm-to-fork continuum and various environmental and socioeconomic factors [5]. In this dynamic context, the integration of crop classification using deep learning emerges as a significant consideration, providing advanced tools for precise and efficient analysis within the intricate web of factors influencing the agrifood system, supported by numerous studies dedicated to the development of advanced tools [6,7]. Hence, food security, quality, and safety, now more than ever, are depending on the entire food supply chain from early production to market accessibility. A food-borne disease is referred to as food contamination, on account of the presence of hazardous contaminants that can cause human body illnesses. They are divided into three groups, namely biological, chemical, and physical contaminants, based on the pollutant and the process by which they enter the food product [8], for example, during crop cultivation, due to contaminated soil (animal manure or chemical fertilizers), as well as the water used for irrigation of produce (groundwater, recovered rainfall, surface water, or re-utilized wastewater) [9,10]. Food contamination represents a big challenge because of its large impact, being related to the whole food supply chain. In order to protect consumers from unsafe foods, standards are required to establish a monitoring system to reduce chemical and microbiological food contamination [11], including all food chain participants, such as farmers, processors, transporters, retailers, and consumers [4]. Even though there are many guidelines to follow, only a fraction of them is followed. In many cases, individuals neglect simple practices, such as proper hand-washing methods and the good use of gloves, which leads to serious food poisoning [12].

It was said that the transmission of viruses through food matrices is less likely compared to direct person-to-person contact, respiratory droplets or contaminated surfaces. The COVID-19 pandemic has forced us to think about the technological preparedness to use these smart technologies, which avoid human-to-human and human-to-food contact during food processing. Hence, there is a need for technological innovation combining analytical strategies with AI tools, taking into consideration economic and feasibility challenges [13]. Artificial intelligence (AI) is primordial for technological advances, developing computational tools to provide sustainable solutions for food security and safety [14]. There are potential applications of machine vision systems in addition to analytical strategies for more accurate and lower-cost techniques for contaminant detection in food [15].

The present work proposes a unique methodology that was developed after a careful investigation of the literature on how to conduct a high-quality comprehensive review. For the data collection, which is detailed later on, the search process is carried out in three major databases: Scopus, Web of Science, and IEEE Xplore. Notwithstanding the fewer databases implicated, the proposed approach is based on executing multiple query searches in each one, leading to an in-depth and exhaustive investigation.

The remainder of this paper is organized as follows. In Section 2, we start by investigating related and/or similar works to our topic of research. Section 3 holds the section in which research questions, search procedures, and paper selection settings are described. Section 4 presents and discusses relevant information extracted from collected papers according to our research questions. In Section 5, we highlight further details concerning gaps in research and challenges. Finally, Section 6 draws conclusions, final remarks, and our suggested future research directions.

2. Background

Spectroscopic measurement methods are versatile tools applied across diverse scientific domains. They play crucial roles in pharmaceutical quality control, environmental monitoring, materials science, medical diagnostics, forensic analysis, agricultural research and much more. The adaptability and precision of these methods make them invaluable for characterizing composition, structure, and properties in various scientific disciplines. Recent advances have emerged in chemical analytical strategies, depending on the variety of studied food matrices. There is a constant quest to find the most suitable technique to investigate certain aspects of a compound and ascertain its consistency or structure.

2.1. Spectroscopy for Food Safety and Quality

Spectroscopic methods serve as analytical tools to identify food's composition, germs, pests, diseases, and adulteration [16]. The following listing presents the most common techniques within the literature: 1. infrared spectroscopy, 2. Raman spectroscopy, 3. nuclear magnetic resonance (NMR) spectroscopy,4. ultraviolet–visible spectroscopy (UV-vis). Infrared spectroscopy using Fourier transform infrared (FTIR) is often employed. The mid-infrared (MIR) region covers an area between 4000 and 400 cm^{-1} [17]. A soil's composition, characteristics, and organic matter may all be found via MIR spectroscopy. Also, the diffuse reflectance infrared Fourier transform (DRIFT) method may identify the chemical characteristics of humus and soil. Moreover, attenuated total reflectance (ATR) can identify organic materials in soil [18]. Anisidine, which is produced during the oxidation of food, is frequently measured using UV-visible spectroscopy to assess the quality of oil [19]. Moreover, fluorescence is a property shared by a large number of microorganisms, including their colonies, making it simple to identify any bacteria by looking at their fluorescence spectra [20,21]. Tryptophan, riboflavin, and lumichrome are three different forms of fluorophores found in yogurt, and their presence enables fluorescence spectroscopy to assess the yogurt's quality [22]. Honey is a substance produced by bees from floral nectar, containing phenolic chemicals that are byproducts of the phenolic acid found in the flower, and as it is packaged and transported, its qualities alter [23]. Fluorescence spectroscopy can quantify the concentration of phenolic chemicals. During processing and storage, mycotoxins and fungus are found in grains using MIR spectroscopy [24]. On the other hand, carbohydrates' structure can change during storage, especially in the presence of water, which can be identified using Raman spectroscopy. This technique characterizes and quantifies the lipid content of foods [25]. NMR spectroscopy can monitor the ripening, drying, and adulteration of food components as well as determining the genotype responsible for a certain phenotype of the grapes used to create wine. As a result, NMR spectroscopy may offer information on mixtures of metabolites [26]. Infrared or Raman fingerprints are the outcomes of observations made on a large number of objects or samples with a wide variety of characteristics (variables, like the absorbance at various wave-numbers or wave-number shifts) from a monochromatic light source for FTIR and Raman, respectively [27]. Therefore, since the exponential rise in computing power and the capacity to gather, store, and analyze enormous volumes of data, machine learning (ML) systems can improve the potential to extrapolate information from complicated spectrum data.

2.2. Integrating AI Tools for Food Safety and Quality Analysis

The development of analytical strategies for food and beverage assessments is crucial for ensuring food safety and public health. Various technologies, including imaging, odor, and taste, have been developed [28]. Spectroscopic techniques, such as infrared spectroscopy and Raman spectroscopy, have proven to be rapid and nondestructive for microorganism detection [29]. Magnetic surface-enhanced Raman scattering nanoprobes show high specificity in separating and detecting multiple pathogens in complex food matrices [30]. The usage of electronic noses, coupled with data acquisition cards and classification methods, enhances success rates in ensuring the originality of saffron products [28]. The integration of more effective algorithms is essential for spectral data processing and microorganism reference database building [29].

Hyperspectral imaging (HSI) systems, including computer vision systems (CVS), are widely applied in the food industry for nonintrusive quality control [28]. HSI covers various food-processing phases, providing the ability to control the quality and safety of processed foods [31]. Challenges lie in the classification, especially for crops within the same family, and the high cost of these technologies [31]. Algorithms and chemometric methods should focus on reducing the dimensionality of data and improving computational efficiency while enhancing performance and robustness [31].

Terahertz spectroscopy techniques, coupled with machine learning tools, ensure quality and security inspection of agricultural products and food [32]. Classical methods of spectral preprocessing, such as smoothing, standard normal variate, and Fourier transformation, can be integrated into multivariate calibration steps for more efficiency [33]. In the quantification of honey adulteration, spectroscopy and hyperspectral imaging, when coupled with machine learning models and optic fiber sensors, provide fast and nondestructive detection [34].

AI development in data mining has made significant breakthroughs, particularly with the application of deep learning (DL) in the analysis of spectral data from food and agricultural products [35]. DL approaches offer a less laborious yet more precise method for this purpose. Combining infrared spectroscopy (IRS) and hyperspectral imaging (HSI) techniques with AI tools shows potential in advancing the quality evaluation of cereals, which are among the top consumed crops globally [36]. The integration of convolutional neural networks (CNNs) in the qualitative and quantitative analysis of spectra involves extracting micro- and macro-features through multiple convolution and pooling layers. DL-spectroscopic sensing techniques have demonstrated promising results in the quality evaluation of food and agro-products, encompassing identification, geographical origin detection, adulteration recognition, bruise detection, and component content prediction for crops [35].

The application of CNNs helps avoid secondary workloads, although challenges persist, such as determining optimal network scale, selecting parameters, addressing overfitting, and enhancing model interpretability—a current dilemma in AI research [33]. Table 1 highlights recent and high-quality papers that couple analytical strategies with machine learning approaches across diverse food safety and quality purposes.

Table 1. Latest applications combining AI tools with analytical approaches for food safety and quality.

Analytical Approach	AI Tool	Problematic	Ref.
Spectroscopy	Python-based portable system using Jetson TX2 Module	Food classification of four classes of coffee and purées	[37]
Near-infrared spectroscopy	Block sparse Bayesian learning (BSBL) with fast marginalized likelihood maximization (FMLM)	Computational cost reduction for calculating the inverse of a large matrix containing absorption peak information	[38]
Impedance spectroscopy	A fuzzy logic model applied on the parameters extracted from distribution of relaxation times (DRT)	Meat-based food classification according to its freshness for different types of muscles	[39]
TeraHertz (THz) spectroscopy and chemometrics	Interval partial least squares (iPLS) for optimizing the THz frequency and other preprocessing techniques combined with extreme learning machine (ELM), genetic algorithm support vector machine (GA-SVM), and artificial bee colony algorithm support vector machine (ABC-SVM) for decision making	Three typical soybean origins' identification	[40]
Fourier transform infrared (FTIR) spectroscopy	FTIR and multispectral imaging (MSI) coupled with support vector machines (SVM) for regression	Meat quality assessment, specifically minced pork patties stored under modified atmosphere packaging (MAP) conditions, by estimating the microbial population	[41]
Raman spectroscopy	A single convolutional neural network (CNN) model development where hyperparameters, activity functions, and loss functions were optimized	Spectral data preprocessing simplification	[42]
Dielectric spectroscopy	Principal component analysis (PCA) for preprocessing and four models, namely support vector machine—SVM, K-nearest neighbor—KNN, linear discriminant—LD and quadratic discriminant—QD, for classification purposes	Discrimination between three citrus juices in order to develop new technologies to identify adulteration	[43]

3. Methodology

Writing a quality review paper is a crucial step in one's research project; it helps in clarifying the state of knowledge, explaining apparent contradictions, identifying needed research, and creating a consensus where none existed before [44]. In the literature, there are three major types of review papers: comprehensive (including systematic reviews), semisystematic, and integrative, depending on its purpose, research questions, search strategy, and data analysis. However, they all generally aim at resolving conceptual ambiguities by providing an integrated, synthesized overview of the current state of the art, as well as presenting research insights, existing gaps, and future research directions [45]. A good literature review must offer both depth and rigor; hence, demonstrating an appropriate strategy for selecting articles and capturing data and insights is crucial [46]. To this end, we propose our own methodology for a comprehensive review: *TriScope Keywords-based Synthesis*, which can be further generalized to a *MultiScope Keywords-based Synthesis* methodology, as represented in Figure 1. The following section describes in detail the process and methodology that we developed to execute this study, taking into consideration the guidelines of writing a review paper [45,46].

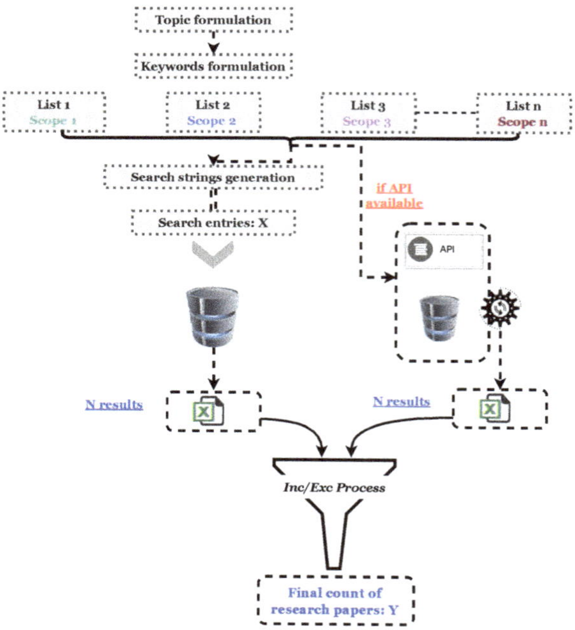

Figure 1. Generalized representation of our *MultiScope Keywords-based Synthesis* methodology.

3.1. Topic Formulation

This study reviews the current advancements in incorporating AI tools to address challenges associated with analytical procedures and advanced chemical methods. These approaches are applied to prevent food contamination within the food supply chain. For this matter, we propose the following questions:

- What are the analytical strategies that are mainly used for crop-food safety and which techniques were dominantly incorporated?
- What are the AI-based tools that were embodied to ensure crop-food safety?
- Did these AI-based systems prove to be beneficial in research and industry? To what degree they proved to be explainable or/and interpretable?

Depending on the type of questions and the targeted scientific audience, the collected papers are analyzed in different manners, either pursuing a qualitative or quantitative analysis. In our case, and while this topic combines three different fields of research, we focused on both quantitative and qualitative analysis, in order to allow people from either fields to comprehend the general outline of this work. In order to answer these questions, an efficient and in-depth analysis of the available literature regarding the addressed topic has been generated and fully explained in the following section.

3.2. Study Design

Academic articles are available through a plethora of scholarly databases. Web Of Science (WOS), Scopus, and IEEE Xplore contain, combined, over 254 million records with a variety of influential academic journals, and are highly recognized by the international academic community. They cover a wide variety of topics, where many papers from different publishers are visible, and they allow access to diverse databases, with the possibility of filtering the search based on personalized search criteria. This paper combines the above three databases as the main source of articles. In order to answer our proposed questions related to our topic of research, we followed a thorough search methodology as visualized in Figure 2, considering the three above-mentioned databases for paper retrieval. The whole process of data collection, including keyword identification, search string generation, and inclusion/exclusion, was performed from October 2022 to January 2023.

These processes were conducted using two approaches, one for the WOS and IEEE Xplore databases, which depended mainly on manual search, and an API-based approach for the Scopus database (see Figures 2 and 3). Figure 2 summarizes the whole process of data collection for all three databases. All searches targeted papers of the same timeline, which is from 1 January 2000 to 30 September 2022. We aimed to explore the evolution of this issue over an extended period, allowing for a comprehensive investigation and analysis.

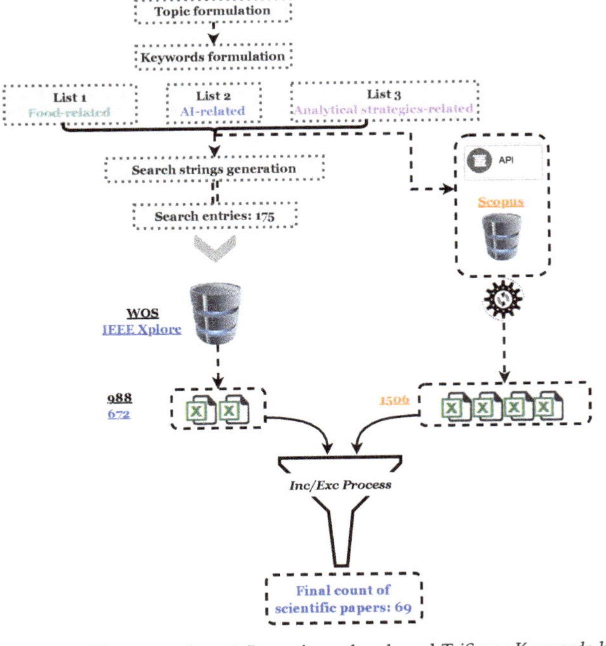

Figure 2. The general workflow of our developed *TriScope Keywords-based Synthesis* for conducting this comprehensive review.

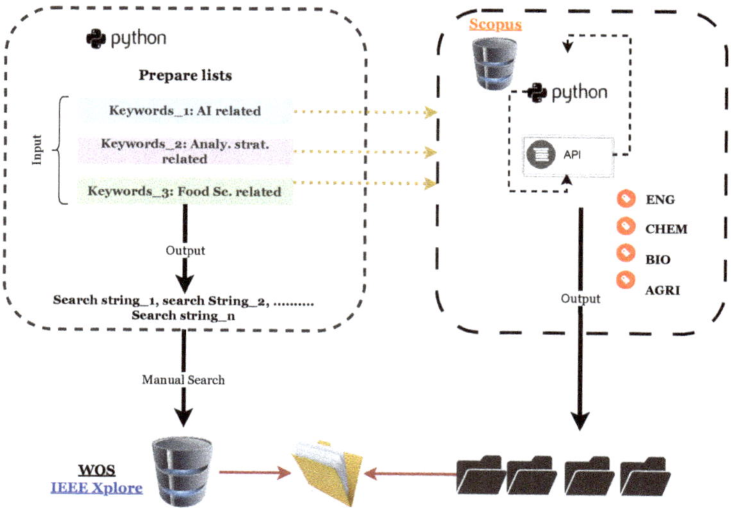

Figure 3. Data collection process, using API for the Scopus databse and manual searching for Web of Science and IEEE Xplore databases.

3.2.1. WOS and IEEE Xplore Data Collection

The first step is the presearch phase. Firstly, we collected the most convenient and in-domain concepts and technical words to describe this topic by investigating the most relevant literature and executing manual preliminary searches on the Google Scholar database, in order to define our keywords. The list of keywords that were used during our search process, divided into three groups:

- Food: food, foodborne, crop, cereal, and toxin.
- Analytical Strategies: analytical strategies, biochemistry, chemical analysis, spectroscopy, omics, immunosensor, and biosensor.
- AI tools: artificial intelligence, machine learning, deep learning, neural networks, and computer vision.

These keywords were then divided into three columns, which represent the three main parts of this research: the first one for AI, the second one for analytical strategies, and the third one for food safety keywords. Secondly, we organized and ran a simple algorithm on Python, which was able to produce all possible combinations between these words from the three columns, to have as an out-put a three-word search string entry, stored in a xlxs. format file. The order of these words isn't important in this case, and the final query string is a combination of three words from three lists of keywords corresponding to food, AI tools and chemical analytical strategies. The final count of these generated search string was 175 query string, written as follows:

$$Search\ String_1 = \text{"}Artificial\ intelligence\text{"} + \text{"}Analytical\ strategies\text{"} + \text{"}cereal\text{"}$$
$$Search\ String_2 = \text{"}Artificial\ intelligence\text{"} + \text{"}Analytical\ strategies\text{"} + \text{"}crop\text{"}$$
$$Search\ String_3 = \text{"}Artificial\ intelligence\text{"} + \text{"}Analytical\ strategies\text{"} + \text{"}food\text{"}$$
$$Search\ String_4 = \text{"}Artificial\ intelligence\text{"} + \text{"}Analytical\ strategies\text{"} + \text{"}foodborne\text{"}$$
$$Search\ String_5 = \text{"}Artificial\ intelligence\text{"} + \text{"}Analytical\ strategies\text{"} + \text{"}toxin\text{"}$$
$$Search\ String_6 = \text{"}Artificialintelligence\text{"} + \text{"}Biochemistry\text{"} + \text{"}cereal\text{"}$$
$$\vdots$$
$$Search\ String_{175} = \text{"}Neural\ networks\text{"} + \text{"}spectroscopy\text{"} + \text{"}toxin\text{"}$$

These search queries were inserted as described previously for each search in the WOS database, but they were converted to another form using the AND Boolean operator and the ALL METADATA filter in order to execute the manual search in IEEE Xplore, which leads to:

$$\begin{gathered} \textit{Search String}_1 = (\textit{“All Metadata”} : \textit{Artificial intelligence})\text{AND} \\ (\textit{“All Metadata”} : \textit{Analytical strategies})\text{AND}(\textit{“All Metadata”} : \textit{cereal}) \end{gathered}$$

Along with these specific keyword combinations, we used database filters to display results based on some specific criteria, which are the timeline (2000–2022) and paper status (only published papers).

In both databases, a total of 175 searches were executed manually between the 30th of October 2022 and the 4th of November 2022 (Figure 3). The outcome of each search resulted in a number of results, ranging between 0–63 results for IEEE Xplore and 0–235 results for WOS. Each search outcome was stored as a link in a xls. format file with its according search string, sorted from the highest to the lowest, in order to eliminate search strings that did not generate any results. These links needed to be briefly inspected in a short period to avoid any research papers being added on the databases. Table 2 highlights the six search strings that provided the highest numbers of results for both databases (more than 30 results).

Table 2. Search strings with the highest number of results (>30) for Web of Science and IEEE Xplore.

IEEE Xplore	WOS
"neural networks" + "spectroscopy" + "food": **63**	"Machine learning" + "spectroscopy" + "food": **235**
"neural networks" + "chemical analysis" + "food": **58**	"neural networks" + "spectroscopy" + "food": **160**
"Artificial intelligence" + "chemical analysis" + "food": **57**	"computer vision" + "spectroscopy" + "food": **111**
"Machine learning" + "spectroscopy" + "food": **55**	"Deep learning" + "spectroscopy" + "food": **77**
"Artificial intelligence" + "spectroscopy" + "food": **50**	"Machine learning" + "spectroscopy" + "crop": **69**
"Machine learning" + "chemical analysis" + "food": **43**	"Artificial intelligence" + "Analytical strategies" + "food": **37**

3.2.2. Scopus Data Collection

To retrieve data from Scopus database, an open access Application Protocol Interface (API) was used in Python. Three string lists were defined, corresponding to Food, Chemical Analytical strategies and AI tools keywords as mentioned before. Subsequently, a list of four subjects was considered, containing Engineering (ENG), Chemistry (CHEM), Biology (BIO), and Agriculture (AGRI). To formulate one specific query, we looped over all possible combinations of keywords formatted in accordance to the standard Scopus advanced search. Then, we retrieved corresponding data for each query in the four considered subjects. Then, we stored the obtained metadata in four .xlsx files corresponding to the considered subjects (see Figure 3).

3.2.3. Inclusion and Exclusion Process

The three databases resulted in a total number of **3166** results as raw data, which went through the same scanning process. The inclusion/exclusion (Inc/Exc) phase consisted of three major subphases, each one bases its outcome on distinguished and precise criteria of selection:

- **Inc/Exc 1:** A first selection based on the relevance of these papers to the topic, after thoroughly reading the title, key words, and abstract sections, respectively. The number of papers included were **30**, **166**, and **242** for IEEE XPLORE, WOS, and SCOPUS databases, respectively.
- **Inc/Exc 2:** A further evaluation of these collected papers was conducted based on their original language and availability online, also removing duplicates, which ended with a total of **109** papers included.
- **Inc/Exc 3:** A full reading process was thoroughly executed to decide which of these papers are the most relevant to our topic of research, leaving **69** papers.

The final count was 69 research papers that have met all the search criteria, and were further organized in MENDELEY https://www.mendeley.com. This software will serve as a tool for storing and analyzing these papers. Figure 4 summarizes in details the process of Inc/Exc phase.

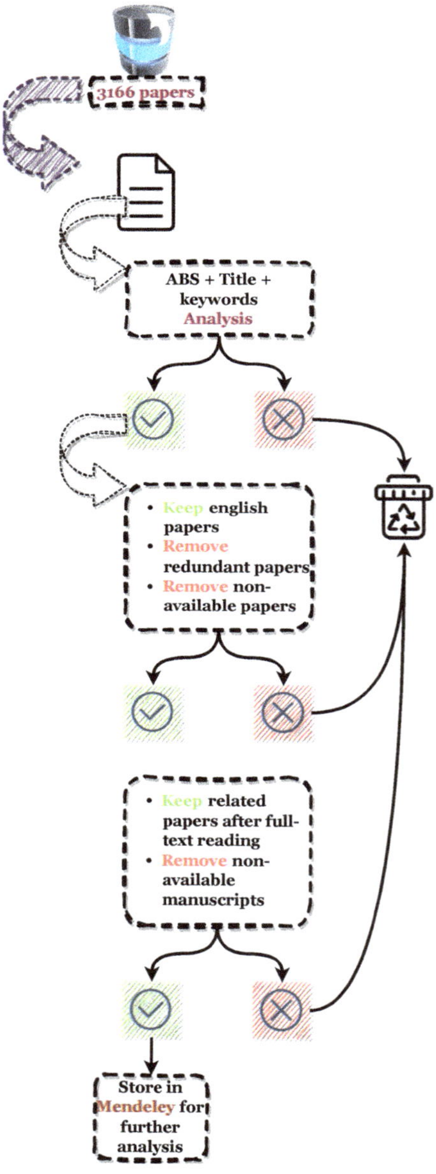

Figure 4. Diagram illustrating the inclusion–exclusion process and criteria of selection.

4. Data Description and Analysis

As discussed in Section 3, our methodology output is 69 papers, including articles and reviews, as detailed in Table 3. Figure 5 also showcases a map of these papers with possible connections and links, highlighting first authors and the according year of publication, as well as numbers of citations for each one. This map was produced using the free online tool ConnectedPapers https://www.connectedpapers.com/, after manually uploading our collected papers. The following data results representation will focus on quantity and quality analysis, as set forth in the upcoming parts.

Table 3. This table represents all collected 69 papers, where A stands for Articles and R stands for Review papers.

Title	First Author	Publisher	Type	Year	Ref.
Detection of defects on selected apple cultivars using hyperspectral and multispectral image analysis	Mehl	ASABE	A	2001	[47]
Differentiation and detection of microorganisms using Fourier Transform infrared photoacoustic spectroscopy	Irudayaraj	Elsevier	A	2002	[48]
Rapid detection of foodborne microorganisms on food surface using Fourier transform Raman spectroscopy	Yang	Elsevier	A	2003	[49]
Differentiation of food pathogens using FTIR and artificial neural networks	Gupta	ASABE	A	2005	[50]
Identification and quantification of foodborne pathogens in different food matrices using FTIR spectroscopy and artificial neural networks	Gupta	ASABE	A	2006	[51]
Applications of Artificial Neural Networks (ANNs) in Food Science	Huang	Taylor & Francis	R	2007	[52]
Supervised pattern recognition in food analysis	Berrueta	Elsevier	R	2007	[53]
Cortical Networks Grown on Microelectrode Arrays as a Biosensor for Botulinum Toxin	Scarlatos	Wiley-Blackwell	A	2008	[54]
Detecting single Bacillus spores by surface enhanced Raman spectroscopy	He	Springer	A	2008	[55]
Self-organizing algorithm for classification of packaged fresh vegetable potentially contaminated with foodborne pathogens	Siripatrawan	Elsevier	A	2008	[56]
Raman Spectroscopy-Compatible Inactivation Method for Pathogenic Endospores	Stöckel	American society for Microbiology	A	2010	[57]
Application of Hyperspectral Imaging in Food Safety Inspection and Control: A Review	Feng	Taylor & Francis	R	2012	[58]
Characterization of food spoilage fungi by FTIR spectroscopy	Shapaval	Wiley-Blackwell	A	2013	[59]
Detection of aflatoxin contaminated figs using Near-Infrared (NIR) reflectance spectroscopy	Güneş	IEEE	A	2013	[60]
Hyperspectral and multispectral imaging for evaluating food safety and quality	Qin	Elsevier	R	2013	[61]
Analytical techniques combined with chemometrics for authentication and determination of contaminants in condiments: A review	Reinholds	Elsevier	R	2015	[62]
Applications of computer vision for assessing quality of agri-food products: a review of recent research advances	Ma	Taylor & Francis	R	2016	[63]
Data mining derived from food analyses using non-invasive/non-destructive analytical techniques; determination of food authenticity, quality & safety in tandem with computer science disciplines	Ropodi	Elsevier	R	2016	[64]
Image analysis operations applied to hyperspectral images for non-invasive sensing of food quality—A comprehensive review	ElMasry	Science Direct	R	2016	[65]
Early Warning Modeling and Application based on Analytic Hierarchy Process Integrated Extreme Learning Machine	Geng	IEEE	A	2017	[66]
Feasibility of Non-Destructive Internal Quality Analysis of Pears by Using Near-Infrared Diffuse Reflectance Spectroscopy	Shen	IEEE	A	2017	[67]
FT-IR Hyperspectral Imaging and Artificial Neural Network Analysis for Rapid Identification of Pathogenic Bacteria	Lasch	American Chemical Society	A	2018	[68]
An Approach for the Development of a Sensing System to Monitor Contamination in Stored Grain	Kaushik	IEEE	A	2019	[69]
Application of Deep Learning in Food: A Review	Zhou	Wiley-Blackwell	R	2019	[70]

Table 3. *Cont.*

Title	First Author	Publisher	Type	Year	Ref.
Machine learning algorithms for the automated classification of contaminated maize at regulatory limits via infrared attenuated total reflection spectroscopy	Öner	Wageningen Academic publishers	A	2019	[71]
Quantitative assessment of zearalenone in maize using multivariate algorithms coupled to Raman spectroscopy	Guo	Elsevier	A	2019	[72]
Raman Spectroscopy Classification of Foodborne Pathogenic Bacteria Based on PCA-Stacking Model	Wan-dan	IEEE	A	2019	[73]
Rapid determination of aflatoxin B1 concentration in soybean oil using terahertz spectroscopy with chemometric methods	Liu	Elsevier	A	2019	[74]
Terahertz Spectroscopy Determination of Benzoic Acid Additive in Wheat Flour by Machine Learning	Sun	Springer	A	2019	[75]
An Overview on the Applications of Typical Non-linear Algorithms Coupled With NIR Spectroscopy in Food Analysis	Zareef	Springer	R	2020	[76]
Application of deep learning and near infrared spectroscopy in cereal analysis	Le	Elsevier	A	2020	[77]
Arcobacter Identification and Species Determination Using Raman Spectroscopy Combined with Neural Networks	Wang	American Society for Microbiology	A	2020	[78]
Deep learning networks for the recognition and quantitation of surface-enhanced Raman Spectroscopy	Weng	The Royal Society of Chemistry	A	2020	[79]
Development of Machine Learning & Edge IoT Based Non-destructive Food Quality Monitoring System using Raspberry Pi	Sahu	IEEE	A	2020	[80]
Emerging techniques for determining the quality and safety of tea products: A review	Yu	Wiley-Blackwell	R	2020	[81]
Machine learning applications to non-destructive defect detection in horticultural products	Nturambirwe	Science Direct	R	2020	[82]
Multi-view Learning for Subsurface Defect Detection in Composite Products: a Challenge on Thermographic Data Analysis	Wu	IEEE	A	2020	[83]
On-line prediction of hazardous fungal contamination in stored maize by integrating Vis/NIR spectroscopy and computer vision	Shen	Elsevier	A	2020	[84]
Optical detection of aflatoxins B in grained almonds using fluorescence spectroscopy and machine learning algorithms	Bertania	Elsevier	A	2020	[85]
Achieving a robust Vis/NIR model for microbial contamination detection of Persian leek by spectral analysis based on genetic, iPLS algorithms and VIP scores	Rahi	Elsevier	A	2021	[86]
Application of Artificial Intelligence in Food Industry—a Guideline	Mavani	Springer	R	2021	[87]
Applications of THz Spectral Imaging in the Detection of Agricultural Products	Ge	MDPI	R	2021	[88]
Bioimpedance data statistical modelling for food quality classification and prediction	Rivola	IEEE	R	2021	[89]
Characterisation and Classification of Foodborne Bacteria Using Reflectance FTIR Microscopic Imaging	Xu	MDPI	A	2021	[90]
Combining optical spectroscopy and machine learning to improve food classification	Magnus	Elsevier	A	2021	[91]
Deep Learning for Rapid Identification of Microbes Using Metabolomics Profiles	Wang	MDPI	A	2021	[92]
Hyperspectral image processing for the identification and quantification of lentiviral particles in fluid samples	Gómez-González	Nature Publishing Group	A	2021	[93]
Identification of the apple spoilage causative fungi and prediction of the spoilage degree using electronic nose	Guo	Wiley-Blackwell	A	2021	[94]
Investigation of nonlinear relationship of surface enhanced Raman scattering signal for robust prediction of thiabendazole in apple	Li	Elsevier	A	2021	[95]
Metaheuristic Optimization to Improve Machine Learning in Raman Spectroscopic based Detection of Foodborne Pathogens	Vakilian	IEEE	A	2021	[96]

Table 3. *Cont.*

Title	First Author	Publisher	Type	Year	Ref.
Microwave Sensing for Food Safety: a Neural Network Implementation	Ricci	IEEE	A	2021	[97]
Non-destructive detection of foreign contaminants in toast bread with near infrared spectroscopy and computer vision techniques	Yin	Springer	A	2021	[98]
Raman spectroscopy combined with machine learning for rapid detection of food-borne pathogens at the single-cell level	Yan	Elsevier	A	2021	[99]
Recent advances in assessing qualitative and quantitative aspects of cereals using nondestructive techniques: A review	Zareef	Elsevier	R	2021	[100]
Trace Identification and Visualization of Multiple Benzimidazole Pesticide Residues on Toona sinensis Leaves Using Terahertz Imaging Combined with Deep Learning	Nie	MDPI	A	2021	[101]
A Novel Method for Carbendazim High-Sensitivity Detection Based on the Combination of Metamaterial Sensor and Machine Learning	Yang	MDPI	A	2022	[102]
Advances in Machine Learning and Hyperspectral Imaging in the Food Supply Chain	Kang	Springer	R	2022	[103]
Component spectra extraction and quantitative analysis for preservative mixtures by combining terahertz spectroscopy and machine learning	Yan	Elsevier	A	2022	[104]
Design of Food Safety Supervision System in the Background of Big Data	Zhang	IEEE	R	2022	[105]
Detection and quantification of peanut contamination in garlic powder using NIR sensors and machine learning	Rady	Academic Press Inc.	A	2022	[106]
Domain Adaptation for In-Line Allergen Classification of Agri-Food Powders Using Near-Infrared Spectroscopy	Bowler	MDPI	A	2022	[107]
Investigation of reflectance, fluorescence, and Raman hyperspectral imaging techniques for rapid detection of aflatoxins in ground maize	Kim	Elsevier	A	2022	[108]
Low-Resolution Raman Spectroscopy for the detection of contaminant species in algal bioreactors	Adejimi	Elsevier	A	2022	[109]
Machine learning-based typing of Salmonella enterica O-serogroups by the Fourier-Transform Infrared (FTIR) Spectroscopy-based IR Biotyper system	Cordovana	Elsevier	A	2022	[110]
Markov Transition Field Combined with Convolutional Neural Network Improved the Predictive Performance of Near-Infrared Spectroscopy Models for Determination of Aflatoxin B1 in Maize	Wang	MDPI	A	2022	[111]
Potential application of hyperspectral imaging in food grain quality inspection, evaluation and control during bulk storage	Aviara	Elsevier	R	2022	[112]
Recent Advances and Applications of Rapid Microbial Assessment from a Food Safety Perspective	Pampoukis	MDPI	R	2022	[113]
Recent Progress in Spectroscopic Methods for the Detection of Foodborne Pathogenic Bacteria	Hussain	MDPI	R	2022	[114]
Spectroscopy and imaging technologies coupled with machine learning for the assessment of the microbiological spoilage associated to ready-to-eat leafy vegetables	Manthou	Elsevier	A	2022	[115]

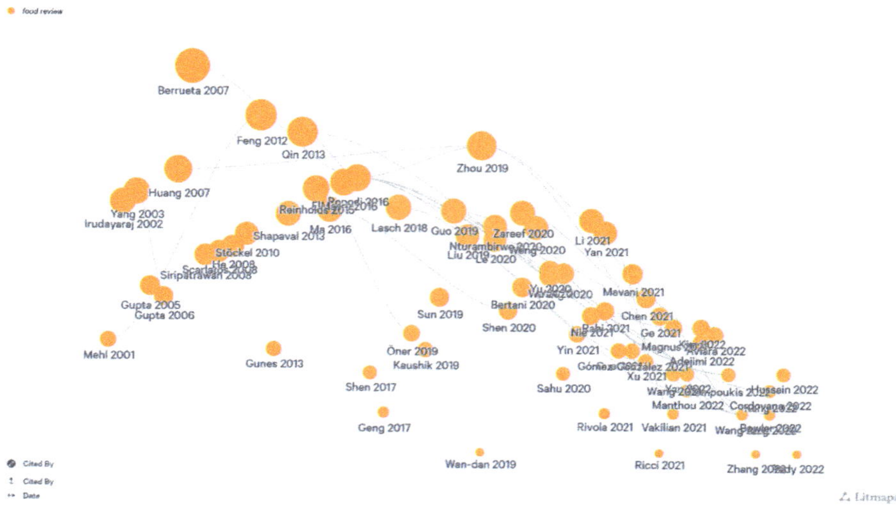

Figure 5. A map of all our collected papers. Each node represents a paper with its first author and year of publication. The size of a node represents the number of citations. Please refer to Table 3 for more details.

4.1. *Quantitative Analysis*

Although we explored a larger period of time, from 1 January 2000 to 30 September 2022, most of our collected papers were between 2019 and 2022. As shown in Figure 6, these papers were generally published in 14 different publishers, which are well classified globally and highly recognised.

Most of the journals are highly ranked, with a quartile of 1 (Q1), which are divided as article papers and review papers; some of these papers, mainly those published by IEEE, are conference papers, as in Figure 7.

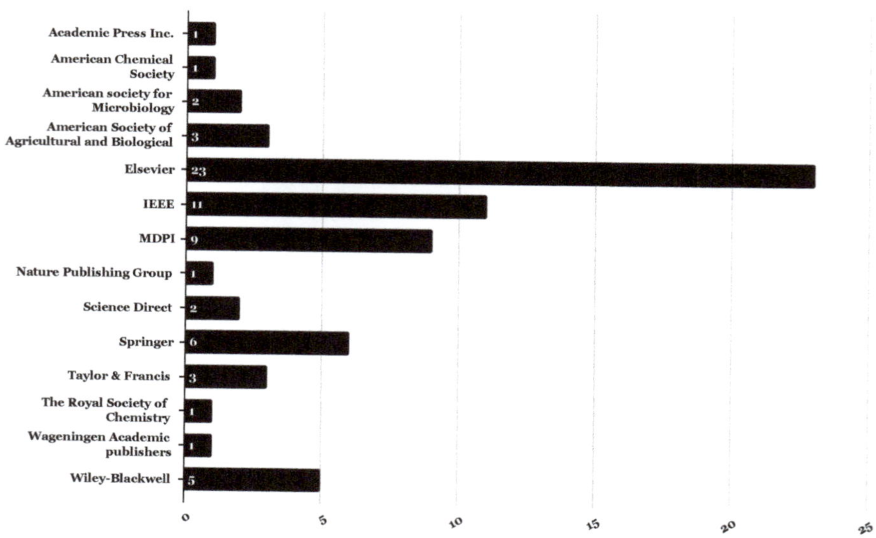

Figure 6. Count of published papers per publisher name.

Journal	Papers
Analyst	1
Analytical Chemistry	1
Applied Engineering in Agriculture	1
Applied and Environmental Microbiology	2
Biosensors	1
Biosystems Engineering	2
Comprehensive Reviews in Food Science and Food Safety	2
Critical Reviews in Food Science and Nutrition	3
Food Chemistry	3
Food Control	3
Food Engineering Reviews	3
Food Measurement and Characterization	1
Food Process Engineering	1
Foods	1
International journal of Molecular Science	1
International journal of food Microbiology	1
Journal of Agriculture and Food Research	1
Journal of Applied Microbiology	1
Journal of Chromatography A	1
Journal of Food Composition and Analysis	2
Journal of Food Engineering	1
Journal of Food Science	1
Journal of Infrared, Millimeter, and Terahertz Waves	1
Journal of Microbiological Methods	1
Journal of Molecular Structure	1
Materials	1
Metabolites	1
Molecular Structure	1
Molecules	1
Photonics	1
Postharvest Biology and Technology	1
Proceedings - 2013 International Conference on Electronics, Computer and Computation (ICECCO)	1
Proceedings - 2017 Intelligent Systems Conference (IntelliSys)	1
Proceedings - 2017 International Conference on Modelling, Identification and Control (ICMIC)	1
Proceedings - 2019 International Conference on Intelligent Informatics and Biomedical Sciences (ICIIBMS)	1
Proceedings - 2019 International Conference on Signal Processing and Integrated Networks (SPIN)	1
Proceedings - 2020 International Conference on Computing, Power and Communication Technologies (GUCON)	1
Proceedings - 2021 IEEE Conference on Antenna Measurements & Applications (CAMA)	1
Proceedings - 2021 International Conference of Signal Processing and Intelligent Systems (ICSPIS)	1
Proceedings - IEEE International Conference on Artificial Intelligence in Everything	1
Proceedings - IEEE International Symposium on Circuits and Systems	1
Science of total Environement	1
Scientific Reports	1
Sensing and Instrumentation for Food Quality and Safety	1
Sensors	2
Sensors and Actuators B chemical	1
Spectrochimica Acta - Part A: Molecular and Biomolecular Spectroscopy	2
Talanta	1
Transactions of the ASABE	2
Transactions on Industrial Informatics	1
Trends in Food Science and Technology	2
Vibrational Spectroscopy	1
World Mycotoxin Journal	1

Figure 7. A table chart depicting the number of papers published per journal.

On the other hand, Figure 8 highlights the distribution of article papers only, in a color-coded format; each color refers to a field of study. Although our keywords were mainly extracted from three different fields of research, after the paper collection, another distribution emerged. These fields are Artificial Intelligence (including computer vision, machine learning, etc.), Analytical Strategies, Biological Sciences, and Food Sciences.

We will discuss these papers' contents while trying to answer our research questions (RQs), which are considered as the qualitative analysis of our results. Firstly, we divide the discussion into two parts, one for review papers and all related types, and the second one for article papers with experiments and applications. Then, the outcome from discussing these review papers will help as a reference while discussing the articles.

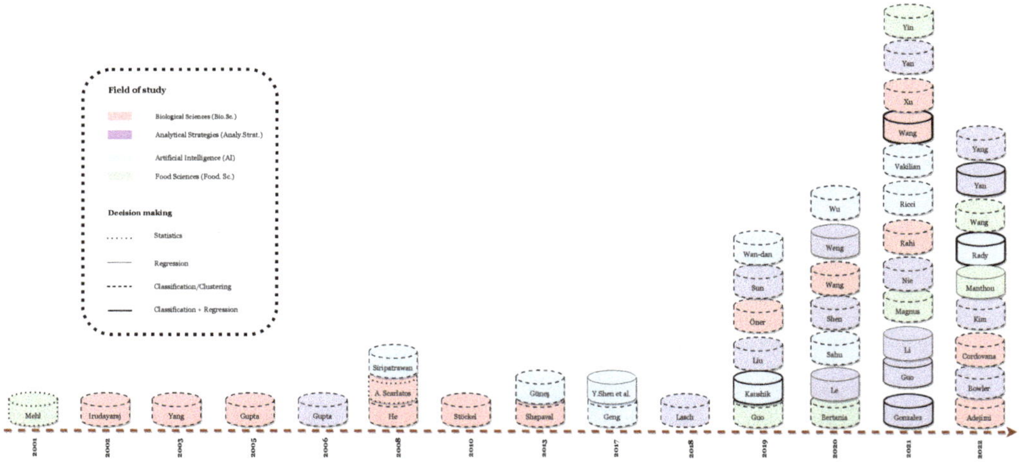

Figure 8. A diagram representing the chronological (from 2001 to 2022) distribution of the resulting articles. Each data-point represents an article, the colors are according to which field of study, and the lines are related to the decision-making objective Mehl, 2001 [47]; Irudayaraj, 2002 [48]; Yang, 2003 [49]; Gupta, 2005 [50]; Gupta, 2006 [51]; He, 2008 [55]; A. Scarlatos, 2008 [54]; Siripatrawan, 2008 [56]; Stöckel, 2010 [57]; Günes, 2013 [60]; Shapaval, 2013 [59]; Geng, 2017 [66]; Y.Shen, 2017 [67]; Lasch, 2018 [68]; Guo, 2019 [72]; Kaushik, 2019 [69]; Liu, 2019 [74]; Öner, 2019 [71]; Sun, 2019 [75]; Wan-dan, 2019 [73]; Bertania, 2020 [85]; Le, 2020 [77]; Sahu, 2020 [80]; Shen, 2020 [84]; Wange, 2020 [78]; Weng, 2020 [79]; Wu, 2020 [83]; Gonzalez, 2021 [93]; Guo, 2021 [94]; Li, 2021 [95]; Magnus, 2021 [91]; Nie, 2021 [101]; Rahi, 2021 [86]; Ricci, 2021 [97]; Vakilian, 2021 [96]; Wang, 2021 [92]; Xu, 2021 [90]; Yan, 2021 [99]; Yin, 2021 [98]; Adejimi, 2022 [109]; Bowler, 2022 [107]; Cordovana, 2022 [110]; Kim, 2022 [108]; Manthou, 2022 [115]; Rady, 2022 [106]; Wang, 2022 [111]; Yan, 2022 [104]; Yang, 2022 [102].

4.2. Qualitative Analysis

Over the last decade, there has been a rapid development of artificial intelligence (AI) tools for the nondestructive evaluation of food and agricultural products. However, a key aspect of the emerging AI tools is the preprocessing phase, where acquired data are adequately prepared for further analysis and decision making using, generally, machine learning models. Through our analysis process of our data papers, new keywords and concepts emerged, for a diversity of techniques and technologies, regarding analytical strategies where spectroscopy and chemometrics were the most used in articles and discussed in review papers (see Figure 9). The commonality among these collected papers is their focus on exploring innovative approaches and technologies applied to food sciences. Their aim is to solve challenges related to food safety, quality assessment, authenticity verification, and inspection processes. In the following sections, we will discuss articles and review papers separately, following the three RQs defined in the Methodology section.

4.3. Review Papers

The review papers report recent advances regarding the application of various AI technologies and methodologies in the field of food sciences. These papers discuss the use of different techniques in AI, such as artificial neural networks (ANNs), computer vision, data mining, deep learning, machine learning, and statistical modeling combined with recent analytical strategies, such as hyperspectral imaging, chemometrics, nondestructive analytical techniques, THz spectral imaging, bioimpedance and spectroscopic methods for food analysis, quality assessment, safety inspection, authenticity determination, and defect detection. Assessments of food products are divided into two pillars: safety and quality parameters, and since our concern is on safety, we focused on extracting information related to this matter, specifically for crop-based food.

Figure 9. A map highlighting subject-related words and concepts that were strongly present within the literature, such as Spectroscopy (Graphical design generated using free-online tool Wordart https://wordart.com/create, accessed on 23 August 2023).

4.3.1. Analytical Strategies and AI as Nondestructive Tools for Crop-Food Safety

These review papers are commonly discussing the usage of analytical techniques and consider them as revolutionary, mostly for being nondestructive Techniques in food analysis, where the key word "nondestructive" was stated plainly in these works [62–64,76,82,88,89,100,112]. These techniques allow for the assessment of crop-based food safety without structural damage. The overall focus of these papers focuses on both aspects of food analysis: safety and quality [52,61,70,112]. Nonetheless, we will inspect only technologies and methodologies that have been implemented in food safety analysis, thus detecting contaminants, assessing microbial risks, determining the presence of pathogens, and evaluating their safety and readiness to be manufactured, stored, transported, or directly consumed. Many examples emerged, such as reviewing methods for determination of contaminants in spices and herbs [62], such as biological contaminants [113,114], caused by foodborne pathogenic microorganisms. These samples are subject to analytical analysis, which then, depending on the objective of the study and which aspect of food safety to tackle, will produce data.

The most referred-to analytical strategies in solving this matter were spectral imaging (SI) technologies [61,113], and spectroscopy-based technologies [52,61,63,64,76,81,82,88,89,100,113,114].

Different techniques have emerged in spectral imaging, like hyperspectral imaging (HSI) [58,61,65,70,82,103,112], for example, near-infrared hyperspectral imaging [112], as well as multispectral imaging [58,61,64,65,113] and THz spectral imaging [88]. In spectroscopy, we list ultraviolet-visible spectroscopy (UV-vis), near-infrared spectroscopy, Fourier transform infrared (FT-IR) spectroscopy, Raman spectroscopy, THz spectroscopy, THz time-domain spectroscopy(THz-TDS), laser-induced breakdown spectroscopy (LIBS), surface-enhanced Raman spectroscopy sensors (SERS), nanoenzymes, and modified chromatographic techniques that were mostly coupled with AI tools [76,81,88,100,112–114]. On the other hand, electroanalytical methods have risen to mimic the human senses by using sensor arrays and pattern recognition systems, known as E-nose, E-tongue, and E-eye [81]. They are often combined with other spectral techniques for real-time detection and higher accuracy results, such as near-infrared spectroscopy (NIRS) [87]. The E-nose is destined to emulate the human olfactory system, hence avoiding exposure to dangerous chemical and biological hazards due to inhaling or skin leisure [82]. Many applications have been studied, like evaluating shelf life by detecting grain off-odor due to microbial spoilage for wheat and barley-based foods [52]. Furthermore, electrochemical impedance spectroscopy (EIS) has been receiving more attention research lately, especially when applied on biological tissues, such as crop-based food membranes. EIS or bioimpedance technology is optimal for surface contaminant detection, with its capacity to detect defects [89].

Several papers discuss the importance of data analysis and processing for food safety. Chemometrics [81,100], machine learning [87,88], computer vision [63,113], and data mining [64,70] are applied to extract meaningful information, detect patterns, and make predictions based on the collected data. Machine learning techniques are summarized in Table 4, with the most common techniques gathered from some relevant review papers related to crop-food safety.

Table 4. Chemometrics and ML approaches for crop-based food safety and quality in some relevant review papers.

Ref.	Crop-Food	Preprocessing Steps	Decision Model
[81]	Tea	Standard normal variate (SNV) and multiplicative scatter corrections	PCA, KNN, KPCA, ANN, HCA, BPNN, PLS, CPNN, SPA, PNN, ELM, LDA, SVM, S-LDA, LVQ, KLDA, MLP, RBF, RF
[103]	Apple, wolfberry, lettuce, pear, green plum, peach, strawberry, Brassica, jujube, lettuce and chives	LLE, LE, ISOMAP and MDS, SNE and t-SNE	PLSR, KNN, LDA, NB, DT, SVR, SVM, RF, LSSVM, LWR, FNN, ResNet, CNN, and DNN
[76]	Grain products, forages, oil, fruits, vegetables, sugarcane seeds, coffee, tea, spices, black/green tea, grapes, apples, wheat flour, rice and barley	Multivariate calibration of spectral data, standard normal variate transformation (SNV), multiplicative scatter correction (MSC), smoothing, derivative, wavelet transforms (WT), and orthogonal signal correction (OSC)	ANN, BP-ANN, GA-ANN, RBFNN, AdaBoost, SVM, LA, ELM,SLFN, LS-SVR, SVM, linear, radial basis function (RBF), normalized polynomial, sigmoid, Gaussian RBF, and string kernels
[62]	Condiments, spices, and herbs	PCA, HCA, parallel factor analysis (PARAFAC), MPLS, PLS, ANOVA, t-test, straight line subtraction (SLS), constant offset elimination (COE), and minimum–maximum normalization (MMN)	ANN, kNN, PLS regression, PLS-DA, HCA, PCA, LDA, k-means cluster analysis (KM-CA), and DA
[87]	Fruits	computer vision systems	Adaptive Neuro Fuzzy Inference System (ANFIS)
[70]	Vegetables	CNN	ResNet-152, AlexNet-SVM classifier, and hybrid CNN-SSAE
[58]	Apples, cucumbers, spinach, and wheat	PCA, PLSR	PLSDA, ANN, LD, and PCA

The best description of the purpose of chemometrics was in a recent work by Yu et al. [81], quoting exactly "Chemometrics, the art of extracting chemically relevant information from data produced in chemical experiments". We can also distinguish between chemometrics and sensometrics, respectively, studying the relationship between the measured chemical parameters and the state of the object by statistical or mathematical methods or studying the link between sensory parameters and the internal characteristics of the object via similar approaches as the first one [70]. Spectra processing and model establishment are two main aspects of chemometrics; each contains multiple analysis methods [100]. These systems that are based on sensors equipped with chemosensitive materials for molecular recognition [113] can also be joined with other techniques for data analysis in electronic sensing [81]. Chemometrics were usually incorporated within data analysis in electronic, E-nose, E-tongue, and E-eye [100] and within spectra processing and model establishment when combining AI tools with spectroscopic techniques. There is application potential of deep learning in chemometrics and sensometrics toward food [70], for example, real-time toxins quantification and detection in cereals by combining nondestructive techniques with chemometrics, such as spectroscopy-based techniques [100]. They were also joined when applied in condiment analysis methods for the determination of contaminants in spices and herbs [62]. Final data acquired by HSI systems were described as "hypercube" for their 3D characteristics: two for spatial coordinates and one for spectral values [58].

4.3.2. Interdisciplinary Approaches in Crop-Food Safety

The review papers often involve interdisciplinary approaches, combining knowledge and techniques from fields such as food science, computer science, spectroscopy, image analysis, and statistical modeling. These collaborations aim to leverage the strengths of different disciplines to address complex challenges in food analysis and safety. Firstly, image correction before implementing chemometric algorithms is necessary to reduce the noise to enhance the signal-to-noise ratio by either spectral or image analysis, Secondly, data preparation and preprocessing use chemometric methods, such as principal component analysis (PCA) and partial least square regression (PLSR), to be fed for model application [58]. Microbiological applications, for mycotoxin detection, for example, include combining nondestructive methods with statistical methods such as PLS at different stages of the food supply chain; harvesting and transporting, due to many climate conditions and agricultural processes [62]; and the detection of the total aflatoxin content using fluorescence fingerprints (FF) in combination with PLSR [62]. FT-NIR data were often analyzed using the PLS regression with various preprocessing techniques, such as straight-line subtraction (SLS), constant offset elimination(COE), and minimum–maximum normalization (MMN). The classification of hyperspectral images of possibly contaminated chili was proposed by using PCA and SANN as a classifier [62]. Supervised ML approaches (classification and regression) are used for the detection of endogenous component content, pesticide residue, microplastic, and heavy metal contamination in crops [103]. THz-TDS can be combined with SVMs for classification purposes, scoring good results [81]. However, this work can be criticized for mentioning various spectral dimensionality reduction techniques for feature extraction and selection, like LLE, LE, ISOMAP, and MDS in addition to SNE and t-SNE, whereas those techniques are not mentioned in the literature in any shape or form [103]. PCR, PLS, and MLR are best to check the linearity, while qualitative techniques are best for non-linearity detection, such as partial residual plot (PRP), residual plot (RP), e-PC (AVP/PaRP), and Mallows' augmented partial residual plot (APaRP). On the other hand, quantitative techniques are mostly referred to for statistical methods such as ANOVA for lack of fit (LOF) for univariate calibration mode, or the most basic tests to assess serial correlation: the Durbin–Watson test and run test [103]. The fuzzy logic technique is used in the food industry, in food modeling, control, and classification, and in addressing food-related problems by managing human reasoning in linguistic terms. FL has been proven to successfully maintain the quality of the foods, and it acts as a prediction tool and control system for food production processes [87]. Deep learning has been introduced into the food field by analyzing RGB images and spectra images of food, as the data analysis tool to solve the problems and challenges in the food domain. Food recognition and classification, such as CNN and image analysis, has been the most commonly used pattern in food recognition and classification. Numerous popular CNN architectures for image processing were cited, including AlexNet, the visual geometry group network (VGG), GoogLeNet, and the residual neural network (ResNet), as well as the usage of hybrid models, such as radial basis function (RBF) kernel-based SVM with ResNet-152, coupling fine-tuned AlexNet with a binary SVM classifier [70].

4.3.3. Analytical Strategies and Recent Technologies for Food Safety

Some papers highlight the use of automation and advanced technologies to streamline food analysis processes, improve efficiency, and enable rapid assessment. This includes the integration of computer vision, machine learning, and artificial intelligence techniques for automated defect detection and quality evaluation. Table 4 highlights some relevant reviews reporting chemometrics and ML applications for food safety and quality.

Analytical strategies deploying AI tools can be described as knowledge-based systems with computer programs using knowledge from different sources for food safety problems, such as expert systems, knowledge-based artificial intelligence, and knowledge-based engineering [87]. Integrating various sensors with AI-based methods has been increasing in food industries over the past few years, for example, E-nose-based systems to detect defects and contamination in crop foods. The classification and differentiation of different fruits have also been determined by using E-nose. CVS usage can be a handful for image processing and pattern recognition combined with near-infrared spectroscopy (NIRS) in the food industry for accurate and precise results [87]. Combining ANNs-sensors in real-time applications, like E-nose and E-tongue for real-time detection, scores faster and higher accuracy results [87]. Fecal contamination on fruits and vegetables—apples, for example—using reflectance-fluorescence, is coupled with band ratio, threshold, separation algorithms, and PCA. Defects like bruises and lesions are considered symptoms of possible contamination if caused by a biological agent. Hyperspectral imaging is then used to inspect defects coupled with many tools, such as PLSDA, ANN, LDA, and PCA [58].

Overall, the common points among these review papers revolve around the application of emerging technologies, nondestructive techniques, data analysis, food safety, quality assessment, automation, and interdisciplinary approaches within the field of food science.

4.4. Article Papers

As depicted in Figure 8, the obtained 49 research papers focus on the identification, and in some cases quantification, of food-borne pathogens, be they bacteria, fungi, or yeast. The employed machine learning (ML) workflow mostly consists of two major steps, namely *data preparation*, including data acquisition, data cleaning, and data transformation; and *decision making* including, tuning, training, and testing of one or multiple machine learning models.

The results are best categorized by how data were gathered. However, in most cases, data acquisition stems from a chemical analytical approach, mostly spectroscopy-based approaches, resulting in *structured data* in the form of numerical values or *unstructured data* in the form of images. Thus, the results are reported following the taxonomy outlined in Section 2.1. Each subsection that follows gathers the research papers that use the specific spectroscopic approach, and its variants if any, combined with various ML methods for food safety purposes.

4.4.1. Raman Spectroscopy

In earlier studies, Raman spectroscopy was used in combination with classical machine learning approaches to enhance food pathogen detection. In [57], principal component analysis (PCA) was applied in combination with linear discriminate analysis (LDA), support vector machines (SVM), and a simple artificial neural network (ANN) architecture for classification of *Bacillus anthracis* endospores and their products. Also, Raman Spectroscopy spectra were preprocessed using a first derivative (D1) and Savitzky–Golay smoothing function and second (2)-order polynomial to enhance the spectra's resolutions and eliminate baseline and linear slope effects, respectively [72]. The resulting signal was used as input for partial least squares (PLS), synergy interval partial least squares (siPLS), and ant colony optimization-siPLS (ACO-siPLS) to quantify the level of maize contamination by zearalenone. On the other hand, after applying the Savitzky–Golay smoothing function on the spectral information, a stacking approach combined with K-nearest neighbors (KNN) and support vector machines (SVM) was proposed in [116] to distinguish between the *Escherichia coli* and *Brucellasuis* vaccine. Later studies investigated the use of complex ANN-based models to enhance classification and quantification performances. In [78], a simple PCA approach was used to categorize three genera (Arcobacter, Campylobacter, and Helicobacter), then a one-dimensional convolution neural network (1D-CNN) and a fully connected artificial neural network (ANN) were used to classify 18 of Arcobacter species and quantify their ratio in the bacterial mixture [78]. In another study, raw Raman spectroscopy was used as an input to the fully connected ANN, genetic algorithm-ANN (GA-ANN), and particle swarm optimization-ANN (PSO-ANN) to distinguish between 12 strains of five bacteria genera, including (*Escherichia, Listeria, Vibrio, Shigella,* and *Salmonella*) [96].

In addition, in the cases where faster sampling, higher resolution, and better signal-to-noise ration are needed, Fourier transform (FT)-Raman is often used with promising results. In an older work, ref. [49] proposed a first attempt to use FT-Raman for a nondestructive characterization and differentiation of six different microorganisms, including the pathogen *Escherichia coli* on whole apples. PCA and canonical variate (CV) plot and cluster the data in two-dimensional scatter-plot.

Furthermore, a more sensitive version of Raman spectroscopy, namely surface-enhanced Raman spectroscopy (SERS), can be used in cases where sensitivity, selectivity, and rapid analysis are needed. In an older work [55], SERS was used to detect and discriminate among five Bacillus spores (*B. cereus, B. cereus, B. cereus, B. subtilis,* and *B. stearothermophilus*). Standard data preparation techniques were conducted, such as normalization, binning, smoothing, and second-derivative transformation, before using PCA and hierarchical cluster analysis (HCA) to cluster the data into three groups. However, a recent comparative study [79] demonstrates the benefit of using deep learning models such as CNNs, fully convolutional networks (FCN), and principal component analysis networks (PCANet) to determine their abilities to measure pirimiphos-methyl in wheat extract in the two input forms of one-dimensional vector or two-dimensional matrix, as opposed to classical ML methods such as random forests (RF), KNN, and SVM.

On the other hand, single-cell Raman spectroscopy (SCRS) has also been used in a recent work [99] to discriminate between 23 common strains from seven genera of food-borne bacteria (*Escherichia, Listeria, Staphylococcus, Cronobacter, Vibrio, Shigella,* and *Salmonella*) at the single cell level. Kernel PCA (KPCA) was used for nonlinear feature extraction, followed by a decision tree (DT) algorithm, with promising results.

In addition, low-resolution Raman spectroscopy (LRRS) has also been used in combination with SVM discriminate analysis to rapidly identify harmful cyanobacterial species and quantify their presence, demonstrating the potential of LRRS technology for the real-time detection of contaminant species within microalgal bioreactors.

4.4.2. Visible and Near-Infrared Spectroscopy

Visibe/near-infrared spectroscopy (Vis/NIR) is of valuable usage for food safety, especially when a broad spectral range analysis is needed. It has been used in combination with LDA and partial least square regression (PLSR) for online detection of *Aspergillus* and *Fusarium* contamination in stored maize [86]. LDA was used for classification results of maize samples according to species of infected fungal strain and infection level. PLSR was used for the respective determination of colony counts. Also, in [84], Vis/NIR was considered with PCA in combination with various data preprocessing techniques, such as moving average (MA), multiplicative scatter correction (MSC), standard normal variate (SNV), first derivative (D1), second derivative (D2), and Savitzky–Golay (SG), and their combinations were considered based on PCA results. The prepared data were the input to the PLS-DA algorithm, as well as three of the variable selection-based variants (iPLSDA, GA-PLSDA, and VIP-PLSDA), to detect contaminants in Persian leek.

On the other hand, recent studies explored the use of near-infrared spectroscopy (NIR) coupled with computer vision in food safety applications. In [106], they used low-cost NIR sensors (NIRONE S2 and S2.5) to acquire spectra with wavelengths 1550–1950 nm and 2000–2450 nm. Standard preprocessing approaches were applied, such as autoscaling, mean centering, Savitzky–Golay, standard normal variate (SNV), and first and second derivatives. The objective is garlic powder contamination, especially allergen contamination such as peanut powder. Thus, five classification models were used (KNN, SVM, LDA, PLS-DA, and DT), to first detect the presence of peanut powder in garlic powder, then identify whether the presence was high (2–20%) or low (0.01–1%). In the latter two cases, a PLSR-based regression model was developed to quantify the presence concentration. Similarly, a deep learning-based approach was proposed in [107] for the in-line allergen classification of agrifood powders, combining domain-adversarial neural networks (DANN) and semisupervised generative adversarial neural networks (SGANN). In addition, two DL approaches were proposed in [98,111], namely a two-dimensional Markov transition field-CNN (2D-MTF-CNN) and a modified version of FCN (U-net) to monitor the aflatoxin B1 (AFB1) content in maize and identify food foreign contaminants (metallic iron, polypropylene plastic, and hair) on the surface of bread, respectively.

NIR spectroscopy in the diffuse reflectance mode has been used [67] for soluble solid content and total acid content analysis of fruits. After reducing the dimension of the acquired data to the leading three principal components, the backpropagation neural network (BPNN) and generalized regression neural network (GRNN) were proposed and compared in predicting the values of SSC and TAC in three cultivars of pears.

Furthermore, earlier works [50,51] investigate the usage of Fourier transform infrared spectroscopy (FTIR) coupled with complex ANN methods to identify five pathogenic bacteria (*Enterococcus faecium*, *Salmonella* Enteritidis, *Bacillus cereus*, *Yersinia enterocolitis*, and *Shigella*) [50], as well as to quantify four food pathogens (*E. coli O26*, *Salmonella* Typhimurium, *Yersinia enterocolitica*, and *Shigella boydii*) [51]. In both works, probabilistic neural networks (PNNs) were developed and tested. FTIR has been used for aflatoxin contamination detection in figs [60]. The forward feature selection (FFS) method was used to reduce features in spectra space, followed by three linear classifiers, namely the linear discriminant classifier (LDC), logarithmic linear classifier (LOGLC), and quadratic discriminant classifier (QDC). Also, it was coupled with smoothing and standard normal variate (SNV) for noise removal, followed by SVM and PLSDA to characterize and differentiate between Bacillus subtilis and Escherichia coli cell suspensions in food spoilage context [90]. The authors of [71] have also showed the benefits of using FTIR combined with five classification approaches, namely adaptive boosting (AdaBoost), random forests (RF), SVM, and multilayer perceptrons (MLPs) for the automated classification of contaminated maize. In a latter study [110], the FTIR spectroscopy-based IR biotyper system was used to successfully classifying a total of 958 characterized Salmonella enterica isolates (25 serogroups and 138 serovars). PCA and LDA were applied for visualization porpuses, followed by SVM and ANNs for classification. Also, an earlier attempt to combine FTIR photoacoustic spectroscopy (FTIR-PAS) with unsupervised analysis (PCA and canonical variate analysis) was explored in [48] to identify various bacteria (*Lactobacillus casei*, *Bacillus cereus*, and *Escherichia coli*), fungi (*Aspergillus niger* and *Fusarium verticilliodes*) and yeast (*Saccharomyces cerevisiae*) on apple surface, with promising results. However, to the knowledge of the authors, no subsequent research was conducted using this technique.

4.4.3. Time-Domain Spectroscopy (THz–TDS)

The time-domain spectroscopy (THz–TDS) system can also be used in conjunction with ML approaches for food safety. Due to its spectral range, sensitivity, and chemical specificity, it is an appealing alternative for data acquisition. In [102], they proposed a novel metamaterial sensor to analyze multiresonance dips in spectra obtained using time-domain spectroscopy (THz–TDS) coupled with mean shift (MS) to investigate the presence of Carbendazim's residue in crops. Also, soybean oil contamination by *aflatoxin B1* (AFB1) was investigated using the same spectroscopic method [74]. Student stochastic neighborhood embedding (t-SNE) dimensionality reduction (DR) methods were applied, followed by the backpropagation neural network (BPNN), least squares SVM (LS-SVM), and RF for classification purposes. As opposed to identification and classification problems, time-domain spectroscopy (THz–TDS) systems were used in tandem multiple linear regressions (MLR), PLS, and LS-SVM for measuring benzoic acid (BA) additives in wheat flour [75]. In a recent study, [104] used TPS systems with singular value decomposition (SVD), non-negative matrix factorization (NMF), self-modeling mixture analysis (SMMA), and support vector machines for regression (SVR) as preventative measures, that is, to identify and quantify the right components of preservatives (sorbic acid, potassium sorbate, and sodium benzoate) to precisely enhance the antimicrobial effect for merchandise safety.

4.4.4. Fluorescence Spectroscopy

Fluorescence spectroscopy, which focuses on the measurement of the emitted light (fluorescence) from a sample after it has been excited by an external energy source, can also be used to acquire data for food contamination analysis. A recent work [85], presents a methodology for the analysis of fluorescence spectra of slurred almonds under 375 nm wavelength excitation for optical detection of *Aflatoxins B* in grained almonds. This application is different than conventional classification frameworks in the sense that labels of contaminated or not contaminated are generated based on multiple threshold values; thus, a single sample could have a vector of labels corresponding to each threshold value. This is referred to as multiexpert learning [117]. In this context, support vector machines (SVM) were used with a majority vote. On the other hand, other techniques of analyzing materials have been used, particularly based on the interaction of microwaves with samples. Unlike spectroscopy, microwave sensing systems operate in the microwave-frequency range, typically from hundreds of MHz to tens of GHz. Recently, a microwave sensing approach coupled with ANNs for food safety was proposed in [97], exploiting the dielectric contrast between the potential intrusion and the surrounding matter of packaged food. In addition, nuclear magnetic resonance (NMR) spectroscopy, which is a prominent analytical approach, was applied in a recent work [92] to distinguish between pathogenic and nonpathogenic bacteria, using PCA and ANNs.

4.4.5. Nuclear Magnetic Resonance (NMR)

Nuclear magnetic resonance (NMR) can be of great importance in the food safety context due to its ability to provide detailed molecular information in a nondestructive manner. However, compared to the above categories, NMR is somewhat less used in the literature. Only one study [92] explored the use of NMR spectra in combination with artificial neural networks to distinguish between 10 different microbial strains such as *Bacillus*, *Salmonella*, and *Yersinia*, among others in food based on metabolite profiles. For this type of data, prior preparation is necessary, like scaling spectra to the internal concentration standard (TSP), removing noise bins, standardization, and PCA for visualization. The proposed ANN model, with hidden layers of 800 neurons, ReLu functions, and two dense layers, was compared with two classical machine learning algorithms (RF and SVM), with slight differences in performance.

4.4.6. Hyperspectral Imaging

Optical spectroscopic techniques have been recently exploited for food safety as an advanced promising tool. Due to the enhanced spectral resolution and improved spacial information, hyperspectral imaging, combined with AI tools, is increasingly appealing for food safety applications. The first application was as early as 2002 [47], where hyperspectral imaging was used to characterize spectral features, and multispectral imaging combined with PCA was used for the detection of defects on three apple cultivars, namely Golden Delicious, Red Delicious, and Gala. Recently, an ANN-based method using FTIR-hyperspectral data was proposed in 2018 for the rapid and cost-effective diagnosis of pathogenic bacteria [68]. Also, in [83], active thermography and an infrared camera were used to capture thermal images for analysis using a multiview learning-based autoencoder (MVAE) for defect detection.

On the other hand, a study used visible and NIR (VNIR) hyperspectral imaging to determine the presence of viral particles in a fluid suspension as well as on a surface upon complete evaporation of its water content [93]. In contrast with the previous studies, which proved the effectiveness of using HSI to detect the presence of fungi or bacteria, detecting the presence of viruses, which are two orders of magnitude below fungi and bacteria, can be crucial to prevent the spread of plant diseases. The acquired VNIR-HSI data were used in combination with PLSDR and FFNN to detect the presence of viruses in addition to the analysis of averaged spectra for quantifcation of the viral load [93].

4.4.7. Comparative Studies

In the light of recent developments and applications of machine learning in food safety, comparative studies are of great importance to understand and assess the viability of the techniques. In [108], a thorough comparative study was conducted using four different hyperspectral imaging techniques (fluorescence, VNIR, SWIR and Raman) combined with five classification approaches (LDA, LSVM, QDA, and QSVM), in addition to four simple preprocessing techniques. Also, the study in [115] provides a comparative assessment of spectroscopy-based techniques and machine learning approaches for evaluating the microbiological spoilage of ready-to-eat leafy vegetables (baby spinach and rocket). In this study, Fourier-transform infrared (FTIR), near-infrared (NIR), visible (VIS) spectroscopy, and multispectral imaging (MSI) were used, with only two regression algorithms, namely SVR and PLSR.

4.4.8. Electroanalytical Methods

The data used in [80] were gathered using the daily inspection of food safety obtained from the Analysis and Testing Institute of one province in China, which includes manufacturer information, origin information, product information, production dates, inspection items, and inspection results, among others. The aim of this study is to first build an AHP-ELM model to distinguish between three types of food safety risks, namely heavy metal pollution, chemical contaminant pollution, and pathogenic bacteria pollution, then construct an AHP-ELM-based model for food safety warning. In an earlier study [54], microelectrode arrays (MEAs) obtained from Multichannel Systems GmBH were used to measure the electrical activity of cortical networks grown in a controlled environment (in vitro). The MEAs measure the electrical spikes of the lyophilized botulism toxin (BoNT A) group for further statistical analysis, in contrast with a controlled group without toxins. Furthermore, the use of cortical networks as a biosensor for botulinum toxin offers several advantages. First, the networks can be grown in a controlled environment, allowing for reproducible experiments. Second, the networks can be easily manipulated and modified to study different aspects of toxin detection. Third, the electrical signals from the network can be recorded noninvasively, minimizing the potential harm to the tissue. This approach provides a promising tool for potential applications in toxin detection in food safety research.

On the other hand, few studies explored the potential for a real-time monitoring sensing system to provide early means to assess food safety [69,80]. A brief study [69] proposed a real-time environment monitoring sensing system to monitor stored grain's condition and potentially increase its shelf-life. First, sensors record the values of temperature, moisture content, and CO_2, among others, and store the data in an Ardinuo data logger periodically. Consequently, the stored data are communicated to a web server using WI-FI. Second, preprocessing techniques are applied in the microprocessor to reduce the inevitable noise generated by the different environment sensors. This is due to long-term drifts, temporary electrical errors, and the effects of nearby sensors. Finally, a classification approach can be used to distinguish between four classes, namely no spoilage, early spoilage, severe spoilage, and early insect infestation. A more recent study [80] proposed an edge IoT and machine learning-based approach for food quality monitoring systems to avoid food waste, which reaches 50% for root crops, fruits, and vegetables. First, three sensors are in place to gather environmental data, namely gas, temperature, and humidity sensors, after which any missing entry is replaced with the median value of the feature. Then, four classification algorithms are considered to categorize the results into three classes (fresh, semifresh, and spoiled), namely linear SVM, RBF-SVM, logistic regression, and random forests (RF). To ensure real-time monitoring, the results are communicated to the user through an Android application.

5. Research Gaps and Challenges

Spectral imaging, along with spectroscopy-based techniques, can be considered as alternatives for conventional methods in order to solve food safety-related problems [58]. For instance, applying AI has been rising since 2015, and is expected to stay on the same path for the next 10 years. In fact, combining analytical technologies with AI and data analysis tools, along with developing sensing systems, is expected to have great potential for the agriculture and food industry [64,88]. However, many of these techniques are still at the research and development stages [81], Sensor applications, for example, and developing robust algorithms are still areas for further exploration [64]. It appears that the driving factors to execute more innovation within a certain applied field, and if we are talking about AI in the food industry, are mainly the Industrial Revolution and the need for intelligent systems, hence smart factory development [87]. AI tool applications in the food industry have been intense in 2020 mostly, where researchers were regarded as carrying out more research work using the AI method, which might be linked to the recent pandemic situation due to the COVID-19 virus. Many industrial corporations are now open to AI applications as an alternative for human workers, for cheaper and better outcomes.

There has been much confusion within the literature, especially when it comes to food safety and quality concepts, as well as concerning soilborne and foodborne infections in crop foods. These misconceptions were mostly present in papers classified as analytical techniques or AI-oriented journals. On the other hand, food technology and/or biology-oriented papers tend to the nonadequate referring to AI tools, maybe due to miscommunication between these research areas. Generally speaking, we may suggest that research-wise, this topic might stumble accross these challenges:

- The complexity and diversity of food matrices, which may require different AI models and parameters for different food products and contaminants.
- The lack of standardization and validation of AI methods and data, which may affect the accuracy, reliability, and comparability of the results.
- The potential risks and uncertainties of AI applications in agriculture, such as environmental impacts, socioeconomic impacts, cyber-attacks, biases, and errors, which may affect food security, sustainability, and resilience.
- The need for interdisciplinary collaboration and stakeholder engagement, which may involve challenges such as communication barriers, knowledge gaps, cultural differences, and conflicting interests.

5.1. *Big Data*

The complexity of food matrices that represent samples in experiments (Figure 10) is an important challenge. Their diversity leads to a variety of data types, hence various protocols and methods. Contaminant detection, for example, laboratory detection, emphasizes accuracy, while field detection emphasizes efficiency and portability [81]. Spectral data can be very noisy and susceptible to continuous change [53]. Big data are a major challenge, where food safety relies on data analysis and the processing of information coming from a variety of resources, such as IoT and information-based systems. At the macro level, data analysis impacts market operations and government policies; at the micro level, it is called the smart supervision of food safety. Creating a food traceability system can establish a full correlation between all resources from different stages of the food supply chain. It can be represented as a food safety management platform based on big data, forming a data warehouse, with a unified and robust data management system, for digital food safety supervision [105]. In fact, large data are required to develop cutting-edge technologies, but difficulties arise, such as confusion between data fitting and modeling, and inadequate decision making. Also, there is the curse of dimensionality, data leakage, limited real datasets to experiment on, and low productivity [113].

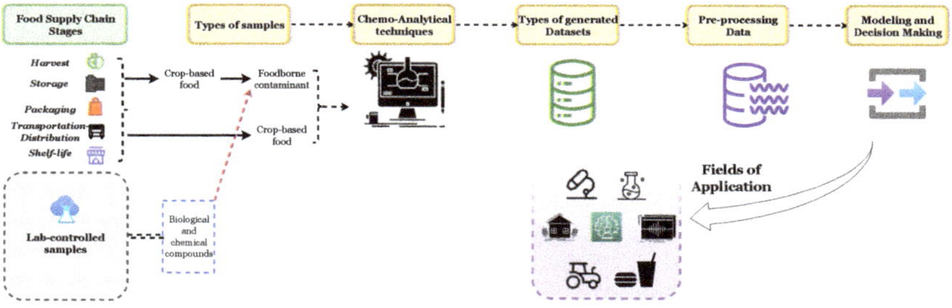

Figure 10. Information and data workflow according to case studies within our collected papers.

5.2. Learning Methods, XAI, and Interpretability

Limited model performances are also a major issue, along with the need for combining this analytical methods with robust nonlinear algorithms [76]. Also, various general guidelines are in place to objectively assess a machine learning model performance, in particular the use of cross-Validation approaches (also known as K-fold cross-validation) to avoid different problems such as over-fitting. However, as depicted in Figure 11c, only 33% of the article papers used cross-validation techniques, with the number of folds ranging between 4 and 10 (see Figure 11d). Also, the evaluation metrics used to validate the model performance are of upmost importance. As shown in Figure 11a, accuracy is the most used metric. In addition, over 42% of research papers only used one metric and an additional 27% only used two metrics to assess the model performance.

Nondestructive techniques in food safety, especially when it comes to spectral imaging, are the objective detection of defects in plants. However, many challenges arise, such as being restricted to a limited number of attributes; there is a need for more efficient image acquisition and processing, in HSI or NIR, for example. The challenge that remains is that of the standardization of techniques; each reported study is more or less limited to a specific instrumental parameter, studying a single food item and particular defect, while using a different learning algorithm or a different validation process [82]. Limitations and challenges of rapid methods are mainly technical but sometimes are depending on the right choice of data analysis techniques, including the most studied features, without losing information [113]. In most cases of spectral data analysis, it is hard to acknowledge the effect of variable combinations and adding interpretability in decision making.

5.3. Real-World Applications

The robustness and transferability of models is one of the main challenges and one of the limitations that restrict the wide-scale application of spectroscopic techniques [69,80,81]. Most of these research papers are only investigations of the feasibility of applications of these approaches. More research should be conducted in routine analysis and real-time implementation [62]. The practical application of emerging techniques, such as the E-nose, still needs to overcome a series of problems, such as the robustness and transitivity of models [81]. Not every mentioned technology seems to be at the same development level in theory as those in laboratories of the food industry; many are still at the infancy level [113], for example, in safety cereal processing, which needs to be further investigated [100]. There should also be more focus on establishing open-source databases and accessible analytical methods for more applications [81].

To our knowledge, there are only a few published papers involving spectral sensing coupled with AI tools, especially for crop-based food safety, although many pieces of research related to food safety and quality, in general, have been executed. Additionally, most studies combing analytical methods with AI tools were from contributors who were mainly scholars of computer sciences and/or related fields, where in-depth safety problem solving is not the main objective of the study, but rather general or specific feature engineering [69,70,80].

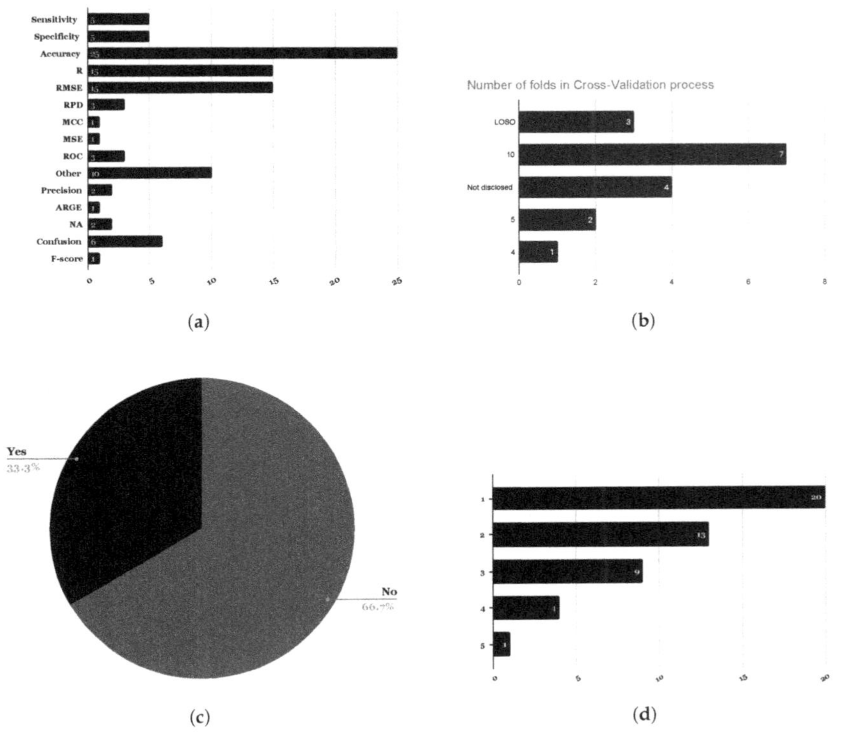

Figure 11. Statistics about evaluation metrics and the use of K-fold cross-validation in the research papers. (**a**) Number of times each evaluation metric is used in the 49 research papers. (**b**) Count of evaluation metrics used per research paper. (**c**) Percentage of research papers that uses K-fold cross-validation for model validation. (**d**) The number of folds used in K-fold.

6. Conclusions and Future Work

This work conducts a qualitative as well a quantitative evaluation of research works regarding our research topic. Furthermore, it proposes a unique methodology to search the literature exhaustively and come up with relevant conclusions and suggestions for further research. This methodology can be further applied in upcoming comprehensive review papers or for searching the literature in emerging areas of research. We executed many searches with many keywords, which was necessary to perform an exhaustive search in the literature to obtain as many papers as possible. Nonetheless, we remarked that most papers were majorly discussing quality assessment issues, mostly, food origins and/or adulteration, using classification and/or regression methods. Also, there was a high tendency to confuse safety and quality measures, which are clearly separate. Since we opted to accept reviews that are remotely related to our specific topic, these review papers were discussing a diversity of points, including our target problem, with more details directed towards meat products, and quality assessment measures, more than safety matters. Food safety, like any other real-life problem, has either been a principal topic in research projects or simply an application to test decision-making tools. It is apparent that there are few applications, and even studies, directed towards food safety when it comes to its contamination prevention, where most studies focused mainly on animal products, like milk and meat; hence, very few studies were regarding crop-based food. To this end, machine learning methods are of crucial importance. Analytical techniques and analytical strategies are completely different concepts: the first one is related to one field of studies involving for example spectral techniques; the second one represents an intersection between many areas of research, hence diverse technologies, as it was discussed and referred to in many case studies papers and review papers, like systems for detecting food defects involving spectral imaging, chemometrics for preprocessing, and computer vision for image analysis and decision making.

There have been many gaps while searching the literature. We at this moment encourage more work to be conducted regarding this problem, for a variety of reasons. First, food safety-security is a global call into question, congregating multiple areas of research and industry. In our study, we investigated crop-food safety while considering all steps of the food supply chain. There are many challenges in optimizing the whole process, where employing AI tools would be highly beneficial. Food contamination prevention is a real-world challenge, with multiple hazards at stake. Our focus was mainly on crop-based foods, regarding the whole process, starting from healthy soil and fresh seeds to grow these crop foods to the marketed food product.

As discussed previously, different contaminants present serious problems, particularly biological contaminants. The first step in ensuring both food security and food safety is growing a healthy crop. On one hand, crop diseases have multiple impacts, decreasing the yield, which leads to insufficient food staples, defecting yield products and causing unsafe food for both humans and animals. Also, regarding farm animals, once they ingest diseased foods, all animal-based foods are susceptible to being unsafe for consumption. Furthermore, there is an apparent gap between AI studies' development and their applications, not only at the industrial level but at the research level as well, particularly in crop and agricultural sciences. Henceforth, upcoming research should be directed towards developing cutting-edge technologies in precision agriculture, by executing more studies in the fields of crop sciences combined with machine learning and computer vision tools. Our future work will be focusing mainly on solving problems related to inspecting resistant crops to biological contaminants in arid and semiarid regions as an important aspect of food safety. Working on drought-resistant crops and combining AI tools is one of the major challenges facing research globally, and African countries specifically. This problem falls within the area of phytopathology, crop improvement, and machine learning. Our quest is to work on tasks related to crop classification while adding interpretability in deep learning and complex ML techniques based on image processing.

Author Contributions: Conceptualization, H.R.; methodology, H.R.; software, H.R. and Y.A.; validation, D.H.P.-O.; formal analysis, H.R. and Y.A.; writing—original draft preparation, H.R., Y.A. and A.T.; data curation, H.R.; visualization, H.R.; writing—review and editing, H.R. and Y.A.; supervision, D.H.P.-O. All authors have read and agreed to the published version of the manuscript.

Funding: This research received no external funding.

Institutional Review Board Statement: Not applicable.

Informed Consent Statement: Not applicable.

Data Availability Statement: Data is contained within the article.

Acknowledgments: We are deeply grateful to all those who played a role in the success of this paper. We would like to thank Kaoutar Benghzial, Sami Bamansour, Mouad Elhamdi and Leandro Lorente for their invaluable input and support throughout the research process. Their insights and expertise were instrumental in shaping the direction of this paper. We are also greatful for the continous support of the SDAS research group.

Conflicts of Interest: The authors declare no conflict of interest.

References

1. Garcia, S.N.; Osburn, B.I.; Jay-Russell, M.T. One health for food safety, food security, and sustainable food production. *Front. Sustain. Food Syst.* **2020**, *4*, 1. [CrossRef]
2. Hargreaves, S.M.; Rosenfeld, D.L.; Moreira, A.V.B.; Zandonadi, R.P. Plant-based and vegetarian diets: An overview and definition of these dietary patterns. *Eur. J. Nutr.* **2023**, *62*, 1109–1121. [CrossRef]

3. Matthews, P.J.; Ghanem, M.E. Perception gaps that may explain the status of taro (*Colocasia esculenta*) as an "orphan crop". *Plants People Planet* **2021**, *3*, 99–112. [CrossRef]
4. Neufeld, L.M.; Hendriks, S.; Hugas, M. Healthy Diet: A Definition for the United Nations Food Systems Summit 2021. In *A Paper from the Scientific Group of the UN Food Systems Summit*; Springer International Publishing: Cham, Switzerland, 2021.
5. FAO. *Thinking about the Future of Food Safety—A Foresight Report*; FAO: Rome, Italy, 2022. [CrossRef]
6. Mohanty, S.P.; Hughes, D.P.; Salathé, M. Using deep learning for image-based plant disease detection. *Front. Plant Sci.* **2016**, *7*, 1419. [CrossRef]
7. Raki, H.; González-Vergara, J.; Aalaila, Y.; Elhamdi, M.; Bamansour, S.; Guachi-Guachi, L.; Peluffo-Ordoñez, D.H. Crop Classification Using Deep Learning: A Quick Comparative Study of Modern Approaches. In Proceedings of the International Conference on Applied Informatics, Serdang, Malaysia, 18–19 May 2022; Springer: Cham, Switzerland, 2022; pp. 31–44.
8. Kamala, K.; Kumar, V.P. Food products and food contamination. In *Microbial Contamination and Food Degradation*; Elsevier: Amsterdam, The Netherlands, 2018; pp. 1–19.
9. Bintsis, T. Microbial pollution and food safety. *AIMS Microbiol.* **2018**, *4*, 377. [CrossRef]
10. Turek, C.; Stintzing, F.C. Stability of essential oils: A review. *Compr. Rev. Food Sci. Food Saf.* **2013**, *12*, 40–53. [CrossRef]
11. Clapp, J.; Moseley, W.G.; Burlingame, B.; Termine, P. The case for a six-dimensional food security framework. *Food Policy* **2021**, *106*, 102164. [CrossRef]
12. Chatterjee, A.; Abraham, J. Microbial contamination, prevention, and early detection in food industry. In *Microbial Contamination and Food Degradation*; Elsevier: Amsterdam, The Netherlands, 2018; pp. 21–47.
13. Chitrakar, B.; Zhang, M.; Bhandari, B. Improvement strategies of food supply chain through novel food processing technologies during COVID-19 pandemic. *Food Control* **2021**, *125*, 108010. [CrossRef]
14. Surampalli, R.Y.; Zhang, T.C.; Goyal, M.K.; Brar, S.K.; Tyagi, R. *Sustainability: Fundamentals and Applications*; John Wiley & Sons: Hoboken, NJ, USA, 2020.
15. Teena, M.; Manickavasagan, A.; Mothershaw, A.; El Hadi, S.; Jayas, D. Potential of machine vision techniques for detecting fecal and microbial contamination of food products: A review. *Food Bioprocess Technol.* **2013**, *6*, 1621–1634. [CrossRef]
16. Craig, A.P.; Franca, A.S.; Irudayaraj, J. Surface-enhanced Raman spectroscopy applied to food safety. *Annu. Rev. Food Sci. Technol.* **2013**, *4*, 369–380. [CrossRef]
17. Brousseau, J.L.; Vidon, S.; Leblanc, R. Investigation of the chemical nature of two-dimensional polymerized octadecyltrimethoxysilane Langmuir films by inelastic electron tunneling spectroscopy. *J. Chem. Phys.* **1998**, *108*, 7391–7396. [CrossRef]
18. Ojanen, P.; Mäkiranta, P.; Penttilä, T.; Minkkinen, K. Do logging residue piles trigger extra decomposition of soil organic matter? *For. Ecol. Manag.* **2017**, *405*, 367–380. [CrossRef]
19. Świeca, M.; Sęczyk, Ł.; Gawlik-Dziki, U.; Dziki, D. Bread enriched with quinoa leaves—The influence of protein–phenolics interactions on the nutritional and antioxidant quality. *Food Chem.* **2014**, *162*, 54–62. [CrossRef]
20. Aghayee, S.; Benadiba, C.; Notz, J.; Kasas, S.; Dietler, G.; Longo, G. Combination of fluorescence microscopy and nanomotion detection to characterize bacteria. *J. Mol. Recognit.* **2013**, *26*, 590–595. [CrossRef]
21. Guo, R.; McGoverin, C.; Swift, S.; Vanholsbeeck, F. A rapid and low-cost estimation of bacteria counts in solution using fluorescence spectroscopy. *Anal. Bioanal. Chem.* **2017**, *409*, 3959–3967. [CrossRef]
22. Christensen, J.; Becker, E.M.; Frederiksen, C. Fluorescence spectroscopy and PARAFAC in the analysis of yogurt. *Chemom. Intell. Lab. Syst.* **2005**, *75*, 201–208. [CrossRef]
23. Gismondi, A.; De Rossi, S.; Canuti, L.; Novelli, S.; Di Marco, G.; Fattorini, L.; Canini, A. From *Robinia pseudoacacia* L. nectar to Acacia monofloral honey: Biochemical changes and variation of biological properties. *J. Sci. Food Agric.* **2018**, *98*, 4312–4322. [CrossRef]
24. Su, W.H.; He, H.J.; Sun, D.W. Non-destructive and rapid evaluation of staple foods quality by using spectroscopic techniques: A review. *Crit. Rev. Food Sci. Nutr.* **2017**, *57*, 1039–1051. [CrossRef]
25. Grundas, S.; Stępniewski, A. *Advances in Agrophysical Research*; BoD–Books on Demand: Norderstedt, Germany, 2013.
26. Kim, H.K.; Choi, Y.H.; Verpoorte, R. NMR-based metabolomic analysis of plants. *Nat. Protoc.* **2010**, *5*, 536–549. [CrossRef]
27. Lindgren, F.; Rännar, S. Alternative partial least-squares (PLS) algorithms. In *3D QSAR in Drug Design: Recent Advances*; Springer: Dordrecht, The Netherlands, 1998; pp. 105–113.
28. Tan, W.K.; Husin, Z.; Yasruddin, M.L.; Ismail, M.A.H. Recent technology for food and beverage quality assessment: A review. *J. Food Sci. Technol.* **2022**, *60*, 1681–1694. [CrossRef]
29. Wang, K.; Pu, H.; Sun, D.W. Emerging spectroscopic and spectral imaging techniques for the rapid detection of microorganisms: An overview. *Compr. Rev. Food Sci. Food Saf.* **2018**, *17*, 256–273. [CrossRef]
30. Wang, Y.; Ravindranath, S.; Irudayaraj, J. Separation and detection of multiple pathogens in a food matrix by magnetic SERS nanoprobes. *Anal. Bioanal. Chem.* **2011**, *399*, 1271–1278. [CrossRef] [PubMed]
31. Liu, Y.; Pu, H.; Sun, D.W. Hyperspectral imaging technique for evaluating food quality and safety during various processes: A review of recent applications. *Trends Food Sci. Technol.* **2017**, *69*, 25–35. [CrossRef]
32. Jiang, Y.; Li, G.; Ge, H.; Wang, F.; Li, L.; Chen, X.; Lu, M.; Zhang, Y. Machine learning and application in terahertz technology: A review on achievements and future challenges. *IEEE Access* **2022**, *10*, 53761–53776. [CrossRef]
33. Wang, H.P.; Chen, P.; Dai, J.W.; Liu, D.; Li, J.Y.; Xu, Y.P.; Chu, X.L. Recent advances of chemometric calibration methods in modern spectroscopy: Algorithms, strategy, and related issues. *TrAC Trends Anal. Chem.* **2022**, *153*, 116648. [CrossRef]

34. Al-Awadhi, M.A.; Deshmukh, R.R. A Review on Modern Analytical Methods for Detecting and Quantifying Adulteration in Honey. In Proceedings of the 2021 International Conference of Modern Trends in Information and Communication Technology Industry (MTICTI), Sana'a, Yemen, 4–6 December 2021; IEEE: Piscataway, NJ, USA, 2021; pp. 1–6.
35. Zhang, X.; Yang, J.; Lin, T.; Ying, Y. Food and agro-product quality evaluation based on spectroscopy and deep learning: A review. *Trends Food Sci. Technol.* **2021**, *112*, 431–441. [CrossRef]
36. An, D.; Zhang, L.; Liu, Z.; Liu, J.; Wei, Y. Advances in infrared spectroscopy and hyperspectral imaging combined with artificial intelligence for the detection of cereals quality. *Crit. Rev. Food Sci. Nutr.* **2022**, *63*, 9766–9796. [CrossRef] [PubMed]
37. Buzura, L.; Budileanu, M.L.; Potarniche, A.; Galatus, R. Python based portable system for fast characterisation of foods based on spectral analysis. In Proceedings of the 2021 IEEE 27th International Symposium for Design and Technology in Electronic Packaging (SIITME), Timisoara, Romania, 27–30 October 2021; IEEE: Piscataway, NJ, USA, 2021; pp. 275–280.
38. Pan, T.; Wu, C.; Chen, Q. Sparse Reconstruction Using Block Sparse Bayesian Learning With Fast Marginalized Likelihood Maximization for Near-Infrared Spectroscopy. *IEEE Trans. Instrum. Meas.* **2021**, *71*, 2500410. [CrossRef]
39. Nouri, H.; Guermazi, M.; Kallel, A.Y.; Hao, W.; Kanoun, O. Meat Freshness Assessment based on Impedance Spectroscopy and Distribution of Relaxation Times (DRT). In Proceedings of the 2022 International Workshop on Impedance Spectroscopy (IWIS), Chemnitz, Germany, 27–30 September 2022; IEEE: Piscataway, NJ, USA, 2022; pp. 41–45.
40. Wei, X.; Zhu, S.; Zhou, S.; Zheng, W.; Li, S. Identification of soybean origin by terahertz spectroscopy and chemometrics. *IEEE Access* **2020**, *8*, 184988–184996. [CrossRef]
41. Fengou, L.C.; Mporas, I.; Spyrelli, E.; Lianou, A.; Nychas, G.J. Estimation of the microbiological quality of meat using rapid and non-invasive spectroscopic sensors. *IEEE Access* **2020**, *8*, 106614–106628. [CrossRef]
42. Shen, J.; Li, M.; Li, Z.; Zhang, Z.; Zhang, X. Single convolutional neural network model for multiple preprocessing of Raman spectra. *Vib. Spectrosc.* **2022**, *121*, 103391. [CrossRef]
43. Trigoso, T.C.; Cruz, J.O.; Miñano, H.A.; Silupu, W.C. Application of machine learning in the discrimination of citrus fruit juices: Uses of dielectric spectroscopy. In Proceedings of the 2020 IEEE Engineering International Research Conference (EIRCON), Lima, Peru, 21–23 October 2020; IEEE: Piscataway, NJ, USA, 2020; pp. 1–4.
44. Suter, G.W. Review papers are important and worth writing. *Environ. Toxicol. Chem.* **2013**, *32*, 1929–1930. [CrossRef] [PubMed]
45. Palmatier, R.W.; Houston, M.B.; Hulland, J. Review articles: Purpose, process, and structure. *J. Acad. Mark. Sci.* **2018**, *46*, 1–5. [CrossRef]
46. Snyder, H. Literature review as a research methodology: An overview and guidelines. *J. Bus. Res.* **2019**, *104*, 333–339. [CrossRef]
47. Mehl, P.; Chao, K.; Kim, M.; Chen, Y. Detection of defects on selected apple cultivars using hyperspectral and multispectral image analysis. *Appl. Eng. Agric.* **2002**, *18*, 219.
48. Irudayaraj, J.; Yang, H.; Sakhamuri, S. Differentiation and detection of microorganisms using Fourier transform infrared photoacoustic spectroscopy. *J. Mol. Struct.* **2002**, *606*, 181–188. [CrossRef]
49. Yang, H.; Irudayaraj, J. Rapid detection of foodborne microorganisms on food surface using Fourier transform Raman spectroscopy. *J. Mol. Struct.* **2003**, *646*, 35–43. [CrossRef]
50. Gupta, M.J.; Irudayaraj, J.M.; Debroy, C.; Schmilovitch, Z.; Mizrach, A. Differentiation of food pathogens using FTIR and artificial neural networks. *Trans. ASAE* **2005**, *48*, 1889–1892. [CrossRef]
51. Gupta, M.J.; Irudayaraj, J.M.; Schmilovitch, Z.; Mizrach, A. Identification and quantification of foodborne pathogens in different food matrices using FTIR spectroscopy and artificial neural networks. *Trans. ASABE* **2006**, *49*, 1249–1255. [CrossRef]
52. Huang, Y.; Kangas, L.J.; Rasco, B.A. Applications of artificial neural networks (ANNs) in food science. *Crit. Rev. Food Sci. Nutr.* **2007**, *47*, 113–126. [CrossRef]
53. Berrueta, L.A.; Alonso-Salces, R.M.; Héberger, K. Supervised pattern recognition in food analysis. *J. Chromatogr. A* **2007**, *1158*, 196–214. [CrossRef] [PubMed]
54. Scarlatos, A.; Cadotte, A.; DeMarse, T.; Welt, B. Cortical networks grown on microelectrode arrays as a biosensor for botulinum toxin. *J. Food Sci.* **2008**, *73*, E129–E136. [CrossRef]
55. He, L.; Liu, Y.; Lin, M.; Mustapha, A.; Wang, Y. Detecting single Bacillus spores by surface enhanced Raman spectroscopy. *Sens. Instrum. Food Qual. Saf.* **2008**, *2*, 247–253. [CrossRef]
56. Siripatrawan, U. Self-organizing algorithm for classification of packaged fresh vegetable potentially contaminated with foodborne pathogens. *Sens. Actuators B Chem.* **2008**, *128*, 435–441. [CrossRef]
57. Stockel, S.; Schumacher, W.; Meisel, S.; Elschner, M.; Rosch, P.; Popp, J. Raman spectroscopy-compatible inactivation method for pathogenic endospores. *Appl. Environ. Microbiol.* **2010**, *76*, 2895–2907. [CrossRef]
58. Feng, Y.Z.; Sun, D.W. Application of hyperspectral imaging in food safety inspection and control: A review. *Crit. Rev. Food Sci. Nutr.* **2012**, *52*, 1039–1058. [CrossRef]
59. Shapaval, V.; Schmitt, J.; Møretrø, T.; Suso, H.; Skaar, I.; Åsli, A.; Lillehaug, D.; Kohler, A. Characterization of food spoilage fungi by FTIR spectroscopy. *J. Appl. Microbiol.* **2013**, *114*, 788–796. [CrossRef]
60. Güneş, A.; Kalkan, H.; Durmuş, E.; Bütükcan, M.B. Detection of aflatoxin contaminated figs using Near-Infrared (NIR) reflectance spectroscopy. In Proceedings of the 2013 International Conference on Electronics, Computer and Computation (ICECCO), Ankara, Turkey, 7–9 November 2013; IEEE: Piscataway, NJ, USA, 2013; pp. 123–126.
61. Qin, J.; Chao, K.; Kim, M.S.; Lu, R.; Burks, T.F. Hyperspectral and multispectral imaging for evaluating food safety and quality. *J. Food Eng.* **2013**, *118*, 157–171. [CrossRef]

62. Reinholds, I.; Bartkevics, V.; Silvis, I.C.; van Ruth, S.M.; Esslinger, S. Analytical techniques combined with chemometrics for authentication and determination of contaminants in condiments: A review. *J. Food Compos. Anal.* **2015**, *44*, 56–72. [CrossRef]
63. Ma, J.; Sun, D.W.; Qu, J.H.; Liu, D.; Pu, H.; Gao, W.H.; Zeng, X.A. Applications of computer vision for assessing quality of agri-food products: A review of recent research advances. *Crit. Rev. Food Sci. Nutr.* **2016**, *56*, 113–127. [CrossRef]
64. Ropodi, A.; Panagou, E.; Nychas, G.J. Data mining derived from food analyses using non-invasive/non-destructive analytical techniques; determination of food authenticity, quality & safety in tandem with computer science disciplines. *Trends Food Sci. Technol.* **2016**, *50*, 11–25.
65. ElMasry, G.M.; Nakauchi, S. Image analysis operations applied to hyperspectral images for non-invasive sensing of food quality—A comprehensive review. *Biosyst. Eng.* **2016**, *142*, 53–82. [CrossRef]
66. Geng, Z.; Zhao, S.; Zhu, Q.; Han, Y.; Xu, Y.; He, Y. Early warning modeling and application based on analytic hierarchy process integrated extreme learning machine. In Proceedings of the 2017 Intelligent Systems Conference (IntelliSys), London, UK, 7–8 September 2017; IEEE: Piscataway, NJ, USA, 2017; pp. 738–743.
67. Shen, Y.; Tian, J.; Li, L.; Wu, Y.; Li, L. Feasibility of non-destructive internal quality analysis of pears by using near-infrared diffuse reflectance spectroscopy. In Proceedings of the 2017 9th International Conference on Modelling, Identification and Control (ICMIC), Kunming, China, 10–12 July 2017; IEEE: Piscataway, NJ, USA, 2017; pp. 31–36.
68. Lasch, P.; Stämmler, M.; Zhang, M.; Baranska, M.; Bosch, A.; Majzner, K. FT-IR hyperspectral imaging and artificial neural network analysis for identification of pathogenic bacteria. *Anal. Chem.* **2018**, *90*, 8896–8904. [CrossRef]
69. Kaushik, R.; Singhai, J. An approach for the development of a sensing system to monitor contamination in stored grain. In Proceedings of the 2019 6th International Conference on Signal Processing and Integrated Networks (SPIN), Noida, India, 7–8 March 2019; IEEE: Piscataway, NJ, USA, 2019; pp. 880–884.
70. Zhou, L.; Zhang, C.; Liu, F.; Qiu, Z.; He, Y. Application of deep learning in food: A review. *Compr. Rev. Food Sci. Food Saf.* **2019**, *18*, 1793–1811. [CrossRef]
71. Öner, T.; Thiam, P.; Kos, G.; Krska, R.; Schwenker, F.; Mizaikoff, B. Machine learning algorithms for the automated classification of contaminated maize at regulatory limits via infrared attenuated total reflection spectroscopy. *World Mycotoxin J.* **2019**, *12*, 113–122. [CrossRef]
72. Guo, Z.; Wang, M.; Wu, J.; Tao, F.; Chen, Q.; Wang, Q.; Ouyang, Q.; Shi, J.; Zou, X. Quantitative assessment of zearalenone in maize using multivariate algorithms coupled to Raman spectroscopy. *Food Chem.* **2019**, *286*, 282–288. [CrossRef]
73. Wan-dan, Z.; Ru-jin, S.; Cheng-wei, W.; Qian-xue, L.; Zhi-ping, X. Raman Spectroscopy Classification of Foodborne Pathogenic Bacteria Based on PCA-Stacking Model. In Proceedings of the 2019 International Conference on Intelligent Informatics and Biomedical Sciences (ICIIBMS), Shanghai, China, 21–24 November 2019; IEEE: Piscataway, NJ, USA, 2019; pp. 304–307.
74. Liu, W.; Zhao, P.; Wu, C.; Liu, C.; Yang, J.; Zheng, L. Rapid determination of aflatoxin B1 concentration in soybean oil using terahertz spectroscopy with chemometric methods. *Food Chem.* **2019**, *293*, 213–219. [CrossRef]
75. Sun, X.; Zhu, K.; Liu, J.; Hu, J.; Jiang, X.; Liu, Y.; Gong, Z. Terahertz spectroscopy determination of benzoic acid additive in wheat flour by machine learning. *J. Infrared Millim. Terahertz Waves* **2019**, *40*, 466–475. [CrossRef]
76. Zareef, M.; Chen, Q.; Hassan, M.M.; Arslan, M.; Hashim, M.M.; Ahmad, W.; Kutsanedzie, F.Y.; Agyekum, A.A. An overview on the applications of typical non-linear algorithms coupled with NIR spectroscopy in food analysis. *Food Eng. Rev.* **2020**, *12*, 173–190. [CrossRef]
77. Le, B.T. Application of deep learning and near infrared spectroscopy in cereal analysis. *Vib. Spectrosc.* **2020**, *106*, 103009. [CrossRef]
78. Wang, K.; Chen, L.; Ma, X.; Ma, L.; Chou, K.C.; Cao, Y.; Khan, I.U.; Gölz, G.; Lu, X. Arcobacter identification and species determination using Raman spectroscopy combined with neural networks. *Appl. Environ. Microbiol.* **2020**, *86*, e00924-20. [CrossRef]
79. Weng, S.; Yuan, H.; Zhang, X.; Li, P.; Zheng, L.; Zhao, J.; Huang, L. Deep learning networks for the recognition and quantitation of surface-enhanced Raman spectroscopy. *Analyst* **2020**, *145*, 4827–4835. [CrossRef]
80. Sahu, B.; Tiwari, A.; Raheja, J.; Kumar, S. Development of Machine Learning & Edge IoT Based Non-destructive Food Quality Monitoring System using Raspberry Pi. In Proceedings of the 2020 IEEE International Conference on Computing, Power and Communication Technologies (GUCON), Greater Noida, India, 2–4 October 2020; IEEE: Piscataway, NJ, USA, 2020; pp. 449–455.
81. Yu, X.L.; Sun, D.W.; He, Y. Emerging techniques for determining the quality and safety of tea products: A review. *Compr. Rev. Food Sci. Food Saf.* **2020**, *19*, 2613–2638. [CrossRef]
82. Nturambirwe, J.F.I.; Opara, U.L. Machine learning applications to non-destructive defect detection in horticultural products. *Biosyst. Eng.* **2020**, *189*, 60–83. [CrossRef]
83. Wu, H.; Zheng, K.; Sfarra, S.; Liu, Y.; Yao, Y. Multiview learning for subsurface defect detection in composite products: A challenge on thermographic data analysis. *IEEE Trans. Ind. Inform.* **2020**, *16*, 5996–6003. [CrossRef]
84. Shen, F.; Huang, Y.; Jiang, X.; Fang, Y.; Li, P.; Liu, Q.; Hu, Q.; Liu, X. On-line prediction of hazardous fungal contamination in stored maize by integrating Vis/NIR spectroscopy and computer vision. *Spectrochim. Acta Part A Mol. Biomol. Spectrosc.* **2020**, *229*, 118012. [CrossRef]
85. Bertani, F.; Businaro, L.; Gambacorta, L.; Mencattini, A.; Brenda, D.; Di Giuseppe, D.; De Ninno, A.; Solfrizzo, M.; Martinelli, E.; Gerardino, A. Optical detection of aflatoxins B in grained almonds using fluorescence spectroscopy and machine learning algorithms. *Food Control* **2020**, *112*, 107073. [CrossRef]

86. Rahi, S.; Mobli, H.; Jamshidi, B.; Azizi, A.; Sharifi, M. Achieving a robust Vis/NIR model for microbial contamination detection of Persian leek by spectral analysis based on genetic, iPLS algorithms and VIP scores. *Postharvest Biol. Technol.* **2021**, *175*, 111413. [CrossRef]
87. Mavani, N.R.; Ali, J.M.; Othman, S.; Hussain, M.; Hashim, H.; Rahman, N.A. Application of artificial intelligence in food industry—A guideline. *Food Eng. Rev.* **2022**, *14*, 134–175. [CrossRef]
88. Ge, H.; Lv, M.; Lu, X.; Jiang, Y.; Wu, G.; Li, G.; Li, L.; Li, Z.; Zhang, Y. Applications of THz spectral imaging in the detection of agricultural products. *Photonics* **2021**, *8*, 518. [CrossRef]
89. Rivola, M.; Ibba, P.; Lugli, P.; Petti, L. Bioimpedance data statistical modelling for food quality classification and prediction. In Proceedings of the 2021 IEEE International Symposium on Circuits and Systems (ISCAS), Daegu, Republic of Korea, 22–28 May 2021; IEEE: Piscataway, NJ, USA, 2021; pp. 1–5.
90. Xu, J.L.; Herrero-Langreo, A.; Lamba, S.; Ferone, M.; Scannell, A.G.; Caponigro, V.; Gowen, A.A. Characterisation and classification of foodborne bacteria using reflectance FTIR microscopic imaging. *Molecules* **2021**, *26*, 6318. [CrossRef]
91. Magnus, I.; Virte, M.; Thienpont, H.; Smeesters, L. Combining optical spectroscopy and machine learning to improve food classification. *Food Control* **2021**, *130*, 108342. [CrossRef]
92. Wang, D.; Greenwood, P.; Klein, M.S. Deep Learning for Rapid Identification of Microbes Using Metabolomics Profiles. *Metabolites* **2021**, *11*, 863. [CrossRef]
93. Gomez-Gonzalez, E.; Fernandez-Muñoz, B.; Barriga-Rivera, A.; Navas-Garcia, J.M.; Fernandez-Lizaranzu, I.; Munoz-Gonzalez, F.J.; Parrilla-Giraldez, R.; Requena-Lancharro, D.; Guerrero-Claro, M.; Gil-Gamboa, P.; et al. Hyperspectral image processing for the identification and quantification of lentiviral particles in fluid samples. *Sci. Rep.* **2021**, *11*, 16201. [CrossRef]
94. Guo, Z.; Guo, C.; Sun, L.; Zuo, M.; Chen, Q.; El-Seedi, H.R.; Zou, X. Identification of the apple spoilage causative fungi and prediction of the spoilage degree using electronic nose. *J. Food Process Eng.* **2021**, *44*, e13816. [CrossRef]
95. Li, H.; Hassan, M.M.; Wang, J.; Wei, W.; Zou, M.; Ouyang, Q.; Chen, Q. Investigation of nonlinear relationship of surface enhanced Raman scattering signal for robust prediction of thiabendazole in apple. *Food Chem.* **2021**, *339*, 127843. [CrossRef]
96. Vakilian, K.A. Metaheuristic Optimization to Improve Machine Learning in Raman Spectroscopic-based Detection of Foodborne Pathogens. In Proceedings of the 2021 7th International Conference on Signal Processing and Intelligent Systems (ICSPIS), Tehran, Iran, 29–30 December 2021; IEEE: Piscataway, NJ, USA, 2021; pp. 1–5.
97. Ricci, M.; Vasquez, J.A.T.; Turvani, G.; Sirena, I.; Casu, M.R.; Vipiana, F. Microwave Sensing for Food Safety: A Neural Network Implementation. In Proceedings of the 2021 IEEE Conference on Antenna Measurements & Applications (CAMA), Antibes Juan-les-Pins, France, 15–17 November 2021; IEEE: Piscataway, NJ, USA, 2021; pp. 444–447.
98. Yin, J.; Hameed, S.; Xie, L.; Ying, Y. Non-destructive detection of foreign contaminants in toast bread with near infrared spectroscopy and computer vision techniques. *J. Food Meas. Charact.* **2021**, *15*, 189–198. [CrossRef]
99. Yan, S.; Wang, S.; Qiu, J.; Li, M.; Li, D.; Xu, D.; Li, D.; Liu, Q. Raman spectroscopy combined with machine learning for rapid detection of food-borne pathogens at the single-cell level. *Talanta* **2021**, *226*, 122195. [CrossRef]
100. Zareef, M.; Arslan, M.; Hassan, M.M.; Ahmad, W.; Ali, S.; Li, H.; Ouyang, Q.; Wu, X.; Hashim, M.M.; Chen, Q. Recent advances in assessing qualitative and quantitative aspects of cereals using nondestructive techniques: A review. *Trends Food Sci. Technol.* **2021**, *116*, 815–828. [CrossRef]
101. Nie, P.; Qu, F.; Lin, L.; He, Y.; Feng, X.; Yang, L.; Gao, H.; Zhao, L.; Huang, L. Trace identification and visualization of multiple benzimidazole pesticide residues on toona sinensis leaves using terahertz imaging combined with deep learning. *Int. J. Mol. Sci.* **2021**, *22*, 3425. [CrossRef] [PubMed]
102. Yang, R.; Li, Y.; Zheng, J.; Qiu, J.; Song, J.; Xu, F.; Qin, B. A Novel Method for Carbendazim High-Sensitivity Detection Based on the Combination of Metamaterial Sensor and Machine Learning. *Materials* **2022**, *15*, 6093. [CrossRef] [PubMed]
103. Kang, Z.; Zhao, Y.; Chen, L.; Guo, Y.; Mu, Q.; Wang, S. Advances in Machine Learning and Hyperspectral Imaging in the Food Supply Chain. *Food Eng. Rev.* **2022**, *14*, 596–616. [CrossRef]
104. Yan, H.; Fan, W.; Chen, X.; Wang, H.; Qin, C.; Jiang, X. Component spectra extraction and quantitative analysis for preservative mixtures by combining terahertz spectroscopy and machine learning. *Spectrochim. Acta Part A Mol. Biomol. Spectrosc.* **2022**, *271*, 120908. [CrossRef] [PubMed]
105. Zhang, C.; Zhao, W.; Dong, L. Design of Food Safety Supervision System in the Background of Big Data. In Proceedings of the 2022 International Conference on Artificial Intelligence in Everything (AIE), Lefkosa, Cyprus, 2–4 August 2022; IEEE: Piscataway, NJ, USA, 2022; pp. 413–418.
106. Rady, A.; Watson, N. Detection and quantification of peanut contamination in garlic powder using NIR sensors and machine learning. *J. Food Compos. Anal.* **2022**, *114*, 104820. [CrossRef]
107. Bowler, A.L.; Ozturk, S.; Rady, A.; Watson, N. Domain Adaptation for In-Line Allergen Classification of Agri-Food Powders Using Near-Infrared Spectroscopy. *Sensors* **2022**, *22*, 7239. [CrossRef]
108. Kim, Y.K.; Baek, I.; Lee, K.M.; Qin, J.; Kim, G.; Shin, B.K.; Chan, D.E.; Herrman, T.J.; Cho, S.k.; Kim, M.S. Investigation of reflectance, fluorescence, and Raman hyperspectral imaging techniques for rapid detection of aflatoxins in ground maize. *Food Control* **2022**, *132*, 108479. [CrossRef]
109. Adejimi, O.E.; Ignat, T.; Sadhasivam, G.; Zakin, V.; Schmilovitch, Z.; Shapiro, O.H. Low-Resolution Raman Spectroscopy for the detection of contaminant species in algal bioreactors. *Sci. Total Environ.* **2022**, *809*, 151138. [CrossRef]

110. Cordovana, M.; Mauder, N.; Join-Lambert, O.; Gravey, F.; LeHello, S.; Auzou, M.; Pitti, M.; Zoppi, S.; Buhl, M.; Steinmann, J.; et al. Machine learning-based typing of Salmonella enterica O-serogroups by the Fourier-Transform Infrared (FTIR) Spectroscopy-based IR Biotyper system. *J. Microbiol. Methods* **2022**, *201*, 106564. [CrossRef]
111. Wang, B.; Deng, J.; Jiang, H. Markov Transition Field Combined with Convolutional Neural Network Improved the Predictive Performance of Near-Infrared Spectroscopy Models for Determination of Aflatoxin B1 in Maize. *Foods* **2022**, *11*, 2210. [CrossRef]
112. Aviara, N.A.; Liberty, J.T.; Olatunbosun, O.S.; Shoyombo, H.A.; Oyeniyi, S.K. Potential application of hyperspectral imaging in food grain quality inspection, evaluation and control during bulk storage. *J. Agric. Food Res.* **2022**, *8*, 100288. [CrossRef]
113. Pampoukis, G.; Lytou, A.E.; Argyri, A.A.; Panagou, E.Z.; Nychas, G.J.E. Recent advances and applications of rapid microbial assessment from a food safety perspective. *Sensors* **2022**, *22*, 2800. [CrossRef] [PubMed]
114. Hussain, M.; Zou, J.; Zhang, H.; Zhang, R.; Chen, Z.; Tang, Y. Recent Progress in Spectroscopic Methods for the Detection of Foodborne Pathogenic Bacteria. *Biosensors* **2022**, *12*, 869. [CrossRef]
115. Manthou, E.; Karnavas, A.; Fengou, L.C.; Bakali, A.; Lianou, A.; Tsakanikas, P.; Nychas, G.J.E. Spectroscopy and imaging technologies coupled with machine learning for the assessment of the microbiological spoilage associated to ready-to-eat leafy vegetables. *Int. J. Food Microbiol.* **2022**, *361*, 109458. [CrossRef]
116. Shi, R.; Xia, F.; Zeng, W.; Qu, H. Raman spectroscopic classification of foodborne pathogenic bacteria based on PCA-Stacking model. *Laser Optoelectron. Prog.* **2019**, *56*, 043003.
117. Umaquinga-Criollo, A.C.; Tamayo-Quintero, J.D.; Moreno-García, M.N.; Aalaila, Y.; Peluffo-Ordóñez, D.H. Developments on Support Vector Machines for Multiple-Expert Learning. In Proceedings of the Intelligent Data Engineering and Automated Learning–IDEAL 2021: 22nd International Conference, IDEAL 2021, Proceedings 22, Manchester, UK, 25–27 November 2021; Springer: Berlin/Heidelberg, Germany, 2021; pp. 587–598.

foods

MDPI

Review

The Application of Artificial Intelligence and Big Data in the Food Industry

Haohan Ding [1,2], Jiawei Tian [2], Wei Yu [3], David I. Wilson [4], Brent R. Young [3], Xiaohui Cui [1,5,*], Xing Xin [1], Zhenyu Wang [6] and Wei Li [2]

[1] Science Center for Future Foods, Jiangnan University, Wuxi 214122, China; dinghaohan@jiangnan.edu.cn (H.D.); x.xin@jiangnan.edu.cn (X.X.)
[2] School of Artificial Intelligence and Computer Science, Jiangnan University, Wuxi 214122, China; z1370196418@gmail.com (J.T.); cs_weili@jiangnan.edu.cn (W.L.)
[3] Department of Chemical & Materials Engineering, University of Auckland, Auckland 1010, New Zealand; w.yu@auckland.ac.nz (W.Y.); b.young@auckland.ac.nz (B.R.Y.)
[4] Electrical and Electronic Engineering Department, Auckland University of Technology, Auckland 1010, New Zealand; diwilson@aut.ac.nz
[5] School of Cyber Science and Engineering, Wuhan University, Wuhan 430072, China
[6] Jiaxing Institute of Future Food, Jiaxing 314050, China; zhenyuwang@whu.edu.cn
* Correspondence: xcui@whu.edu.cn

Abstract: Over the past few decades, the food industry has undergone revolutionary changes due to the impacts of globalization, technological advancements, and ever-evolving consumer demands. Artificial intelligence (AI) and big data have become pivotal in strengthening food safety, production, and marketing. With the continuous evolution of AI technology and big data analytics, the food industry is poised to embrace further changes and developmental opportunities. An increasing number of food enterprises will leverage AI and big data to enhance product quality, meet consumer needs, and propel the industry toward a more intelligent and sustainable future. This review delves into the applications of AI and big data in the food sector, examining their impacts on production, quality, safety, risk management, and consumer insights. Furthermore, the advent of Industry 4.0 applied to the food industry has brought to the fore technologies such as smart agriculture, robotic farming, drones, 3D printing, and digital twins; the food industry also faces challenges in smart production and sustainable development going forward. This review articulates the current state of AI and big data applications in the food industry, analyses the challenges encountered, and discusses viable solutions. Lastly, it outlines the future development trends in the food industry.

Keywords: artificial intelligence; big data; food industry; emerging technologies; smart sensors

1. Introduction

Artificial intelligence (AI) is a set of technologies that simulate human intelligence, which allows computers to imitate aspects of human thinking and behavior to achieve autonomous learning, reasoning, planning, and decision-making. The core of artificial intelligence is machine learning (ML), deep learning, natural language processing, computer vision, and other technologies, which can be applied to a variety of fields and industries.

The term 'big data' refers to the large and diverse collection of data [1] typically generated by a plethora of sensors or mobile devices, scraped from the internet and other sources, and includes structured and unstructured data, such as text, images, and videos. These data collections are usually characterized by high speed, high density, and high dimensionality, so they need to be stored, processed, and analyzed using bespoke technologies, often now referred to as 'big data technologies'. In the food industry, big data analytics can help predict market demand, optimize the supply chain, improve food safety and quality, and bring more opportunities and competitive advantages to businesses [2].

Artificial intelligence and big data, as two important technologies that complement each other, play a key role in the food industry and are making a significant impact on innovation and development [3].

Currently, artificial intelligence is widely used in healthcare [4], finance, transportation, manufacturing, and the gaming and processing industries [5,6]. The application of AI in these industries can improve efficiency, reduce costs, and improve service quality.

Artificial intelligence technologies have also been widely used in the food sector. Expert systems, fuzzy logic systems, ANFIS (adaptive neuro-fuzzy inference system) technology, big data, blockchain, and smart sensors are applied to food classification, production development, marketing, supervision, food quality improvement, and supply chain management [7]. The technology has improved food safety and quality [8]. Now, AI technology can also help food companies achieve intelligent production and management, which includes food quality testing, food production control, and food safety monitoring [9].

In addition to artificial intelligence, big data is also an important support that cannot be ignored in the food industry. Big data analytics technology enables companies to extract valuable information from huge data sets and gain insight into key information such as market trends and consumer preferences. Through big data analysis, food companies can better grasp market demand, optimize supply chains, reduce waste, and improve production efficiency. Big data technology can also be combined with other technologies, such as blockchain technology. Blockchain can ensure the traceability of food provenance and quality, transparency from farm to fork, and provide consumers with more reliable product information.

With the continuous progress of technology and the expansion of application scenarios, the application prospects of artificial intelligence and big data in the food industry will be broader. They will become an important means for the transformation and upgrading of the food industry and enhance its competitiveness, bringing more opportunities and challenges to the food industry [10,11].

According to Bradford's Law [12], selected core journals are regarded as important sources of information and represent the latest scientific and technological advances. In terms of the journal selection of review papers, this review searched the Web of Science database and studied the relevant literature on the application of artificial intelligence (AI) and big data (BD) in the field of food. We systematically refined the retrieved literature with keyword selection (search expression: TS = (food (detection or processing)) and (big data or (artificial intelligence and algorithms)) and year (last 5 years), extracted more than 900 articles, and finally selected more than 150 articles related to artificial intelligence, big data, and food or agriculture to ensure access to the latest research results. Then, by drawing a word cloud (as shown in Figure 1 below), we conducted keyword cluster analysis on relevant articles in the Web of Science database. In Figure 1, high-frequency words are visually highlighted to help readers quickly perceive important information.

The paper is organized as follows. In Section 1, we begin with an introduction to the basic conceptual topics of artificial intelligence and big data, giving an overview of the current state of the food industry. In Section 2, we delve into the application of big data analytics in the food sector, highlighting the potential uses of blockchain technology in terms of security and supply and demand. In Section 3, we discuss the various applications of AI in the food industry and explore its future developments and challenges.

1.1. The Early Situation in the Food Industry

In the early stages of the development of the food industry [13], there was a focus on developing more efficient hardware, equipment, and new processing techniques to improve the efficiency of food production and processing. These technological innovations were essential to meet the growing demand for food and to ensure food quality.

Figure 1. Word Cloud Generated Based on Keyword Retrieval in the Web of Science Database.

During the nascent stages of the food industry, improvements in agricultural tools and techniques were major initiatives to increase the yield and quality of crops. As agricultural production evolved, machinery and equipment such as stone mills and presses were gradually introduced to make processing more efficient and to maintain the freshness and quality of food. At the same time, food processing techniques were improved, and methods such as curing, drying, and smoking were developed to extend the shelf life of food and facilitate storage and transport.

In addition to the development of hardware and processing technology, the effective connection of the entire food chain was also an important concern for the early food industry. The lack of close collaboration and information sharing between raw material producers, processors, and retailers led to inefficiencies in supply chain management, generating excessive inventory and food waste. In order to solve this problem, the food industry has started to establish a more comprehensive supply chain management system [14]. Through the application of information technology, data exchange and information sharing between different segments have been improved, enabling more efficient logistics and inventory management. Meanwhile, through measures such as standardization and quality certification, cooperation and trust between different links in the food industry chain have been enhanced to ensure that the origin and quality of food can be traced.

In summary, the food industry has significantly improved the efficiency of food production and processing from its early stages through the development of more efficient hardware, equipment, and new processing technologies to meet the growing demand for food. At the same time, through the establishment of a more comprehensive supply chain management system, the entire food industry chain has realized effective connection and collaboration, from the production of raw materials to end-users, laying a solid foundation for the development and prosperity of the food industry.

1.2. The Current State of the Food Industry

The food industry currently covers a wide range of areas, including food service, food processing, and food retailing [3] and related industries, including agriculture, fisheries, and livestock. It is a large global industry that involves the entire food value chain from production to consumption. Food provides consumers with increasingly diverse choices and demands for safe, secure food. Among them, some new technologies are widely used in the food industry, including artificial intelligence, big data, 3D printing technology, and blockchain technology. The application of these new technologies can improve productivity, reduce waste, improve food quality and safety, increase mass demand, and retain food consumers [15].

1.3. The Importance of Food Safety

Food safety issues have always been a focus of attention. Establishing a safety traceability system is key to ensuring food safety and generating revenue for food supply chain components [16–19]. The food safety cloud [20] is the intersection of food safety work with big data, information technology, and the internet to achieve data and information technology for food quality traceability and identification and to help government departments predict food safety-sensitive information and identify periodic and trending food safety key issues promptly using big data analysis. Meanwhile, it can also provide consumers with information about food's origin, composition, personalized nutrition programs, standards, and other individualized services and dietary structure advice to help consumers choose safe food and nutritious food, not only to eat safely but also to eat healthily. In the field of food safety [21], machine-learning technology has shown significant improvement in detecting potential food contamination sources. By employing a supervised learning algorithm, we can train the model to analyze multiple variables in the production process, such as temperature, humidity, and chemical composition. This model can identify patterns associated with past food safety incidents and detect anomalies in real-time production. For example, image analysis using convolutional neural networks (CNN) [22] can accurately identify tiny defects or foreign objects on food surfaces, which can help prevent substandard products from entering the market. In the field of food quality [23–25], the application of deep learning technology has become the frontier of improving the precision of product quality monitoring. Using sequential models such as recurrent neural networks (RNN) [26] and short-term memory networks (LSTM) [27], multiple parameters can be monitored in real-time on the production line, and fine classification of product appearance and quality characteristics can be achieved. For example, by training convolutional neural networks on high-resolution images, small color changes or shape defects can be detected, thereby improving the overall quality level of the product.

1.4. Digital Transformation in the Food Industry

As technology continues to advance, the food industry is undergoing a digital transformation [28]. Automated production and smart manufacturing technologies have improved production efficiency, while big data analytics have provided better insight into consumer needs. In addition, blockchain technology enhances the transparency and traceability of the supply chain, providing security for food safety. At the same time, emerging technologies such as 3D-printed food and virtual reality also open up new possibilities for future food innovation. These digital trends provide a solid foundation for the application of artificial intelligence and big data, and this article will further explore big data, blockchain, and artificial intelligence and their key roles and advantages in the food industry.

2. Big Data in the Food Industry

Big data refers to data sets that are huge in scale, of many types, and difficult to process [29–31]. It arises from the rapid advancement of computer technology, the Internet of Things, cloud computing, mobile internet, and other technologies. The development of data has driven the progress of science and technology, and the huge amount of data provides new opportunities and challenges for data analysis. Big data generally has 5 V characteristics [32]. Volume refers to the scale of big data. The characteristic of big data is that the amount of data is very large, far exceeding the capacity of traditional data processing methods. Velocity is the speed at which data are generated and transmitted. Variety is the variety of types and formats of data. Value is the ability to extract useful information and insights from big data. Veracity is the accuracy and credibility of the data. This kind of data is very difficult to manage and analyze using traditional databases or data processing tools and requires distributed computing and storage technologies. Data technology covers the collection, storage, management, analysis, and visualization of data. However, big data technology differs from small data technology in that big data technology needs to deal with larger data sets, which require more powerful data collection, storage, management, analysis, and visualization capabilities. Through big data technology, people can better understand customer needs, market trends, product performance, and other information to provide better decision support and market insight for enterprises. In the food field, big data, on the other hand, can provide massive food-related information and analysis results to provide decision support for food companies. Applications of big data in the food field include data analysis, food traceability, food nutrition, and health [33].

Big data has an extremely important position in modern society [34]. It can be widely used in business, government, healthcare, transportation, finance, culture and entertainment, and scientific research. For example, Torre–Bastida et al. [35] discussed the application of big data in the field of transportation and mobility, including urban traffic management, intelligent transportation systems, travel mode analysis, and traffic congestion prediction. The analysis of big data [36] can provide unexpected information and insights that can help companies make business decisions, increase productivity, improve product quality, and optimize the customer experience. The application of big data is particularly evident in the food industry [37]. For example, Zuheros et al. [38] used big data to propose a decision model based on artificial intelligence sentiment analysis that analyzes and evaluates restaurant reviews to recommend restaurants to users. Delanoy and Kasztelnik [39] believe that big data analytics can enable factories that produce food to improve food safety and security. With the development of the computer industry, artificial intelligence and big data technology have also developed significantly. When artificial intelligence and big data are combined, synergies can be created. Artificial intelligence can process big data through techniques such as machine learning, deep learning, natural language processing, and computer vision to discover patterns and laws in the data, thus helping people to better understand the data and grasp the information. These intelligent technologies have had a profound impact on people's lives and brought a lot of convenience. Especially in the food industry [40], these two technologies have achieved remarkable results in food flavor matching, food safety, and food testing [41–43].

2.1. Applications of Big Data in the Food Industry

With progress and economic development, people's requirements for food have become more stringent. People are now more concerned about food flavor and food nutrition [44]. Traditional food science and technology fail to meet the specific food needs of each region and each individual. The emerging artificial intelligence and big data analytics technologies have brought revolutionary changes to the food industry in terms of supply chain optimization and food safety. The internet, combined with artificial intelligence and big data, is the future direction of food safety, and big data is an important factor in food safety [45]. For example, Kazama and Sugimoto [46] have used big data technology to form a neural network that converts food from one country's recipes to another's recipes based on its ingredients and composition, completing the transformation of regional food styles and making it easier for people to taste cuisines from exotic places. Kalra et al. [47] built a big data-based nutritional assessment system that allows people to give higher priority to the nutritional content of the food on their plates and calculate the nutrition of food through recipe analysis.

In addition, Whitehouse et al. [48] used big data technology (surveillance cameras) to monitor kitchen operations, improving food safety and providing greater peace of mind for guests sitting at the table. Not only from the food point of view, AI and big data technologies can also advise diners to choose quality restaurants and help restaurant managers make the best and most rational decisions. Lee et al. [49] developed a predictive model of restaurant reviews based on big data that is useful for customer decision-making and to help restaurant managers make the best and most rational decisions. The integration of AI and big data brings benefits to small-scale food producers or localized food systems, as described by Ajit Maru in 2018 [50], using AI and big data to design tools and applications tailored to their specific circumstances and capabilities to make data-driven agriculture [51] friendlier to them. Through real-time monitoring and feedback, local food producers can adjust their production strategies more flexibly to better meet local market needs. In addition, by establishing a blockchain-based traceability system, small-scale producers can provide more transparent and traceable product information, increasing consumer trust in local products, thereby promoting the sustainable development of local food systems. Figure 2 illustrates the different applications of big data in the food sector, including food safety and traceability, consumer insight and market analysis, product optimization and quality control, and innovation and sustainability. Figure 3 also illustrates five aspects of big data security in the food industry and the processing model of big data.

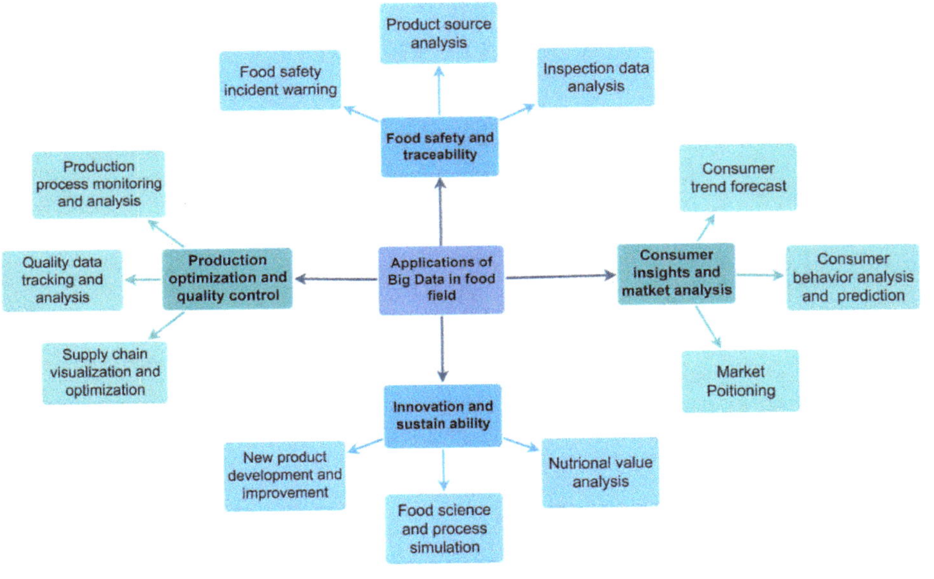

Figure 2. Applications of big data in the field of food safety and traceability, consumer insights and market analysis, production optimization and quality control, and innovation and sustainability.

2.1.1. Application of Personalized Marketing and Recommendation System

Personalized marketing is a customized marketing strategy and recommended content based on consumers' interests, preferences, and behavioral characteristics. For example, Liao et al. [52] explored how marketing campaigns in social networks can be analyzed with big data to understand user behavior, preferences, and interaction patterns. The application of personalized marketing and a recommendation system can be realized through big data analysis and artificial intelligence algorithms.

With the help of big data analytics, companies can understand consumers' purchase history, preferences, and behaviors to provide personalized product recommendations, pricing strategies, and promotions that enhance consumers' buying experience and loyalty [53].

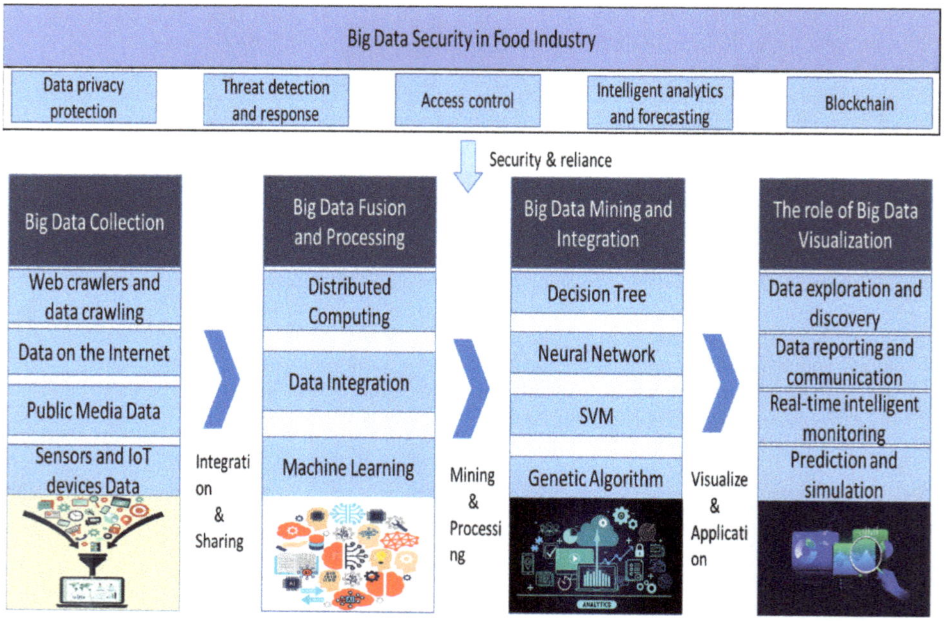

Figure 3. Five aspects of big data security in the food industry and the processing model of big data.

A recommendation system [54] uses algorithms and models to recommend food products that suit consumers' tastes and needs based on their preferences and similar consumer behaviors, increasing purchase conversion rates and sales.

2.1.2. Consumer Behavior Analysis and Forecasting

Big data analytics can help companies better understand consumer behavior patterns, decision-making processes, and purchase motivations and, thus, predict trends in consumer behavior and changes in preferences [55]. Through consumer behavior analysis, companies can identify the characteristics and needs of different consumer groups and optimize product positioning, marketing, and channel strategies. Predicting consumer behavior can help companies make more accurate inventory management, production planning, and supply chain decisions and improve operational efficiency and the ability to meet consumer demand. Through the application of personalized marketing and recommendation systems and consumer behavior analysis and prediction, food companies can better understand consumers, provide personalized products and services, enhance consumer satisfaction and loyalty, and, thus, gain an advantage in the market competition.

2.1.3. The Utilization of Big Data Analytics in Supply Chain Management

Big data analytics holds a significant position in food supply chain management [56]. By collecting and analyzing large amounts of supply chain data, companies can gain insights into inventory, procurement, production, distribution, and more. Big data analytics can help companies optimize supply chain processes, improve operational efficiency, reduce inventory costs, and decrease transportation time. With the technology and tools of big data analytics, enterprises [57] can conduct real-time monitoring and forecasting of the supply chain, identify potential risks and bottlenecks, and take appropriate measures to adjust and optimize. The application of AI and big data in the food supply chain has also proposed specific solutions to supply chain inefficiencies and food waste. As Joshi pointed out in 2020, through advanced predictive analysis of machine learning [58], accurate prediction of demand fluctuations can be achieved, and then real-time adjustment of optimal inventory levels can be achieved through intelligent inventory management systems. In addition, big data analysis plays a role in optimizing production plans [59], integrating information such as market trends, raw material supply, and production capacity to ensure the precise matching of production and market demand. Real-time monitoring and feedback mechanisms improve the visibility and responsiveness of the supply chain through IoT (Internet of Things) sensors and advanced data analytics, helping to identify and resolve problems in a timely manner. These advanced applications provide strong support for improving supply chain efficiency and reducing food waste.

2.1.4. Application of Forecasting Models and Machine Learning Algorithms in Demand Forecasting

Demand forecasting [60] is a critical component of decision-making in the food industry. By forecasting changes and trends in consumer demand, companies can plan production and supply rationally and avoid overstock or out-of-stock situations. Forecasting models and machine learning algorithms can use historical sales data, market trends, and other relevant factors to build accurate demand forecasting models. By analyzing and learning from the data, these models and algorithms can automatically identify and capture demand patterns, make forecasts, and provide targeted recommendations and decision support.

Through the application of big data analytics in supply chain management and predictive models and machine learning algorithms in demand forecasting, food companies can more accurately understand the supply chain situation and consumer demand, achieve matching of supply and demand, and improve operational efficiency and customer satisfaction. This will help optimize supply chain management [61] and production planning and enhance the competitiveness of companies.

2.2. The Bottleneck of Big Data Applications for the Food Industry

Big data technology is now extensively employed in the food sector [62], including many aspects of production, sales, and consumption. Although big data technology can bring many advantages, such as increasing production efficiency, improving marketing strategies [63], and enhancing product quality, there are also some disadvantages. These include data privacy issues, quality control problems, and bias problems. In the process of big data analysis, a great deal of data must be collected and analyzed, including consumers' personal information, behaviors, and preferences. These data may be illegally accessed and misused, posing a risk of privacy leakage to consumers. Although big data analytics can help producers better understand consumer preferences and needs, it may sometimes lead to producers overemphasizing market demand at the expense of product quality and safety, thus posing a health risk to consumers. Big data analytics is based on historical data and is used to make predictions and analyses, so there may be some bias. If producers make decisions based on data analysis alone, without considering the impact of other factors, the final decision may be wrong or risky.

Although big data technology can facilitate the development and progress of the food industry, it must be applied carefully to avoid negative effects such as data privacy issues [64] and quality control problems. Meanwhile, the government should strengthen regulation and legislation to protect the legitimate rights of consumers.

Blockchain technology is considered a viable solution to address the challenges and implications of big data [65]. 'Blockchain combined with Big Data' is considered to be an effective solution for sharing data. The changes brought by blockchain technology on the food supply chain include food security, food safety, food integrity, smallholder support, monitoring and management, etc. It also employs technologies for verifying and digitally signing documents, verifies and tracks ownership of intellectual property and proprietary systems, enables smart contracts [66], and tracks patient health records. Decentralized ledger solutions leveraging blockchain technology can also be linked with smart contracts and decentralized applications [67]. As De Filippi explained in 2016 [68], the decentralized nature of blockchain ensures distributed storage of data, eliminating a single point of attack and, thus, reducing the risk of data being hacked or tampered with. Each block contains information from the previous block, forming an immutable chain structure that provides a reliable guarantee of data transparency and integrity. Secondly, the blockchain's smart contract capabilities can be used to formulate and enforce data access rights [69]. With smart contracts, it is possible to specify which parties have the right to access, modify, or share specific types of data. This method ensures that data access can be traced and controlled and improves the management level of data privacy.

2.3. Blockchain Technology

Blockchain technology is one of the emerging technologies currently receiving a lot of attention, and it brings many benefits to the agri-food supply chain [70]. Blockchain, derived from Bitcoin, is a decentralized database recorded in the form of cryptographic blocks for the execution and sharing of every transaction or digital occurrence in a public ledger, whose information can be authenticated at any time in the future. The core features of blockchain are data immutability and decentralization, which makes the information recorded more authentic and reliable and effectively solves the problem of mutual distrust. In the field of food traceability, blockchain can achieve traceable supply chain management by assigning a unique digital identifier to each food product [71,72]. This includes the growing conditions, batch number, and expiration date of the food. This will not only help reduce food waste but also help consumers assess the ecological footprint of their food and manage the redistribution of extra food. In addition, food authentication in the food supply chain [73] (such as food ingredients, production date, production source, processing, and technology used) has been a focus of attention, and traditional authentication methods often have problems such as fraud, while the use of blockchain technology can effectively avoid such problems [71,74].

Lin et al. [75] proposed a framework for building a food traceability system with smart contracts built upon blockchain technology and the Ethereum platform. The framework is explicitly applied to the food industry, using blockchain technology to ensure transparency and safety in food production and supply chains. Mohan et al. [76] proposed a model that combines existing food quality systems and technologies at all stages of the supply chain and uses blockchain technology to propose a solution for food (chicken) tracking. Kim and Laskowski [73] used a blockchain modeling approach that integrates IoT devices for capturing and sharing data from supply chain sources. The study also addresses the issue of the existence of common data standards at different stages of the supply chain when using decentralized blockchain networks. On the other hand, Kumar and Iyengar [13], built a rice supply chain system built upon blockchain technology that aims to improve security in the management process. This case is directly relevant to the food industry, using blockchain technology to solve security issues in the rice supply chain. However, it is also important to recognize that not all blockchain applications are equally focused on food safety issues, while some may focus on other aspects such as finance and logistics.

Blockchain technology, with its reliable, secure, distributed-based nature, provides a monitored and managed full-chain solution in the agri-food supply chain from the production of raw materials to the store shelves, involving producers, consumers, suppliers, and regulators, and greatly improving food traceability [77,78]. In addition, with smart contract technology, manufacturers can further reduce costs and enhance the overall efficiency of the manufacturing industry. It is also important to mention that the potential transparency offered by blockchain technology can also help facilitate the development of a reputation-based trading system. In the context of blockchain systems applied to food supply chains, the communication and processes are typically represented as one-way. Figure 4 illustrates the process of the food supply chain in a blockchain system and the behavior of different participants, and Table 1 illustrates how blockchain can guarantee transparency and traceability in the food supply chain, improve food safety and product quality, and enhance consumer trust in products.

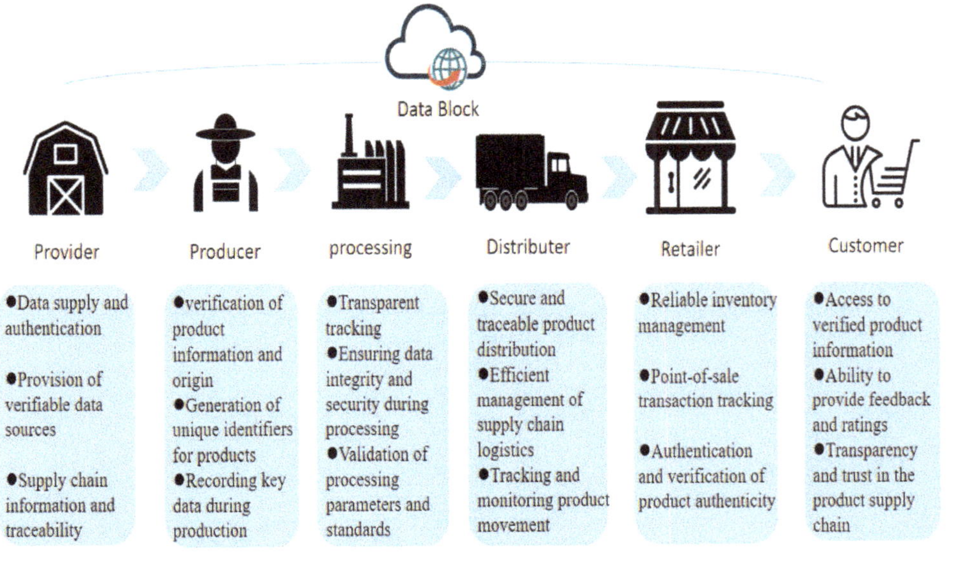

Figure 4. The process of the food supply chain in a blockchain system and the specific behaviors of different participants.

Table 1. The application of blockchain in the supply chain of agricultural products.

Products	Objectives	Projects/Companies Involved
Beer	Tracking the entire production process of beer to reveal its relevant ingredients. (Downstream is the first company to apply blockchain technology to beer.)	Downstream Brewing Company [79]
Beef	Implement blockchain technology to detect its supply chain process and prevent food fraud.	BeefLedger Corporation [80]
Grain	Identify the entire supply chain.	Agri-Digital [81]
Mango	Guarantee the traceability of the mango production chain.	IBM, Wal-Mart, Nestle, etc. [82]
High fructose corn syrup	Supervision and management.	The Coca-Cola Company
Chicken	Ensure its traceability.	Gogochicken, OriginTrail Inc. [83]
Food waste	Monitoring and management, waste forecasting.	Plastic Bank, Agora Technology Labs
Rice	Supervision and to ensure the quality of rice during transportation.	"Agri-Food Blockchain" Project [13]
Milk	Traceability to prevent food fraud in the dairy production process.	"Agri-Food Blockchain" Project

However, blockchain technology also faces future challenges and issues to be addressed [84]. With the integration of an increasing number of components into blockchain systems, such as RFID, smart sensors, smart robots, biometric data, IoT, and big data, the underlying logic and implementation may become more complex. In addition, although blockchain technology is already widely applied in food supply chains around the world, operation and maintenance costs are increasing, which may limit the entry of new suppliers into the market. In addition, large companies may adopt privately licensed blockchain technology, which may lead to giants and oligopolies monopolizing the market. In addition, how to improve the regulatory system related to blockchain technology is still an issue that needs to be further debated. Currently, there is no consensus among top policymakers and experts on how to properly apply blockchain technology and cryptocurrency transactions.

3. Artificial Intelligence in the Food Sector

When we talk about artificial intelligence in food, we are in a field full of cutting-edge technology and limitless potential. With the rapid development of science and technology, artificial intelligence has profoundly affected all levels of the food industry. From expert systems to fuzzy logic systems, from ANFIS technology to NIRS technology to CVS technology and the application of artificial intelligence and sensors in the food industry, these innovations are bringing unprecedented changes to food production, quality control, and safety [85,86]. Compared with big data, these technologies have unique characteristics and application scenarios, and each plays a unique role in solving problems and optimizing processes.

3.1. Knowledge-Based Expert Systems in the Food Industry

In the food industry, expert systems, as a knowledge-based artificial intelligence technology [87,88], can simulate the thinking style and knowledge structure of experts to achieve automated decision-making and problem-solving [89]. The system can use the knowledge and experience of domain experts, combined with techniques such as machine learning and natural language processing, to build a system with certain reasoning capabilities to automate the solution of complex problems. For example, expert systems can be used in applications such as assessing food quality, detecting food safety problems, and optimizing production processes.

In addition, some expert systems can continuously learn and improve [90]. As data and knowledge accumulate, the system can update and optimize itself, thereby improving its problem-solving accuracy and efficiency. A knowledge-based system, alternatively referred to as an expert system, is a computer application that integrates a large number of problem solutions associated with a particular domain. The system is capable of mimicking the decision-making process of human experts. Typically, it comprises six components: the human–computer interface, knowledge base, inference engine, interpreter, comprehensive database, and knowledge acquisition module, among which the most important are the knowledge base and the reasoning machine. The knowledge base stores a large number of facts, objects, cases, and rule conditions, which are expressed in the form of "IF-THEN" statements. The reasoning machine can perform specific operations on this knowledge and generate solutions to various problems with the help of human experts.

Since food safety is a top priority in the food production industry, expert systems have been applied in the production of food, quality testing, and food risk assessment [91]. In addition, expert systems are also used in the food industry, such as web-based expert systems for diagnosing pests and diseases of banana plants [92], intelligent expert system databases [93] for automatic control of product quality indexes, and expert systems based on fuzzy logic models [94] applied to the coffee industry.

The Future and Challenges of Expert Systems

In the coming years, we predict knowledge-based expert systems will be combined with techniques such as fuzzy logic and neural networks for advanced control processes in food processing, control modeling, and multivariate and nonlinear processes. In particular, the ability of fuzzy logic to handle uncertainty in hybrid fuzzy expert systems may bring significant advantages. In the food field, the application of knowledge-based expert systems has the potential to help companies reduce production costs, improve productivity, optimize product quality, and enhance market competitiveness. In the 1990s, due to the lack of hardware and software conditions, expert systems did not reach their full potential. Today, with the rapid development of hardware and software and the progress of artificial intelligence and machine learning, expert systems are highly practical and have been applied to engineering, medicine [95], business, food, and other fields; it is an integral part of artificial intelligence [96].

3.2. Fuzzy Logic Systems

Traditional knowledge-based general expert systems have a distinct disadvantage in that they are unable to handle problems that are beyond the scope of their knowledge database. When the system is confronted with problems that are not included in the knowledge base, the rule-based system will not be able to provide deterministic results. To solve the problems of traditional expert systems in dealing with uncertain and fuzzy data, Zadeh first introduced the concept of fuzzy logic in 1965. A fuzzy logic system is an artificial intelligence technique based on the principles of fuzzy logic [97–99]; its main use is to deal with problems involving fuzzy concepts. Unlike traditional logic systems, fuzzy logic systems allow the truth value of propositions to take any value between 0 and 1, thus allowing better handling of problems with fuzzy boundaries. In addition, fuzzy set theory provides a convenient means of dealing with uncertainty and transforming expert knowledge into computer-processable quantitative functions. Traditional analytical and statistical methods often struggle to deal with expert knowledge, but complex mathematical relationships are not required in building fuzzy logic applications. Fuzzy models can be expressed in easily comprehensible linguistic rules, resembling the format of rule-based expert systems. The emergence of fuzzy logic systems introduces the stereotypical judgment and imprecise nature of human judgment and can improve the generalizability of expert systems.

Fuzzy logic translates variable values into a linguistic representation, where the interpretation corresponds to a fuzzy set, and based on these representations, the fuzzy system then proceeds to the next step of judgment. Fuzzification [100] is a process in which explicit values are converted into affiliations, and a fuzzy input set is generated. The correspondence of a membership function fuzzy system is usually between 0 and 1. Fuzzy rules are also called "IF-THEN" rules [101], where IF is the necessary precondition and THEN is the corresponding result. The fuzzy logic system consists of the following steps: first, the actual values are transformed into fuzzy values, and then rules are applied to map these fuzzy inputs to fuzzy outputs. Next, a fuzzy output is obtained by combining the results of the fuzzy rules through statistical methods. Finally, the fuzzy output is transformed into specific actual values through the process of defuzzification. These steps work in concert to form the basic operational flow of a fuzzy logic system. Figure 5 illustrates these steps in the operations of a fuzzy logic controller.

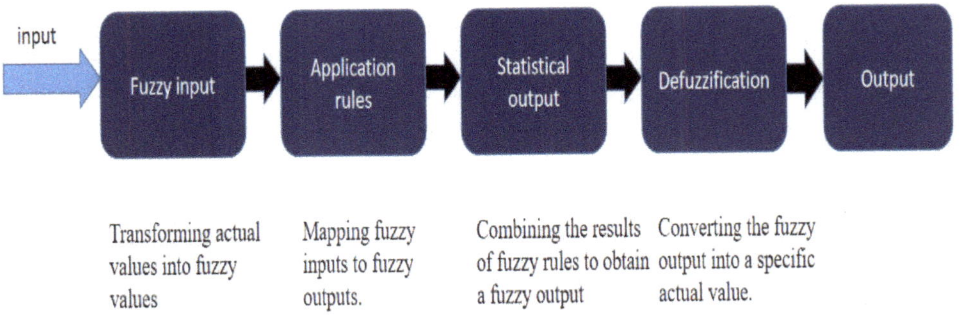

Figure 5. The operational steps of fuzzy logic controller.

Fuzzy logic systems are efficient and simple designs for fast analysis and problem-solving with a highly accurate approach. As a result, these systems have been widely used in industry [94,100,102,103]. In the food industry, fuzzy logic techniques are also applied. For example, Farzaneh et al. [104] proposed an adaptive neuro-fuzzy inference system and applied it to canola oil extraction modeling. Samodro et al. [105] used fuzzy logic techniques to help a coffee roaster maintain the smell and quality of coffee. Yulianto et al. [106] implemented a fuzzy inference system and used it for salt yield estimation. Basak et al. [107] employed fuzzy logic techniques to assess the concentration of essential oil in withered leaves and its efficacy as a preservative for fruit juices. Vivek and Subbarao [108] utilized fuzzy logic techniques in the sensory evaluation of food items. Furthermore, emerging approaches such as fuzzy set theory have proven effective in evaluating the sensory attributes of diverse traditional and innovative foods developed [109] through fortification and modified processing methods.

In fuzzy modeling, linguistic entities such as "unsatisfactory, fair, moderate, good, and very good" are used to describe sensory attributes (including color, aroma, taste, texture, and mouthfeel) of food products obtained through subjective evaluation. Shahbazi et al. [110] proposed a food traceability system based on blockchain machine learning that combines a fuzzy logic traceability system based on shelf life management systems to manipulate perishable foods to address issues regarding food resolution such as lightweight, evaporation, warehouse transactions, or shipment time. With the application of fuzzy logic systems, problems with fuzzy boundaries can be better handled, leading to more accurate and objective decisions.

3.3. Adaptive Neuro-Fuzzy Inference System (ANFIS) Technology

The adaptive neuro-fuzzy inference system [111] is an innovative structure for fuzzy inference systems that seamlessly integrates fuzzy logic and neural networks. It is a technique that adjusts the premise parameters and conclusion parameters using a hybrid algorithm of least squares and backpropagation and can automatically generate IF-THEN rules. It combines the computational learning capability of ANN (artificial neural network) networks with the human-like reasoning capability of a fuzzy logic system. Where ANFIS comes into play is reflected in the fact that, in some cases, it can automatically determine the appropriate parameters of the membership functions and does not need to set the appropriate affiliation functions and their parameters by itself. This is especially the case when we already have a set of inputs and associated output variables and values. Like artificial neural networks (ANN), ANFIS systems can "automatically" adjust their nodes and the connections between them. The five-layer structure of ANFIS consists of a fuzzy layer, a product layer, a normalization layer, a defuzzification layer, and a total output layer. Figure 6 illustrates the network structure of ANFIS.

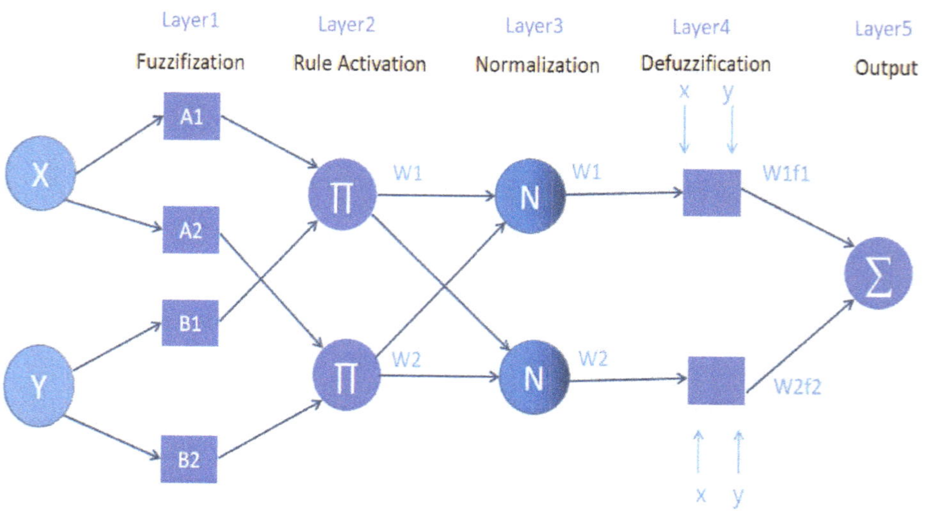

Figure 6. The five-layer structure of the ANFIS network.

In Figure 5, f_i is the output result corresponding to rule i; A1 and B1 are the nonlinear parameters corresponding to rule i; Wi represents the trigger ratio of rule i in all rule bases.

The functions of the nodes in each layer are as follows.

Layer 1: The fuzzification layer, where input variables are transformed into fuzzy sets, and the output represents the degree of membership in the fuzzy set. A1, A2, B1, B2 belong to fuzzy set A. It is responsible for fuzzifying the input features x and y using membership functions to obtain a degree of membership in the range of [0, 1].

Layer 2: The rule activation layer, which implements the operation of the fuzzy set in the antecedent part of each rule. Each node in this layer is fixed and computes the strength of each rule by multiplying the degree of membership of each feature. The output of each node represents the activation strength of a rule, and the node function can be implemented as taking bounded products or strong products.

Layer 3: The normalization layer, which normalizes the activation strengths of each rule to indicate their relative weights in the entire rule base.

Layer 4: The defuzzification layer, the result of the calculation rule, is generally given by a linear combination of input features.

Layer 5: The output layer calculates the exact output by defuzzifying the aggregated outputs of the rules. The single node in this layer computes the sum of all incoming signals as the final output of the system.

With the progress made in fuzzy logic and neural networks, the adaptive neural network fuzzy inference system, as a product combining these two theories, has emerged as a significant research direction in the field of computational intelligence due to the combined benefits of the expressive nature of fuzzy logic and the self-learning capability of neural networks. In the field of food testing, this technique has had an impact [107]. For example, researchers have used the ANFIS model to study the sensory attributes of ice cream [108] and predicted the acceptability of the taste based on the input parameters. The model achieved a minimum error rate of 5.11% and a correlation coefficient of 0.93. In a separate study, another researcher employed the ANFIS model to forecast the quality of virgin olive oil samples [112]. These results show that combining fuzzy inference systems with neural networks, especially in food testing, has great advantages. In addition, Abbaspour-Gilandeh et al. [113] designed a system combining artificial neural networks and ANFIS for predicting the kinetic energy and energy of quince under a hot air dryer, Kaveh [114] designed an ANFIS model for predicting the moisture diffusivity and specific energy consumption of drying potatoes, garlic, and melons under convection hot air dryers, Arabameri et al. [112] utilized an adaptive neuro-fuzzy inference system (ANFIS) to assess and forecast the oxidative stability of virgin olive oil. Mokarram et al. [115] employed an adaptive neuro-fuzzy inference system and multivariate linear regression to estimate the flavor of oranges. Table 2 lists the published ANFIS technology applications in the food industry.

Table 2. Published ANFIS technology applications in the food industry.

Authors	Research Subjects	Expected Goals	Experimental Results
Arabameri et al. [112]	Olive Oil	Prediction of the quality of olive oil samples and determination of the influence of other factors	Highly accurate prediction of olive oil quality and successful prediction of the effects of time, temperature, and phenolics on its stability
Kaveh et al. [114]	Potatoes, garlic, and cantaloupe	Predicted moisture diffusion rate and energy consumption ratio	Successful use of the ANFIS model for accurate prediction of its water content
Mokarram et al. [115]	Orange	Predicting orange flavor	Successful use of the ANFIS model for accurate prediction of orange flavor

Table 2. *Cont.*

Authors	Research Subjects	Expected Goals	Experimental Results
Abbaspour-Gilandeh et al. [113]	Quince	Prediction of kinetic energy and energy of quince under hot air drying	Accurate prediction of kinetic energy of quince using the ANFIS model and multiple linear regression
Kumar et al. [116]	Taro	Optimization of the extraction process of taro	Successful optimization of extraction process of taro bioactive compounds using response surface methodology and ANFIS
Ojediran JO et al. [117]	Yam	Predicting the drying characteristics of yam	Accurate prediction of drying characteristics of yam slices in convective hot air desiccant using ANFIS

3.4. Near-Infrared Spectroscopy Technology Combined with Artificial Intelligence

Near-infrared spectroscopy (NIR) is a non-invasive analytical technique based on optical principles that can be used for rapid and accurate analysis and detection of the chemical and physical properties of substances, which uses the absorption and scattering properties of the substance molecules in the near-infrared spectral region to analyze the chemical composition, structure, and properties of substances by detecting the intensity of the absorbed and scattered light. NIR technology can be used for rapid and accurate analysis of food composition and quality indicators [118]. Compared with previous techniques, NIR technology has the advantages of no chemical substances, fast and accurate results, non-destructive, low cost, and resource-saving, thus becoming a viable alternative to traditional techniques. In the past few years, the fusion of artificial intelligence algorithms and NIR technology has achieved effective monitoring of food internal quality detection and disease conditions, mainly using methods such as least squares and multiple linear regression. In the food field, NIR technology has been widely applied in food quality inspection and quality control. For example, the NIRS system combined with artificial intelligence technology can classify and detect food products, and NIR technology can detect mechanical damage to mangoes [119]. In 2020, Curto et al. [120] used artificial neural network-based near-infrared spectroscopy to precisely forecast the sensory characteristics of cheese; this technology provides an efficient and reliable quality assessment tool, but its application is not limited to cheese, it is also applicable to the quality control of other foods. Gunaratne et al. [121] used NIR spectroscopy and machine learning modeling for chocolate quality assessment; it provides a scientific basis for improving the quality and consistency of chocolate products. At the same time, this technology also provides a reference for other food composition analyses and quality assessments. Qiao et al. [122] used a hyperspectral imaging system to assess pork quality and marbling levels; it provides a new perspective on food quality control. This technology not only has a wide application in meat quality but also can be used for quality monitoring of other agricultural products, bringing new possibilities to the agricultural field. Alshejari et al. [123] designed an intelligent system for decision support in identifying meat spoilage through the analysis of multispectral images; it also provides an innovative solution to other food freshness and preservation problems.

In conclusion, NIR technology is fast, accurate, non-invasive, and low-cost, which can enhance the quality and market competitiveness of food products. With the continuous development and popularization of artificial intelligence technology, the application prospects of NIRS technology in the food field will be broader.

3.5. Application of Computer Vision Systems in the Food Industry

Computer vision is an artificial intelligence technology that utilizes image processing and pattern recognition techniques to automate the analysis and interpretation of visual data, such as images and videos. It aims to simulate human vision by enabling computers and related devices to understand digital images and videos, extract meaningful information, and make informed decisions. It involves the development of sophisticated algorithms, including traditional methods and deep learning approaches, to enable computers to perceive, analyze, and comprehend visual data in a manner similar to humans. The underlying principle of computer vision is to enable a computer to process and comprehend images at the pixel level, utilizing specialized software algorithms to retrieve, process, and interpret visual information effectively. Its main task is to acquire images and videos and process them with a computer to obtain three-dimensional information about the corresponding scene, including image segmentation, image classification, and image detection. The earliest experiments with computer vision began in the 1950s when it was used to parse typed and handwritten text. At that time, the analysis procedure was relatively simple and required a lot of manual operations, where the operator had to provide data samples manually for analysis. The traditional manual operation made it difficult to provide large amounts of data, and coupled with limited computing power, the error rate of such analysis was quite high. Today, we do not lack powerful computing power. Cloud computing, coupled with powerful algorithms, can help us solve even the most complex problems. However, it is not only the combination of new hardware and advanced algorithms that are driving the development of computer vision technology; it is also the vast amount of publicly available visual data that we generate every day that is driving the technology. Computer vision has a wide range of applications, including but not limited to medical image detection, autonomous driving, facial recognition, agriculture, and food inspection. In the food field, computer vision technology has been widely used [124]. Computer vision can be used in food production lines for automated inspection and quality control and can also be applied to food quality inspection. Lopes et al. [125] developed a computer vision classification system for barley flour that incorporates spatial pyramid segmentation integration. Phate et al. [126] employed a computer vision system to cluster the ANFIS weighing model for sweet orange (*Citrus limetta*); Nadim et al. [127] applied image processing techniques to quality control in mushrooms; Villaseñor-Aguilar et al. [128] used an artificial vision system for sweet pepper for ripeness assessment for quality control. Among the various deep learning neural networks, convolutional neural networks have been applied to computer vision, and it is particularly suitable for handling various tasks in the field of computer vision, such as image classification, object detection, and semantic segmentation. With advances in machine learning, especially the development of convolutional neural networks, collected food pictures can be further processed to classify food. Figure 7 illustrates the use of a convolutional neural network (CNN) in computer vision to automatically extract food shape features for food recognition. Table 3 lists published computer vision applications in the food industry.

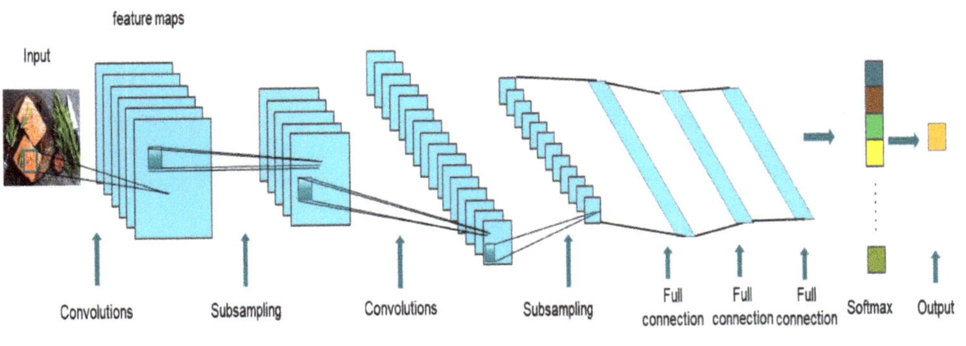

Figure 7. Automatic extraction of food shape features for food recognition using convolutional neural networks (CNN) in computer vision.

Table 3. Published computer vision applications in the food industry.

Authors	Research Subjects	Objectives	Experimental Results
Lopes et al. [125]	Barley flour	Forecast for barley flour	Classification using spatial pyramid segmentation method, the final prediction with SVM is 95%
Siswantoro et al. [129]	Eggs	Predicting egg volume	Successfully predicted egg volume with ANN model with a 97.38% success rate
Villager-Aguilar et al. [128]	Sweet pepper	Predicting the ripening status of bell peppers	Successfully developed an artificial vision system using CVS and ANN/FL to predict the ripeness of bell peppers with a maximum accuracy of 88% for FL and 100% for ANN
Bakhshipour et al. [130]	Iranian black tea and green tea	Classification of black and green teas in Iran	Successful classification of both with REP decision trees
Mazen et al. [131]	Banana	Predicting the ripening of bananas	Successfully used SVM and ANN algorithms to accurately predict the ripening level of bananas with an accuracy of 98%
Wan et al. [132]	Tomato	Predicting the ripeness of fresh tomatoes	Accurate detection of tomato ripeness with ANN algorithm with 99% accuracy
Markande et al. [133]	Potatoes	Grade classification of potatoes	A combination of CVS technology and fuzzy logic system successfully classifies potatoes and reduces costs
Garcia et al. [134]	Vegetable seeds	Sorting vegetable seeds	Successful classification of spinach seeds and cabbage seeds with ANN technology
Ozkan et al. [135]	Dry beans	Classification of different types of dry bean seeds	Successful classification of dry bean seeds with SVM, DT, ANN, and KNN algorithms
Zareiforoush et al. [136]	Rice	Grading the quality of rice	Successfully developed a system to grade rice quality with 97% accuracy

3.6. *Artificial Intelligence Combined with Smart Sensors for Real-Time Inspection in the Food Industry*

In today's food industry, guaranteeing product quality and safety holds utmost significance. With the ongoing advancements in artificial intelligence and smart sensor technology, it is now feasible to monitor and assess the quality and safety of food products in real-time. Artificial intelligence analyzes large amounts of data through machine learning algorithms to predict possible problems and identify and solve food safety issues promptly. Meanwhile, smart sensors can monitor parameters such as temperature, humidity, pressure, and odor in real-time to detect potential problems. The application of these technologies enables the food industry to more effectively ensure product quality and safety, bringing better health and quality of life to consumers.

The integration of artificial intelligence and smart sensors [137] allows for real-time monitoring and early warning to ensure food safety. Once the sensor detects an abnormality, the system immediately notifies the operator and takes appropriate action to avoid food safety problems. In addition, the sensor data can be analyzed to better understand the characteristics of product quality and process changes so that the production process and product quality can be continuously optimized. The effectiveness and practicality of this technology application open up a wider scope of development for the contemporary food industry and provide a reliable guarantee for safeguarding public health.

Smart sensors are the core foundation of artificial intelligence technologies and can be divided into different types, such as physical sensors, chemical sensors, and biological sensors. Artificial intelligence technologies that can enhance sensor systems encompass a variety of methodologies, such as knowledge-based systems, fuzzy logic, automatic knowledge acquisition, neural networks, genetic algorithms, case-based reasoning, and ambient intelligence. These intelligent techniques contribute to the capabilities of sensor systems, enabling them to acquire, process, and interpret data more effectively. These artificial intelligence technologies can be widely used in scenarios such as small sensor systems and single sensors.

The appropriate utilization of artificial intelligence technology can enhance the competitiveness of sensor systems and applications. One common approach involves integrating sensors for real-time detection, such as electronic nose (E-nose), electronic tongue (E-tongue) [138], and machine learning. These sensor integrations are often combined with technologies such as artificial neural networks to enable advanced data analysis, pattern recognition, and decision-making capabilities. This integration enables the development of intelligent sensor systems that can effectively analyze complex data and make accurate real-time assessments in various domains [138]. The first of these technologies is used to monitor food products in real-time, achieve high accuracy, and issue alerts in a short time so that timely action can be taken to safeguard food safety and quality [139]. The application of these technologies can not only improve the efficiency and productivity of the food industry but also provide the public with safer and healthier food.

In recent years, there has been significant progress in the implementation of smart sensors incorporating artificial intelligence techniques in the food sector. For example, McVey et al. [140] developed a spectroscopy-based smart sensor. Ndisya et al. [141] used an optical sensor technology based on hyperspectral imaging to assess the quality variation of purple-speckled yam slices in dry hot air and successfully developed a predictive model. This research is a clear example of applications in the food industry, but the same hyperspectral imaging technology can also be used in agriculture to monitor plant growth and predict disease. In addition, Abedi-Firoozzah et al. [142] applied kale anthocyanins in biosensors and food packaging. Sanaeifar et al. [143] used an electronic nose to classify, detect, and identify different types of fruits as well as to detect defective parts in food products. Feng et al. [144] designed an electronic nose that can be used to detect spoilage of food products such as meat and fish. The electronic tongue is an intelligent sensor capable of analyzing and identifying food through the sense of taste. It can identify different chemical substances, both organic and inorganic, through eight built-in electronic sensors. Electronic tongue technology is used in a wide range of applications in the food sector, including food traceability, food freshness, food quality grading, and quality monitoring during food production. For example, Wadehra and Pati [145] detail the application of electronic tongues in different food production and processing processes, such as fruit juices, dairy products, vegetables, and fruits. The electronic tongue technology is also capable of quickly distinguishing between chicken breeds and product quality and will, therefore, become an important tool for the evaluation of chicken meat and its product quality. An electronic tongue system developed in China uses bare metal electrodes such as gold, silver, and platinum as working electrodes and can be used in combination with suitable data analysis methods for quality differentiation of wine, liquor, tea, meat, and cow milk, as well as qualitative and quantitative analysis of food-borne pathogenic bacteria and pesticides. Through sensor technology and the Internet of Things, key parameters and indicators in the food production process can be automatically collected and recorded, reducing errors from manual intervention and providing a more accurate and comprehensive database. These data can be used for production process optimization, quality control, and food safety, helping companies achieve intelligent production and refined management. Table 4 shows a comparison of conventional laboratory instruments, electronic nose, electronic tongue, computer vision, and sensory analysis. Where '√' represents meeting this condition, and '×' represents not meeting it.

Table 4. Comparison of conventional laboratory instruments, electronic nose, electronic tongue, computer vision, and sensory analysis.

Feature	Conventional Laboratory Instruments	Electronic Nose	Electronic Tongue	Computer Vision	Sensory Analysis
Fast detection	√	√	√	√	×
Low-cost analysis	√	√	√	√	×

Table 4. *Cont.*

Feature	Conventional Laboratory Instruments	Electronic Nose	Electronic Tongue	Computer Vision	Sensory Analysis
Chemical free analysis	×	×	×	×	×
Objectivity	√	√	√	√	×
Non-destructive measurement	√	√	×	√	√
Sample pre-treatment	×	×	√	×	×
simple	√	√	√	√	×
Single operator	√	√	√	√	×
Permanent data storage	√	√	√	√	√

4. Future Trends and Challenges for Artificial Intelligence Applications in the Food Field

In recent years, the food industry has undergone rapid and dramatic changes thanks to the food industry revolution [146]. With the development of artificial intelligence, the future of the food industry will be based on technologies such as smart agriculture, robotic agriculture, drones, 3D printing, and digital twins [147–150]. Meanwhile, thanks to robotics and automation in the sustainable food industry [151], the food manufacturing industry is shifting from traditional manual production to an automated production phase. Within this change, packaging, warehousing, distribution, marketing, and consumer service are also moving toward automation. However, the emergence of these new models places higher demands on the workforce of the food industry. In addition, food safety issues have become a major international concern [152,153]. The development of IoT technology [154,155] can help solve this problem as it can identify products and trace them back to every step of food production and processing. The concept of Industry 4.0 for food processing represents an improvement in the quality and safety of processed foods in the current digital age by leveraging the Fourth Industrial Revolution (called Industry 4.0) technologies. Industry 4.0 for food processing technology has gained a lot of attention in recent years, revolutionizing and transforming many aspects of the food industry, including food processing and food inspection. The introduction of Industry 4.0 for food processing to the food industry has led to an increase in the utilization of artificial intelligence in the food sector, which addresses the various issues that arise in the food sector through innovative approaches. Industry 4.0 for food processing [156,157] integrating technology or intelligent systems into traditional industries enhances the safety and quality of processed food, standardizes production processes, reduces production costs and time, conserves energy and resources, and minimizes food loss and waste. Figure 8 introduces the elements related to AI and big data in Industry 4.0 for food processing.

More and more researchers are now conducting in-depth research in the food sector, more AI methods will emerge, and the amount of AI applications will continue to rise in a straight line. Here are some of the latest and most cutting-edge AI technologies being applied to the food industry, covering synthetic food, food production and energy efficiency management, supply chain management, sales forecasting, assisted cooking, and personalized nutrition. Compared to traditional naturally derived ingredients, machine learning and biotechnology can be used to create new synthetic foods using technologies similar to 3D printing, such as meat-free plant-based meat and egg-free cakes. By using the Internet of Things, deep learning and computational vision technologies can minimize waste energy and material resources, making production processes more efficient and environmentally friendly. At the same time, AI can also enhance the quality of food production by automatically controlling details of the production process, such as temperature, humidity, and ventilation, based on monitored environmental parameters. AI technologies can intelligently track large supply chains to ensure food traceability and improve corporate response to crises by monitoring risk factors such as weather and natural disasters that may affect food production and distribution, which is called supply chain management. Sales forecasting refers to feeding large amounts of data into AI models (such as the recently popular ChatGPT 4.0 model [158,159]), which can accurately predict consumer buying behavior. For example, factors such as food recipes, tastes, and weather drive consumer demand based on consumer buying habits and calendar events to promote sales to achieve personalized sales forecasts. The use of computer vision technology can help chefs design dishes and improve ingredient recognition and the quality of dishes, which is called assisted cooking. Nowadays, with the widespread popularity of smart bracelets, weight scales, and sports watches, users' physical data can be accessed by these devices, and they can also be combined with artificial intelligence technology to analyze users' physical conditions and provide more personalized health management solutions.

Figure 8. Key technology elements related to Industry 4.0 for food processing.

4.1. *Future Development Direction and Outlook*

4.1.1. Application Prospects of Emerging Technologies

Applications for emerging technologies include augmented reality, virtual reality, blockchain, and edge computing. Augmented reality (AR) and virtual reality (VR) technologies can provide immersive food experiences that help consumers better understand the origin, preparation process, and nutritional information of products, enhancing consumer engagement and trust. Blockchain technology can provide food traceability and trustworthiness, ensuring transparency and security in the food supply chain. Consumers can track information on the origin, processing, and transportation of food to ensure the quality and traceability of products. Edge computing pushes data processing and analysis to edge devices close to the data source, allowing real-time data processing and decision-making at places such as food production sites or farms, improving production efficiency and quality control.

4.1.2. Possible Directions for Innovation and Improvement

Possible directions for innovation and improvement cover the application of deep learning and reinforcement learning, as well as multimodal data fusion. By further developing deep learning and reinforcement learning algorithms, the prediction and decision-making capabilities of AI systems in the food industry can be improved to better address production, quality control, and risk management issues. Multimodal data fusion technology will integrate multiple data sources (such as images, text, sensor data, etc.) to provide more comprehensive food information and analysis results to help decision-makers better understand and manage food production and supply chains.

4.1.3. Exploration of Feasibility and Sustainability Issues

With the increasing amount of data in the food industry, it has become particularly important to protect the privacy and security of data for consumers and businesses. Appropriate data protection policies and technical measures are needed to guarantee the secure storage and transmission of data. In addition, training professionals with AI and big data analysis skills are key, and there is also a need to popularize and promote related technologies so that more businesses and farmers can benefit from AI and big data applications. As we advance AI and big data applications, it is imperative to intensify our focus on mitigating their adverse environmental impacts. Proactively exploring eco-friendly data center solutions and optimizing energy consumption strategies becomes paramount to ensure a sustainable trajectory of our progress within the natural world. These directions and issues provide an outlook, but the sustainability of actual developments and solutions requires further research and practice.

4.1.4. Future Challenges Ahead

In the future, the food industry will face numerous challenges. The production process generates a vast amount of data. Industry 4.0 for food processing plays a role in capturing these data, and big data plays a crucial role in collecting and processing these data, which poses a challenge for pre-processing. AI technology is used to evaluate and analyze the data produced by the food industry, but the data processed by AI often need to be manually labeled. As for blockchain, it is a relatively new and advanced technology that may encounter technical problems, cost issues, and data security and authenticity issues. AI in the food industry also faces challenges such as data privacy and security, technical complexity, and integration with traditional methods. In terms of data privacy and security, new technologies may face potential data breach risks when processing large amounts of consumer information [160]. In addition, due to the technical complexity, the maintenance and updating of AI systems may require highly specialized skills, which may limit its adoption by small businesses or farmers. When integrated with traditional methods, production processes may be disrupted or uncoordinated due to technical differences. When adapting to actual production conditions, AI systems may be affected by environmental changes. For example, certain technologies may suffer from reduced performance in extreme weather conditions, resulting in reduced productivity. In addition, if the algorithm relies on large-scale data sets that fail to adequately account for local factors, applications in specific regions may produce inaccurate results. Therefore, we need to pay more attention to these potential problems and take appropriate measures to address or mitigate the impact of these threats and vulnerabilities. The implementation of artificial intelligence (AI) and big data technologies in the food industry faces a number of limitations and obstacles, especially in terms of technical complexity and integration with existing systems. First, technical complexity, which involves the understanding and application of advanced algorithms, models, and data processing techniques, requires highly specialized technical teams. In addition, as Anagnostopoulos pointed out in 2016 [161], data in the food industry are often fragmented and heterogeneous, and consistency and standardization in processing these data is also a challenge. In this process, ensuring the quality and integrity of these data is particularly important. Integration with existing systems is another prominent issue [162].

Many food businesses already have large information management systems with traditional management approaches, and integrating new AI and big data technologies into these systems requires careful planning and effective change management. Differences in technology architecture, inconsistent data formats, and acceptance of new technologies are also challenges in the integration process. In addition, for some traditional enterprises, the change in culture and organizational structure is also an aspect that cannot be ignored in the implementation process, and there may be situations such as incompatibility with existing technologies. Figure 9 illustrates the relationship between food safety and big data, blockchain, and AI.

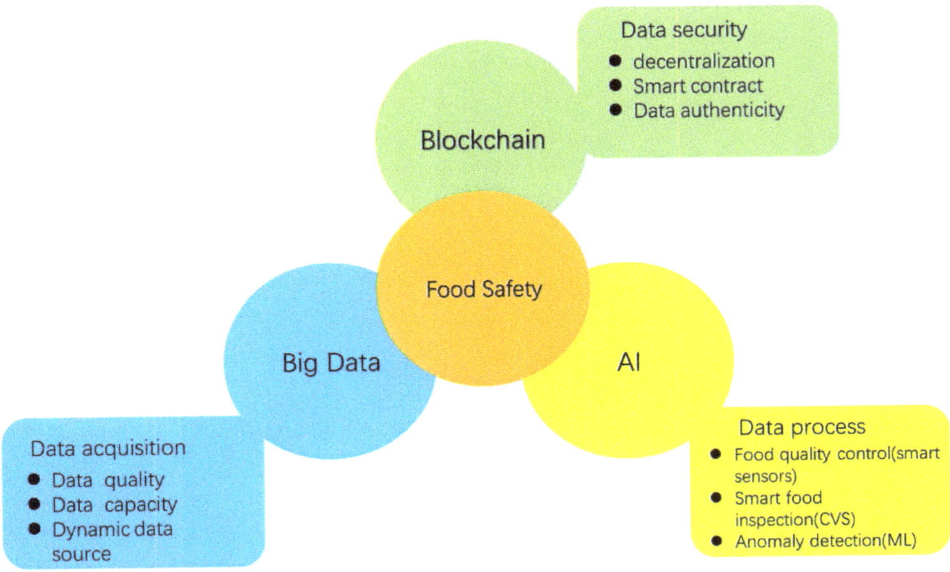

Figure 9. The relationship between food safety, big data, blockchain, and AI.

The application of artificial intelligence and big data in the food field can not only enhance production efficiency and product quality but also improve consumer experience and food safety. The emergence of emerging technologies derived from the combination of AI and big data, such as predictive analytics, automated production, intelligent quality inspection, personalized nutrition and food design, and intelligent marketing and personalized recommendation, will change the business model of the food industry, improve agricultural production efficiency, and provide a more traceable food supply chain. Future developments will need to focus on data privacy and security, talent development and technology diffusion, and environmental sustainability [163] to ensure the viability and sustainability of AI and big data in the food industry.

By leveraging the full potential of artificial intelligence and big data, the food industry can embrace a more efficient, safer, and sustainable development to provide better food products and services to consumers. This will play an important role in the future development of the global food industry.

5. Conclusions

This paper explores the application of big data and artificial intelligence (AI) in the food industry, critically analyzing the challenges these technologies face and the innovative solutions they offer. The research in this review shows that AI and big data have great potential and importance for the food industry. The application of these technologies has brought significant benefits to the food field, such as increased productivity, reduced costs, optimized supply chain management, and improved product quality and safety. However, we also note some of the challenges AI faces in the food industry, such as data privacy and security, technical complexity, and integration with traditional methods. With the continuous progress of technology, we can explore more advanced deep learning algorithms to improve the accuracy of detection in the food field, the combination of blockchain and big data as well as artificial intelligence to ensure the security of the data, and at the same time, the emergence of Industry 4.0 has accelerated the development of various technologies in the food industry, which is expected to realize a large number of industrial applications. In addition, the close integration of AI and big data will drive the food industry toward intelligence operations, sustainability, and innovation.

Author Contributions: H.D.: Writing—original draft; funding acquisition; writing—review and editing. J.T.: Writing—original draft; writing—review and editing. W.Y.: Writing—review and editing; formal analysis. D.I.W.: Writing—review and editing; data curation. B.R.Y.: Writing—review and editing; supervision. X.C.: Writing—review and editing; conceptualization. X.X.: Writing—review and editing. Z.W.: Writing—review and editing. W.L.: Writing—review and editing. All authors have read and agreed to the published version of the manuscript.

Funding: This research was funded by the National Key Research and Development Program of China (2022YFF1101100) and the Fundamental Research Funds for the Central Universities (JUSRP123053).

Institutional Review Board Statement: Not applicable.

Informed Consent Statement: Not applicable.

Data Availability Statement: Not applicable.

Acknowledgments: This work was supported by the National Key Research and Development Program of China (2022YFF1101100) and the Fundamental Research Funds for the Central Universities (JUSRP123053).

Conflicts of Interest: There are no conflict of interest to declare.

References

1. Akoka, J.; Comyn-Wattiau, I.; Laoufi, N. Research on Big Data—A systematic mapping study. *Comput. Stand. Interfaces* **2017**, *54*, 105–115. [CrossRef]
2. Schroeder, R. Big data business models: Challenges and opportunities. *Cogent Soc. Sci.* **2016**, *2*, 1166924. [CrossRef]
3. Misra, N.; Dixit, Y.; Al-Mallahi, A.; Bhullar, M.S.; Upadhyay, R.; Martynenko, A. IoT, big data, and artificial intelligence in agriculture and food industry. *IEEE Int. Things J.* **2020**, *9*, 6305–6324. [CrossRef]
4. Sahu, M.; Gupta, R.; Ambasta, R.K.; Kumar, P. Artificial intelligence and machine learning in precision medicine: A paradigm shift in big data analysis. *Prog. Mol. Biol. Transl. Sci.* **2022**, *190*, 57–100. [CrossRef] [PubMed]
5. Ramakrishna, R.R.; Abd Hamid, Z.; Zaki, W.M.D.W.; Huddin, A.B.; Mathialagan, R. Stem cell imaging through convolutional neural networks: Current issues and future directions in artificial intelligence technology. *PeerJ* **2020**, *8*, e10346. [CrossRef] [PubMed]
6. Serazetdinova, L.; Garratt, J.; Baylis, A.; Stergiadis, S.; Collison, M.; Davis, S. How should we turn data into decisions in AgriFood? *J. Sci. Food Agric.* **2019**, *99*, 3213–3219. [CrossRef] [PubMed]
7. Ben Ayed, R.; Hanana, M. Artificial Intelligence to Improve the Food and Agriculture Sector. *J. Food Qual.* **2021**, *2021*, 7. [CrossRef]
8. Corney, D. Food bytes: Intelligent systems in the food industry. *Br. Food J.* **2002**, *104*, 787–805. [CrossRef]
9. Qian, C.; Murphy, S.; Orsi, R.; Wiedmann, M. How can AI help improve food safety? *Annu. Rev. Food Sci. Technol.* **2023**, *14*, 517–538. [CrossRef]
10. Camarena, S. Artificial intelligence in the design of the transitions to sustainable food systems. *J. Clean. Prod.* **2020**, *271*, 122574. [CrossRef]
11. Mavani, N.R.; Ali, J.M.; Othman, S.; Hussain, M.; Hashim, H.; Rahman, N.A. Application of artificial intelligence in food industry—A guideline. *Food Eng. Rev.* **2022**, *14*, 134–175. [CrossRef]
12. Brookes, B.C. Theory of the Bradford law. *J. Doc.* **1977**, *33*, 180–209. [CrossRef]
13. Kumar, M.V.; Iyengar, N.C.S. A framework for Blockchain technology in rice supply chain management. *Adv. Sci. Technol. Lett.* **2017**, *146*, 125–130. [CrossRef]
14. Zhong, R.; Xu, X.; Wang, L. Food supply chain management: Systems, implementations, and future research. *Ind. Manag. Data Syst.* **2017**, *117*, 2085–2114. [CrossRef]

15. Jagtap, S.; Bader, F.; Garcia-Garcia, G.; Trollman, H.; Fadiji, T.; Salonitis, K. Food logistics 4.0: Opportunities and challenges. *Logistics* **2020**, *5*, 2. [CrossRef]
16. van Asselt, E.D.; Van der Fels-Klerx, H.; Marvin, H.; Van Bokhorst-van de Veen, H.; Groot, M.N. Overview of food safety hazards in the European dairy supply chain. *Compr. Rev. Food Sci. Food Saf.* **2017**, *16*, 59–75. [CrossRef]
17. Aung, M.M.; Chang, Y.S. Traceability in a food supply chain: Safety and quality perspectives. *Food Control* **2014**, *39*, 172–184. [CrossRef]
18. Fung, F.; Wang, H.S.; Menon, S. Food safety in the 21st century. *Biomed. J.* **2018**, *41*, 88–95. [CrossRef]
19. Janjarasskul, T.; Suppakul, P. Active and intelligent packaging: The indication of quality and safety. *Crit. Rev. Food Sci. Nutr.* **2018**, *58*, 808–831. [CrossRef]
20. Chen, K.; Tan, H.; Gao, J.; Lu, Y. Big data based design of food safety cloud platform. *Appl. Mech. Mater.* **2014**, *536*, 583–587. [CrossRef]
21. Hassoun, A.; Jagtap, S.; Garcia-Garcia, G.; Trollman, H.; Pateiro, M.; Lorenzo, J.M.; Trif, M.; Rusu, A.V.; Aadil, R.M.; Simat, V.; et al. Food quality 4.0: From traditional approaches to digitalized automated analysis. *J. Food Eng.* **2023**, *337*, 16. [CrossRef]
22. Aguilar, E.; Bolaños, M.; Radeva, P. Exploring food detection using CNNs. In Proceedings of the Computer Aided Systems Theory–EUROCAST 2017: 16th International Conference, Las Palmas de Gran Canaria, Spain, 19–24 February 2017; Revised Selected Papers, Part II 16. Springer: Berlin/Heidelberg, Germany, 2018; pp. 339–347.
23. Ding, H.; Wilson, D.I.; Yu, W.; Young, B.R. An investigation of the relative impact of process and shape factor variables on milk powder quality. *Food Bioprod. Process.* **2021**, *126*, 62–72. [CrossRef]
24. Ding, H.; Wilson, D.I.; Yu, W.; Young, B.R. Assessing and Quantifying the Surface Texture of Milk Powder Using Image Processing. *Foods* **2022**, *11*, 1519. [CrossRef] [PubMed]
25. Ding, H.; Wilson, D.I.; Yu, W.; Young, B.R.; Cui, X. Application of Three-Dimensional Digital Photogrammetry to Quantify the Surface Roughness of Milk Powder. *Foods* **2023**, *12*, 967. [CrossRef] [PubMed]
26. Zhong, J.; Sun, L.; Zuo, E.; Chen, C.; Chen, C.; Jiang, H.; Li, H.; Lv, X. An ensemble of AHP-EW and AE-RNN for food safety risk early warning. *PLoS ONE* **2023**, *18*, e0284144. [CrossRef] [PubMed]
27. Do, H.-D.; Kim, D.-E.; Lam, M.B.; Chung, W.-Y. Self-powered food assessment system using LSTM network and 915 MHz RF energy harvesting. *IEEE Access* **2021**, *9*, 97444–97456. [CrossRef]
28. Dayıoğlu, M.A.; Turker, U. Digital transformation for sustainable future-agriculture 4.0: A review. *J. Agric. Sci.* **2021**, *27*, 373–399. [CrossRef]
29. Chen, J.; Chen, Y.; Du, X.; Li, C.; Lu, J.; Zhao, S.; Zhou, X. Big data challenge: A data management perspective. *Front. Comput. Sci.* **2013**, *7*, 157–164. [CrossRef]
30. Li, Z.-H.; Wang, G.-R.; Zhou, A.-Y. Research progress and trends of big data from a database perspective. *Comput. Eng. Sci./Jisuanji Gongcheng Yu Kexue* **2013**, *35*, 19–30. [CrossRef]
31. Mohamed, A.; Najafabadi, M.K.; Wah, Y.B.; Zaman, E.A.K.; Maskat, R. The state of the art and taxonomy of big data analytics: View from new big data framework. *Artif. Intell. Rev.* **2020**, *53*, 989–1037. [CrossRef]
32. Géczy, P. Big data characteristics. *Macrotheme Rev.* **2014**, *3*, 94–104. [CrossRef]
33. Sravanthi, K.; Reddy, T.S. Applications of big data in various fields. *Int. J. Comput. Sci. Inf. Technol.* **2015**, *6*, 4629–4632. [CrossRef]
34. Du, X.; Lu, W.; Zhang, F. History, Present, and Future of Big Data Management Systems. *J. Softw.* **2019**, *30*, 127–141. [CrossRef]
35. Torre-Bastida, A.I.; Del Ser, J.; Lana, I.; Ilardia, M.; Bilbao, M.N.; Campos-Cordobes, S. Big Data for transportation and mobility: Recent advances, trends and challenges. *Iet Intell. Transp. Syst.* **2018**, *12*, 742–755. [CrossRef]
36. Chen, J.; Ding, Z.; Gao, X. Survey of Big Data Hot Techniques. *J. Beijing Univ. Technol.* **2017**, *43*, 358–367. [CrossRef]
37. Guo, P.; Wang, K.; Luo, A.; Xue, M. Computational Intelligence for Big Data Analysis: Current Status and Future Prospect. *J. Softw.* **2015**, *26*, 3010–3025. [CrossRef]
38. Zuheros, C.; Martínez-Cámara, E.; Herrera-Viedma, E.; Herrera, F. Sentiment analysis based multi-person multi-criteria decision making methodology using natural language processing and deep learning for smarter decision aid. Case study of restaurant choice using TripAdvisor reviews. *Inf. Fusion* **2021**, *68*, 22–36. [CrossRef]
39. Delanoy, N.; Kasztelnik, K. Business Open Big Data Analytics to Support Innovative Leadership Decision in Canada. *Bus. Ethics Leadersh.* **2020**, *4*, 56–74. [CrossRef]
40. Zhou, Q.; Zhang, H.; Wang, S. Artificial intelligence, big data, and blockchain in food safety. *Int. J. Food Eng.* **2021**, *18*, 1–14. [CrossRef]
41. Li, W.; Gu, C.; Chen, J.; Ma, C.; Zhang, X.; Chen, B.; Wan, S. DLS-GAN: Generative Adversarial Nets for Defect Location Sensitive Data Augmentation. *IEEE Trans. Autom. Sci. Eng.* **2023**, *33*, 1–17. [CrossRef]
42. Ding, H.; Li, B.; Boiarkina, I.; Wilson, D.I.; Yu, W.; Young, B.R. Effects of morphology on the bulk density of instant whole milk powder. *Foods* **2020**, *9*, 1024. [CrossRef] [PubMed]
43. Ding, H.; Yu, W.; Boiarkina, I.; Depree, N.; Young, B.R. Effects of morphology on the dispersibility of instant whole milk powder. *J. Food Eng.* **2020**, *276*, 109841. [CrossRef]
44. Schakel, S.F.; Buzzard, I.M.; Gebhardt, S.E. Procedures for estimating nutrient values for food composition databases. *J. Food Compos. Anal.* **1997**, *10*, 102–114. [CrossRef]
45. Marvin, H.J.; Janssen, E.M.; Bouzembrak, Y.; Hendriksen, P.J.; Staats, M. Big data in food safety: An overview. *Crit. Rev. Food Sci. Nutr.* **2017**, *57*, 2286–2295. [CrossRef] [PubMed]

46. Kazama, M.; Sugimoto, M.; Hosokawa, C.; Matsushima, K.; Varshney, L.R.; Ishikawa, Y. A neural network system for transformation of regional cuisine style. *Front. ICT* **2018**, *5*, 14. [CrossRef]
47. Kalra, J.; Batra, D.; Diwan, N.; Bagler, G. Nutritional profile estimation in cooking recipes. In Proceedings of the 2020 IEEE 36th International Conference on Data Engineering Workshops (ICDEW), Dallas, TX, USA, 14–20 April 2020; IEEE: Piscataway, NJ, USA, 2020; pp. 82–87.
48. Whitehouse, S.; Yordanova, K.; Paiement, A.; Mirmehdi, M. Recognition of unscripted kitchen activities and eating behaviour for health monitoring. In Proceedings of the 2nd IET International Conference on Technologies for Active and Assisted Living (TechAAL 2016), London, UK, 24–25 October 2016; IET: London, UK, 2016; pp. 1–6.
49. Lee, M.; Kwon, W.; Back, K.-J. Artificial intelligence for hospitality big data analytics: Developing a prediction model of restaurant review helpfulness for customer decision-making. *Int. J. Contemp. Hosp. Manag.* **2021**, *33*, 2117–2136. [CrossRef]
50. Maru, A.; Berne, D.; De Beer, J.; Ballantyne, P.; Pesce, V.; Kalyesubula, S.; Fourie, N.; Addison, C.; Collett, A.; Chaves, J. Digital and data-driven agriculture: Harnessing the power of data for smallholders. *F1000Research* **2018**, *7*, 525. [CrossRef]
51. Kamble, S.S.; Gunasekaran, A.; Gawankar, S.A. Achieving sustainable performance in a data-driven agriculture supply chain: A review for research and applications. *Int. J. Prod. Econ.* **2020**, *219*, 179–194. [CrossRef]
52. Liao, S.-H.; Yang, C.-A. Big data analytics of social network marketing and personalized recommendations. *Soc. Netw. Anal. Min.* **2021**, *11*, 1–19. [CrossRef]
53. Mariani, M.M.; Wamba, S.F. Exploring how consumer goods companies innovate in the digital age: The role of big data analytics companies. *J. Bus. Res.* **2020**, *121*, 338–352. [CrossRef]
54. Verma, J.P.; Patel, B.; Patel, A. Big data analysis: Recommendation system with Hadoop framework. In Proceedings of the 2015 IEEE International Conference on Computational Intelligence & Communication Technology, Ghaziabad, India, 13–14 February 2015; IEEE: Piscataway, NJ, USA, 2015; pp. 92–97.
55. Le, T.M.; Liaw, S.-Y. Effects of pros and cons of applying big data analytics to consumers' responses in an e-commerce context. *Sustainability* **2017**, *9*, 798. [CrossRef]
56. Ji, G.; Hu, L.; Tan, K.H. A study on decision-making of food supply chain based on big data. *J. Syst. Sci. Syst. Eng.* **2017**, *26*, 183–198. [CrossRef]
57. Rabl, T.; Sadoghi, M.; Jacobsen, H.-A.; Gómez-Villamor, S.; Muntés-Mulero, V.; Mankowskii, S. Solving big data challenges for enterprise application performance management. *arXiv* **2012**, arXiv:1208.4167. [CrossRef]
58. Joshi, A.; Khosravy, M.; Gupta, N. Machine learning for predictive analysis. *Proc. ICTIS* **2020**, *13*, 225. [CrossRef]
59. Cheng, Y.; Chen, K.; Sun, H.; Zhang, Y.; Tao, F. Data and knowledge mining with big data towards smart production. *J. Ind. Inf. Integr.* **2018**, *9*, 1–13. [CrossRef]
60. Hofmann, E.; Rutschmann, E. Big data analytics and demand forecasting in supply chains: A conceptual analysis. *Int. J. Logist. Manag.* **2018**, *29*, 739–766. [CrossRef]
61. Djekic, I.; Sanjuán, N.; Clemente, G.; Jambrak, A.R.; Tonda, A. Review on environmental models in the food chain-Current status and future perspectives. *J. Clean. Prod.* **2018**, *176*, 1012–1025. [CrossRef]
62. Bronson, K.; Knezevic, I. Big Data in food and agriculture. *Big Data Soc.* **2016**, *3*, 2053951716648174. [CrossRef]
63. Ducange, P.; Pecori, R.; Mezzina, P. A glimpse on big data analytics in the framework of marketing strategies. *Soft Comput.* **2018**, *22*, 325–342. [CrossRef]
64. Mehmood, A.; Natgunanathan, I.; Xiang, Y.; Hua, G.; Guo, S. Protection of big data privacy. *IEEE Access* **2016**, *4*, 1821–1834. [CrossRef]
65. Deepa, N.; Pham, Q.-V.; Nguyen, D.C. A survey on blockchain for big data: Approaches, opportunities, and future directions. *Future Gener. Comput. Syst.* **2022**, *131*, 209–226. [CrossRef]
66. Khan, S.N.; Loukil, F.; Ghedira-Guegan, C.; Benkhelifa, E.; Bani-Hani, A. Blockchain smart contracts: Applications, challenges, and future trends. *Peer–Peer Netw. Appl.* **2021**, *14*, 2901–2925. [CrossRef] [PubMed]
67. Puthal, D.; Malik, N.; Mohanty, S.P.; Kougianos, E.; Yang, C. The blockchain as a decentralized security framework [future directions]. *IEEE Consum. Electron. Mag.* **2018**, *7*, 18–21. [CrossRef]
68. De Filippi, P. The interplay between decentralization and privacy: The case of blockchain technologies. *J. Peer Prod. Issue* **2016**, *14*, 508–527.
69. Macrinici, D.; Cartofeanu, C.; Gao, S. Smart contract applications within blockchain technology: A systematic mapping study. *Telemat. Inform.* **2018**, *35*, 2337–2354. [CrossRef]
70. Ahmed, S.; ten Broek, N. Blockchain could boost food security. *Nature* **2017**, *550*, 43. [CrossRef]
71. Feng, H.; Wang, X.; Duan, Y.Q.; Zhang, J.; Zhang, X.S. Applying blockchain technology to improve agri-food traceability: A review of development methods, benefits and challenges. *J. Clean. Prod.* **2020**, *260*, 15. [CrossRef]
72. Jaison, F.; Ramaiah, N.S. A survey on traceability in food safety system using blockchain. *J. Discret. Math. Sci. Cryptogr.* **2022**, *25*, 793–799. [CrossRef]
73. Kim, H.M.; Laskowski, M. Toward an ontology-driven blockchain design for supply-chain provenance. *Intell. Syst. Account. Financ. Manag.* **2018**, *25*, 18–27. [CrossRef]
74. Demestichas, K.; Peppes, N.; Alexakis, T.; Adamopoulou, E. Blockchain in Agriculture Traceability Systems: A Review. *Appl. Sci.* **2020**, *10*, 4113. [CrossRef]

75. Lin, Q.; Wang, H.; Pei, X.; Wang, J. Food safety traceability system based on blockchain and EPCIS. *IEEE Access* **2019**, *7*, 20698–20707. [CrossRef]
76. Mohan, T. Improve Food Supply Chain Traceability Using Blockchain. Master's Thesis, The Pennsylvania State University, State College, PA, USA, 2018.
77. Mao, D.H.; Hao, Z.H.; Wang, F.; Li, H.S. Innovative Blockchain-Based Approach for Sustainable and Credible Environment in Food Trade: A Case Study in Shandong Province, China. *Sustainability* **2018**, *10*, 3149. [CrossRef]
78. Smith, B.G. Developing sustainable food supply chains. *Philos. Trans. R. Soc. B-Biol. Sci.* **2008**, *363*, 849–861. [CrossRef] [PubMed]
79. Ariemma, L.; De Carlo, N.; Pennino, D.; Pizzonia, M.; Vitaletti, A.; Zecchini, M. Blockchain for the supply chain of the Italian craft beer sector: Tracking and discount coupons. *Int. J. Parallel Emergent Distrib. Syst.* **2023**, *16*, 1–34. [CrossRef]
80. Cao, S.; Powell, W.; Foth, M.; Natanelov, V.; Miller, T.; Dulleck, U. Strengthening consumer trust in beef supply chain traceability with a blockchain-based human-machine reconcile mechanism. *Comput. Electron. Agric.* **2021**, *180*, 105886. [CrossRef]
81. Lucena, P.; Binotto, A.P.; Momo, F.d.S.; Kim, H. A case study for grain quality assurance tracking based on a Blockchain business network. *arXiv* **2018**, arXiv:1803.07877. [CrossRef]
82. Kshetri, N. Blockchain and the economics of food safety. *It Prof.* **2019**, *21*, 63–66. [CrossRef]
83. Bai, C.; Sarkis, J. A supply chain transparency and sustainability technology appraisal model for blockchain technology. *Int. J. Prod. Res.* **2020**, *58*, 2142–2162. [CrossRef]
84. Zheng, Z.; Xie, S.; Dai, H.-N.; Chen, X.; Wang, H. Blockchain challenges and opportunities: A survey. *Int. J. Web Grid Serv.* **2018**, *14*, 352–375. [CrossRef]
85. Marvin, H.J.; Bouzembrak, Y.; Van der Fels-Klerx, H. Digitalisation and Artificial Intelligence for sustainable food systems. *Trends Food Sci. Technol.* **2022**, *120*, 344–348. [CrossRef]
86. Li, W.; Chen, J.; Cao, J.; Ma, C.; Wang, J.; Cui, X.; Chen, P. EID-GAN: Generative adversarial nets for extremely imbalanced data augmentation. *IEEE Transac. Ind. Inform.* **2022**, *19*, 3208–3218. [CrossRef]
87. Kaula, R.; Lander, L.C. A Module-Based Conceptual-Framework For Large-Scale Expert-Systems. *Ind. Manag. Data Syst.* **1995**, *95*, 15–23. [CrossRef]
88. Yang, X.; Zhu, D.-Q.; Sang, Q.-B. Research and prospect of expert system. *Jisuanji Yingyong Yanjiu/Appl. Res. Comput.* **2007**, *24*, 4–9. [CrossRef]
89. Valavanis, K.P.; Kokkinaki, A.I.; Tzafestas, S.G. Knowledge-Based (Expert) Systems In Engineering Applications—A Survey. *J. Intell. Robot. Syst.* **1994**, *10*, 113–145. [CrossRef]
90. Myers, W. Introduction to expert systems. *IEEE Intell. Syst.* **1986**, *1*, 100–109. [CrossRef]
91. Filter, M.; Appel, B.; Buschulte, A. Expert systems for food safety. *Curr. Opin. Food Sci.* **2015**, *6*, 61–65. [CrossRef]
92. Vasquez, R.P.; Aguilar-Lasserre, A.A.; Lopez-Segura, M.V.; Rivero, L.C.; Rodriguez-Duran, A.A.; Rojas-Luna, M.A. Expert system based on a fuzzy logic model for the analysis of the sustainable livestock production dynamic system. *Comput. Electron. Agric.* **2019**, *161*, 104–120. [CrossRef]
93. Blagoveshchenskiy, I.; Blagoveshchenskiy, V.; Besfamilnaya, E.; Sumerin, V. Development of databases of intelligent expert systems for automatic control of product quality indicators. In *Journal of Physics: Conference Series*; IOP Publishing: Bristol, UK, 2020; p. 012019.
94. Hernández-Vera, B.; Aguilar Lasserre, A.A.; Gastón Cedillo-Campos, M.; Herrera-Franco, L.E.; Ochoa-Robles, J. Expert system based on fuzzy logic to define the production process in the coffee industry. *J. Food Process Eng.* **2017**, *40*, e12389. [CrossRef]
95. Saibene, A.; Assale, M.; Giltri, M. Expert systems: Definitions, advantages and issues in medical field applications. *Expert Syst. Appl.* **2021**, *177*, 114900. [CrossRef]
96. Wagner, W.P. Trends in expert system development: A longitudinal content analysis of over thirty years of expert system case studies. *Expert Syst. Appl.* **2017**, *76*, 85–96. [CrossRef]
97. Castillo, O.; Amador-Angulo, L.; Castro, J.R.; Garcia-Valdez, M. A comparative study of type-1 fuzzy logic systems, interval type-2 fuzzy logic systems and generalized type-2 fuzzy logic systems in control problems. *Inf. Sci.* **2016**, *354*, 257–274. [CrossRef]
98. Eerikainen, T.; Linko, P.; Linko, S.; Siimes, T.; Zhu, Y.-H. Fuzzy-Logic and Neural-Network Applications In Food-Science And Technology. *Trends Food Sci. Technol.* **1993**, *4*, 237–242. [CrossRef]
99. Li, H. Output-back fuzzy logic systems and equivalence with feedback neural networks. *Chin. Sci. Bull.* **2000**, *45*, 592–596. [CrossRef]
100. Farzaneh, V.; Ghodsvali, A.; Bakhshabadi, H.; Ganje, M.; Dolatabadi, Z.; Carvalho, I.S. Modelling of the Selected Physical Properties of the Fava Bean with Various Moisture Contents UsingFuzzy Logic Design. *J. Food Process Eng.* **2017**, *40*, 9. [CrossRef]
101. Zadeh, L.A. The calculus of fuzzy if-then rules. In Proceedings of the Fuzzy Engineering toward Human Friendly Systems—Proceedings of the International Fuzzy Engineering Symposium, Leipzig, Germany, 20–22 September 2023; pp. 11–12.
102. Mahadevappa, J.; Gross, F.; Delgado, A. Fuzzy logic based process control strategy for effective sheeting of wheat dough in small and medium-sized enterprises. *J. Food Eng.* **2017**, *199*, 93–99. [CrossRef]
103. Tanima, C.; Madhusweta, D. Sensory evaluation of aromatic foods packed in developed starch based films using fuzzy logic. *Int. J. Food Stud.* **2015**, *4*, 29–48. [CrossRef]
104. Farzaneh, V.; Bakhshabadi, H.; Gharekhani, M.; Ganje, M.; Farzaneh, F.; Rashidzadeh, S.; Carvalho, I.S. Application of an adaptive neuro_fuzzy inference system (ANFIS) in the modeling of rapeseeds' oil extraction. *J. Food Process Eng.* **2017**, *40*, 8. [CrossRef]

105. Samodro, B.; Mahesworo, B.; Suparyanto, T.; Atmaja, D.B.S.; Pardamean, B. Maintaining the quality and aroma of coffee with fuzzy logic coffee roasting machine. In *IOP Conference Series: Earth and Environmental Science*; IOP Publishing: Bristol, UK, 2020; p. 012148.
106. Yulianto, T.; Komariyah, S.; Ulfaniyah, N. Application of fuzzy inference system by Sugeno method on estimating of salt production. In *AIP Conference Proceedings*; AIP Publishing: Melville, NY, USA, 2017.
107. Basak, S. The use of fuzzy logic to determine the concentration of betel leaf essential oil and its potency as a juice preservative. *Food Chem.* **2018**, *240*, 1113–1120. [CrossRef]
108. Vivek, K.; Subbarao, K.; Routray, W.; Kamini, N.; Dash, K.K. Application of fuzzy logic in sensory evaluation of food products: A comprehensive study. *Food Bioprocess Technol.* **2020**, *13*, 1–29. [CrossRef]
109. Pelsmaeker, S.; Gellynck, X.; Delbaere, C.; Declercq, N.; Dewettinck, K. Consumer-driven product development and improvement combined with sensory analysis: A case-study for European filled chocolates. *Food Qual. Prefer.* **2015**, *41*, 20–29. [CrossRef]
110. Shahbazi, Z.; Byun, Y.-C. A procedure for tracing supply chains for perishable food based on blockchain, machine learning and fuzzy logic. *Electronics* **2020**, *10*, 41. [CrossRef]
111. Azeem, M.F.; Hanmandlu, M.; Ahmad, N. Generalization of adaptive neuro-fuzzy inference systems. *IEEE Trans. Neural Netw.* **2000**, *11*, 1332–1346. [CrossRef] [PubMed]
112. Arabameri, M.; Nazari, R.R.; Abdolshahi, A.; Abdollahzadeh, M.; Mirzamohammadi, S.; Shariatifar, N.; Barba, F.J.; Mousavi Khaneghah, A. Oxidative stability of virgin olive oil: Evaluation and prediction with an adaptive neuro-fuzzy inference system (ANFIS). *J. Sci. Food Agric.* **2019**, *99*, 5358–5367. [CrossRef] [PubMed]
113. Abbaspour-Gilandeh, Y.; Jahanbakhshi, A.; Kaveh, M. Prediction kinetic, energy and exergy of quince under hot air dryer using ANNs and ANFIS. *Food Sci. Nutr.* **2020**, *8*, 594–611. [CrossRef] [PubMed]
114. Kaveh, M.; Sharabiani, V.R.; Chayjan, R.A.; Taghinezhad, E.; Abbaspour-Gilandeh, Y.; Golpour, I. ANFIS and ANNs model for prediction of moisture diffusivity and specific energy consumption potato, garlic and cantaloupe drying under convective hot air dryer. *Inf. Process. Agric.* **2018**, *5*, 372–387. [CrossRef]
115. Mokarram, M.; Amin, H.; Khosravi, M.R. Using adaptive neuro-fuzzy inference system and multiple linear regression to estimate orange taste. *Food Sci. Nutr.* **2019**, *7*, 3176–3184. [CrossRef] [PubMed]
116. Kumar, V.; Sharma, H.K. Process optimization for extraction of bioactive compounds from taro (Colocasia esculenta), using RSM and ANFIS modeling. *J. Food Meas. Charact.* **2017**, *11*, 704–718. [CrossRef]
117. Ojediran, J.O.; Okonkwo, C.E.; Adeyi, A.J.; Adeyi, O.; Olaniran, A.F.; George, N.E.; Olayanju, A.T. Drying characteristics of yam slices (*Dioscorea rotundata*) in a convective hot air dryer: Application of ANFIS in the prediction of drying kinetics. *Heliyon* **2020**, *6*, 3. [CrossRef]
118. Weeranantanaphan, J.; Downey, G.; Allen, P.; Sun, D.-W. A review of near infrared spectroscopy in muscle food analysis: 2005–2010. *J. Near Infrared Spectrosc.* **2011**, *19*, 61–124. [CrossRef]
119. Rivera, N.V.; Gomez-Sanchis, J.; Chanona-Perez, J. Early detection of mechanical damage in mango using NIR hyperspectral images and machine learning. *Biosyst. Eng.* **2014**, *122*, 91–98. [CrossRef]
120. Curto, B.; Moreno, V.; Garcia-Esteban, J.A.; Blanco, F.J.; Gonzalez, I.; Vivar, A.; Revilla, I. Accurate Prediction of Sensory Attributes of Cheese Using Near-Infrared Spectroscopy Based on Artificial Neural Network. *Sensors* **2020**, *20*, 3566. [CrossRef] [PubMed]
121. Gunaratne, T.M.; Gonzalez Viejo, C.; Gunaratne, N.M.; Torrico, D.D.; Dunshea, F.R.; Fuentes, S. Chocolate quality assessment based on chemical fingerprinting using near infra-red and machine learning modeling. *Foods* **2019**, *8*, 426. [CrossRef] [PubMed]
122. Qiao, J.; Ngadi, M.O.; Wang, N.; Gariepy, C.; Prasher, S.O. Pork quality and marbling level assessment using a hyperspectral imaging system. *J. Food Eng.* **2007**, *83*, 10–16. [CrossRef]
123. Alshejari, A.; Kodogiannis, V.S. An intelligent decision support system for the detection of meat spoilage using multispectral images. *Neural Comput. Appl.* **2017**, *28*, 3903–3920. [CrossRef]
124. Kakani, V.; Nguyen, V.H.; Kumar, B.P.; Kim, H.; Pasupuleti, V.R. A critical review on computer vision and artificial intelligence in food industry. *J. Agric. Food Res.* **2020**, *2*, 12. [CrossRef]
125. Lopes, J.F.; Ludwig, L.; Barbin, D.F.; Grossmann, M.V.E.; Barbon, S. Computer Vision Classification of Barley Flour Based on Spatial Pyramid Partition Ensemble. *Sensors* **2019**, *19*, 2953. [CrossRef] [PubMed]
126. Phate, V.R.; Malmathanraj, R.; Palanisamy, P. Clustered ANFIS weighing models for sweet lime (*Citrus limetta*) using computer vision system. *J. Food Process Eng.* **2019**, *42*, 16. [CrossRef]
127. Nadim, M.; Ahmadifar, H.; Mashkinmojeh, M. Application of image processing techniques for quality control of mushroom. *Casp. J. Health Res.* **2019**, *4*, 72–75. [CrossRef]
128. Villaseñor-Aguilar, M.-J.; Bravo-Sánchez, M.-G.; Padilla-Medina, J.-A.; Vázquez-Vera, J.L.; Guevara-González, R.-G.; García-Rodríguez, F.-J.; Barranco-Gutiérrez, A.-I. A maturity estimation of bell pepper (*Capsicum annuum* L.) by artificial vision system for quality control. *Appl. Sci.* **2020**, *10*, 5097. [CrossRef]
129. Siswantoro, J.; Hilman, M.; Widiasri, M. Computer vision system for egg volume prediction using backpropagation neural network. In *IOP Conference Series: Materials Science and Engineering*; IOP Publishing: Bristol, UK, 2017; p. 012002.
130. Bakhshipour, A.; Zareiforoush, H.; Bagheri, I. Application of decision trees and fuzzy inference system for quality classification and modeling of black and green tea based on visual features. *J. Food Meas. Charact.* **2020**, *14*, 1402–1416. [CrossRef]
131. Mazen, F.M.; Nashat, A.A. Ripeness classification of bananas using an artificial neural network. *Arab. J. Sci. Eng.* **2019**, *44*, 6901–6910. [CrossRef]

132. Wan, P.; Toudeshki, A.; Tan, H.; Ehsani, R. A methodology for fresh tomato maturity detection using computer vision. *Comput. Electron. Agric.* **2018**, *146*, 43–50. [CrossRef]
133. Bhagat, N.B.; Markande, S. Automatic grading of potatoes with fuzzy logic. In Proceedings of the 2016 Online International Conference on Green Engineering and Technologies (IC-GET), Coimbatore, India, 19 November 2016; IEEE: Piscataway, NJ, USA, 2016; pp. 1–4.
134. Garcia, J.A.A.; Arboleda, E.R.; Galas, E.M. Identification of visually similar vegetable seeds using image processing and fuzzy logic. *Int. J. Sci. Technol. Res.* **2020**, *9*, 4925–4928. [CrossRef]
135. Koklu, M.; Ozkan, I.A. Multiclass classification of dry beans using computer vision and machine learning techniques. *Comput. Electron. Agric.* **2020**, *174*, 105507. [CrossRef]
136. Zareiforoush, H.; Minaei, S.; Alizadeh, M.R.; Banakar, A. A hybrid intelligent approach based on computer vision and fuzzy logic for quality measurement of milled rice. *Measurement* **2015**, *66*, 26–34. [CrossRef]
137. Yu, Z.; Jung, D.; Park, S.; Hu, Y. Smart traceability for food safety. *Crit. Rev. Food Sci. Nutr.* **2022**, *62*, 905–916. [CrossRef]
138. Di Rosa, A.R.; Leone, F.; Cheli, F.; Chiofalo, V. Fusion of electronic nose, electronic tongue and computer vision for animal source food authentication and quality assessment—A review. *J. Food Eng.* **2017**, *210*, 62–75. [CrossRef]
139. Jafarzadeh, S.; Nooshkam, M.; Zargar, M.; Garavand, F.; Ghosh, S.; Hadidi, M.; Forough, M. Green synthesis of nanomaterials for smart biopolymer packaging: Challenges and outlooks. *J. Nanostruct. Chem.* **2015**, *24*, 1–24. [CrossRef]
140. McVey, C.; Elliott, C.T.; Cannavan, A.; Kelly, S.D.; Petchkongkaew, A.; Haughey, S.A. Portable spectroscopy for high throughput food authenticity screening: Advancements in technology and integration into digital traceability systems. *Trends Food Sci. Technol.* **2021**, *118*, 777–790. [CrossRef]
141. Ndisya, J.; Gitau, A.; Mbuge, D.; Arefi, A.; Badulescu, L.; Pawelzik, E.; Hensel, O.; Sturm, B. Vis-NIR Hyperspectral Imaging for Online Quality Evaluation during Food Processing: A Case Study of Hot Air Drying of Purple-Speckled Cocoyam (*Colocasia esculenta* (L.) Schott). *Processes* **2021**, *9*, 1804. [CrossRef]
142. Abedi-Firoozjah, R.; Yousefi, S.; Heydari, M.; Seyedfatehi, F.; Jafarzadeh, S.; Mohammadi, R.; Rouhi, M.; Garavand, F. Application of Red Cabbage Anthocyanins as pH-Sensitive Pigments in Smart Food Packaging and Sensors. *Polymers* **2022**, *14*, 1629. [CrossRef] [PubMed]
143. Sanaeifar, A.; ZakiDizaji, H.; Jafari, A.; de la Guardia, M. Early detection of contamination and defect in foodstuffs by electronic nose: A review. *TrAC Trends Anal. Chem.* **2017**, *97*, 257–271. [CrossRef]
144. Feng, H.; Zhang, M.; Liu, P.; Liu, Y.; Zhang, X. Evaluation of IoT-enabled monitoring and electronic nose spoilage detection for salmon freshness during cold storage. *Foods* **2020**, *9*, 1579. [CrossRef] [PubMed]
145. Wadehra, A.; Patil, P.S. Application of electronic tongues in food processing. *Anal. Methods* **2016**, *8*, 474–480. [CrossRef]
146. Defelice, S.L. The Nutraceutial Revolution—Its Impact On Food-Industry R-And-D. *Trends Food Sci. Technol.* **1995**, *6*, 59–61. [CrossRef]
147. Cheein, F.A.A.; Carelli, R. Agricultural Robotics Unmanned Robotic Service Units in Agricultural Tasks. *IEEE Ind. Electron. Mag.* **2013**, *7*, 48–58. [CrossRef]
148. Maddikunta, P.K.R.; Hakak, S.; Alazab, M.; Bhattacharya, S.; Gadekallu, T.R.; Khan, W.Z.; Pham, Q.V. Unmanned Aerial Vehicles in Smart Agriculture: Applications, Requirements, and Challenges. *IEEE Sens. J.* **2021**, *21*, 17608–17619. [CrossRef]
149. Quan, H.Y.; Zhang, T.; Xu, H.; Luo, S.; Nie, J.; Zhu, X.Q. Photo-curing 3D printing technique and its challenges. *Bioact. Mater.* **2020**, *5*, 110–115. [CrossRef]
150. Tao, F.; Xiao, B.; Qi, Q.L.; Cheng, J.F.; Ji, P. Digital twin modeling. *J. Manuf. Syst.* **2022**, *64*, 372–389. [CrossRef]
151. Zhuang, Y.; Wang, W.; Yun, W. Status and development of robotic control technologies based on networks. *Robot* **2002**, *24*, 276–282. [CrossRef]
152. Griffith, C.J.; Livesey, K.M.; Clayton, D. The assessment of food safety culture. *Br. Food J.* **2010**, *112*, 439–456. [CrossRef]
153. Mauer, W.A.; Kaneene, J.B.; DeArman, V.T.; Roberts, C.A.; Miller, R.; Pong, L.; Dickey, T.E. Ethnic-food safety concerns: An online survey of food safety professionals. *J. Environ. Health* **2006**, *68*, 32–38. [CrossRef] [PubMed]
154. Milic, D.C.; Dujmenovic, I.; Peko, M. An Approach to the Application of the Internet of Things in Logistics. *Teh. Glas.-Technol. J.* **2023**, *17*, 134–140. [CrossRef]
155. Tudora, E.; Tirziu, E.; Gheorghe-Moisii, M. Internet of Things in energy: Technologies, applications, challenges and future trends. *Rom. J. Inf. Technol. Autom. Control.-Rev. Romana Inform. Autom.* **2021**, *31*, 45–58. [CrossRef]
156. Ismail, A.A.; Hassan, R. Technical Competencies in Digital Technology towards Industrial Revolution 4.0. *J. Technol. Educ. Train.* **2019**, *11*, 56–62. [CrossRef]
157. Kim, S.S.; Kim, S. Impact and prospect of the fourth industrial revolution in food safety: Mini-review. *Food Sci. Biotechnol.* **2022**, *31*, 399–406. [CrossRef] [PubMed]
158. Niszczota, P.; Rybicka, I.; Kamini, N. The credibility of dietary advice formulated by ChatGPT: Robo-diets for people with food allergies. *Nutrition* **2023**, *112*, 112076. [CrossRef]
159. Paul, J.; Ueno, A.; Dennis, C. ChatGPT and consumers: Benefits, Pitfalls and Future Research Agenda. *Int. J. Consum. Stud.* **2022**, *13*, 1213–1225. [CrossRef]
160. Cheng, L.; Liu, F.; Yao, D. Enterprise data breach: Causes, challenges, prevention, and future directions. *Wiley Interdiscip. Rev. Data Min. Knowl. Discov.* **2017**, *7*, e1211. [CrossRef]

161. Anagnostopoulos, I.; Zeadally, S.; Exposito, E. Handling big data: Research challenges and future directions. *J. Supercomput.* **2016**, *72*, 1494–1516. [CrossRef]
162. Massaro, A.; Calicchio, A.; Maritati, V.; Galiano, A.; Birardi, V.; Pellicani, L.; Gutierrez Millan, M.; Dalla Tezza, B.; Bianchi, M.; Vertua, G. A case study of innovation of an information communication system and upgrade of the knowledge base in industry by ESB, artificial intelligence, and big data system integration. *Int. J. Artif. Intell. Appl. (IJAIA)* **2018**, *9*, 27–43. [CrossRef]
163. Frank, B. Artificial intelligence-enabled environmental sustainability of products: Marketing benefits and their variation by consumer, location, and product types. *J. Clean. Prod.* **2021**, *285*, 125242. [CrossRef]

foods

MDPI

Perspective

The Application of Multi-Parameter Multi-Modal Technology Integrating Biological Sensors and Artificial Intelligence in the Rapid Detection of Food Contaminants

Longlong Zhang [1,2], Qiuping Yang [2,3] and Zhiyuan Zhu [2,*]

[1] Key Laboratory of Intelligent Manufacturing Technology (Shantou University), Ministry of Education, Shantou 515063, China
[2] College of Electronic Engineering, Southwest University, Chongqing 400715, China
[3] Hubei Key Laboratory of Food Nutrition and Safety, Huazhong University of Science and Technology, Wuhan 430030, China
* Correspondence: zyuanzhu@swu.edu.cn

Abstract: Against the backdrop of continuous socio-economic development, there is a growing concern among people about food quality and safety. Individuals are increasingly realizing the critical importance of healthy eating for bodily health; hence the continuous rise in demand for detecting food pollution. Simultaneously, the rapid expansion of global food trade has made people's pursuit of high-quality food more urgent. However, traditional methods of food analysis have certain limitations, mainly manifested in the high degree of reliance on personal subjective judgment for assessing food quality. In this context, the emergence of artificial intelligence and biosensors has provided new possibilities for the evaluation of food quality. This paper proposes a comprehensive approach that involves aggregating data relevant to food quality indices and developing corresponding evaluation models to highlight the effectiveness and comprehensiveness of artificial intelligence and biosensors in food quality evaluation. The potential prospects and challenges of this method in the field of food safety are comprehensively discussed, aiming to provide valuable references for future research and practice.

Keywords: artificial intelligence; biosensors; feature extraction; machine vision; data analysis

1. Introduction

In the contemporary food industry, globalization and technological advancement are driving the increasing demand for food safety and quality among consumers [1]. This trend becomes particularly significant under rapid economic and social development, with a growing concern for individual health issues. Food safety, as an indispensable element in safeguarding human health, has garnered widespread attention, encompassing the physical properties, nutritional value, and preventive measures of food [2]. Meanwhile, the rapid growth in food trade has made high-quality food a key factor in market competition [3]. However, food is a complex entity composed of multiple components, making comprehensive quality and safety assessment crucial [4]. Currently, the applicability and completeness of food safety assessment have not been fully established, lacking corresponding risk assessment systems and technological frameworks [5]. Moreover, the assessment methods are cumbersome, time-consuming, and costly, with subjective biases and insufficient scientific bases, making large-scale food quality monitoring and assessment impractical [6]. Therefore, to ensure food quality and safety, continuous monitoring of product quality characteristics during transportation and storage is paramount. One method for food safety prevention is to utilize conventional technologies to detect food quality during transportation, such as chromatography [7], ultraviolet detection [8], or fluorescence techniques [9], combined with separation techniques [10]. However, these analytical methods have limitations. Since these analyses are performed post-production, they may lead to contaminated food already entering the market. Moreover, they are labor-intensive, intricate, and costly, demanding a substantial number of samples and proficient operators.

In this context, biosensors, as rapid, sensitive, and highly specific analytical devices, offer new possibilities for addressing food safety issues. Biological sensors in the food industry can be analogized to infrared photodetectors [11], audiovisual photodetectors [12], synergistic graphene photodetectors [13], potential fluctuation engineering-enhanced graphene photodetectors [14], and ultra-high-gain short-wave infrared detectors [15] used in the fields of communication and detection. These photodetectors enhance information acquisition and processing in communication and detection. Similarly, biological sensors in the food industry, through real-time monitoring and precise analysis, enhance food safety and quality control. Biosensors utilize biological receptors (such as enzymes, antibodies, etc.) combined with sensors to generate measurable signals through specific physicochemical transformations, thereby determining the molecular content in samples [16]. The selection of biosensors and sensors depends on the characteristics of the sample and the type of measurable characteristic considered. Biological receptors represent the key components of biosensors, responding only to specific analytes and not to other biological receptors present in the analytical sample. Compared to traditional chemical detection methods, biosensors offer advantages such as rapid response, ease of operation, and low cost, thus being widely applied in various stages of food production, processing, and sale. Through biosensors, real-time monitoring of food quality and safety is achievable, enabling prompt detection of issues and actions to ensure food safety and health.

However, due to the diversity and complexity of food safety issues, a single biosensor may not cover all potential food safety hazards, necessitating comprehensive assessment with other technologies and means [17]. In recent years, innovative technologies such as artificial intelligence have been introduced into the field of food safety, proposing various advanced food quality inspection strategies, including AI methods [18–20], red and purple spectroscopy, and biosensors [21–25]. Among them, the combination of artificial intelligence and biosensors has given rise to a new detection mode, which utilizes complex AI algorithms to transform information obtained by sensors, such as physical, chemical, biological, environmental, or identity-related parameters, into decision processes with enhanced accuracy and intelligence. Currently, AI biosensors have been widely applied in continuous glucose monitoring [26] and rapid pathogen detection [27], and they have made significant progress in the development of wearable electronic devices [28]. The continuous development and application of these technologies provide new avenues and possibilities for comprehensive monitoring and analysis of food quality and nutrition.

This paper aims to explore the latest advances in artificial intelligence and biosensors in acquiring external indicators and internal characteristics of food products. By comprehensively reviewing key research on convolutional neural networks and biosensors in the field of food quality perception and classification, this paper aims to highlight the synergistic role of artificial intelligence and biosensors in the perception and classification processes of food quality and to describe how biosensors and deep learning are combined to comprehensively assess food quality.

2. The Role of Biosensors in Food Safety Assessment

The role of biosensors in food safety assessment cannot be overlooked. The monitoring and analysis of contaminants in food are crucial steps to ensure food safety and public health. Jordi et al. [29] suggested that biosensors are analytical devices that can serve as an alternative to traditional methods for detecting pathogenic bacteria in food (Figure 1a). These sensors consist of biological components, such as antibodies, aptamers, or single-stranded DNA (ssDNA) receptors, that interact with target analytes (e.g., the pathogens depicted in the figure). The sensing layer, coupled with suitable transducers like carbon nanotubes, graphene, or nanoparticles, generates measurable signals from these biorecognition events. Various signal detection methods, including impedimetric, amperometric, and voltammetric techniques, allow biosensors to produce responses proportional to the concentration of the target analytes. This technology can be used for the pre-treatment and detection of pathogens in various foods. In label-free biosensors, the sensor directly generates a signal from the biorecognition event between the target analyte and the receptor. Conversely, label-based biosensors require a secondary molecule or aggregate (label) to generate a measurable signal, as the biorecognition event alone does not produce one. Figure 1 illustrates a typical label-based biosensor, where one receptor (e.g., an antibody or aptamer) is immobilized on the biosensor to capture the target analyte. Subsequently, another receptor (e.g., labeled ssDNA) binds to the captured analyte, and an appropriate molecular label produces the signal. These sensors convert biological reactions into assessable and convertible signals [30] and, with the support of modern technology, translate biological signals into electrical signals. Biosensors comprise molecular recognition and transduction components, converting biological signals into electrical signals through modern technology. Their structure includes a biological membrane and physical or chemical transducers used for the selective determination of analytes, thereby detecting viruses in food, water, environment, and agriculture [31]. By coating synthetic nanomaterials on biosensors, sensing electrodes can be modified to enhance sensing parameters, and signals can be conveyed to individuals through optical, piezoelectric, electrochemical, etc., means [32]. These signals are ultimately converted into electrical signals indicating the presence of biochemical targets, accessible via cloud-based data access. The outcomes of biosensors can be presented in graphical, tabular, or mathematical research formats [33].

The pioneering glucose oxidase (GOx) biosensor debuted in 1962, laying the foundation for biosensor technology [34]. Despite subsequent advancements since the 1960s, GOx sensors remain prevalent, representing the most favored type. These sensors offer advantages such as light weight, ease of use, portability, high detection limits, small analyte volumes required, and applicability in complex biological fluids. However, challenges they face include issues where surface topological structures fail to exhibit sufficient sensitivity and unique associations with target biologics, as well as challenges posed by pH values and ionic strength of biological analytes, leading to varied reactions [35]. These challenges can be addressed by integrating artificial intelligence (AI) technology with biosensors. Next-generation biosensor arrays incorporate AI algorithms, enabling them to exhibit higher specialization, selectivity, responsiveness, and consistency. These sensor arrays integrate solid-state and surface mechanical knowledge with integrated circuits, bioengineering, and data processing. With AI assistance, they can more accurately identify biological analytes, enhancing sensor performance and reliability [36]. Therefore, attention should now be directed towards the latest advancements in this field and outlining directions for the future development of AI-based biosensors.

In addition to enzyme biosensors [37], other types of biosensors such as optical and piezoelectric sensors have also been applied in the detection of toxins and chemicals in food production. Neethirajan et al. [38] illustrated the complexity of food contamination, highlighting common food contaminants and their interrelationships (Figure 1b). Different types of contaminants intertwine, making food safety monitoring complex and diverse. Safeguarding food safety requires comprehensive consideration and response to these different types of contaminants, taking effective measures to reduce their impact on human health and ensure the health and sustainability of the food supply chain [39]. Phumlani Tetyana et al. [40] proposed that the principle of each biosensor is based on molecular recognition, making its characterization crucial (Figure 1c). Effective characterization of biosensors requires comprehensive consideration of factors such as endogenous compounds and contaminants in food; therefore, the adoption of appropriate experimental methods and technological means is necessary. Amidst the prevailing challenges in food safety assessment, the integration of artificial intelligence (AI) with biosensors has assumed paramount importance [41]. Key attributes, including sensitivity, selectivity, stability, detection limit, reproducibility, response time, and range or linearity [42], are not only pivotal for the standalone performance of biosensors but also hold pivotal roles in their amalgamation with AI. This integration allows for the precise identification and interpretation of signals from biosensors, enabling the swift detection and accurate analysis of harmful substances in food [43]. AI algorithms heavily depend on the stable and consistent data furnished by sensors, highlighting the indispensability of sensor stability for ensuring algorithmic robustness and reliability. Low detection limits and high reproducibility enhance data quality, allowing AI models to predict and analyze the presence and concentration of harmful substances in food samples with greater accuracy [44]. The response time directly influences the real-time operational efficacy of AI systems, a crucial aspect for timely detection of food safety issues. Lastly, by optimizing range or linearity, AI models can adeptly adapt to data spanning varying concentration ranges, thereby augmenting the precision and credibility of food safety assessments. The amalgamation of these attributes is poised to propel technological innovation and advancement in the realm of food safety, furnishing indispensable support for public health.

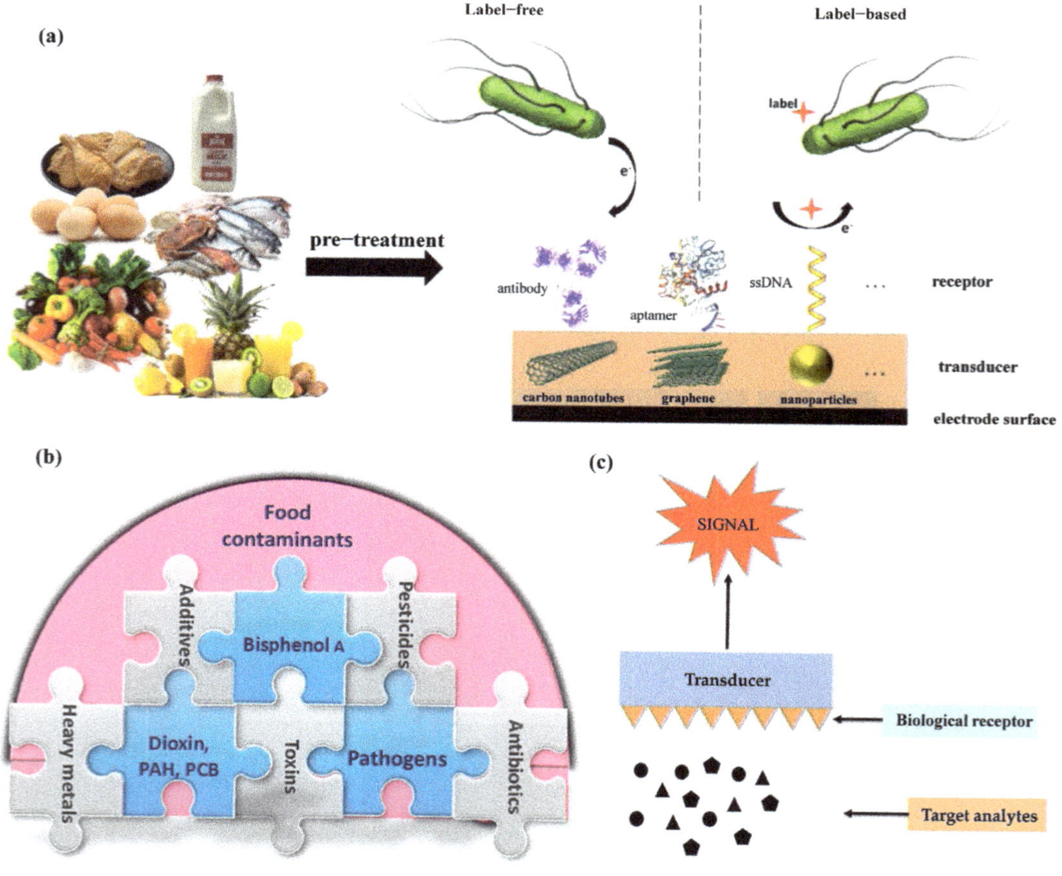

Figure 1. Application and evaluation of biological sensors in food. (**a**) Diagram illustrating biological sensor with close contact between sensor and receptor [29]. (**b**) Predominant food contaminants in food industries [38]. (**c**) Fundamentals of biosensors [40].

3. Multi-Modal Model Based on Artificial Intelligence and Biosensors

Integrating intelligent recognition methods that combine visual signals and bioelectric signals represents a pioneering direction in the field of food quality perception and classification [45]. This approach integrates computer vision with biosensor technology, aiming to overcome the inherent limitations of single-mode perception and human dependency in traditional methods [46]. At its core lies the fusion model structure [47], which comprises a feature extraction fusion stage and a perception scoring stage. As depicted in Figure 2a, in the feature extraction fusion stage, food image data capturing samples of varying qualities are generated and fed into a convolutional neural network (CNN) as input. Through multiple stacked convolutional and pooling layers, the CNN progressively extracts complex and abstract representations of input food images, forming one-dimensional feature vectors containing attributes such as color, texture, and overall appearance. Simultaneously, relevant substance information is transformed into sequential representations by biosensors. Then, feature vectors extracted from both image feature vectors and bioelectric signal sequences are concatenated as input to train the Deep Neural Network (DNN) model. Constructing the structured mathematical framework of the DNN model through extensive datasets, neural network weights and biases embedded within are trained. In the forward propagation algorithm, a series of linear operations are executed using multiple weight coefficient matrices and bias vectors, while activation functions facilitate the transformation from linear to nonlinear (Figure 2b). The selection of activation functions plays a crucial role in network learning and expressing nonlinear relationships, with commonly used activation functions including ReLU and Sigmoid functions (Figure 2c). Forward propagation transmits input data through hidden layers, ultimately generating the network's output to achieve classification, regression, and other tasks. Upon completing forward propagation, the difference between predicted output and actual output is measured by a loss function, and weights and biases are updated using gradient descent methods to optimize the model parameters.

On the other hand, the primary objective of computer vision is to automate visual decision making by simulating human cognitive capabilities [48]. By learning from extensive datasets, algorithms in computer vision can make precise, efficient, and complex judgments [49]. A mature method within this field uses CNNs, which implement convolution through absorbing convolutional operations. The structure of CNNs, as depicted in Figure 2d, comprises input, convolution, pooling, and fully connected layers, among others, focusing on extracting information from input images. This information encompasses different aspects such as the texture and color of the images. Convolutional layers extract these attributes through convolution operations. This process entails traversing convolution filters over input images with defined strides, conducting cross-correlation computations to generate valuable feature maps [50]. Pooling layers play a crucial role in selecting the most significant features from convolutional layers. The overall configuration of convolution and pooling layers is illustrated in Figure 2e.

In the process of food evaluation and grading, there is no necessity to quantitatively acquire specific values of food attributes such as color and texture. Rather, reliance on neural networks for internal analysis is crucial [51]. Convolutional and pooling layers are fundamental building blocks in CNNs for extracting features from input images. Convolutional layers perform local feature extraction through convolution operations, capturing edges, textures, shapes, and other local features. Pooling layers reduce the spatial dimensions of feature maps through downsampling and aggregation operations, enhancing network efficiency and generalization. Multiple convolution kernels typically expand single feature maps into multiple smaller-sized feature maps, compensating for feature loss resulting from size reduction during convolution and pooling processes, essentially trading quantity for quality. Parallel learning and capturing of different types of features by multiple convolution kernels, parameter sharing to reduce the number of parameters, and sparse connections to reduce computational costs characterize the CNN. The translation invariance and feature compression properties of pooling layers make the network insensitive to positional changes, enhancing model robustness. By alternately using convolutional and pooling layers, CNN gradually extracts more abstract and advanced features, providing robust image analysis and recognition capabilities for subsequent food quality perception and grading tasks [52]. Currently, CNNs have been employed to address various challenges in the food domain, including food recognition, calorie estimation, foreign object detection, and maturity assessment [53]. However, leveraging computer vision techniques to solve challenges in the food domain inevitably relies on visual image features, which may have limitations in cases requiring maturity detection and quality assessment [54]. Relying solely on surface analysis of food appearance for detection and classification may result in incomplete evaluations. Therefore, there is an urgent need to explore a comprehensive approach that integrates internal and external factors to assess food properties more comprehensively.

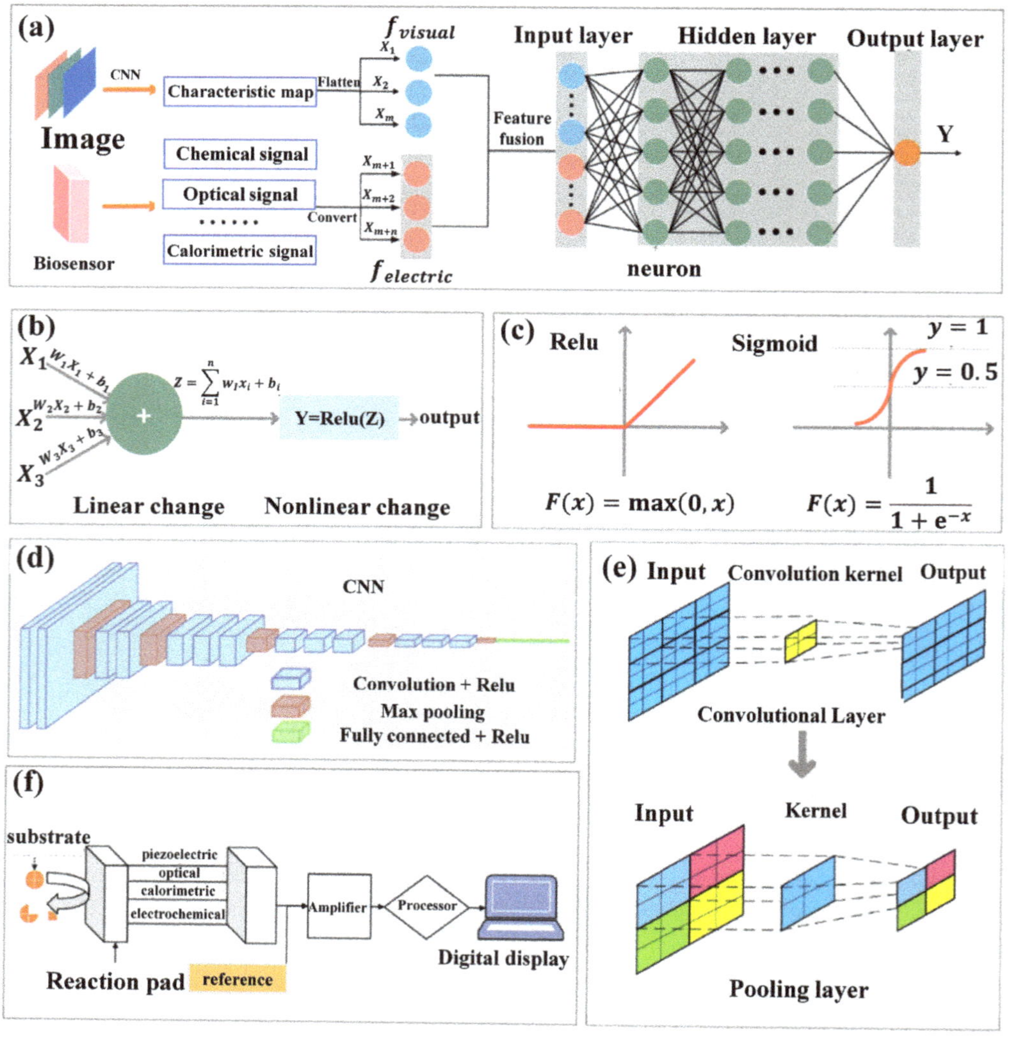

Figure 2. (**a**) Overall structure of the fusion model. (**b**) Structure and function of neurons. (**c**) Common nonlinear activation functions. (**d**) Overall structure of convolutional neural networks. (**e**) Convolution and pooling operations. (**f**) General composition of biosensors.

As illustrated in Figure 2f, biosensors automatically analyze intrinsic food components through a series of systematic processes. Firstly, an automated sampling and preprocessing system ensures precise acquisition of food samples while minimizing potential variations due to human intervention. Subsequently, bioactive substances are selected based on the characteristics of target components to induce specific reactions that generate measurable signals. These signals are then detected and recorded using specialized sensor devices (such as optical, electrochemical, or bioluminescent sensors). Finally, advanced data analysis algorithms interpret and quantify these signals to determine the concentration of target components in the food matrix. By integrating the capabilities of automated sampling, carefully selected bioactive substances, signal detection, real-time monitoring, and remote control, biosensors achieve efficient, accurate, and streamlined analysis of intrinsic food components, reducing reliance on cumbersome manual procedures [55].

Building upon the principles of biosensors and artificial intelligence technology, researchers have developed various types of AI biosensors to detect specific substances. For instance, biosensors utilizing glucose [56], vitamins [57], tyramine [58], bisphenol A [59], and L-glutamate [60] have been developed to measure trace elements in food. Changes in the content of trace elements in food may be caused by various factors, such as nutrient-poor soil, adverse environmental conditions, or contamination during additive application or processing. The integration of artificial intelligence and biosensors brings significant advantages. Artificial intelligence can accelerate the process of data analysis, improving the sensitivity and accuracy of biosensors. By utilizing machine learning algorithms, artificial intelligence can identify complex patterns and trends, thereby enhancing the detection and analysis of trace element content [61]. Junjie, et al. [62] explained that biological sensors simulate the human olfactory system for rapid detection and identification of mixed odors. Miguel Peris et al. [63] proposed the use of specialized chemical sensors to mimic human odors, producing unique odor signatures when interacting with gas mixtures. These signatures are then compared with standard odor profiles to identify the mixed components. The biosensor system primarily consists of three parts (see Figure 3A(a)): (1) a sample processing system, (2) a detection system, and (3) a data processing system. This device has extensive applications in the quality assessment of agricultural products and food. The core of this system is a sensor array, where the excitation of sensor elements produces signals that generate unique responses to specific odor samples (odor fingerprints). This type of biosensor, made from biological materials, converts biological signals into electrical or optical signals. The biosensor, simulating the human taste system, comprises three parts: (i) an autosampler (optional), (ii) a set of chemically selective sensors, and (iii) signal processing algorithm software (see Figure 3A(b)). Biosensors emulate the human senses of smell and taste; electronic noses and tongues mimic this process by detecting samples through sensor arrays. Subsequently, a data processing system analyzes these data and identifies the samples, enabling high-sensitivity and accurate qualitative and quantitative determination of chemicals in complex samples. Magdalena Śliwińska et al. [64] introduced a sensor-simulated taste system consisting of three key components: a sample distribution chamber or optional automatic sample distributor, a sensor array with varying selectivities, and software for data processing designed to mimic the functionality of the brain's image recognition systems. This innovative system enables the classification of chemical odors in liquid samples, as illustrated in Figure 3B. Facure et al. [65] prepared four types of nanocomposite thin-film sensor arrays by drop-casting solutions of rGO, PPy-rGO, PEDOT: PSS-rGO, and PEDOT: PSS-rGO-AuNPs onto gold interdigitated electrodes (IDEs). The sensing units were characterized by impedance spectroscopy and other methods, and the data were processed by Principal Component Analysis (PCA) (Figure 3C). Through the fine-tuning of design parameters of the sensors, especially for complex signals generated by sensor arrays, such as material selection, sensor structure, and detection methods, more reliable and continuous monitoring of trace elements can be achieved. This optimization not only significantly extends the lifespan of sensors but also improves their stability, making them more effectively applicable in practical food detection. Therefore, the combination of artificial intelligence and biosensors can provide a more comprehensive, accurate, and efficient solution for food safety assessment, aiding consumers in assessing the adequacy of their nutritional needs and assisting regulatory agencies in monitoring and managing food safety.

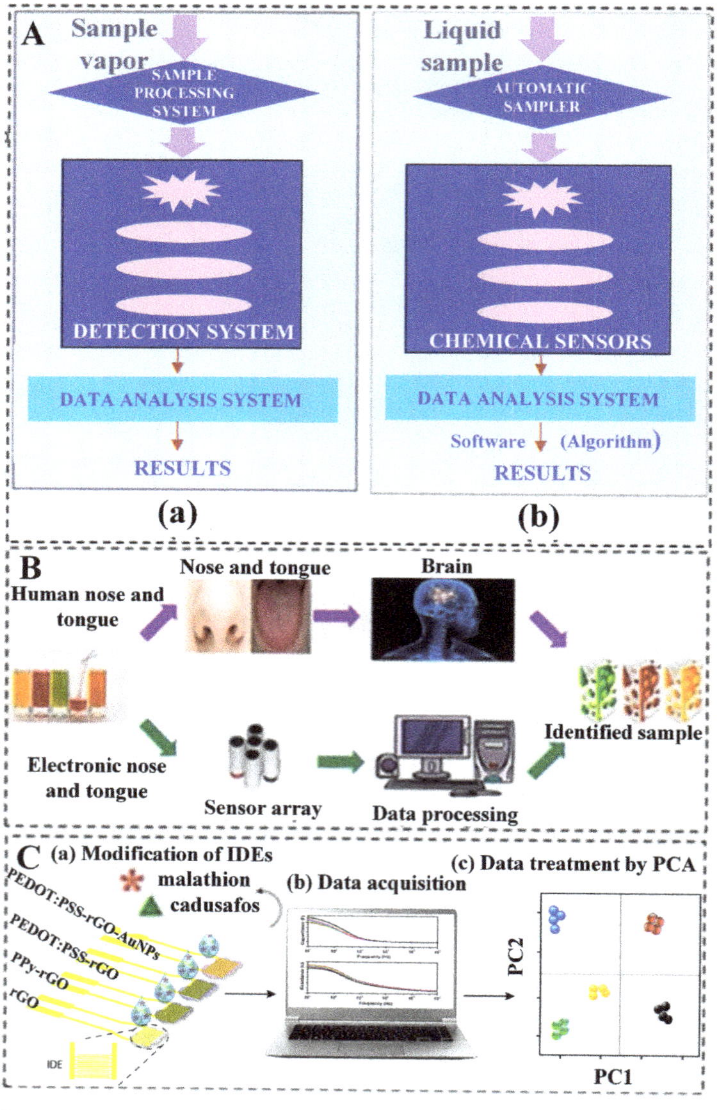

Figure 3. The application of artificial intelligence and biosensors in the food domain. (**A**) Schematic diagram of biosensor simulation (**a**) The simulation of the human olfactory system (**b**) The human gustatory system by biosensors [63] (reproduced from [63], with permission from Elsevier, 2016). (**B**) A comparison of the working principles of the gustatory system and the olfactory system [64] (reprinted (adapted) with permission from [64]. Copyright 2014 American Chemical Society). (**C**) A schematic diagram of an intelligent biosensor system based on graphene hybrid nanocomposites, used for detecting pesticide residues in agricultural products and statistical techniques for processing data collected by biosensors (reproduced (adapted) from [65], with permission from Elsevier, 2022) [65].

4. Discussion

The integration of biosensors and artificial intelligence (AI) offers several distinct advantages for systematically assessing food quality, including automation, efficiency enhancement, and cost-effectiveness [66]. However, several obstacles need to be addressed to make this approach feasible for food quality assessment. One key barrier involves establishing reliable and fully operational sensing systems capable of effectively detecting and analyzing various components present in different types of food. Additionally, careful consideration is required when selecting appropriate biosensors that align with the unique nutritional characteristics of different food categories [67]. Another significant issue revolves around relying on extensive statistical data to construct accurate assessment models, necessitating dedicated hardware resources. To address these challenges, advanced technologies such as chip integration, cloud computing, 5G networks, and other innovative solutions can be utilized effectively [68]. Combining biosensors with microchip integration has enabled the miniaturization and portability of sensing devices, facilitating real-time monitoring of on-site food quality [69]. Progress in cloud computing technology and big data analytics allows sensor data to be seamlessly uploaded to the cloud for efficient processing and storage. This enables rapid analysis of large-scale food samples, ensuring the maintenance of superior quality and safety standards. The advanced capabilities of 5G networks, including high-speed and low-latency attributes, establish robust and efficient communication infrastructure, enhancing the reliability and timely dissemination of sensor data for real-time monitoring purposes [70]. In addition to the aforementioned factors, integrating other complementary solutions such as the Internet of Things (IoT), artificial intelligence (AI), and big data analytics further strengthens the potential of biosensor-driven food quality assessment systems. The connectivity provided by the IoT enhances the functionality of sensors, enabling remote monitoring and real-time data collection. Moreover, harnessing AI methods and tapping into the power of big data analytics aids in the identification and analysis of potential issues as well as emerging trends within food samples [71]. Overall, these technological advancements contribute to significantly enhancing the efficiency and intelligence of biosensor-driven food quality assessment systems, ensuring the maintenance of excellent food quality and safety standards.

The fusion of biosensors and artificial intelligence methods in food quality assessment offers multiple advantages. Firstly, this technology achieves high-precision detection by utilizing biosensors for real-time monitoring of trace elements and harmful substances in food, coupled with AI for data processing and analysis. Secondly, the combination of biosensors and artificial intelligence enables rapid detection through real-time data collection, efficient data processing, and analysis, thereby quickly generating assessment results and promptly addressing risks and issues. Additionally, AI methods can predict and provide alerts for potential food quality issues by analyzing extensive data and historical information, thereby identifying abnormal patterns and potential risks. Moreover, the integration of biosensors and artificial intelligence enables real-time monitoring and remote access. Biosensors can collect and transmit data at any time, connected to the cloud via the Internet of Things, while AI algorithms enable remote monitoring, timely handling of anomalies, and providing remote support and solutions. In summary, the fusion of biosensors and artificial intelligence methods brings significant advantages to food quality assessment, including high-precision detection, rapid detection speed, predictive capabilities, real-time monitoring, and remote access. These advantages play a crucial role in ensuring food safety and safeguarding consumer health.

5. Conclusions

This paper comprehensively explores the synergistic integration of artificial intelligence and biosensors in food quality assessment, emphasizing their advantages and limitations. Through in-depth analysis, we highlight the importance of multi-modal feature fusion for an innovative food quality perception approach. This approach combines unique features extracted by biosensors with advanced artificial intelligence techniques for subsequent processing. Based on the research analysis in this paper, we believe that researchers in the field of food quality assessment can tailor their methods according to the specific food categories they study. By fully leveraging multiple biosensors and effectively integrating bioelectric signals with image features, researchers can significantly enhance the comprehensiveness and accuracy of food quality perception and classification, thereby improving the overall assessment process. This integrated approach not only helps address the inherent limitations of single-modal perception and human dependence but also provides important reference and inspiration for future research in the field of food quality assessment.

Author Contributions: Conceptualization, L.Z. and Z.Z; methodology, L.Z. and Q.Y.; validation, L.Z., Q.Y. and Z.Z.; data curation, L.Z.; writing—original draft preparation, L.Z. and Z.Z.; writing—review and editing, L.Z. and Z.Z.; visualization, L.Z. and Q.Y.; supervision, Z.Z. All authors have read and agreed to the published version of the manuscript.

Funding: This work was supported by Key Laboratory of Intelligent Manufacturing Technology (Shantou University), Ministry of Education (No. STME2024003), and open funding from Hubei Key Laboratory of Food Nutrition and Safety (Huazhong University of Science and Technology) (FNS-HBKL2023A04).

Institutional Review Board Statement: Not applicable.

Informed Consent Statement: Not applicable.

Data Availability Statement: No new data were created or analyzed in this study. Data sharing is not applicable to this article.

Conflicts of Interest: The authors declare no conflict of interest.

References

1. Nayak, R.; Waterson, P. Global Food Safety as a Complex Adaptive System: Key Concepts and Future Prospects. *Trends Food Sci. Technol.* **2019**, *91*, 409–425. [CrossRef]
2. Gallo, M.; Ferrara, L.; Calogero, A.; Montesano, D.; Naviglio, D. Relationships between Food and Diseases: What to Know to Ensure Food Safety. *Food Res. Int.* **2020**, *137*, 109414. [CrossRef] [PubMed]
3. Al-shami, H.A.; Abdullah, S. Halal Food Industry Certification and Operation Challenges and Manufacturing Execution System Opportunities. A Review Study from Malaysia. *Mater. Today Proc.* **2023**, *80*, 3607–3614. [CrossRef]
4. Mihafu, F.D.; Issa, J.Y.; Kamiyango, M.W. Implication of Sensory Evaluation and Quality Assessment in Food Product Development: A Review. *Curr. Res. Nutr. Food Sci. J.* **2020**, *8*, 690–702. [CrossRef]
5. Qian, J.; Ruiz-Garcia, L.; Fan, B.; Robla Villalba, J.I.; McCarthy, U.; Zhang, B.; Yu, Q.; Wu, W. Food Traceability System from Governmental, Corporate, and Consumer Perspectives in the European Union and China: A Comparative Review. *Trends Food Sci. Technol.* **2020**, *99*, 402–412. [CrossRef]
6. Cosby, A.; Manning, J.; Fogarty, E.; Wilson, C.; Islam, N.; Trotter, M. *Assessing Real Time Tracking Technologies to Integrate with Identification Methods and National Traceability Requirements*; Report; CQUniversity: Melbourne, Australia, 2021. [CrossRef]
7. Li, S.; Tian, Y.; Jiang, P.; Lin, Y.; Liu, X.; Yang, H. Recent Advances in the Application of Metabolomics for Food Safety Control and Food Quality Analyses. *Crit. Rev. Food Sci. Nutr.* **2021**, *61*, 1448–1469. [CrossRef]
8. Singh, H.; Bhardwaj, S.K.; Khatri, M.; Kim, K.-H.; Bhardwaj, N. UVC Radiation for Food Safety: An Emerging Technology for the Microbial Disinfection of Food Products. *Chem. Eng. J.* **2021**, *417*, 128084. [CrossRef]
9. Wu, Y.; Sun, J.; Huang, X.; Lai, W.; Xiong, Y. Ensuring Food Safety Using Fluorescent Nanoparticles-Based Immunochromatographic Test Strips. *Trends Food Sci. Technol.* **2021**, *118*, 658–678. [CrossRef]
10. Choi, J.R.; Yong, K.W.; Choi, J.Y.; Cowie, A.C. Emerging Point-of-Care Technologies for Food Safety Analysis. *Sensors* **2019**, *19*, 817. [CrossRef]
11. Fu, J.; Guo, Z.; Nie, C.; Sun, F.; Li, G.; Feng, S.; Wei, X. Schottky Infrared Detectors with Optically Tunable Barriers beyond the Internal Photoemission Limit. *Innovation* **2024**, *5*, 100600. [CrossRef]
12. Fu, J.; Nie, C.; Sun, F.; Li, G.; Shi, H.; Wei, X. Bionic Visual-Audio Photodetectors with in-Sensor Perception and Preprocessing. *Sci. Adv.* **2024**, *10*, eadk8199. [CrossRef]
13. Jiang, H.; Fu, J.; Wei, J.; Li, S.; Nie, C.; Sun, F.; Wu, Q.Y.S.; Liu, M.; Dong, Z.; Wei, X.; et al. Synergistic-Potential Engineering Enables High-Efficiency Graphene Photodetectors for near- to Mid-Infrared Light. *Nat. Commun.* **2024**, *15*, 1225. [CrossRef]
14. Jiang, H.; Wei, J.; Sun, F.; Nie, C.; Fu, J.; Shi, H.; Sun, J.; Wei, X.; Qiu, C.-W. Enhanced Photogating Effect in Graphene Photodetectors via Potential Fluctuation Engineering. *ACS Nano* **2022**, *16*, 4458–4466. [CrossRef]

15. Jiang, H.; Wang, M.; Fu, J.; Li, Z.; Shaikh, M.S.; Li, Y.; Nie, C.; Sun, F.; Tang, L.; Yang, J.; et al. Ultrahigh Photogain Short-Wave Infrared Detectors Enabled by Integrating Graphene and Hyperdoped Silicon. *ACS Nano* **2022**, *16*, 12777–12785. [CrossRef] [PubMed]
16. Perumal, V.; Hashim, U. Advances in Biosensors: Principle, Architecture and Applications. *J. Appl. Biomed.* **2014**, *12*, 1–15. [CrossRef]
17. Schröder, M.J.A. Food Composition. In *Food Quality and Consumer Value: Delivering Food That Satisfies*; Schröder, M.J.A., Ed.; Springer: Berlin/Heidelberg, Germany, 2003; pp. 101–136. [CrossRef]
18. Beć, K.B.; Grabska, J.; Huck, C.W. Principles and Applications of Miniaturized Near-Infrared (NIR) Spectrometers. *Chemistry* **2021**, *27*, 1514–1532. [CrossRef]
19. Savoia, S.; Albera, A.; Brugiapaglia, A.; Di Stasio, L.; Ferragina, A.; Cecchinato, A.; Bittante, G. Prediction of Meat Quality Traits in the Abattoir Using Portable and Hand-Held near-Infrared Spectrometers. *Meat Sci.* **2020**, *161*, 108017. [CrossRef]
20. Teye, E.; Amuah, C.L.Y.; McGrath, T.; Elliott, C. Innovative and Rapid Analysis for Rice Authenticity Using Hand-Held NIR Spectrometry and Chemometrics. *Spectrochim. Acta Part A Mol. Biomol. Spectrosc.* **2019**, *217*, 147–154. [CrossRef] [PubMed]
21. Thakur, M.S.; Ragavan, K.V. Biosensors in Food Processing. *J. Food Sci. Technol.* **2013**, *50*, 625–641. [CrossRef]
22. Mustafa, F.; Andreescu, S. Nanotechnology-Based Approaches for Food Sensing and Packaging Applications. *RSC Adv.* **2020**, *10*, 19309–19336. [CrossRef]
23. Yousefi, H.; Su, H.-M.; Imani, S.M.; Alkhaldi, K.; Filipe, C.D.M.; Didar, T.F. Intelligen t Food Packaging: A Review of Smart Sensing Technologies for Monitoring Food Quality. *ACS Sens.* **2019**, *4*, 808–821. [CrossRef] [PubMed]
24. Zhang, D.; Yu, S.; Wang, X.; Huang, J.; Pan, W.; Zhang, J.; Meteku, B.E.; Zeng, J. UV Illumination-Enhanced Ultrasensitive Ammonia Gas Sensor Based on (001)TiO_2/MXene Heterostructure for Food Spoilage Detection. *J. Hazard. Mater.* **2022**, *423*, 127160. [CrossRef] [PubMed]
25. Parmar, R.R.; Jain, K.R.; Modi, C.K. Unified Approach in Food Quality Evaluation Using Machine Vision. In *Advances in Computing and Communications*; Abraham, A., Mauri, J.L., Buford, J.F., Suzuki, J., Thampi, S.M., Eds.; Springer: Berlin/Heidelberg, Germany, 2011; pp. 239–248. [CrossRef]
26. Jin, X.; Cai, A.; Xu, T.; Zhang, X. Artificial Intelligence Biosensors for Continuous Glucose Monitoring. *Interdiscip. Mater.* **2023**, *2*, 290–307. [CrossRef]
27. Yi, J.; Wisuthiphaet, N.; Raja, P.; Nitin, N.; Earles, J.M. AI-Enabled Biosensing for Rapid Pathogen Detection: From Liquid Food to Agricultural Water. *Water Res.* **2023**, *242*, 120258. [CrossRef] [PubMed]
28. Zhang, Y.; Hu, Y.; Jiang, N.; Yetisen, A.K. Wearable Artificial Intelligence Biosensor Networks. *Biosens. Bioelectron.* **2023**, *219*, 114825. [CrossRef] [PubMed]
29. Riu, J.; Giussani, B. Electrochemical Biosensors for the Detection of Pathogenic Bacteria in Food. *TrAC Trends Anal. Chem.* **2020**, *126*, 115863. [CrossRef]
30. Grieshaber, D.; MacKenzie, R.; Vörös, J.; Reimhult, E. Electrochemical Biosensors—Sensor Principles and Architectures. *Sensors* **2008**, *8*, 1400–1458. [CrossRef]
31. Parihar, A.; Choudhary, N.K.; Sharma, P.; Khan, R. MXene-Based Aptasensor for the Detection of Aflatoxin in Food and Agricultural Products. *Environ. Pollut.* **2023**, *316*, 120695. [CrossRef] [PubMed]
32. Karakuş, S.; Baytemir, G.; Taşaltın, N. Digital Colorimetric and Non-Enzymatic Biosensor with Nanoarchitectonics of Lepidium Meyenii-Silver Nanoparticles and Cotton Fabric: Real-Time Monitoring of Milk Freshness. *Appl. Phys. A* **2022**, *128*, 390. [CrossRef]
33. Bhardwaj, T. Review on biosensor technologies. *Int. J. Adv. Res. Eng. Technol.* **2015**, *6*, 36–62.
34. Clark, L.C.; Lyons, C. Electrode Systems for Continuous Monitoring in Cardiovascular Surgery. *Ann. N. Y. Acad. Sci.* **1962**, *102*, 29–45. [PubMed]
35. Ma, J.; Guan, Y.; Xing, F.; Eltzov, E.; Wang, Y.; Li, X.; Tai, B. Accurate and Non-Destructive Monitoring of Mold Contamination in Foodstuffs Based on Whole-Cell Biosensor Array Coupling with Machine-Learning Prediction Models. *J. Hazard. Mater.* **2023**, *449*, 131030. [CrossRef] [PubMed]
36. Huang, W.-C.; Wei, C.-D.; Belkin, S.; Hsieh, T.-H.; Cheng, J.-Y. Machine-Learning Assisted Antibiotic Detection and Categorization Using a Bacterial Sensor Array. *Sens. Actuators B Chem.* **2022**, *355*, 131257. [CrossRef]
37. Verma, M.L. Nanobiotechnology Advances in Enzymatic Biosensors for the Agri-Food Industry. *Environ. Chem. Lett.* **2017**, *15*, 555–560. [CrossRef]
38. Neethirajan, S.; Ragavan, V.; Weng, X.; Chand, R. Biosensors for Sustainable Food Engineering: Challenges and Perspectives. *Biosensors* **2018**, *8*, 23. [CrossRef] [PubMed]
39. Cazelles, K.; Zemlak, T.S.; Gutgesell, M.; Myles-Gonzalez, E.; Hanner, R.; Shear McCann, K. Spatial Fingerprinting: Horizontal Fusion of Multi-Dimensional Bio-Tracers as Solution to Global Food Provenance Problems. *Foods* **2021**, *10*, 717. [CrossRef]
40. Tetyana, P.; Morgan Shumbula, P.; Njengele-Tetyana, Z. Biosensors: Design, Development and Applications. In *Nanopores*; Ameen, S., Shaheer Akhtar, M., Shin, H.-S., Eds.; IntechOpen: London, UK, 2021. [CrossRef]
41. Zhao, Y.; Wang, X.; Sun, T.; Shan, P.; Zhan, Z.; Zhao, Z.; Jiang, Y.; Qu, M.; Lv, Q.; Wang, Y.; et al. Artificial intelligence-driven electrochemical immunosensing biochips in multi-component detection. *Biomicrofluidics* **2023**, *17*, 041301. [CrossRef] [PubMed]
42. Burcu Aydın, E.; Aydın, M.; Kemal Sezgintürk, M. Biosensors and the Evaluation of Food Contaminant Biosensors in Terms of Their Performance Criteria. *Int. J. Environ. Anal. Chem.* **2020**, *100*, 602–622. [CrossRef]

43. Milovanovic, M.; Žeravík, J.; Obořil, M.; Pelcová, M.; Lacina, K.; Cakar, U.; Petrovic, A.; Glatz, Z.; Skládal, P. A Novel Method for Classification of Wine Based on Organic Acids. *Food Chem.* **2019**, *284*, 296–302. [CrossRef]
44. Lu, Z.; Gong, Y.; Shen, C.; Chen, H.; Zhu, W.; Liu, T.; Wu, C.; Sun, M.; Su, G.; Wang, X.; et al. Portable, Intelligent MIECL Sensing Platform for Ciprofloxacin Detection Using a Fast Convolutional Neural Networks-Assisted Tb@Lu_2O_3 Nanoemitter. *Food Chem.* **2024**, *444*, 138656. [CrossRef]
45. Shi, Y.; Wang, X.; Borhan, M.S.; Young, J.; Newman, D.; Berg, E.; Sun, X. A Review on Meat Quality Evaluation Methods Based on Non-Destructive Computer Vision and Artificial Intelligence Technologies. *Food Sci. Anim. Resour.* **2021**, *41*, 563–588. [CrossRef]
46. Deng, Z.; Wang, T.; Zheng, Y.; Zhang, W.; Yun, Y.-H. Deep Learning in Food Authenticity: Recent Advances and Future Trends. *Trends Food Sci. Technol.* **2024**, *144*, 104344. [CrossRef]
47. Lahat, D.; Adali, T.; Jutten, C. Multimodal Data Fusion: An Overview of Methods, Challenges, and Prospects. *Proc. IEEE* **2015**, *9*, 1449–1477. [CrossRef]
48. Green, T.M.; Ribarsky, W.; Fisher, B. Building and Applying a Human Cognition Model for Visual Analytics. *Inf. Vis.* **2009**, *8*, 1–13. [CrossRef]
49. Lin, Y.; Ma, J.; Sun, D.-W.; Cheng, J.-H.; Zhou, C. Fast Real-Time Monitoring of Meat Freshness Based on Fluorescent Sensing Array and Deep Learning: From Development to Deployment. *Food Chem.* **2024**, *448*, 139078. [CrossRef] [PubMed]
50. Dönmez, E. Enhancing Classification Capacity of CNN Models with Deep Feature Selection and Fusion: A Case Study on Maize Seed Classification. *Data Knowl. Eng.* **2022**, *141*, 102075. [CrossRef]
51. Icagic, S.D.; Kvascev, G.S. A Smart Alcoholmeter Sensor Based on Deep Learning Visual Perception. *Sensors* **2022**, *22*, 7394. [CrossRef]
52. Liu, Y.; Pu, H.; Sun, D.-W. Efficient Extraction of Deep Image Features Using Convolutional Neural Network (CNN) for Applications in Detecting and Analysing Complex Food Matrices. *Trends Food Sci. Technol.* **2021**, *113*, 193–204. [CrossRef]
53. Maurya, P.; Singh, N.P. Mushroom Classification Using Feature-Based Machine Learning Approach. In *Proceedings of the 3rd International Conference on Computer Vision and Image Processing, Prayagraj, India, 4–6 December 2020*; Chaudhuri, B.B., Nakagawa, M., Khanna, P., Kumar, S., Eds.; Springer: Singapore, 2020; pp. 197–206. [CrossRef]
54. Zhu, L.; Spachos, P.; Pensini, E.; Plataniotis, K.N. Deep Learning and Machine Vision for Food Processing: A Survey. *Curr. Res. Food Sci.* **2021**, *4*, 233–249. [CrossRef]
55. Accelerated Estimation of Coffee Sensory Profiles Using an AI-Assisted Electronic Tongue for IFSET—IBM Research. Available online: https://research.ibm.com/publications/accelerated-estimation-of-coffee-sensory-profiles-using-an-ai-assisted-electronic-tongue (accessed on 10 June 2024).
56. Wang, J. Electrochemical Glucose Biosensors. *Chem. Rev.* **2008**, *108*, 814–825. [CrossRef]
57. Sempionatto, J.R.; Khorshed, A.A.; Ahmed, A.; De Loyola E Silva, A.N.; Barfidokht, A.; Yin, L.; Goud, K.Y.; Mohamed, M.A.; Bailey, E.; May, J.; et al. Epidermal Enzymatic Biosensors for Sweat Vitamin C: Toward Personalized Nutrition. *ACS Sens.* **2020**, *5*, 1804–1813. [CrossRef] [PubMed]
58. Sánchez-Paniagua López, M.; Redondo-Gómez, E.; López-Ruiz, B. Electrochemical Enzyme Biosensors Based on Calcium Phosphate Materials for Tyramine Detection in Food Samples. *Talanta* **2017**, *175*, 209–216. [CrossRef] [PubMed]
59. Marchesini, G.R.; Meulenberg, E.; Haasnoot, W.; Irth, H. Biosensor Immunoassays for the Detection of Bisphenol A. *Anal. Chim. Acta* **2005**, *528*, 37–45. [CrossRef]
60. Md Muslim, N.Z.; Ahmad, M.; Heng, L.Y.; Saad, B. Optical Biosensor Test Strip for the Screening and Direct Determination of L-Glutamate in Food Samples. *Sens. Actuators B Chem.* **2012**, *161*, 493–497. [CrossRef]
61. Hu, Q.; Wang, S.; Duan, H.; Liu, Y. A Fluorescent Biosensor for Sensitive Detection of *Salmonella* Typhimurium Using Low-Gradient Magnetic Field and Deep Learning via Faster Region-Based Convolutional Neural Network. *Biosensors* **2021**, *11*, 447. [CrossRef] [PubMed]
62. Zhang, J.; Huang, H.; Song, G.; Huang, K.; Luo, Y.; Liu, Q.; He, X.; Cheng, N. Intelligent Biosensing Strategies for Rapid Detection in Food Safety: A Review. *Biosens. Bioelectron.* **2022**, *202*, 114003. [CrossRef]
63. Peris, M.; Escuder-Gilabert, L. Electronic Noses and Tongues to Assess Food Authenticity and Adulteration. *Trends Food Sci. Technol.* **2016**, *58*, 40–54. [CrossRef]
64. Śliwińska, M.; Wiśniewska, P.; Dymerski, T.; Namieśnik, J.; Wardencki, W. Food Analysis Using Artificial Senses. *J. Agric. Food Chem.* **2014**, *62*, 1423–1448. [CrossRef] [PubMed]
65. Facure, M.H.M.; Mercante, L.A.; Mattoso, L.H.C.; Correa, D.S. Detection of Trace Levels of Organophosphate Pesticides Using an Electronic Tongue Based on Graphene Hybrid Nanocomposites. *Talanta* **2017**, *167*, 59–66. [CrossRef]
66. Mavani, N.R.; Ali, J.M.; Othman, S.; Hussain, M.A.; Hashim, H.; Rahman, N.A. Application of Artificial Intelligence in Food Industry—A Guidelin. *Food Eng. Rev.* **2021**, *14*, 134–175. [CrossRef]
67. Zhang, Z.; Lou, Y.; Guo, C.; Jia, Q.; Song, Y.; Tian, J.-Y.; Zhang, S.; Wang, M.; He, L.; Du, M. Metal–Organic Frameworks (MOFs) Based Chemosensors/Biosensors for Analysis of Food Contaminants. *Trends Food Sci. Technol.* **2021**, *118*, 569–588. [CrossRef]
68. Paul, P.K. Aspects of Biosensors with Refers to Emerging Implications of Artificial Intelligence, Big Data and Analytics: The Changing Healthcare–A General Review. In *Next Generation Smart Nano-Bio-Devices*; Dutta, G., Biswas, A., Eds.; Springer Nature: Singapore, 2023; pp. 1–18. [CrossRef]
69. Mi, F.; Hu, C.; Wang, Y.; Wang, L.; Peng, F.; Geng, P.; Guan, M. Recent Advancements in Microfluidic Chip Biosensor Detection of Foodborne Pathogenic Bacteria: A Review. *Anal. Bioanal. Chem.* **2022**, *414*, 2883–2902. [CrossRef] [PubMed]

70. Verma, D.; Singh, K.R.; Yadav, A.K.; Nayak, V.; Singh, J.; Solanki, P.R.; Singh, R.P. Internet of Things (IoT) in Nano-Integrated Wearable Biosensor Devices for Healthcare Applications. *Biosens. Bioelectron. X* **2022**, *11*, 100153. [CrossRef]
71. Ding, H.; Tian, J.; Yu, W.; Wilson, D.I.; Young, B.R.; Cui, X.; Xin, X.; Wang, Z.; Li, W. The Application of Artificial Intelligence and Big Data in the Food Industry. *Foods* **2023**, *12*, 4511. [CrossRef] [PubMed]

MDPI AG
Grosspeteranlage 5
4052 Basel
Switzerland
Tel.: +41 61 683 77 34

Foods Editorial Office
E-mail: foods@mdpi.com
www.mdpi.com/journal/foods